Handbook of Medicinal Plants

FOOD PRODUCTS PRESS®

Crop Science

Amarjit S. Basra, PhD
Senior Editor

The Lowland Maya Area: Three Millennia at the Human-Wildland Interface edited by A. Gómez-Pompa, M. F. Allen, S. Fedick, and J. J. Jiménez-Osornio

Biodiversity and Pest Management in Agroecosystems, Second Edition by Miguel A. Altieri and Clara I. Nicholls

Plant-Derived Antimycotics: Current Trends and Future Prospects edited by Mahendra Rai and Donatella Mares

Concise Encyclopedia of Temperate Tree Fruit edited by Tara Auxt Baugher and Suman Singha

Landscape Agroecology by Paul A Wojkowski

Concise Encylcopedia of Plant Pathology by P. Vidhyaskdaran

Molecular Genetics and Breeding of Forest Trees edited by Sandeep Kumar and Matthias Fladung

Testing of Genetically Modified Organisms in Foods edited by Farid E. Ahmed

Fungal Disease Resistance in Plant: Biochemistry, Molecular Biology, and Genetic Engineering edited by Zamir K. Punja

Plant Functional Genomics edited by Dario Leister

Immunology in Plant Health and Its Impact on Food Safety by P. Narayanasamy

Abiotic Stresses: Plant Resistance Through Breeding and Molecular Approaches edited by M. Ashraf and P. J. C. Harris

Teaching in the Sciences: Learner-Centered Approaches edited by Catherine McLoughlin and Acram Taji

Handbook of Industrial Crops edited by V. L. Chopra and K. V. Peter

Handbook of Medicinal Plants edited by Zohara Yaniv and Uriel Bachrach

Durum Wheat Breeding: Current Approaches and Future Strategies edited by Conxita Royo, M. M. Nachit, N. Di Fonzo, J. L. Araus, W. H. Pfeiffer, and G. A. Slafer

Handbook of Statistics for Teaching and Research in Plant and Crop Science by Usha Rani Palaniswamy and Kodiveri Muniyappa Palaniswamy

Handbook of Microbial Fertilizers edited by M. K. Rai

Eating and Healing: Traditional Foods As Medicine edited by Andrea Pieroni and Lisa Leimar Price

Physiology of Crop Production by N. K. Fageria, V. C. Baligar, and R. B. Clark

Plant Conservation Genetics edited by Robert J. Henry

Introduction to Fruit Crops by Mark Rieger

Sourcebook for Intergenerational Therapeutic Horticulture: Bringing Elders and Children Together by Jean M. Larson and Mary Hockenberry Meyer

Agriculture Sustainability: Principles, Processes, and Prospects by Saroja Raman

Introduction to Agroecology: Principles and Practice by Paul A. Wojtkowski

Handbook of Molecular Technologies in Crop Disease Management by P. Vidhyasekaran

Handbook of Precision Agriculture: Principles and Applications edited by Ancha Srinivasan

Dictionary of Plant Tissue Culture by Alan C. Cassells and Peter B. Gahan

Handbook of Medicinal Plants

Zohara Yaniv, PhD
Uriel Bachrach, PhD
Editors

CRC Press is an imprint of the
Taylor & Francis Group, an informa business

First published 2005 by Haworth Press,Inc

Published 2010 by CRC Press
Taylor & Francis Group
6000 Broken Sound Parkway NW, Suite 300
Boca Raton, FL 33487-2742

ISBN-13: 978-1-56022-995-7 (pbk)

PUBLISHER'S NOTE
This book has been published solely for educational purposes and is not intended to substitute for the medical advice of a treating physician. Medicine is an ever-changing science. As new research and clinical experience broaden our knowledge, changes in treatment may be required. While many potential treatment options are made herein, some or all of the options may not be applicable to a particular individual. Therefore, the author, editor, and publisher do not accept responsibility in the event of negative consequences incurred as a result of the information presented in this book. We do not claim that this information is necessarily accurate by the rigid scientific and regulatory standards applied for medical treatment. **No warranty, expressed or implied, is furnished with respect to the material contained in this book. The reader is urged to consult with his/her personal physician with respect to the treatment of any medical condition.**

Cover design by Lora Wiggins.

Library of Congress Cataloging-in-Publication Data

Handbook of medicinal plants / Zohara Yaniv, Uriel Bachrach, editors.
p. ; cm.
Includes bibliographical references and index.
ISBN 1-56022-994-2 (hard : alk. paper) — ISBN 1-56022-995-0 (soft : alk. paper)
1. Materia medica, Vegetable—Handbooks, manuals, etc. 2. Medicinal plants—Handbooks, manuals, etc. 3. Traditional medicine—Handbooks, manuals, etc.
[DNLM: 1. Plants, Medicinal. 2. Medicine, Traditional. 3. Phytotherapy—methods. QV 766 H236 2005] I. Yaniv, Zohara. II. Bachrach, Uriel.

RS164.H273 2005
615'.321—dc22

2004012055

CONTENTS

ABOUT THE EDITORS

Zohara Yaniv, PhD, received her master's degree in biochemistry from The Hebrew University in Jerusalem and her PhD in plant biochemistry from Columbia University. Dr. Yaniv worked at the Boyce Thompson Institute for Plant Research in New York. Since 1978, she has been a senior scientist at the Department of Medicinal Plants at the Institute of Field Crops in the Agricultural Research Center of Israel. She was appointed professor at Bar Ilan University, Ramat Gan, and the Faculty of Agriculture at The Hebrew University, Jerusalem. She is a former visiting scientist at the University of Paris; the National Institutes of Health in Bethesda, Maryland; Lubbock University in Texas; and the Weitzmann Institute of Science in Rehovoth, Israel. Dr. Yaniv has also been a visiting professor at the University of Vienna, Austria, and Shanghai University, China. For six summers, she conducted an international course in ethnobotany at MAICH, Chania, Crete.

Dr. Yaniv's scientific work includes research in ethnobotany of the Middle East with special emphasis on the Holy Land. This research resulted in numerous publications and a book entitled *Medicinal Plants of the Holy Land,* published in Hebrew and English (2000). She has been an invited speaker at various international meetings on medicinal plants. For the past ten years she has served as the head of The Department of International Collaboration of the Agricultural Research Center of Israel.

Uriel Bachrach, PhD, is a professor emeritus of Molecular Biology at the Hebrew University-Hadassah Medical School in Jerusalem. Dr. Bachrach is the former head of the Department of Molecular Biology, President of the Israel Society of Biochemistry, and Advisor in Biotechnology to the Prime Minister's Office of the State of Israel. He is a former visiting professor at the University of California, San Francisco; Northwestern University; the University of Addis Ababa, Ethiopia; and Bologna University in Italy. He is a former visiting scientist at the National Institutes of Health in Bethesda, Maryland, as well as Columbia University in New York, the University of Manchester in the United Kingdom, and Deutsche Krebs Forschung Zentrum (DKFZ) in Heidelberg, Germany.

Dr. Bachrach has published more than two hundred scientific volumes, holds three patents, and was editor or co-editor of nine books, including two volumes of *The Physiology of Polyamines.* He wrote *Function of Naturally Occurring Polyamines,* published in 1973. He received an honorary doctorate in medicine from the University of Bologna and a medal from the Polish Society of Physiology. He was also awarded a medal and honorary professorship from the Shandong Academy of Medical Science of the People's Republic of China. Dr. Bachrach organized several international meetings, including the Gordon Conferences and a FEBS meeting held in Jerusalem in 1980.

CONTRIBUTORS

Emanuela Appetiti has been with the Botany Department of the Natural History Museum, Smithsonian Institution, Washington, DC, since 2000 as a research collaborator and is currently a scientific program specialist. An independent scholar working on Australian Aboriginal traditional medicine, she graduated in sociology, with a specialization in cultural anthropology, in 1987, from the University "La Sapienza" in Rome, Italy. She is a member of the Washington Academy of Sciences, the National Coalition of Independent Scholars, the Botanical Society of Washington, and the American Anthropological Association. In the past, she was a research fellow of the Istituto Italiano per gli Studi Filosofici (Italian Institute for Philosophical Studies). She is currently preparing an annotated bibliography on Australian Aboriginal uses of medicinal plants. In addition, she is for the third year a co-organizer of the Earthwatch Expedition Medicinal Plants of Antiquity, a research program based in Rome, Italy.

Claudio Galli is professor of pharmacology at the University of Milan, where he studies natural antioxidants and essential fatty acids. He has been the president of ISSFAL (International Society for the Study of Fatty Acids and Lipids) and on the Board of Directors of the Italian Pharmacology Society.

Simona Grande graduated in Biology and is currently research associate at the Department of Pharmacological Sciences, University of Milan.

Michael Heinrich is a professor and the head of the Centre for Pharmacognosy and Phytotherapy, The School of Pharmacy, University of London. He is a pharmacognosist, biologist (University of Freiburg 1989/ 1996), and an anthropologist (MA, Wayne State University, 1982), with many years of research experience in a multitude of aspects of medicinal and food plants as well as the interface of cultural and natural sciences. He has a particular interest in the biomedical and cultural bases of medicinal plant use in Lowland Mexico, the Mediterranean, and other countries. He is a lead author of *Fundamentals of Pharmacognosy and Phytotherapy.*

Peter John Houghton, B. Pharm., PhD F.R.P. Pharm. S., F.R.S.C.C. Chem, is a professor in pharmacognosy in the Department of Pharmacy, King's College London. He received his PhD in pharmacognosy from London University and bachelor's degree from Chelsea College. He has published more than 200 papers and articles on medicinal plants and is on the committees of several international societies and several editorial boards. His research is focused on combining biological tests with chemical isolation to elucidate the compounds responsible for the activity of traditional medicinal plants, especially those used for wound healing, cancer, and for CNS conditions associated with ageing.

Zohar Kerem, PhD, is a lecturer at the Institute of Biochemistry, Food Science and Nutrition, The Faculty of Agricultural, Food and Environmental Quality Sciences, The Hebrew University of Jerusalem. He received his PhD in agricultural biotechnology and his master's degree from The Hebrew University of Jerusalem and a bachelor's degree from Tel Aviv University. He specializes in the chemistry of natural products and their antioxidant activities in biological systems. He has authored several journal articles on the mechanisms of biological oxidation and their inhibition.

Ruth Kutalek, MA, PhD, is an assistant in the Department of Ethnomedicine, Institute for the History of Medicine, Medical University of Vienna. She received her MA and PhD in social anthropology (ethnomedicine) from the University of Vienna. She has conducted research in Tanzania, Ethiopia, and Austria, and has published several articles in the fields of ethnomedicine, ethnopharmacology, ethnomycology, and ethnoentomology.

Marie-Aleth Lacaille-Dubois is a professor of pharmacognosy and director of the research team UMIB, EA 3660 "Molecules of Biological Interest" on the faculty of pharmacy, University of Burgundy, Dijon, France. She received her PhD and habilitation in pharmacognosy at the University of Burgundy, and her master's degree at the University of Paris V. She has received several distinctions from the Faculty of Pharmacy in Paris, and the National Academy of Science and was a fellow of the DAAD (Munich, Germany) and the JSPS (Fukuoka Japan). She has been a visiting scientist and delivers several lectures in Germany, Japan, Brazil, Pakistan, and Cameroon. She is author of seventy original articles, and chapters of books on the topics of isolation, structure elucidation, and biological assays of flavonoids and triterpenoids mainly in the field of cancerology and immunology. She has also been invited to deliver lectures at several international conferences (Portugal, Poland, Pakistan, and Morocco).

Efraim Lewinsohn, PhD, is a senior research scientist in the Department of Vegetable Crops and Genetics, The Newe Yaar Research Center, Agricultural Research Organization, Israel. He received his PhD and MSc in biology from the Weizmann Institute of Science, Rehovot, Israel, and his bachelor's degrees in horticulture from the Hebrew University of Jerusalem in Rehovot, Israel. He specializes in the study of secondary metabolism in plants with special emphasis in the enzymology and molecular biology of aroma formation. He has authored several journal and review articles on topics related to the discovery and biotechnological applications of gene coding for the formation of aroma compounds in plants.

Chang-Xiao Liu is a professor emeritus in Tianjin Institute of Pharmaceutical Research and Tianjin University, and the director of the State Key Laboratory of Pharmacodynamics and Pharmacokinetics at Tianjin Institute of Pharmaceutical Research. He received his Academian from China Academy of Engineering, and has published twelve books and 200 papers.

David Mantle, PhD FRSC FRC Path, is currently visiting lecturer in the School of Biology, University of Newcastle upon Tyne, United Kingdom. His background is in hospital-based chemical pathology. He has published more than 100 research papers in scientific journals on the role of free radicals/antioxidants and proteolytic enzymes/inhibitors in disease pathogenesis, and the use of medicinal plants and nutritional supplements for the prevention and treatment of disease.

Lester A. Mitscher is a Kansas University Distinguished Professor of Medicinal Chemistry and Molecular Biosciences, at Kansas University, Lawrence, Kansas. He received his PhD in organic and physiological chemistry from Wayne State University in 1958 and his BS in pharmacy from the same institution. He has published more than 250 research articles and seven books on the chemistry of pharmaceuticals, consults with various firms and governmental agencies, and is on the editorial boards of eight journals. He has received a number of awards including the Smissman Prize from the American Chemical Society, the Volweiler Prize from the American Association of Colleges of Pharmacy, the Research Achievement Award from the American Pharmaceutical Association, and is a fellow of The American Association for the Advancement of Science.

Daniel E. Moerman is a William E. Stirton Professor of Anthropology at The University of Michigan-Dearborn. He received his PhD in anthropology from the University of Michigan in 1974. His early work related to health emerged during his dissertation research with a rural black population in coastal South Carolina in the early 1970s. Since then, he has done research primarily in medicinal plants (primarily of Native American peoples), and their impact on health. His book *Native American Ethnobotany* (Timber Press, 1998) received the Annual Literature Award of the Council on Botanical and Horticultural Libraries for 2000. In 2002, his book *Meaning, Medicine and the "Placebo Effect"* about the role of meaning in the healing process was published by Cambridge University Press.

Éva Németh, PhD, is a professor at the Department of Medicinal and Aromatic Plants, Faculty of Horticultural Sciences, BUESPA, Budapest, Hungary. She holds an MSc in horticultural sciences and an MSc in genetics and plant breeding. She received her PhD in biology at the Hungarian Academy of Sciences. She is author of more than 200 scientific publications about the biological bases, intraspecific chemical variability, and technological development of medicinal plant production. She is a breeder of twelve medicinal plant varieties.

Graham Pinn is the director of the Dubai London Clinic. He received his DCH from London University, MRNZCGP from New Zealand, RCGP and FRCP from London University. His consultant posts have included New Zealand, the United Kingdom, Australia, West Germany, Seychelles, and the United Arab Emirates. He has published in the fields of gastroenterology, cardiology, tropical medicine, and herbal medicine. In 2003, he published *A Practical Guide for Medical Practitioners*.

Sir Ghillean T. Prance, DPhil, FRS, is former director of the Royal Botanic Gardens, Kew, Endland, and a visiting professor at the University of Reading. He received his DPhil in forest botany from Oxford University. He is author of eighteen books and monographs and more than 450 scientific and general papers on plant taxonomy, ethnobotany, vegetation ecology, and conservation. He received the International Cosmos Prize from Japan in 1993.

Armin Prinz, PhD, MD, is a professor in the Department of Ethnomedicine, Institute for the History of Medicine, Medical University of Vienna. He received his PhD in social anthropology in Vienna and his MD from the University of Vienna. He conducted research in the Republic of the Congo (formerly Zaire), Senegal, Ethiopia, and Tanzania. He has published extensively in the field of ethnomedicine, nutritional anthropology, ethnopharmacology, and travel medicine.

Ya'akov Tadmor was educated as a biochemist and plant geneticist. Dr. Tadmor is senior scientist at the Israeli Agricultural Research Organization and has expertise in the genetics of health-benefiting traits in agricultural products, including the chemical and genetic analyses of these traits.

Alain Touwaide, PhD, is a research associate at the Department of Botany of the National Museum of Natural History, at the Smithsonian Institution in Washington, DC. He obtained a PhD in philosophy and letters from the University of Louvain (Belgium), with a thesis on ancient Greek medicine. A specialist of the history of pharmacology in the different cultures of the Mediterranean area from Antiquity to the Renaissance, he has researched and taught in several universities and research institutions in Europe and the United States, and has published several books and numerous journal articles. In 2003 he received the Award for Achievement in the Behavioral and Human Sciences of the Washington Academy of Sciences, of which he is a life fellow.

D. Francesco Visioli, PhD, holds appointments at the Department of Pharmacological Sciences of the University of Milan and at the College of Pharmacy, Oregon State University. He studies omega-3 fatty acids and natural antioxidants, with particular and active investigation (in vitro and in vivo) of extra virgin olive oil. He has published more than 100 papers.

Professor Dr. h. c. Hildebert Wagner is a Universitäts-Professor, University of Munich, Institut of Pharmacy, Pharmaceutical Biology, Munich, Germany. He was appointed as full professor of pharmacognosy in 1965. He was director of the Institute of Pharmaceutical Biology, Munich, until 1999. He gained Distinguished Visiting Professorship at Columbus, Ohio; was dean of the faculty of chemistry/pharmacy at the Institute of Pharmaceutical Biology, Munich; PhD *honoris causae* of the University of Budapest and the University of Debrecen (Hungary), Dijon (France), and Helsinki, Finland. He was editor of the *International Journal of Phytomedicine* and a member of the advisory boards of *Phytochemistry, Journal of Ethnopharmacology,* and *Journal of Natural Products* (Lloydia). His research included the isolation, structure determination, synthesis, and analysis of biologically and pharmacologically active compounds of medicinal plants, particularly in the fields of alkaloids, heart glycosides, flavonoids, and lignans; drugs with antiviral, antiasthmatic, antiphlogistic, immunostimulating, and adaptogenic activity; and standardization of Chinese drugs. He has published 800 original papers, thirty review articles, and eight books.

Ming-Wei Wang, MD, PhD, is the director of the Chinese National Center for Drug Screening and a professor of pharmacology at Shanghai Institute of Material Medica, Chinese Academy of Sciences. He received his MD from Shanghai First Medical College and a PhD from Cambridge University. He has established a number of biopharmaceutical companies in the United States, the United Kingdom, China, and Hong Kong. He is the inventor of several international new drug patents, and he has authored more than seventy research papers.

Christoph Wawrosch is an assistant professor in the Institute of Pharmacognosy at the University of Vienna. He received his PhD in plant tissue culture of *Achillea* species and his master's degree from the University of Vienna. Research interests include the in vitro propagation of medicinal plants for conservation or selection purposes, and bioreactor-based propagation systems for efficient plant production. He teaches pharmacognosy and plant tissue culture within the pharmacy curriculum at the Institute of Pharmacognosy, University of Vienna.

Richard M. Wilkins, PhD, MS, BSc, MIBiol, is currently a senior lecturer in ecotoxicology in the School of Biology and the deputy director of the Medicinal Plant Research Centre, Newcastle University, United Kingdom. He received his MS and PhD from the University of Washington, Seattle, and worked in both the pharmaceutical and crop protection industries. He published two books and more than eighty papers in scientific journals on toxicology, biochemistry, medicinal plants, and controlled release.

Preface

The almost explosive growth in publications on medicinal plants is continuing with no sign of abatement. For centuries plants have provided humankind with useful, sometimes lifesaving, drugs. In the last century the sources of natural medicine have expanded to include higher and lower plants and marine organisms, as well as animals. The importance of medicinal plants can be realized from their prominence in the over-the-counter and prescription drug market. Recent studies revealed that about one-third of all prescriptions issued in developed countries such as the United States and Canada, as well as the countries of Europe, contained an herb or a purified extract of an active component derived from herbs. In the less developed countries where synthetic drugs are expensive, such as China and the countries of the Far East, herbs can be present in 70 to 90 percent of prescriptions. With the progress of new technologies, new avenues were opened to purify active components from plants, to establish their chemical structure, and even to synthesize and/or modify them chemically.

Since the 1970s the interest in medicinal plants increased tremendously due to the failure of modern medicine to provide effective treatment for chronic diseases. The emergence of drug-resistant bacteria and parasites also stimulated the search for alternative treatments. Plant extracts that have been used for centuries without harmful side effects became a source of hope for drug companies and clinicians.

Because of this explosion of information, it was a perfect time for experts in the field to review the subject. This handbook is divided into five parts: Part I is an introduction. Part II deals with the various traditions in different parts of the world. Part III deals with modern technologies applied in order to separate, identify, and produce active components of plants. Part IV is devoted to the use of medicinal plants for the treatment of different diseases. In many cases the structure of the active component is illustrated and if possible the mode of action of the drug is explained. Part V is a concluding chapter that reflects on the topics discussed in the book as well as considers the future of medicinal plants.

This handbook is intended to furnish the reader with the latest developments in the field of medicinal plants. It is our hope that this handbook will encourage further research and understanding of all aspects of this intriguing and complex field.

Preface

PART I: INTRODUCTION

Chapter 1

Trends and Challenges in Phytomedicine: Research in the New Millennium

Hildebert Wagner

INTRODUCTION

To define and describe the future tasks of phytomedicinal research in the new millennium requires an analysis not only of the current state of development of phytomedicinal research but also of chemosynthetic pharmaceutical research. Both find themselves in a race to develop new medicines, with fewer or no side effects, for therapeutic and preventive application in illnesses for which causality-based treatment has been nonexistent or imperfect.

This analysis is, on one hand, imperative for an improved treatment of diseases. On the other hand, we must also consider what the drug law stipulates in terms of guaranteeing the quality, safety, and effectiveness of the medicines we develop. These requirements might differ between Western nations and the so-called third world or Asia, but a global consensus exists on at least a couple of points: that medicines must become safer and more effective, and that their use in treatment must be rationalized scientifically. These demands will determine to a large degree the content and strategies of future phytomedicinal research. The prerequisites for innovation are favorable to a renaissance in this field.

- Of the 2,000 registered acute and chronic diseases, only 30 percent are presently curable. The rest can be treated only symptomatically to greater or lesser effect. Therefore, a great need exists for new highly effective and causality-based treatments, to which phytotherapy is expected to contribute.
- One advantage of phytotherapy is the availability of a wide group of medicinal drugs and preparations that have been used over the centuries almost exclusively on the basis of empirical evidence. The current availability of high-tech methods allows researchers to optimize the

effectiveness, standardization, and clinical testing of these traditional medicines to meet today's international standards. In addition, a reservoir of around 300,000 plant species exists, of which only about 30 percent have been investigated scientifically, inclusively the herbs and preparations of Chinese, Indian, South American, and African traditional medicines.

- Today's high throughput screening techniques enable us to examine thousands of plant extracts and raw materials via several test models in a relatively short period of time, followed by the isolation and structure elucidation of all pharmacologically active components, even if they are present in a plant in minor quantities.
- These analytical procedures can now be supplemented with molecular-biological test models, which have recently been introduced into plant screening techniques. These allow us to determine a hitherto unknown mechanism of action of a given plant extract or raw material at a molecular level, as well as facilitate the discovery of new indications which offer a causality-based therapy. The synergetic effects of components in extract preparations can also be elucidated using molecular-biological methods.
- Through the use of these techniques for the chemical and pharmacological study of phytopreparations, researchers can for the first time present reproducible and internationally recognized clinical results. More than 400 positive, placebo-controlled, randomized, double-blind studies have been completed for around 20 standardized phytopreparations. These reports are further evidence that the development of phytomedicinal preparations should be of high priority for the near future.
- Two paradigm shifts, exemplified by the following changes in chemotherapeutic strategies, serve as additional confirmation and legitimization of phytomedicinal research and phytotherapy: a gradual renunciation of the long-standing reliance on monotherapy as an ideal, in favor of polychemotherapy and the development of totally new application concepts, through which the evaluation of different pathophysiological processes and accompanying symptoms, rather than the search for single disease-causing agents (e.g., tumor cells or infection), has assumed significance. (See also one example of the new concept on cancer treatment and prevention, Chapter 14 by Mantle and Wilkins, in this book.)

Phytotherapy has long followed and developed both such strategies, although without scientific evidence to support them directly. Now it appears that the therapeutic strategies of phytotherapy and chemotherapy are becoming more more parallel to each other.

Numerous reviews and books have been published since 1985, offering a view of the state of research in phytomedicine, its tasks, and its methods. Among the most important of these works are the following:

- "Phytomedicines in Europe: Their Chemistry and Biological Activity"[1]
- "Natural Products in Drug Discovery and Development"[2]
- "Recent Natural Products Based Drug Development: A Pharmaceutical Industry Perspective"[3]
- "Phytomedicines: Back to the Future"[4]
- "Phytomedicine Research in Germany"[5]
- *Herbal Medicine: A Concise Overview for Professionals*[6]
- *Rational Phytotherapy,* Fourth Edition[7]
- "Phytochemistry and Medicinal Plants"[8]
- *A Desktop Guide to Complementary and Alternative Medicine: An Evidence-Based Approach*[9]

In the following 3 sections, with the use of examples from my own research and that of other laboratories, I report the progress of knowledge in phytomedicine, as well as relevant results from the three research emphases of phytomedicinal research. In the opinion of the author, these same three emphases will, in the next ten years, claim the highest priority:

- The assurance of phytopreparation quality by standardization
- The search for new bioactive compounds
- The phytopharmacological and molecular-biological research and the clinical, pharmacokinetic, and bioavailability studies

Safety is the most important and self-evident precondition for the development of new phytopreparations, and it must be guaranteed through toxicological investigations. These tasks can be achieved only through interdisciplinary and international cooperation.

QUALITY ASSURANCE OF MEDICINAL DRUGS AND PHYTOPREPARATIONS

Most countries have developed structures for the quality control of phytopreparations destined for the internal or export market. Officially recognized drugs are subjected to identity and purity test values and the determination of standard values for the major bioactive constituents of an herbal drug, the assurance of plant conservation, and microbiological purity. Currently, quality code numbering and criteria of national pharmacopeia vary from country to country, but a first step has been made toward a global harmonization of guidelines, by the International Federation of Pharmaceutical Manufacturers Associations (IFPMA).[10] The World Health Organization (WHO) has collected current data on the legal status of traditional medicine and complementary/alternative medicine of 123 member countries.[11]

In addition, some countries and organizations have drawn up their own specific herbal drug monographs (e.g., German Commission E and Kostantin Keller, WHO, ESCOP [European Scientific Cooperative on Phytotherapy]),[12] or are in the process of doing so (e.g., China and Brazil). These monographs contain all stipulated quality standards to be observed in bringing a drug or a combined phytopreparation to the drug market.

The quality proof of an herbal drug begins with definitive authentication and taxonomic assignment. This can be difficult even for taxonomic experts because of the extreme diversity within some genera and the occurrence of inter- and intraspecific variations (polymorphism). Therefore more than a few misidentifications of species have occurred. This handicap can be overcome by applying the new molecular approach of DNA fingerprint analysis, especially the sequencing techniques and restriction analyses (RELPs) combined with (micro) morphological investigations.[13] This method can be carried out with a relatively small quantity of plant material, using independent molecular markers from isolated nuclear and chloroplast DNA. The sequences are alienable (transferable), editable, and analyzable using computer programs for reconstruction of phylogenetic relationships and genetic similarities.

By this method it has been possible to resolve, for example, the long-lasting insecurity about the correct taxonomic assignment of *Phyllanthus amarus* and *P. niruri*. On the basis of sequence analysis obtained from two independent markers of *P. amarus, P. niruri,* and eight taxonomically related species, it was possible to characterize *P. amarus* by species-specific mutations. Furthermore, there is now evidence from a cladistic analysis that *P. fraternus* and *P. abnormis* are the species related most closely to *P. amarus*. The analysis revealed further that *P. niruri* is not genetically linked with *P. amarus,* and that *P. debilis, P. tenellus,* and *P. urinaria* are clearly separate species.[14]

Over the past 15 years, similar DNA analyses of many species with uncertain taxonomic assignment have been performed. Because this method has become a routine analytical tool, it is likely that in the future drug regulation authorities will demand a DNA-analysis-based identity proof for a plant drug when doubts about its correct taxonomic assignment exist. The same method could also be used for an identity assessment if an artificially produced new plant hybrid with extraordinary chemical composition and therapeutic efficacy or a newly detected natural plant species, subspecies, subvariety, or chemical race should be protected by a patent.

If an herbal drug of traditional medicine with a great folkloric reputation and long-term use is to be be further developed to a new phytopreparation which meets present-day international standards of quality, safety, and efficacy, the following steps must be followed:

1. The isolation and structural elucidation of all major constituents of the herbal drug that might be responsible for its overall pharmacological activity and efficacy must occur. This can be performed with or without a bioguided fractionation using the modern chromatographic and spectroscopic techniques available.[15] The most important methods are listed in Box 1.1 (see also Chapter 11). Great progress in this area has been achieved through the application of specific and selective analytical methods, using a combination of high performance liquid chromatography (HPLC) and diode-array technique (online recorded ultraviolet [UV] spectra), or mass spectrometry (MS) and nuclear magnetic resonance (NMR) detection. As far as pharmacological in vitro assays are concerned, aside from enzymatic methods, more and more molecular-biological assays have been integrated into the screening of extracts and isolated compounds.
2. The elucidation of the major active constituents aims primarily at the standardization of an extract or phytopreparation, which is based on the true bioactive compounds of an extract. Because several constituents contribute to the overall pharmacological and therapeutic effects of an extract, the quantitation of one marker compound or a collective of compounds according to a pharmacopoeia or a national monograph will no longer be sufficient. In these cases, the standardization method must aim at the quantitation of two or three major bioactive components, as evaluated by pharmacological investigations.[16]
3. The standardization of a phytopreparation can be hampered and complicated when a multiextract mixture must be handled. To overcome this difficulty, two procedures can be used: the single extracts are standardized or fingerprinted before mixing the individual extracts, or a 3D HPLC fingerprint analysis of the multicomponent extract mixture must be performed as shown in Figure 1.1 for the phytopreparation Sho-seiryu-to (TJ-19) of the Kampo medicine, which contains the ex-

tracts of eight single herbs.[17] In any case, however, for such analytical procedures, the knowledge of all the major bioactive constituents of the individual herbs is a primary precondition. Since 1995 many specific HPLC fingerprints with quantitation of single bioactive components of multiextract mixtures have been developed. General guidelines for the characterization and standardization of plant materials that are used for pharmacological, clinical, or toxicological studies, with examples for three categories of herbal preparations, have been made by Bauer and Tittel.[18]

4. Because the criteria for the proof of quality of an herbal drug differ from country to country, steps should be taken toward a global harmonization of the guidelines, such as those used by the International Federation of Pharmaceutical Manufacturers Associations (IFPMA).[19] This organization represents the global research-based pharmaceutical industry and aims to ensure worldwide standards of safety, quality, and efficacy for new medicines as well as more efficient registration. Accordingly, efforts in this direction have been made and are still in progress by such bodies as the German Commission E, WHO, and ESCOP.[20]

BOX 1.1. High-tech analytical and isolation methods for plan screening and isolation work.

Analysis of plant constituents and accordant metabolites in blood, serum, and urine

- Thin-layer chromatography (TLC), thin-layer electrophoresis (TLE)
- Isotachophoresis (ITP)
- Capillary electrophoresis (CE)
- Capillary electrochromatography (CEC)
- HPLC, gas chromatography (GC)
- HPLC coupled with MS (chemical and ionization or electrospray ionization technique)
- Liquid chromatography (LC) coupled with UV/MS/NMR/Fourier transform ion cyclotron resonance (FT-ICR)

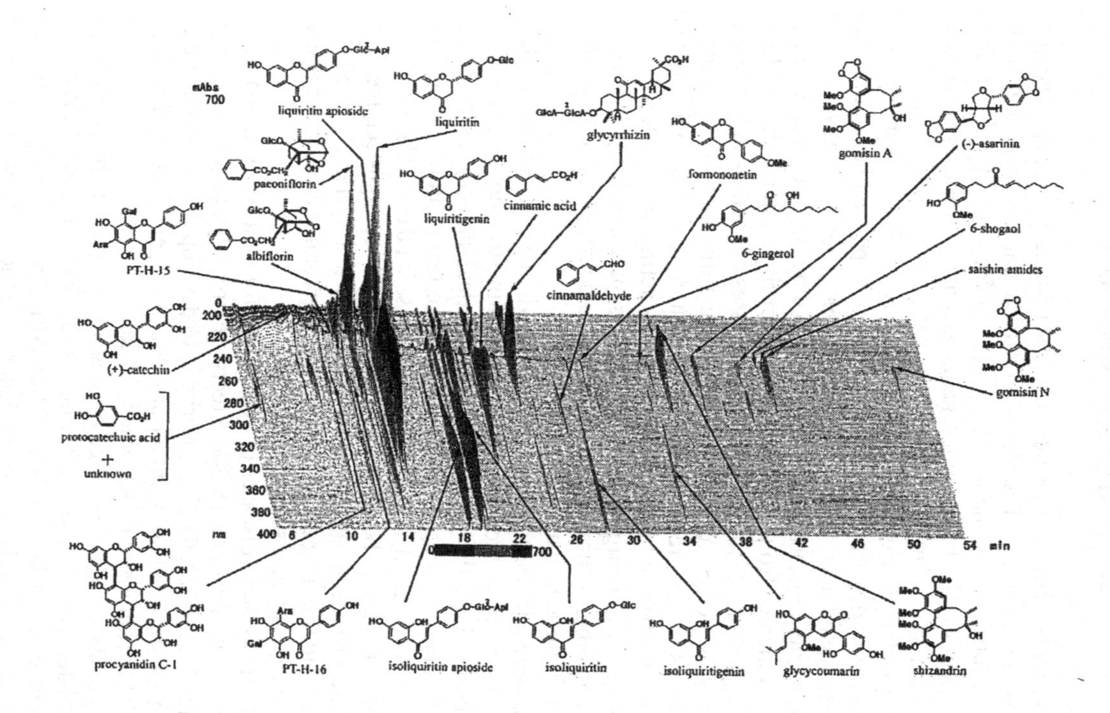

Figure 1.1. 3D HPLC fingerprint analysis of Sho-seiryu-to (TJ-19) extract produced from the eight herbal drugs *Pinellia ternata, Glycyrrhiza glabra, Cinnamomum cassia, Schisandra chinensis, Asarum sieboldii, Paeonia lactiflora, Ephedra sinica*, and *Zingiber officinale*. (*Source:* Amagaya, S., Iizuka, A., Makino, B., Kubo, M., Komatsu, Y., Cheng F.-Ch., Ruo, I.I., Itoh, T., and Teresawa, K., 2001, General pharmacological properties of Sho-seiryu-to (TJ-19) extracts, *Phytomedicine* 8: 338-347.)

SEARCH FOR NEW BIOACTIVE COMPOUNDS AND CHEMICAL MODIFICATION OF LEADING COMPOUNDS

Since the mid-1980s, a great number of screening programs have been initiated worldwide by pharmaceutical industries, universities, and national research institutions with the aim of searching for new bioactive natural products in plants and other natural organisms that are promising enough to be developed into new drugs.

This approach is very demanding for two reasons. First, around 300,000 plant species exist on earth, of which only about 30 percent have been thoroughly investigated, chemically and pharmacologically. Every year, some interesting, highly bioactive compounds detected by comprehensive screening projects emerge into the limelight; some of them later become blockbusters. Second, despite the magnificent successes of chemo- and antibiotic therapy, only 30 percent of about 2,000 existing diseases can be cured at present. Others can be treated only symptomatically or insufficiently and some not at all. The need for efficient, causally acting drugs to treat severe diseases is therefore an urgent challenge. All resources, inclusive of medicinal plants and other natural products which could provide new drugs, must be included in these research programs.

As far as screening procedures are concerned, this task cannot be solved economically by using the classical methods hitherto employed in university laboratories. Over the years, the pharmaceutical industry has developed high throughput screening methods which have been automated and perfected to such an extent that 10,000 samples of isolated or synthesized compounds and extracts can be screened per month. These high-tech methods, however, are promising only when a battery of test systems can be employed. Receptor binding assays have attained great importance because of their high selectivity and sensitivity.[21]

Box 1.2 shows those diseases or syndromes for which new and more effective drugs with fewer or no side effects are urgently needed. In the field of anti-infectious drugs, the first priority is to develop drugs that can be effectively used for combating infections which are multiresistant to presently available chemosynthetics or antibiotics.

In this context, it must be stressed that in many cases bioactive compounds isolated from plants do not have the desired maximal efficiency. This prompted many laboratories to transform leading compounds from plants into structural derivatives with better efficiency, bioavailability, pharmacokinetics, or stability. Attempts have also been made to reduce the toxic potential of a compound or to convert it into a drug with better applicability. Two examples may demonstrate this approach: the well-known betulinic

BOX 1.2. New drugs of high priority wanted worldwide.

Cancer
- Therapeutics and preventives

Cardiovascular disease
- Antihypertonics
- Antiatherosclerotics
- Anti-ischemics (drugs for stroke prevention)

Central nervous system disease
Therapeutics and preventives for
- Alzheimer's disease
- Parkinson's disease

Infectious disease
- Antiviral drugs (e.g., anti-HIV, anti-hepatitis B and C)
- Antiparasitic drugs
- Antifungal drugs

Inflammatory disease
- Antiasthmatics
- Anticolitis drugs
- Antineurodermatitic/antipsoriatic drugs

acid, a constituent of *Betula* sp. and occurring also in many other plants, has been found to exert an antitumor effect via an apoptosis-inducing mechanism.

A French research group, headed by Prof. Lacaille-Dubois (see also Chapter 19), synthesized 38 betulinic acid derivatives with other substituents in positions 3 and C28. The apoptosis-inducing effect in melanoma cells could be increased from an IC_{50} value of 1.1 µg/mL to an IC_{50} value of 0.7 µg/mL.[22] Surprisingly, in a second approach, the researchers found betulinic acid also effective against HIV infection. Substitution of C_3-OH and C-28 COO-group by a dimethylsuccinyl-and a ω-amino-alkanoyl group, respectively, enhanced the anti-HIV activity by a factor of ten (see Figure 1.2). Another example is artemisinin, a sesquiterpenlacton peroxide from the herb *Artemisia annua,* which has been used in China since ancient times for the treatment of malaria. Converting artemisinin into a 6-methyl derivative (arte ether) increased bioavailability after preventional applica-

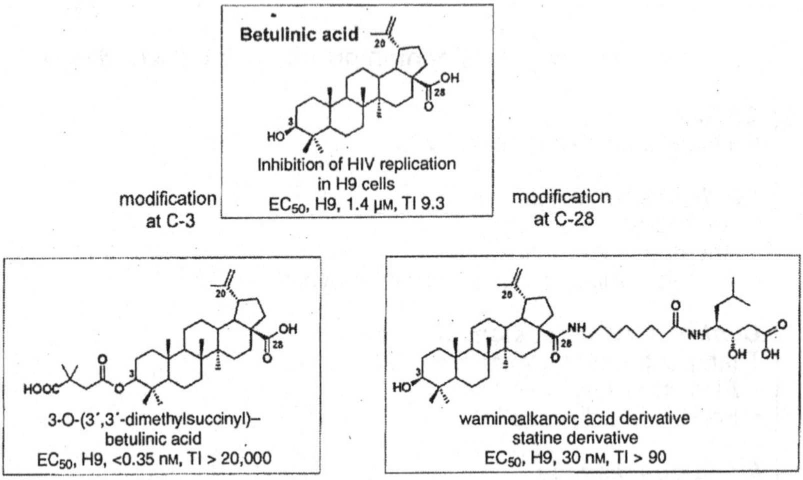

Betulinic acid

Inhibition of HIV replication in H9 cells

EC_{50}, H9, 1.4 μM, TI 9.3

modification at C-3

modification at C-28

3-O-(3′,3′-dimethylsuccinyl)–betulinic acid

EC_{50}, H9, <0.35 nM, TI > 20,000

waminoalkanoic acid derivative statine derivative

EC_{50}, H9, 30 nM, TI > 90

FIGURE 1.2. Anti-HIV activity of betulinic acid and its semisynthetic derivatives. (*Source:* Baglin, I., Mitaine-Offer, A.C., Nour, M., Tan, K., Cavé, C., and Lacaille-Dubois, M.A., 2003, A review of natural and modified betulinic, ursolic, and echinocystic acid derivatives as potential antitumor and anti-HIV agents, *Medicinal Chemistry,* 3, 525-539.)

tion and thereby the efficacy against *Plasmodium falciparum* strains, which have become resistant to most chemosynthetic drugs.[23]

MOLECULAR-BIOLOGICAL SCREENING: OVERALL PHARMACOLOGICAL PROFILES AND SYNERGISTIC EFFECTS

The development and introduction of highly selective and sensitive molecular-biological bioassays into the screening of plant extracts, phytopreparations, and isolated compounds has revolutionized medicinal plant research. These methods operate on a molecular level (e.g., gene and receptor domains, signal transduction level) and provide a much better understanding of the mechanisms of action of a drug. They might hold the key to the development of new causally acting drugs and a more rational phytotherapy. Even the first investigations with these new methods have revealed that the pharmacological potential of many plant extracts and phytopreparations is greater than previously supposed. In addition, detailed pharmacological experiments with single isolated compounds, in comparison to the original extracts or extract fractions, have confirmed the general knowledge that many plant constituents, among them primarily phenolic compounds and terpenoids, exert polyvalent pharmacological effects. This might explain some pharmacological synergistic effects and the phenomenon that very often an extract possesses a much better therapeutic effect than single isolated constituents thereof.

Example 1

In an attempt to rationalize the cardiotonic activity of hawthorn *(Crataegus oxyacantha)* extract preparations, verified by numerous clinical studies,[24] it was found that the procyanidines and flavon-C-glycosides appear to be the main constituents of the herb responsible for pharmacological and therapeutic cardiotonic activity. In addition to pharmacological investigations performed with the extracts, we also found that the fraction of the extract containing procyanidines and flavonoids exhibits in vitro an angiotensin-converting enzyme (ACE) inhibiting effect which, together with the endothelin-dependent, smooth-muscle-relaxing effect found recently in an aorta model, might account for the dilating effect on blood vessels and simultaneous reduction in blood pressure.[25]

Furthermore, the flavonoids act as antioxidants, inhibitors of cyclooxygenase and 5-lipoxygenase, such that additional anti-inflammatory and thrombocyte-aggregation inhibitory effects can be expected. These additional investigations have assessed the flavonoids and flavon-C-glycosides as the major cardiotonic active constituents of hawthorn. They suggest further that, in order to increase the efficacy of hawthorn medication, extracts must be produced which contain both active classes of compounds in enriched form. Therefore, this is also an argument to standardize hawthorn extracts in both classes of compounds. In conclusion it can be stated that the polyvalence of hawthorn constituents now best explains the therapeutic application of hawthorn for the treatment of heart insufficiency grades I and II.

Example 2

Since the pharmacological role of garlic *(Allium sativum)* in the prevention and treatment of cancer and atherosclerosis has received increasing attention, and thorough investigations into the molecular mechanisms of action of garlic compounds are lacking, allicin and ajoene have been investigated in two new in vitro models. The first used an apoptosis-inducing model; the second was performed with inducible nitric oxide synthase (iNOS) from human macrophages.

- The first experiment showed that ajoene induces apoptosis in human leukemic cells, but not in peripheral mononuclear blood cells of healthy donors. Ajoene increased the production of intracellular peroxide (reactive oxygen species—ROS) in dose- and time-dependent fashion, which could be blocked partially by preincubation of human leukemic cells with the antioxidant N-acetyl cysteine. This result suggests that ajoene might induce apoptosis via the stimulation of peroxide production and activation of the nuclear factor κB (see Figure 1.3).[26] This finding is an important step in elucidating the underlying molecular mechanisms of its antitumor action.
- The second experiment showed that allicin and ajoene inhibit the expression of iNOS in activated macrophages (see Figure 1.4).[27] Because it is known that the inflammatory environment in human atherosclerotic lesions results in an expression of iNOS and, subsequently, in the formation of peroxynitrite, thereby aggravating the process, these results may provide an interesting basic contribution with regard to the atherogenic beneficial effects claimed for garlic in atherosclerosis prophylaxis. Meanwhile, it has also been shown that the mistletoe lectin I of *Viscum album* is able to induce the apoptosis of cancer cells, suggesting that aside from the well-known immunostimulating activity, a second molecular biological mechanism for the described antitumor activity of mistletoe preparation has been found.

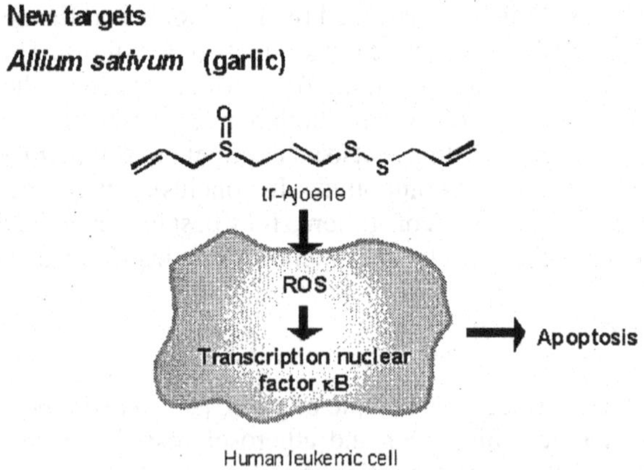

FIGURE 1.3. Mechanism of action of ajoene of garlic on apoptosis induction in human leukemic cells. (*Source:* Dirsch, V.M., Gerbes, A.T., and Vollmar, A.M., 1998, Ajoene, a compound of garlic, induces apoptosis in human leucemic cell species and activation of nuclear factor κB, *Mol. Pharmacol.*, 53: 402-407.)

New targets

***Allium sativum* (garlic)**

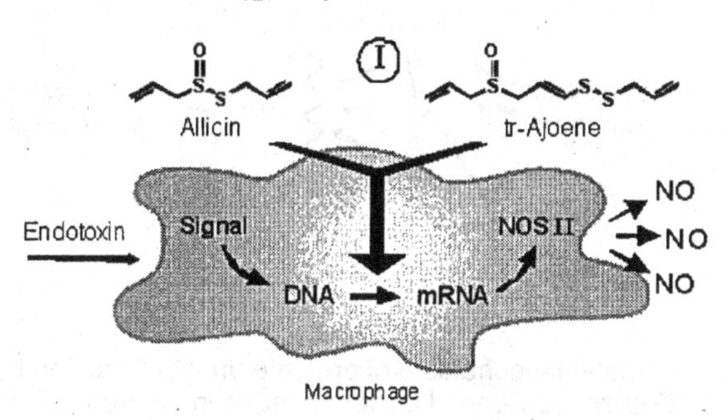

FIGURE 1.4. Allicin and ajoene of garlic inhibit the expression of iNOS in endotoxine activated macrophages. (*Source:* Dirsch, V.M., Kiemer, A.K., Wagner, H., and Vollmar, A.M., 1998, Effect of allicin and ajoene, two compounds of garlic on inducible nitric oxide synthase, *Atherosclerosis,* 139: 333-339.)

With these results, the known thrombocyte-aggregation-inhibiting, anti-inflammatory, triglyceride- and cholesterol-decreasing, antioxidant and antimicrobial polyvalence of garlic has been enlarged by two new pharmacological effects. It can be considered great progress that, with these molecular-biological experiments, a sufficiently plausible scientific explanation for the preventional application of garlic preparation and its possible beneficial effect can be presented for the first time. Similar apoptosis effects have been found for many other plant constituents known for their antitumor activity.

Example 3

Water-alcohol extracts of the root of *Urtica dioica* (stinging nettle root) have long been used and are still in use for the treatment of benign prostatic hyperplasia. Several observational and double-blind studies have assessed its clinical efficacy.[28] The roots contain a mixture of isolectins (urtica dioica agglutin [UDA]) (molecular weight ~ 9,500 Da), which possess N-acetyl-glucosamine specifity.[29] By using a ^{125}I-labeled epidermal growth factor receptor preparation, we were able to show that this UDA binds to the epidermal growth factor (EGF) receptor of epidermal cancer cell line CA 431 in a dose-dependent manner.[30] This could result in a competitive inhibition of EGF-induced proliferation (see Figure 1.5). In addition, the polysaccharides we isolated from the same root showed immunostimulating and anti-

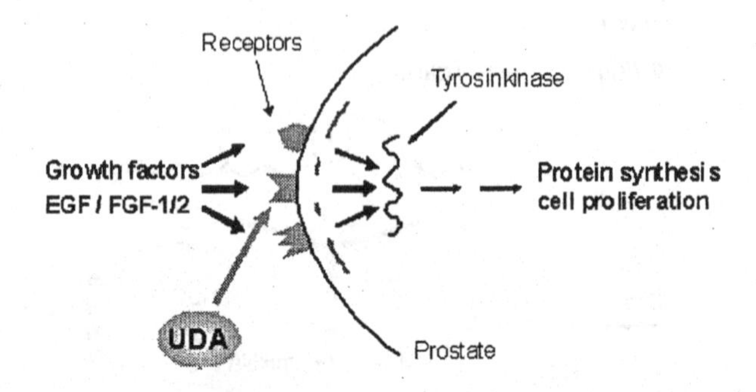

FIGURE 1.5. Putative mechanism of prostate growth inhibition by UDA of *Urtica dioica* root. (*Source:* Schilcher, H., Herbal drugs in the treatment of benign prostatic hyperplasia, in L.D. Lawson and R. Bauer (eds.), *Phytomedicines of Europe,* American Chemical Society, Symposium Series 691, 1998, pp. 62-73.)

inflammatory activity.[31] Because the lectins are very stable against heat and acids and ^{125}I-labeled UDA after oral application could be detected in the serum, it is likely that UDA is resorbed from the gut in a nondegraded form. Therefore, we believe that the well-documented clinical efficacy of *Urtica* preparations is based on a synergism of antiproliferative and anti-inflammatory effects caused by these two water-soluble classes of these compounds.

Example 4

Cannabis *(Cannabis sativa),* used in ancient times as an extract or tincture because of its analgesic, sedative, and anticonvulsive effects, was banned from the drug market after the hallucinogen tetrahydrocannabinol (THC) was isolated from the herb. Cannabis has attracted interest again because of the detection of an endogen cannabinoid system in some parts of the human brain and the immune system, represented by the two cannabinoid receptors CB1 and CB2. Molecular-biological investigations resulted in the identification of arachidonylethanolamine (anandamide) and 2-arachidonylglycerol (ether) as endogen CB ligands. According to current knowledge, this endogenic cannabinoid system plays an important role in the development of memory, in pain transduction and inhibition, in control of appetite, in lactation, in generation of emesis, and as immunomodulator. Among the most conspicuous pharmacological effects of (–)*trans*-Δ9-tetrahydrocannabinol, which recently came onto the market under the name dronabinol, the muscle-relaxant, appetite-stimulating, and analgesic effects are the most interesting.[32]

The idea of using THC alone for the treatment of multiple sclerosis (MS) patients was given up after a comparative study with cannabis extract showed that the extract exhibited a much better antispastic activity than THC alone, as measured in an immunogenic model of MS.[33] More than 1 percent of MS patients take cannabis for amelioration of spasticity and the pain associated with the condition.[34] The reason for this better effect is probably due to the content of cannabidiol (CBD) in the extract, which amplifies the antispastic effect and at the same time reduces the undesirable psychotropic side effects of THC. Cannabis is one of the best examples for synergy effects of several constituents in an extract due to interactions on a molecular level.

Example 5

More screening and isolation efforts are necessary to evaluate the pharmacological profile of multiextract preparations. First results have now been presented from a multidisciplinary research group at the University of Hong Kong. The research chemically and pharmacologically investigated a fixed herbal drug combination that has a great reputation in traditional Chinese medicine (TCM) for the prevention and therapy of stroke.[35] The herbal drug combination consists of the following eight herbal drugs: root of *Salvia miltiorrhiza,* rhizome of Ligustic, cum chuanxiong root of *Paeonia rubra,* root of *Angelica pubescens,* root of *Stephania tetrandra,* Ramulus C. Uncis of *Uncaria,* rhizome of *Gastrodia elata,* and root of *Panax ginseng.* As a result of the detailed chemical and comprehensive pharmacological investigation the researchers were able to assign defined pharmacological effects to the individual herbal drugs and their major bioactive constituents, which in turn had a direct relevance for stroke prevention and therapy.

The TCM drugs were classified into four categories: channel collateral stroke affecting, anti-inflammatory, antithrombotic, and neuroprotective. The therapeutic effects can be described as improving blood circulation, increasing cerebral blood flow, and protecting the brain from ischemic and reperfusion injuries. For the first time, researchers also made an attempt to interpret the yin and yang doctrine of TCM for stroke in terms of Western diagnostic criteria. The clinical efficacy of this TCM herbal drug mixture is explained by a pharmacokinetic or pharmacodynamic synergism. From this last example it can be concluded that in many cases drug combinations are superior in their efficacy to single isolated compounds in positively influencing a complex pathophysiological event.

The idea that a whole or partially purified extract of a plant has advantages over a single isolated constituent is not new but until recently has not been investigated systematically and rationalized. Evidence to support the occurrence of synergy within phytomedicines is accumulating and was reviewed recently by Williamson.[36] Previous results in classical pharmacology, using mixtures of bioactive compounds, have shown that a differentiation between additive and synergistic overadditive or potentiating effects is necessary. If two bioactive substances of a mixture have the same pharmacological target, an additive effect can be expected. If, however, two or more substances of a mixture have different pharmacological targets, a pharmacologically synergistic effect may result which can be greater than expected for the individual substances taken together (provided none of the substances in the mixture exerts an antagonizing effect). For several extracts or fractions of extracts, synergistic effects could be measured which exceeded the effect of single compounds, or mixtures of them at equivalent concentration, by a factor of 50 or more. Such dose-response investigations with mixtures of bioactive pure compounds can be carried out using the isobol methods as proposed by Berenbaum. We have carried out such an experiment using the thrombocyte aggregation assay with mixtures of ginkgolides A and B, two major constituents of *Ginkgo biloba*. As shown by the concave up isobol curve in Figure 1.6,[37] a typical potentiated synergistic effect was recorded. Among the many similar examples, the following combinations of bioactive compounds showed similar results.

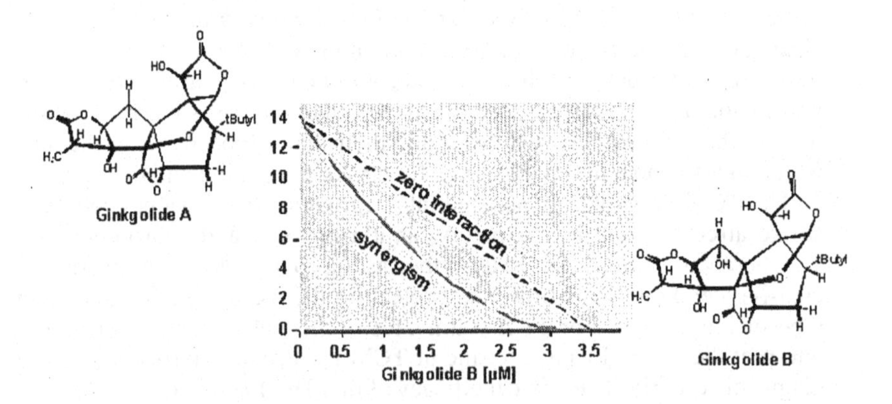

FIGURE 1.6. Synergy effects of a mixture of ginkgoglide A + B in a thrombocyte aggregation inhibiting assay. (*Source:* Williamson, E.M., 2001, *Phytomedicine,* 8: 401-409.)

In an in vitro antimalaria test, the activity of artemisinin from *Artemisia annua* was enhanced by the presence of the flavonoids artemetin and casticin. In another test, pairs of flavonoids taken from genistein, baicalein, hesperetin, naringenin, and quercetin were shown to be synergistic in inhibition of a human breast cancer cell line.

It is evident that this isobol method cannot be applied to herbal extracts or extract mixtures. Here detailed pharmacological investigations with single constituents or mixtures of them in comparison to an extract fraction or the whole extract must be used. In Table 1.1 are listed those herbal drugs for which pharmacological evidence of this potentiated synergistic effects has been assessed. The corresponding major constituents of these herbal drugs, which were used for comparison, are also given. Similar synergistic pharmacological effects have been also reported for mixed-herb formulations such as the combination of nettle, *Urtica dioica,* and pygeum bark, *Pygeum africanum,* which have a good reputation in the treatment of benign prostatic hyperplasia.[38] In this case, the combination inhibits 5-α-reductase and aromatase more significantly than either herb alone.

In contrast to these examples, antagonistic pharmacological effects can also result from the combination of two constituents or extracts. Further examples for agonistic and antagonistic effects within a single herb extract are presented in the next section.

CLINICAL TRIALS, PHARMACOKINETIC STUDIES, AND BIOAVAILABILITY STUDIES

Phytochemists, biologists, and molecular biologists cannot perform clinical trials, but they do have to provide standardized phytochemicals to clinicians. Furthermore, as trained experts in instrumental analysis of herbal drug constituents, they can perform all qualitative and quantitative determinations of chemicals and their metabolites in serum, urine, and other bodily fluids in order to elucidate the bioavailability and pharmacokinetic profile of defined plant constituents of applied plant extracts. Because the efficacy of a phytopharmaceutical in humans can be conclusively proven only by a controlled clinical trial, great efforts have been made since 1995 (especially in Europe and Germany) to fulfill the requirements of the regulatory authorities. The accepted "gold standard" for testing the efficacy of drugs (herbal or synthetic) is the double-blind, randomized, placebo-controlled clinical trial. Depending on the research question, these trials can be conducted against placebo only or also against a reference medication, if an effective drug for the indication in question exists.

TABLE 1.1. Herbal drugs for which potential synergistic effects have been evidenced.

Extract/essential oil	Major constituents
Ginkgo biloba [a]	Gingkolides A + B
Piper methysticum [b]	Yangonin, desmethoxy, dihydromethysticin
Glycyrrhiza glabra [c,d,e]	Glycyrrhizin/isoliquiritin
Cannabis sativa [f]	Tetrahydrocannabinol/cannabidiol
Valeriana officinalis [g]	Valtrate, isovaltrate/valeranone, valerenic acid
Zingiber officinale [h] (essential oil)	α-zingiberene
	β-sesquiphellandrene
	Bisabolene, curcumene

[a]Chung, K.F., McCusker, M., Page, P., Dent, kG., Guinot, P., and Barnes, P.J., 1987, Effect of ginkgolide mixture (BN52063) in antagonizing skin and platelet responses to platelet activating factor in man, *The Lancet* 31: 248-250.
[b]Singh, Y.N. and Blumenthal, M., 1997, Kava: An overview, *Herbalgram,* 39: 33-56.
[c]Cantelli-Forti, G., Maffei, F., Hrelia, P., Bugamelli, F., Bernadi, M., D'Intino, P., Maranesi, M., and Raggi, M.M., 1994, Interaction of licorice on glyzyrrhizin pharmacokinetics, *Environ. Health Perspect.,* 102(Suppl. 9): 65-68.
[d]Kimura, M., Kimuar, I., Guo. X., Luo, B., and Kobayashi, S., 1992, Combined effects of Japanese-Gino medicine Kakkon-to-ka-senkyu-shine and its related combinations and component drugs on adjuvant-induced inflammation in mice, *Phytother. Res.,* 6(4): 209-216.
[e]Miaorong, P. and Jing, L., 1996, Correlativity analysis on detoxifying effect of Radix Glyzyrrhizae on Radix Aconiti Preparata in Sini Decoction. *Proc. 40th Anniversary Conference,* Beijing University of Chinese Medicine, August 28-30. Beijing University Press.
[f]Zuardi, A.W., Shirakawa, I., and Finkelfarb, E., 1982, Action of cannabidiol on the anxiety and other effects produced by delta-9-THC in normal subjects, *Psychopharmacology,* 76: 245-250.
[g]Hölzl, J., The pharmacology and therapeutics of *Valeriana,* in P.J. Houghton (ed.) and R. Hardman (series ed.), *Medicinal and aromatic plants—Industrial profiles,* Vol. 1, *Valerian* (The Netherlands: Harwood Academic Publishers, 1997), pp. 55-57.
[h]Beckstrom-Sternberg, S.M. and Duke, J.A., Potential for synergistic action of phytochemicals in spices, in G. Charalambous (ed.), *Spices, herbs and edible fungi* (Amsterdam: Elsevier, 1994), pp. 210-233.

The specific requirements for a license and marketing authorization for a herbal product in Europe are laid down in Council Directive 75/318 EEC (1975) as amended by Directive 91/507/EEC (1991). In addition, two Committee for Proprietary Medicinal Products (CPMP) guidelines are currently available on herbal medicinal products, dealing with aspects of quality and manufacture of herbal medicinal products 1989/1992.

The efficacy requirements for herbal medicinal products show great divergence in approach among member states in Europe, but efforts are under way to harmonize the directives in the near future.[39] Indicators suggest that in the future three categories of herbal remedies will stand:

- Licensed conventional herbal medicinal products
- Herbal remedies of traditional medicine
- Food or dietary supplements

Clinical trials will be mandatory for the first category only. The requirement of safety, however, is a requisite for all classes of remedies, including dietary supplements.[40]

At present more than 400 mono- and double-blind placebo-controlled clinical trials that meet all international requirements of performance and efficacy have been performed. About 80 percent of the existing studies were done with the 14 herbal drugs listed in Box 1.3. The progress achieved can be summarized as follows:

- The clinical studies performed have made it possible to integrate the investigated phytopreparations into a rational overall concept of general drug therapy.
- The long-existing "indication lyric" of many phytopreparations could be eliminated, the indication for many phytopharmaceuticals could be confined (narrowed), and, for some phytopharmaceuticals, new indications could be detected.
- Surprisingly, about 20 percent of phytopreparations investigated, also tested against chemosynthetic drugs, showed therapeutic equivalence at the same indication (see Table 1.2).
- For many standardized extracts, therapeutic superiority over single isolated compounds thereof could be assessed.

The latter result confirms conclusively some comparative pharmacological investigations with extracts and single compounds which are explained by the action of potentiated synergistic effects. The best evidence for this existing synergism can be demonstrated by a comparison of the amount of

BOX 1.3. Efficacy assessment of herbal drug preparations by double-blind, placebo-controlled clinical trials.

Level 2001:

> 400 controlled mono- and double-blind studies, of which
- ~ 80 percent were monoextract preparations
- ~ 20 percent were combination preparations

Major herbal drugs:

Ginkgo	*Hypericum*	*Crataegus*
Echinacea	*Allium sativum*	*Valeriana*
Piper methysticum	*Sabal*	*Urtica*
Vitex agnuscastus	*Harpagophytum*	*Salix*
Aesculus hipp.	*Silybum marianum*	

TABLE 1.2. Examples of therapeutic equivalence of standardized plant extracts with synthetic drugs, evidenced by comparative placebo-controlled clinical studies.

Herbal extract	Chem. synth. drug	Indication
Crataegus (hawthorn)	Capotopril	Heart insufficiency, I + II NYHA
Hypericum (St. John's wort)	Imipramine Amitriptyline	Mild and moderate depression
Sabal (saw palmetto)	Proscar (finasteride)	Benign prostatic hyperplasia I + II
Hedera helix (ivy)	Ambroxol	Chronic bronchitis
Boswellia (incense)	Sulfasalazin	Crohn's disease

bioactive compounds in a *Hypericum* extract with those of the synthetic psychopharmacon imipramine in a clinical trial.[41]

The procentual reduction of the Hamilton Rating Scale for Depression (HAM-D) score values after six weeks' treatment of patients with moderately severe depression were the same for the *Hypericum* extract given in a dosage of 500 mg extract corresponding to 8-10 mg bioactive compounds (hypericins, hyperforin, xanthones, flavonoids, procyanidins all together) per day, and imipramine given in a concentration of 150 mg per day (see Figure 1.7). This means that 8-10 mg *Hypericum* extract's bioactive constituents together are therapeutically equivalent to 150 mg imipramine. This therapeutic equivalence cannot be explained by an enhanced resorption rate, and thereby improved bioavailability of *Hypericum* constituents, due to some nonbioactive by-products of the *Hypericum* extract.

Meanwhile, clinical evidence is available for the therapeutic superiority of *Salix* root extract over an isolated single constituent thereof, and for the herbal drug combinations valerian/kava kava and ginseng/ginkgo over only one of the two herbal drug combinations.[42] In recent years, these clinical studies have been paralleled by additional bioavailability and pharmacokinetic studies of the major bioactive constituents of several phytopharmaceuticals. These studies have become possible due to improved analytical methods, such as HPLC techniques combined with ion mass spectroscopy, diode-array or colorimetric-array method, the radio immune assay or computer-assisted xenon-positron-tomography to test the effect of administered drugs on blood circulation and neuronal activity in the human brain.[43]

Why do we also need bioavailability and pharmacokinetic studies for phytopharmaceuticals? One reason is to disprove the oft-expressed opinion that phytopharmaceuticals are placebos because no active constituents there-

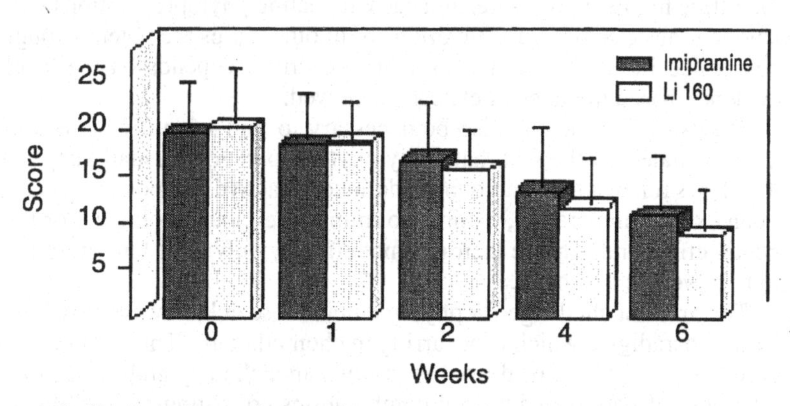

Indication: Moderate neurotic depression
Dosage: *Hypericum:* 3 × 300 mg extract per day ≅ 10 mg hypericins, hyperforin amentoflavone, procyanidins; imipramine: 3 × 25 to 35 mg
Parameter: Hamilton Rating Scale for Depression (HAM-D)

FIGURE 1.7. Comparative double-blind study with *Hypericum* extract and imipramine.

of could be detected in the body after administration. The scientific advantages, however, are obvious. Pharmacological effects can be correlated with clinical efficacy, dose regimes can be optimized, a better application form can be found to improve resorption rate of the phytopreparation from the gut, metabolization and elimination pathways can be elucidated, and possible interactions between herbal constituents and other comedicated drugs can be detected. It is likely that in the near future these studies will become a necessary requirement for clinical trials and at the same time an important contribution to rationalize phytotherapy.

SCOPE AND AIM OF PHYTOMEDICINE IN THE NEAR FUTURE

The three research topics of phytomedicine will be the scope and aim of the near future, regardless of whether phytopharmaceuticals are classified as conventional or traditional drugs. The search for new pharmacologically active compounds for drug development is an important issue but not the only one, as the trend toward using standardized plant extracts of high quality, safety, and efficacy will continue. Therefore, all efforts have to be targeted to reveal the chemical-pharmacological profiles of extracts and fixed combinations and to rationalize their therapeutic application. Whether in the future highly active, safe, and causally acting phytopreparations will be able to replace some synthetic drugs, or in other cases are potent enough to be applied in combination with synthetic drugs, depends on the level of evidence-based therapeutic efficacy achieved.

In this context, it will also be necessary to find scientific explanations and rationales for the fact that many phytopreparations, usually applied in low doses relative to the amount of defined bioactive compounds contained in an extract preparation, exhibit no immediate pharmacological or therapeutic effect and achieve their optimal efficacy only after long-term (three to four weeks) treatment.

The greatest challenge for phytomedicine research, however, will be the shift of paradigms which is occurring in chemotherapy. This change can be described as a withdrawal from monosubstance therapy and a transition to treatment of patients with drug combinations consisting of two, three, or more single drugs. This multichemotherapy has been introduced, for example, in the treatment of AIDS, hypertension, and many other diseases. The second paradigm shift can be defined as a change in the strategy of medication, characterizable as a multitarget therapy. Taking tumor therapy as an example, this new strategy aims to destroy tumor cells not via direct interaction of the drug with the tumor cell cycle but via various other mechanisms which do not damage healthy cells. This medication could be directed, e.g., to induce apoptosis of tumor cells, to inhibit angiogenesis, to stimulate specific and unspecific immune defence mechanisms, to induce the expression of antioncogenes, and to activate the production of cell-protecting proteins (heat shock proteins) (see Figure 1.8). This new, very ambitious therapeutic strategy is still in its infancy, but it is a challenge for phytomedicine research because the attempt to treat diseases according to this strategy is an actually old phytotherapeutic concept.

CONCLUSION

Four major research areas will determine to a large degree the content and strategies of future phytomedicinal research:

1. The high-tech methods for the chemical analysis of plant extracts and standardization
2. The search for bioactive constituents in plants and their use as templates to develop new drugs for diseases which at present cannot be effectively treated
3. The integration of new molecular biological methods into the screening of plant extracts and their constituents
4. The good clinical practice (GCP)-conform studies of the efficacy proof and bioavailability of standardized plant extracts

One major concern will be to investigate the multivalent and multitarget actions of plant constituents and extracts with the aim to understand and rationalize the therapeutic superiority of many plant extracts over single

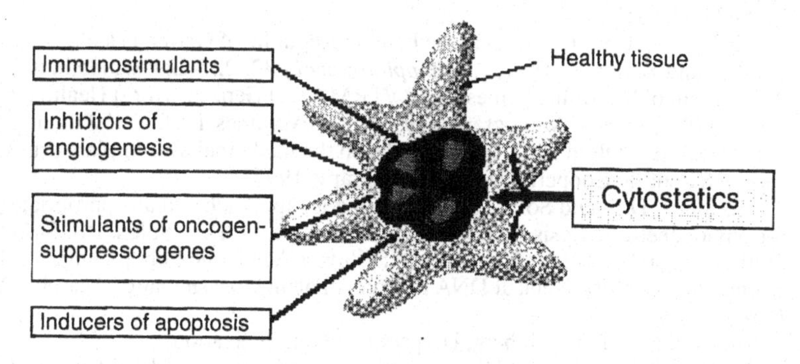

FIGURE 1.8. New strategies for the development of new drugs—cancer therapy example.

isolated constituents. The assessment of the efficacy of phytopreparations by placebo-controlled double-blind trials with bioavailability and pharmacokinetic studies holds the key for a rational and fully accepted phytotherapy.

NOTES

1. Nettleton, D.E., 1996, Phytomedicine in Europe: Their chemistry and biological activity, *Drugs of the Future,* 21(12): 1257-1264.

2. Cragg, G.M., Newman, D.J., and Snader, K.M., 1997, Natural products in drug discovery and development, *J. Nat. Prod.,* 60: 52-60.

3. Shu, Y.-Z., 1998, Recent natural products based drug development: A pharmaceutical industry perspective, *J. Nat. Prod.,* 61: 1053-1071.

4. Tyler, V.E., 1999, Phytomedicines: Back to the future, *J. Nat. Prod.,* 62: 1589-1592.

5. Wagner, H., 1999, Phytomedicine research in Germany, *Environmental Health Perspectives,* 7: 779-781.

6. Ernst, E., *Herbal medicine: A concise overview for professionals* (Oxford: Butterworth-Heinemann, 2000).

7. Schulz, V., Hänsel, R., and Tyler, V., *Rational phytotherapy* (Fourth edition) (Berlin: Springer-Verlag, 2001).

8. Phillipson, J.D., 2001, Phytochemistry and medicinal plants, *Phytochemistry,* 56: 237-243.

9. Ernst, E., Pittler, M.H., Stevinson, C., White, A., and Eisenberg, D. (eds.), *A desktop guide to complementary and alternative medicine: An evidence-based approach* (London: Mosby, Inc., 2001).

10. IFPMA, 1997, Major steps towards global drug regulations (4 ICH Conference in Brussels), *Health Horizons,* 32.

11. WHO, *Legal status of traditional medicine and complementary alternative medicine: A worldwide review,* Geneva: World Health Organization (2001). WHO/EDM/TRM/2001.2.

12. German Commission E at the Federal Republic of German Health Authority (Berlin) and Keller, K, 1991, *J. Ethnopharmacol.,* 32: 225-229; Guidelines for the Assessment of herbal medicines (WHO/TRM/91.4) Geneva: World Health Organization 1991 (State of Monographs 1999/2001) Volumes 1 and 2 (50 monographs published); ESCOP in 2003. Monographs on the medicinal uses of plant drugs, European Scientific Cooperative on Phytotherapy, 1996.

13. Soltis, D.E. and Soltis, P.S., Choosing an approach and an appropriate gene for phylogenetic analysis in molecular systematics, in D.E. Soltis, P.S. Soltis, J.J. Soltis, and J.J. Doyle (eds.), (Dordrecht: Kluwer Academic Publ., 1998), pp. 1-34; Bachmann, K., 1997, Nuclear DNA markers in plant biosystematic research, *Opera Bot.,* 137-148.

14. Heubl, G. and Meinberg, H., personal communication.

15. Wolfender, J.-L. and Hostettmann, K., Applications of liquid chromatography-mass spectrometry in the investigation of medicinal plants, in J.T. Anason, R. Mata, and Romeo (eds.), *Phytochemistry of medicinal plants* (New York: Plenum

Press, 1995), pp. 189-215; TRAC-Trends, Anal. Chem. (in press) Stecher, G., Huck, C.W., Stöggl, W.M., Bonn, G.K. Phytoanalysis—A challenge in phytomics.

16. Bauer, R. and Tittel, G., 1996, Quality assessment of herbal preparations as a precondition of pharmacological and clinical studies, *Phytomedicine,* 2(3): 193-198.

17. Amagaya S., Iizuka, A., Makino, B., Kubo, M., Komatsu, Y., Cheng F.-Ch., Ruo, I.I., Itoh, T., and Teresawa, K., 2001, General pharmacological properties of Sho-seiryu-to (TJ-19) extracts, *Phytomedicine,* 8: 338-347.

18. Bauer and Tittel, 1996, Quality assessment of herbal preparations.

19. IFPMA, 1997, Major steps towards global drug regulations.

20. German Commission E at the Federal Republic of Germany-German Health Authority, 1991; Keller (1991); Guidelines for the assessment of herbal medicines (WHO/TRM/91.4); ESCOP monographs on the medicinal uses of plant drugs, 1996.

21. Bohlin, L. and Bruhn, J.G., 1999, Bioassay methods in natural product research and drug development, *Proceedings of the Phytochemical Society of Europe,* Volume 43 (Dordrecht: Kluwer Academic Publ.).

22. Baglin, I., Mitaine-Offer, A.C., Nour, M., Tan, K., Cavé, C., and Lacaille-Dubois, M.A., 2003, A review of natural and modified betulinic, ursolic and echinocystic acid derivatives as potential antitumor and anti-HIV agents, *Medicinal Chemistry,* 3: 525-539.

23. Phillipson, J.D., 1999, New drugs from plants: It could be yew, *Phytotherapy Res.,* 13: 1-7.

24. Tauchert, M., Block, M., and Hübner, W.D., 1994, Wirksamkeit des Weißdorn-Extraktes LI 132 im Vergleich mit Captopril-Multizentrische Doppelblindstudie bei 192 Patienten mit Herzinsuffizienz im Stadium II nach NYHA, *Münch. Med. Wschr.,* 136(Suppl. 1): 27-33; Tauchert, M., 2002, Efficacy and safety of *Crataegus* extract WS 1442 in comparison to placebo in patients with chronic stubel heart insufficiency at NYHA III, *Am. Heart Journal,* 143(5): 910-915.

25. Wagner, H., Elbl, G., Lotter, H., and Guinea, M., 1991, Evaluations of natural products as inhibitions of angiotensin-I-converting enzyme (ACE), *Pharma. Pharmacol. Lett.,* 1: 15-18; Franck, U., Günther, B., Vierling, W., and Wagner, H., 1996, Investigation of *Cecropia* and *Crataegus* extracts for their angiotensin converting enzyme inhibitory and vasorelaxant activities, *Phytomedicine,* 1: 93.

26. Dirsch, V.M., Gerbes, A.T., and Vollmar, A.M., 1998, Ajone, a compound of garlic, induces apoptosis in human promyclo-leucemie species and activation of nuclear factor κB, *Mol. Pharmacol.,* 53: 402-407.

27. Dirsch, V.M., Kiemer, A.K., Wagner, H., and Vollmar, A.M., 1998, Effect of allicin and ajoene, two compounds of garlic on inducible nitric oxide synthase, *Atherosclerosis,* 139: 333-339.

28. Schilcher, H., Herbal drugs in the treatment of benign prostatic hyperplasia, in L.D. Lawson and R. Bauer (eds.), *Phytomedicines of Europe,* American Chemical Society. Symposium Series 691, 1998.

29. Peumans, W.J., De-Ley, M., and Broehaert, W.F., 1984, An unusual lectin from stinging nettle *(Urtica dioica)* rhizomes, *FEBS Let.,* 177: 99-103.

30. Wagner, H., Willer, F., Samtleben, R., and Boos, G., 1994, Search for the antiprostatic principle of stinging nettle *(Urtica dioica)* roots, *Phytomedicine,* 1: 213-214; Wagner, H., Geiger, W.N., Boos, G., and Samtleben, R., 1995, Studies on

the binding of *Urtica dioica* agglutinin (UDA) and other lectins in an in vitro epidermal growth factor recptor test, *Phytomedicine,* 4: 287-290.

31. Wagner et al., 1994, Search for the antiprostatic principle of stinging nettle.

32. Bruhn, C., 2002, Dronabinol der Wirkstoff im Hanf, *Dtsch. Apoth. Ztg.* 142: 3057-3063.

33. Baker, D., Pryce, G., Croxford, J.L., Brown, P., Pertwee, R.G., Huffman J.W., and Layward, L., 2000, Cannabinoids control spasticity and tremor in a multiple sclerosis model, *Nature,* 404: 84-87.

34. Williamson, E.M. and Evans, F.J., 2000, Cannabinoids in clinical practice, *Drugs,* 60(6): 1305-1314.

35. Gong, X. and Sucher, N.J., 1999, Stroke therapy in traditional Chinese medicine (TCM): Prospects of drug discovery and developments, *Trends Pharmacol. Sci,* 20: 191-196.

36. Williamson, E.M., (2001), Synergy and other interactions in phytomedicines. *Phytomedicine,* 8: 401-409.

37. Berenbaum, M., 1989, What is synergy? *Pharmacol. Rev.,* 41: 93-141; Wagner, H. and Steinke, B., in press, Synergy effects of a mixture of Ginkgolide A + B in a thrombozyte aggregation inhibiting assay, *Phytomedicine.*

38. Hartmann, R.W., Mark, M., and Soldati, F., 1996, Inhibition of a 5α-reductase and aromatase by PHL-00801 (Prostatonin), a comination of PY102 *(Pygeum africanum)* and UR 102 *(Urtica dioica)* extracts, *Phytomedicine,* 3/2: 121-128.

39. Anderson, L.A., Herbal medicinal products: Regulation in the UK and European Union, in E. Ernst (ed.), *Herbal medicine: A concise overview for professionals* (Oxford: Butterworth Heinemann, 2000). pp. 100-113; Loew, D. and Schroedter, A., Quality and standardization of herbal medicinal products, in E. Ernst (ed.), *Herbal medicine: A concise overview for professionals* (Oxford: Butterworth Heinemann, 2000), pp. 59-68.

40. De Smet, P.A.G.M., 1992, Toxicological outlook on the quality assurance of herbal drugs, in P.A.G.M. De Smet, K. Keller, R. Hänsel, and R.F. Chandler (eds.), *Herbal Drugs* Volume 1 (Springer Verlag), pp. 1-72.

41. Woelk, H. (2000), Comparison of St. John's wort and imipramine for treating depression: Randomised controlled trial, *British Medical Journal,* 321: 536-539.

42. Schmidt, B., Ludke, R., Selbmann, H.K., Kotter, I., Tschirdewahn, B., Schaffner, W., and Heide, L. 2001, Efficacy and tolerability of a standardized willow bark extract in patients with ostearthritis randomised, placebo-controlled, double blind clinical trial, *Phytother. Res.,* 15(4): 344-350; Wheatley, D., 2001, Stress induced insomnia treated with kava and valerian, *Human Psychopharmacol.,* 16: 353-356; Scholey, A.B. and Kennedy, D.O., 2002, Acute, dose-dependent cognitive effects of *Ginkgo biloba, Panax ginseng* and their combination in healthy young volunteers: Differential interactions with cognitive demand, *Psychopharmacology,* 17: 35-44.

43. Bhattaram, V.A., Graefe, U., Kohlert, C., Kveit, M., and Derendorf, H., 2002, Pharmacokinetics and bioavailability of herbal medicinal products, *Phytomedicine,* Suppl. 2, pp. 1-33.

PART II: THE USE OF MEDICINAL PLANTS THROUGHOUT HISTORY

Chapter 2

History of Application of Medicinal Plants in China

Chang-Xiao Liu

INTRODUCTION

The Chinese materia medica is an important part of traditional Chinese medicine and Chinese civilization. Chinese traditional medicine rose from mythical medicine to a system of drugs and herbal drugs. The first book on materia medica, *Herbal Classics of the Divine Plowman (Shennong Bencao Jing),* known as "the canon of materia medica," was composed in the first century B.C. and the first century A.D. by an author using the name Shennong, the Holy Farmer. It is well known that China was a leader in the use of medicinal plants, and this distinction has remained and has been even further developed. Medicinal plants today play an outstanding role within the framework of official health services in China. China is endowed with an abundance of local medicinal plants. More than 5,000 plants have been identified as medicinal plants in China. The 1995 edition of *The Chinese Pharmacopoeia* recorded more than 700 kinds of Chinese drugs originating from medicinal plants.

The terms *Chinese traditional and herbal drugs (zhongcaoyao)* are the general designations for medicinal agents used in China that originate from natural products. As China has vast territory and includes many nationalities, Chinese herbal medicines can be classified according to their different usages into three categories: Chinese traditional drugs *(zhongyao),* national minorities drugs *(minzuyao),* and Chinese folk drugs *(cao yao).*

CHINESE TRADITIONAL AND HERBAL DRUGS

Chinese Traditional Drugs

Chinese traditional drugs of Chinese materia medica *(zhongyao)* are used by doctors of Chinese traditional medicine. Their usages are directed by a series of systematic and self-contained theories. The drugs have spread throughout China and have been regarded as the main component of the Chinese herbal medicines. About 500 items are commonly used in Chinese traditional medicines, of which *Panax ginseng, Glycyrrhiza uralensis,* and *Rheum palmatum* are well known.

National Minorities Drugs

National minorities drugs *(minzuyao)* are used by doctors of different national minorities, of which Tibetan, Mongolian, Uygur, and Tai are the most important. These drugs are all closely connected with Chinese traditional medicine but are characterized by factors of nationality and locality. Their use is generally limited to certain districts, usually only within local or ethnic territories. Based on common knowledge, the commonly used Tibetan traditional drugs include about 200 items. National minorities drugs are estimated to number between 500 and 600 items.

Chinese Folk Drugs

Herbal drugs or Chinese folk drugs *(caoyao)* are used throughout China, and they are characterized by application on the basis of trial and error. They are ailment oriented, usually without theoretical direction. Very often the most effective herbal drugs and prescriptions are kept secret, and are accessible only to a lucky few who will teach their own descendants or favorite apprentices.

BRIEF HISTORY OF APPLICATION OF MEDICINAL PLANTS

It is reasonable to assume that early humans, compelled by the drive to find foodstuffs, examined various substances encountered and thus acquired empirical knowledge. Accumulation of this knowledge might have told them that some of the materials were palatable and nutritious while others were disagreeable and caused adverse effects. This information was handed down to their progeny, and the study of nature by humans has grown from there. This process led to the discovery of a group of substances called *drugs,* which showed beneficial effects to the sick. Therefore, discoveries of drugs were closely connected with the search for edible stuffs, as evidenced by numerous writings on materia media at different periods in history.

In Chinese history, many well-known medicinal works recorded a great number of medicinal plants. These works are treasured as a precious cultural heritage. Since the majority of Chinese traditional drugs are of plant origin, this system was once called *herbal.* In Chinese folklore, the discovery of drugs is accredited to the prehistoric emperor Shennong. According to certain texts, it is said that Shennong, a famous originator of traditional Chinese medicine, tried hundreds of herbs and tested 70 poisons a day. This folklore illustrates the difficulties of the "trial-and-error" accumulated clinical experience in the treatment of disease. Thus, the first book on Chinese materia medica is named after him, the *Herbal Classics of the Divine Plowman.* The book is believed to have been written between the first century B.C. and the first century A.D. It represents a systematic account of all knowledge related to the application of drugs in that period in Chinese history. Unfortunately, the original text was lost long ago and what is available today is a script assembled during the Ming and Qing dynasties from quotations contained in many other classics.

The information traced from such records can be broken down into broad categories of particular Chinese dynasties: the Han dynasty and before, the Three Kingdoms epoch, the north-south division epoch, the Tang dynasty, the Song dynasty, the Yuan dynasty, the Ming dynasty, the Qing dynasty, and the New China period.

Han Dynasty and Before (B.C. 206-A.D. 200)

The first record of Chinese medicinal material is the *Herbal Classics of the Divine Plowman,* which is also called *Shennong Herbs (Shennong Bencao Jing).* This work represents a crystallization of the empirical knowledge of the prehistoric and early historical people of China about the healing efficacy of natural drugs from plants, animals, and minerals. The authorship is not known, but it is attributed to Shennong, a mythological figure who is said to have used himself as an experimental subject and to have learned how to help people medically through his experiences. *Shennong Bencao Jing* contains 365 kinds of drugs, among which 18 are repeated. Therefore, the number of different drugs is actually 347, of which 239 are derived from medicinal plants, 65 from animals, and 43 from minerals. The 365 kinds of drugs were divided into three classes: superior, intermediate, and inferior. The superior drugs are strengthening and tonic drugs. They can be administered for prolonged periods without harmful effects. The intermediate drugs are tonic to subjects and effective against diseases with toxicity dependent to dosage. The inferior drugs possess specific therapeutic activity but are toxic and should not be taken for any prolonged course. The pharmacological actions and therapeutic indications of each drug were described in the classical works.

Three Kingdoms (A.D. 220-280) and Western Jin Epochs (A.D. 265-420)

During the three kingdoms and Western Jin epochs (A.D. 220-420) two persons wrote about their experiences in China, where they had encountered Arabian traders with unusual objects. In *Zi-An-Ji* (Collection of man of peace) Sima Sui (A.D. 231-273) mentioned that *Ruta graveolens* was used for keeping silverfish from books. Ji Han (A.D. 263-306) in *Nan-Fang Cao-Mu Zhuang* (Description of plants of the south) recorded *Phoenix dactylifera* and *Lawsonia inermis,* introduced by Arabian merchants, and *Areca catechu,* from tropical Asia.

North-South Division Epoch (A.D. 425-590)

The first increase in recorded knowledge of Chinese material medica occurred during the north-south division epoch. The outstanding recorder was Tao Hong-Jing, a native of southern China (Jiangsu province). He prepared two treatises on materia medica, *Unrecorded Materials of Eminent Physicians,* which is also called *Transactions of Famous Physicians (Ming-Yi Bie-Lu),* and *Commentary on Shennong's Herbal Classics (Shennong Bencao Jing Zhu).* He incorporated his findings of *Unrecorded Materials of Eminent Physicians* into a copy of *Shennong's Herbal.* Tao Hong-Jing's work was quoted repeatedly by Li Shi-Zhen. *Ming-Yi Bie-Lu,* compiled by Tao Hong-Jing (452-536 A.D.), came next to be the most important classical writings on Chinese materia medica as the annotation to *Shennong Herbs.* It specialized in the study of herbs during the Liang dynasty. Tao Hong-Jing expounded *Shennong Herbs* on one hand, and supplemented it with 365 additional medicinal substances on the other. This book doubled the list of known drugs to 730. It described the actions and uses of these drugs and classified them according to their source, e.g., jades, stones, herbs, woods, fruits, vegetables, and cereals. The book in seven volumes is also called *Shennong Bencao Jing Zhu* (Commentary on Shennong herbs).

Tang Dynasty (A.D. 618-907)

Emperor Gao-Zong of the Tang dynasty was concerned with having a complete pharmacopoeia and appointed the minister of works, Li Ji, to revise Tao's *Commentary on Shennong's Herbal Classics.* The complete work consisted of seven volumes and is known as *Ying-Gong Tang Bencao* (Duke Ying's Tang pharmacopoeia). A decade later, Su Gong pointed out the need for a revision of this pharmacopoeia. The Emperor Gao Zong appointed a team of 22 scholars including Su Jing and Li Ji 659 A.D. The complete work, *Tang Bencao* (New Tang pharmacopoeia), consisted of 53 volumes with illustrations. It listed 644 medicinal substances. Since it was sponsored by the Tang dynasty government, it is considered to be the earliest officially published pharmacopoeia in the world.

Chen Cong-Qi completed the *Bencao Shi-Yi* (Unrecorded materia medica) in A.D. 739. He was a thorough researcher and added approximately 400 kinds of drugs not mentioned in the *Tang Bencao.* By the middle of the eighth century, the first materia medica of subtropical and tropical China and adjacent areas of the Indo-Malaysian Peninsula appeared. Li Xun edited *Hai-Yao Bencao* (Register of the medicine imported via the seas) in A.D. 756.

Song Dynasty (A.D. 960-1279)

In the history of Chinese materia medica, the Song dynasty could be considered the golden era for publication of pharmacopoeias. The first emperor, Song Tai-Zu, appointed imperial pharmacist Liu Han and Taoist monk Ma Zhi to lead a team of nine in revising the *Tang Bencao.* Using the works written during the Tang dynasty as references, they organized the material, placed many former names into synonymy, and added 133 new items. It took them seven years to complete the revision, which filled 21 volumes. This work is known as the *Kai-Bao Bencao* (Pharmacopoeia of the Kaibao reign) and was published in A.D. 973.

In A.D. 1058, Song Ren-Zong, the fourth emperor of the Song dynasty, asked the imperial medical team, headed by Zhang Yu-xi and Lin Yi, to revise the 100-year-old *Kai-Bao Bencao.* The working team added 82 new items and revised 17 entries. Known as the *Revised Pharmacopoeia of Giao-You Reign,* it consisted of 20 volumes. In A.D. 1062, a new work, *Tu-Jin Bencao* (An illustrated pharmacopoeia) was compiled, which consisted of 21 volumes. This work has been quoted by all subsequent authors dealing with Chinese materia medica.

Jingshi Zhenglei Beiji Bencao (Classic classified materia medica for emergencies) was compiled by Tang She-Wei, a physician especially proficient in therapeutics who declined the offer of an official appointment in order to devote his life to medical practice and the collection of folk recipes. Its publication at the end of the eleventh century made it the earliest well-preserved book on materia medica. It listed 1,746 medicines with directions for use and preparation, together with many new prescriptions. The book, in 31 volumes, laid a solid foundation for the development of the modern knowledge of herbs.

In A.D. 1108, Emperor Song Hui-Zong received a manuscript on materia medica prepared by Tang Shen-wei. In this manuscript, Tang modified the text of the *Revised Pharmacopoeia of Giao-You Reign* and incorporated the illustrations from *Tu-Jin Bencao*. This combined work Tang had titled *Zheng-Lei Bencao* (Verified identification of materia medica). The emperor accepted the manuscript, which then became known as the *Da-Guan Bencao* (Pharmacopoeia of Da-Guan reign).

Yuan Dynasty (A.D. 1279-1368)

The Yuan dynasty is the period when China was under Mongol rule. Near the end of the Yuan dynasty, Zhu Zhen-Heng prepared *Bencao Yan-Yi Buyi* (A supplement to herbal commentaries).

Ming Dynasty (A.D. 1368-1644)

Chinese medicine was developed during the Ming dynasty. *Dian-Nan Bencao* (Materia medica of southern Yunana) was edited by Lan Mao in 1370, and *Jun-Huang Bencao* (Famine herbal) by Zhu Xiao in 1407. Li Shi-Zhen's (1552-1578) *Bencao Gangmu* (Compendium of materia medica) is the best-known Chinese classic materia medica. However, the most comprehensive work was published in 1596, after Li's death. It ran 52 volumes and listed 1,892 medical substances, with more than 1,000 illustrations and more than 10,000 detailed descriptions of the appearance, properties, method of collection, preparation, and use of each substance. This book is far more than a pharmaceutical text; it is a branch of natural history including botany, zoology, mineralogy, and metallurgy. Li Shi-Zhen (1552-1578) was a great physician and naturalist. He was born in a physician's family. After three unsuccessful attempts at the higher official examinations, he decided to enter his father's profession. He concentrated on medical studies, and his medical skills soon gained wide recognition among his contemporaries. He wrote a dozen medical works.

With Li Shi-Zhen's knowledge of the Chinese classics and his practical experience in diagnosing illnesses and prescribing remedies, he prepared a manual for practitioners of traditional Chinese medicine, the *Bencao Gangmu.* When Li used *gang* and *mu,* he implied arrangement and classification. From the 800 references, he selected 1,518 items; he also provided 374 new entries. For each of the 1,892 items, he gave a recognized name, which he called *gang,* and cited all the synonyms with his commentaries, which are called *mu.* To this framework, he added information on properties, tastes, and efficacy for the treatment of diseases, as well as recipes. The Western world learned about Chinese medicine through the material medica completed during the Ming dynasty, particularly form *Bencao Gangmu.* During the Ming dynasty, a famous treatise on plants and agriculture, *Qun Fang Tu* (Monograph on flowers) was published by Wang Xiang-Jin.

Qing Dynasty (A.D. 1644-1911)

Medicinal plants used in traditional Chinese medicine were recorded largely in ancient herbals and pharmacopoeias before the Qing dynasty. During the Qing dynasty, both medicinal plants employed and the source references had changed. The famous works were *Flower Mirror* by Chen Hua-Zi in 1688, *Englarged Monograph of Flowers* by Liu Hao in 1708, *New Version of Materia Medica* by Wu Yi-Bao in 1757, and *Bencao Gangmu Shiyi* (Supplement to the compendium of materia medica) in 1765.

Bencao Gangmu Shiyi was written by Zhao Xue-Min. It described 716 medical substances that had not been included in Li Shi-Zhen's *Bencao Gangmu.* This brought the total of the drugs recorded in formal literature at that time to 2,608 kinds of drugs. Zhao Xue-Min was a native of Qiantang (now Hangzhou, Zhejiang province), a physician, and a famous pharmacist. He collected and systemized the folk doctor Zhao Bi-Yun's experiences in treating diseases in two books titled *Chuanya Neipian* (Internal treatise on folk medicine) and *Chuanya Waipian* (Extra treatise on folk medicine).

In addition to these books, many other works in Chinese history are well known. Table 2.1 lists some of them.

MEDICINAL PLANTS IN CHINA IN THE PAST CENTURY

In 1911, China was transformed from an empire to a republic. Since 1949, the establishment of New China and the research and application of Chinese medicinal plants produced wide changes. Three modern references are important to mention: *Zhongyao Zhi* (The Chinese materia medica),

TABLE 2.1. Some well-known historical Chinese medical works.

Author(s)	Book	Dynasty	Number of drugs
Chen Cang-Qi	*Supplement to the Herbs*	Tang	—
Wang Hao-Gu	*Materia Medica of Decoction*	Tang	238
Zhu Su	*Herbs for Relief of Famines*	Tang	414
Wang Ang	*Essentials of Materia Medica*	Ming	470
Yuan Shan and Dao Ren	*Nature of the Drugs*	Ming	90
Lei Xiao	*Lei's Methods of Preparing Drugs*	Fifth century	—
Zhuang J-Guang and Miao Xi-Yong	*Complete Handbook of the Preparations Principles of Correct Diet*		200
Hu Si-Hei	*An Illustrated Book of Plants*	Qing	1,714
Wu Qi-Jun	*A Lengthy Compilation of Plants with Illustrations*	Qing	1,848

Zhongyao Dacidian (Encyclopedia of Chinese materia medica), and *Quaguo Zhongcaoyao Huibian* (A compilation of Chinese medicinal herbs).

Encylopedia of Chinese Materia Medica

Zhongyao Dacidian was compiled by Jiangsu College of Chinese Traditional Medicine and published by Shanghai Scientific and Technical Publishers in 1977 in two volumes with an extra appendix. It is the most comprehensive compilation on Chinese material medica to date, with 5,767 items of crude drugs from 4,161 species of plants, 442 species of animals, and 58 kinds of minerals included. For each plant entry, the account includes the recognized Chinese name, its synonyms, and the source material with scientific identification, a morphological description illustrated, distribution, and information on cultivation, harvesting time, and procedure, product characteristics, chemical composition, pharmacology, treatment use, and references.

The Chinese Materia Medica

Zhongyao Zhi was edited by the Institute of Medicinal Plant Development, Chinese Academy of Medical Sciences, and published by People's Health Publishers from 1959 onward in four volumes. This book is intended to be an authoritative reference, describing 494 drugs with meticulous care. The second edition, in six volumes, appeared in 1979. A new edition of this book was completed in 2000 and the English edition was published.

A Compilation of Chinese Medicinal Herbs

Quaguo Zhongcaoyao Huibian was published by People's Health in 1975. It contained two volumes and listed about 2,200 herbal medicines. The names, sources, morphology, environment, cultivation, collection and preparation, chemistry, pharmacology, nature, tastes, uses, indications, and administrations of the medicines are given in detail, along with illustrations of the plants.

Color Atlas of Chinese Herbal Drugs

In 1982, People's Health Publishing Press and Xionghua Publishing House (Japan) published the first volumes of the *Color Atlas of Chinese Herbal Drugs.* It will eventually include 25 volumes and a total of 500 drug items, along with color prints, most of them of medicinal plants.

New Compendium of Chinese Materia Medica

New Compendium of Chinese Materia Medica was edited by Wu, Zhou, and Xiao and published in three volumes. The book includes 7,000 medicinal plants. The first, second, and third volumes appeared in 1988, 1990, and 1991, respectively.

A Pictorial Encyclopedia of Chinese Medical Herbs

A Pictorial Encyclopedia of Chinese Medical Herbs was edited by Xiao. It included 5,000 color photographs of Chinese herbal medicines in ten volumes and was published in 1988 and 1989.

In the course of Chinese history, Chinese medicines rose from mythical substances to a system of herbal medical treatments. Although the materia medica has been recompiled many times, the basic contents have remained the same. This is because the evaluation of a drug was recorded by famous and well-experienced physicians. Many of their famous prescriptions were handed down from generation to generation with only minor alterations. In other words, these herbal agents had to be screened through the clinical experiences of many ages to be effective, with more than a few being remarkably efficient. The process of trial and error gradually eliminated the worthless or less efficient ones. At present, about 600 effective agents are in common use, about 100 of which are at the top of the list for trial in scientific studies in pharmacology, botanical classification, chemical analysis, and clinical evaluation.

Medicinal Plants in The Chinese Pharmacopoeia

As noted, Chinese traditional drugs are composed of materials derived from plants, animals, and minerals. Based on a nationwide survey, at least 80 percent of Chinese traditional and herbal drugs are from plant origin, from five editions of *The Pharmacopoeia of the People's Republic of China.* These drugs identify 7,295 plant species of therapeutic value (see Tables 2.2 and 2.3).[1]

More than 5,000 kinds of Chinese materia medica are used for the treatment of various diseases, and commonly used Chinese materia medica includes about 500 kinds, the number of species of Chinese materia medica recorded in the 1995 edition of the Pharmacopoeia of the People's Republic of China.[2] The drug names, origin, and parts used of Chinese materia medica, originating from Chinese medicinal plants, were listed in the books edited by Liu and Xiao.[3]

TABLE 2.2. The number of species of Chinese materia medica recorded in *The Pharmacopoeia of the People's Republic of China.*

Edition	Species
1953	65
1963	446
1977	882
1985	506
1990	509
1995	522
2000	531

TABLE 2.3. An overview of Chinese medicinal plants.

Origin	**No. of medicinal plant species**
Thallophytes	467
Bryophytes	43
Pteridophytes	455
Gymnosperms	126
Angiosperms	—
Dicotyledons	8,598
Monocotyledons	1,429
Total	11,118

Mainly owing to official government promotion, some important contributions of either their intensive or extensive nature have been published in the more than four decades since the founding of the People's Republic of China. *The Chinese Pharmacopoeia* underwent a major change in format in the five recent editions published in 1977, 1985, 1990, 1995, and 2000. The 1977 edition was issued in two volumes for the first time. The first volume devoted itself entirely to the Chinese materia medica in describing 822 drugs or with their extracts and oils, along with 270 compound formulations. The 1985 edition was also issued in two volumes. The first volume describes 506 individual drugs or their single preparation, and 207 compound preparations. The 1990 edition was also issued in two volumes. The first volume collects 509 single drugs or with their oils and 275 kinds of complex preparations. The 1995 edition collected 522 items of drugs and 398 preparations. The 2000 edition recorded 531 kinds of drugs and 461 preparations.

The plant families, which contain more than 100 medicinal plant species, are listed in Table 2.4, and the plant genera, which contain more than 15 medicinal plant species and with the higher proportion of therapeutic members, are listed in Table 2.5.[4]

This increase is due, on one hand, to the fact that China has a long history in the use of Chinese traditional medicine and, thus, has accumulated vast experiences which have frequently involved conspicuous and unique effects on certain diseases, as well as generally little toxicity and few side effects. On the other hand, owing to the various geographical and ecological conditions, both the flora and medicinal plant resources are very abundant. In addition, the Chinese government has emphasized systematizing Chinese traditional medicine and raising it to a higher level. As a result, the tra-

TABLE 2.4. Families containing more than 100 medicinal plant species.

Family	Medicinal species/ genera	Example of important medicinally used genera
Compositae	778/155	*Artemisia, Senecio, Aster, Saussurea, Atractylodes*
Leguminosae	490/107	*Cassia, Sophora, Crotalaria, Glycyrrhiza, Indigofera, Astragalus*
Ranunculaceae	420/34	*Aconitum, Delphinium, Thalictrum , Clematis, Anemone, Coptis*
Lamiaceae	436/75	*Salvia, Rabdosia, Nepeta, Clinopodium, Scutellaria, Thymus*
Liliaceae	358/46	*Fritillaria, Polygonatum, Smilax, Ophiopogon, Veratrum, Allium*
Rosaceae	360/39	*Crataegus, Prunus, Rosa, Rubus, Potentilla*
Orchidaceae	287/76	*Dendrobium, Gastrodia, Habenaria, Liparis, Bletilla*
Apiaceae	234/55	*Apiaceae, Angelica, Heracleum, Bupleurum, Ferul, Ligusticum, Peucedanum*
Rubiaceae	219/59	*Uncaria, Hedyotis, Oldenlandia, Morinda, Rubia, Gardenia*
Euphorbiaceae	160/39	*Euphorbia, Glochidion, Croton, Mallotus*
Saxifragaceae	155/24	*Astilbe, Bergenia, Saxifraga*
Papaverzceae	135/15	*Corydalis, Meconopsis, Papaver*
Polygonaceae	123/8	*Rheum, Polygonum, Fagopyrum*

ditional medicines and their preparations are widely accepted throughout China today.

From this information, one may conclude that Chinese traditional drugs and herbal drugs have made a great contribution to the health of the Chinese people. They play an important role in modern drug research and development in which new drugs from biological sources are being sought.

DEVELOPMENT OF RESOURCES

Due to the destruction of forests, overgrazing of meadows, expansion of industry, and urbanization, as well as excessive collection of rare and en-

TABLE 2.5. Genera containing more than 15 medicinal species and with higher proportion of therapeutic members.

Genus	Species/ total species	Main ethnopharmacological data
Aconitum	46/167	Anodyne, antirheumatic, dispelling, internal cold, arrow poison, asthmolytic, cardiotonic
Aralia	19/30	General tonic, antirheumatic, promote circulation
Aristolochia	31/39	For various infections, snakebite, treatment of abdominal diseases
Berberis	67/200	Antidysentery, various infections, antipyretic, antidote
Codonopsis	26/50	Tonic, invigorate the functions of the digestive system, for weakness
Corydalis	73/150	Anodyne, treatment of coronary heart diseases, febrile and detoxicant
Delphinium	35/113	Anodyne, antirheumatic
Dendrobium	35/60	Tonic, for febrile diseases with thirst and dry mouth, dry cough, and chronic tidal fever
Hypericum	16/50	Antimicrobial, emonogogue, treatment of hepatitis
Lysimachia	28/50	Invigorating blood circulation and eliminating blood statis, asthmolytic
Rabdosia	30/90	For menstrual disorders
Salvia	41/78	Antimicrobial, anticancer, febrifugal, and detoxicant
Swertia	22/70	Antimicrobial, treatment of coronary heart diseases, also as a tranquilizer
Thalictrum	39/67	Treatment of hepatitis, bitter-tonic, febrifugal, and detoxicant
Gentiana	53/247	Antipyretic, antimicrobial, for various infections, anticancer
Scutellaria	33/102	Antipyretic, antidote, stomachic, anti-inflammatory

dangered plants and animals in the wild, the natural resources of medicinal plants and animals are being reduced day by day. There is an urgent need to draw up the necessary plans for medicinal resource utilization and conservation.

Conservation and Implementation

In spite of abundant and rich natural resources, medicinal plants are not unlimited and must be well protected by law and by the people. Collection of any medicinal plant should be guided by precise knowledge of the species, including its locality, time of maturation, parts to be collected, and conservation needs. Steps should be taken to avoid overexploitation and excessive collection. The gathering of rare and endangered species, such as *Panax ginseng, Coptis chinensis, Gastrodia elata,* and *Paris polyphylla,* should be prohibited.

Gene banks of medicinal plants should be established. At present, quite a number of important Chinese medicinal plants have already been preserved in a gene bank under the auspices of several agricultural institutions.

Introduction and Acclimatization

A considerable number of the plants listed in the ancient Chinese materia medica were of foreign origin, and even today some of the drugs prescribed have to be imported, for example, American ginseng from the United States and Canada, and *Amomum kravanh* from Thailand. To meet the market demand, several botanical gardens in China have initiated the task of introducing these plants for acclimatization, aimed an eventual plantations in farms. The introduction policy also includes those nonofficial crude drugs of wild origin, which have been used by people for practical medication. Some examples of these categories are as follows:

- Of foreign origin: *Panax quinquefolium, Amomum kravanh, Strychnos nux-vomica, Rauwolfia vomitoria, Syzygium aromaticum, Cassia acutifolia, Vinca minor*
- Indigenous species of wild origin: *Schisandra chinensis, Gentiana scabra, Cistanche deserticola, Asarum sieboldii, Scutellaria baicalensis, Bupleurum chinensis, Atractylodes lancea, Anemarrhena asphodeloides, Fritillaria unibracteata, Alpinia oxyphylla*

Cultivation

With the increasing use of medicinal plants in China, it is necessary to cultivate the most commonly used ones to guarantee supplies. About 100 species of medicinal plants are under cultivation, covering some 460,000 hectares (ha). The main plants are shown in Table 2.6.[5] The most important cultivated medicinal plants are *Panax ginseng, Panax notoginseng,*

TABLE 2.6. Commonly cultivated medicinal plants in China.

Plant	Part used
Achyranthes bidentata	Root
Alisma orientale	Rhizome
Allium tuberosum	Seed
Amomum villosum	Fruit
Angelica dahurica	Root
Angelica sinensis	Root
Andrographis paniculata	Plant
Areca catechu	Seed
Artemisia argyi	Leaf
Asparagus cochinchinensis	Root
Aster tataricus	Root
Astragalus mongholicus	Root
Atractylodes macrocephala	Rhizome
Aucklandia lappa	Rhizome
Brassica juncea	Seed
Carthamus tinctorius	Flower
Cassia acutifolia	Leaf
Cassia obtusifolia	Seed
Celosia cristata	Flower
Chaenomeles speciosa	Fruit
Citrus grandis	Exocarp
Citrus aurantium	Fruit
Citrus reticulata	Exocarp
Codonopsis pilosula	Root
Coix lacryma-jobi var. *ma-yuan*	Seed
Coptis chinensis	Rhizome
Coptis deltoidea	Rhizome
Cornus officinalis	Fruit
Corydalis yanhusuo	Rhizome
Crataegus pinnatifida	Fruit
Curcuma aromatica	Rhizome
Curcuma domestica	Rhizome
Curcuma zedoaria	Rhizome
Datura metel	Flower

TABLE 2.6 *(continued)*

Plant	Part used
Datura innoxia	Flower
Dendrobium nobile	Plant
Dioscorea opposita	Rhizome
Dolichos lablab	Seed
Eriobotrya japonica	Leaf
Eucommia ulmoides	Bark
Euphorbia longan	Aril
Eupatorium fortunei	Leaf
Euryale ferox	Seed
Evodia rutaecarpa	Fruit
Fritillaria thunbergii	Bulb
Gardenia jasminoides	Fruit
Gastrodia elata	Rhizome
Ginkgo biloba	Leaf, seed
Gleditsia sinensis	Fruit
Glehnia littoralis	Root
Illicium verum	Fruit
Impatiens balsamina	Seed
Isatis indigotica	Leaf, root
Lilium brownii var. viridulum	Bulb
Lonicera japonica	Flower
Lycium barbarum	Fruit
Magnolia officinalis	Bark
Magnolia biondii	Flower
Melia toosendan	Fruit
Mentha haplocalyx	Root bark
Morus alba	Bark, leaf
Nelumbo nucifera	Leaf
Ophiopogon japonicus	Root
Paeonia lactiflora	Root
Paeonia suffruticosa	Root bark
Panax ginseng	Root
Panax notoginseng	Root
Perilla fructicosa	Leaf, seed

Plant	Part used
Phellodendron amurensis	Bark
Piper nigrum	Fruit
Pogostemon cablin	Leaf
Polygonatum cyrtonema	Rhizome
Poria cocos	Plant
Prunus armeniaca	Seed
Pnunus mume	Fruit
Pnunus persica	Seed
Pseudostellaria heterophylla	Root
Psoralea corylifolia	Fruit
Raphanus sativus	Seed
Rehmannia glutinosa	Rhizome
Ricinus communis	Seed
Salvia miltiorrhiza	Root
Schizonepeta tenuifolia	Plant
Scrophularia ningpoensis	Root
Sesamum indicum	Seed
Sinapis alba	Seed
Sophora japonica	Flower bud
Stephania tetrandra	Root
Trigonella foenum-graecum	Seed
Tussilago farfara	Flower bud
Zingiber officinale	Rhizome
Ziziphus jujuba	Fruit

Astragalus mongholicus, Angelica sinensis, Coptis chinensis, Codonopsis pilosula, Rehmannia glutinosa, Paeonia suffruticosa, Cinnamomum cassia, Amomum villosum, and *Atractylodes macrocephala;* they are all commonly used Chinese traditional drugs.

At the same time, several wild medicinal plants that are needed in vast quantities are *not* being introduced and cultivated, e.g., *Glycyrrhiza uralensis, Rheum palmatum, Cistanche deserticola, Poria cocos,* and *Dioscorea nipponica.*

In addition, modern biological technology has been used. For example, tissue culture is used for propagating *Lithospermum erythrorhizon, Panax quinquefolium, Corydalis yanhusuo, Scopolia tangutica,* and some others.

NOTES

1. Liu, C.X., Xiao, P.G., and Li, D.P., *Modern research and application of Chinese medicinal plants* (Hong Kong: Hong Kong Medical Publishers, 2000), pp. 1-100.

2. Pharmacopoeia Committee of People's Republic of China, *Pharmacopoeia of the People's Republic of China* (Beijing: People's Health Publishers, 1953, 1963, 1975, 1990, 1995, 2000).

3. Liu, Xiao, and Li, 2000, *Modern research and application*; Liu, C.X. and Xiao, P.G., *An introduction to Chinese materia medica* (Beijing: Beijing Medical University and Peking Union Medical College Press, 1993), pp. 1-36; Liu, C.X. and Xiao, P.G., *Chinese medicinal plants* (Tianjin: Tianjin Pharmaceutical Research, 1995), pp. 1-256.

4. Liu, Xiao, and Li, 2000, *Modern research and application.*

5. Liu and Xiao, 1993, *An introduction to Chinese materia medica.*

Chapter 3

Ethnopharmacology of Traditional Chinese Drugs from Medicinal Plants

Chang-Xiao Liu

INTRODUCTION

Traditional Chinese medicine, as practiced today, is still largely based on its unique system of theories. The ethnopharmacological classification, properties, tastes, actions, and indications of traditional Chinese drugs are based on the classical theories. In this chapter I introduce the ethno-pharmacological classification of traditional Chinese drugs recorded in the *Pharmacopoeia of the People's Republic of China,* as well as the origin, properties, actions, and indications of commonly used Chinese traditional drugs. Current reviews classify traditional Chinese drugs, according to classical pharmacology, into 18 groups as follows.[1]

DIAPHORETICS

Diaphoretics are drugs for treating exterior symptom complexes. These drugs dispel pathogenic factors from the exterior of the body. They are divided into two categories according to their properties: warm-pungent or cool-pungent drugs.

The Warm-Pungent Drugs

These drugs are warm and pungent in property. They are usually reserved for the treatment of exterior symptom complexes caused by wind and cold. The main function of warm-pungent drugs is to dispel wind and cold. They are indicated in warm-cold excess exterior syndrome and are accompanied by chills, fever, absence of sweating, headache, thin and white tongue coating, and superficial and rapid pulse. These drugs can be used to treat cough, asthma, edema, and pain caused by invasion of wind-damp.

The commonly used drugs are *Centipedae* (herb), *Cinnamomi* (branch), *Angelicae dahuricae* (root), *Ephedrae* (herb), *Zingiberis recens* (rhizome), *Ligustici* (rhizome), *Magnoliae* (flower), *Perillae* (leaf), *Saposhnikoviae* (root), *Schizonepetae* (herb), and *Xanthii* (fruit).

The Cool-Pungent Drugs

These drugs are cool and pungent in property. They are usually used for the treatment of exterior symptom complexes caused by wind and heat. They are not enough to promote sweating. They are indicated in wind-heat exterior syndrome manifested as high fever, mild chills, thirst, sore throat, scanty or no sweating, thin and yellow tongue coating, and superficial and rapid pulse. Some of these drugs are also effective in promoting the full expression of rashes in measles, and are for use on boils and carbuncles on the exterior. The most commonly used drugs are *Arctii* (fruit), *Bupleuri* (root), *Chrysanthemi* (flower), *Periostracum cicadae* (flower), *Cimicifuga* (rhizome), *Dichroae* (root), *Mori* (leaf), *Menthae* (herb), *Sojae praparatum* (seed), and *Spirodelae* (herb).

ANTITUSSIVE EXPECTORANTS AND ANTIASTHMATICS

Drugs for Resolving Cold-Phlegm

Phlegm-resolving drugs with warm properties are usually used for the treatment of symptom complexes caused by cold phlegm or phlegm-dampness, marked by expectoration of watery thin phlegm or profuse foamy phlegm when the lung is involved; nausea and vomiting when the stomach is involved; and so forth. The commonly used drugs are *Arisaematis* (rhizome), *Inulae* (flower), *Pinelliae* (rhizome), *Platycodi* (root), and *Typhonii* (rhizome).

Drugs for Resolving Heat-Phlegm

Phlegm-resolving drugs with cold properties are usually used for treating symptom complexes caused by expectoration of thick yellow phlegm and other symptoms reflecting heat in the lung. Some of these drugs are also used for the treatment of goiter, scrofula, impairment of consciousness, convulsion, or even mania if these conditions are caused by phlegm. The commonly used drugs are *Adenophorae* (root), *Bambusae* (stalk), *Laminariae seu Eckloniae* (thallus), *Peucedani* (root), *Sargassum, Trichosanthis* (fruit), and *Trichosanthis* (root).

Antitussive Drugs

Antitussives are used for relieving cough. The commonly used drugs are *Armeniaceae amarum* (seed), *Sterculiae lychnophorae* (seed), *Farfarae* (flower), *Rhododendrri daurici* (leaf), *Ginkgo* (seed), *Lilii* (bulb), *Eriobotryae* (leaf), *Syringae* (bark), *Morus radicis* (root bark), *Stemonae* (root), *Fritillaria cirrhosa* (bulb), and *Fritillaria thunbergii* (bulb).

Antiasthmatic Drugs

Antiasthmatics are used for relieving asthmastic symptoms. Most of the antitussives may have some effect on asthma, whereas antiasthmatics also relieve cough. The commonly used drugs are *Aristolochiae* (fruit), *Daturae* (flower), *Physochlainae* (root), *Cymbopogonis* (herb), *Lepidii* (seed), and *Perillae* (fruit). They are usually considered in combination.

DRUGS CLEARING HEAT AND DRYING DAMPNESS

Drugs that clear heat and dry dampness, which are also called antipyretic drugs, mainly remove internal heat. Generally, internal heat syndrome excludes exterior syndrome due to invasion by exogenous pathogenic factors and heat syndrome caused by retention of blood in the interior. Based on the functions and indications, these drugs can be classified as five subgroups: drugs for clearing heat and reducing fire, drugs for clearing heat and dispelling dampness, drugs for clearing heat and cooling blood, drugs for clearing heat and releasing toxins, and drugs for clearing heat caused by yin deficiency. In general, drugs that clear heat are cool in property. This action may impair the normal functions of the spleen and stomach. Therefore, they should be used with caution in those individuals with poor appetite, weakness of the spleen and stomach, or diarrhea. They serve as auxiliary drugs in combination with toxic herbs or with herbs nourishing yin.

Drugs for Clearing Heat and Reducing Fire

Drugs clearing heat and reducing fire are mainly indicated for syndromes of heat in the "Qi" level. These symptoms are due to exogenous pathogenic heat invasion manifested as high fever, sweating, thirst, delirium, irritability, scanty and brown urine, yellow and dry tongue coating, surging and forceful pulse, excessive heat in lungs, excessive heat in the stomach, and excessive heat in the heart. The most commonly used drugs for clinical application in traditional Chinese medicine are *Anemarrhenae* (rhizome), *Phragmitis* (rhizome), *Trichosanthis* (root), *Banbusa* (leaf), *Lophatheri* (herb), *Gardeniae* (fruit), *Prunella* (spice), *Eriocauli* (flower), *Buddlejae* (flower), and *Celosiae* (seed).

Drugs for Clearing Heat and Dispelling Dampness

Drugs that clear heat and dispel dampness are bitter and dry, and are mainly indicated for interior syndromes of excessive heat without consumption of body fluids. The most commonly used drugs for clinical application in traditional Chinese medicine are *Scutellariae* (root), *Coptidis* (rhizome), *Phellodendri* (bark), *Gentianae* (root), and *Sophora flavescentis* (root).

Drugs for Clearing Heat and Cooling Blood

Drugs that cool blood are used to cool excess heat in the blood. The most commonly used drugs for clinical application in traditional Chinese medicine are *Rehmanniae* (root), *Scrophulariae* (root), *Moutan* (bark), *Paeoniae rubra* (root), and *Lithospermi seu Arnebiae* (root).

Drugs for Clearing Heat and Releasing Toxins

Drugs that clear heat and remove toxins are mainly used for syndromes of excessive toxic heat, such as epidemic diseases, toxic dysentery, boils, and carbuncles. The most commonly used drugs for clinical application in traditional Chinese medicine are *Lonicerae* (flower), *Forsythiae* (fruit), *Taraxaci* (herb), *Violae* (herb), *Isatidis* (leaf), *Houttuyniae* (herb), *Dictamni* (bark), *Rhapontici seu Echinopsis* (root), *Baptisia, Pulsatillae* (root), *Portulacae* (herb), and *Andrographitis* (herb).

Drugs for Clearing Heat Caused by Yin Deficiency

Drugs that clear heat caused by deficiency of yin are indicated for syndromes of interior heat due to deficient yin, such as afternoon tidal fever. The most commonly used drugs for clinical application in traditional Chinese medicine are *Artemisia annua* (herb), *Cynachi* (root), *Lycii* (bark), and *Stellariae* (root).

PURGATIVE DRUGS

Purgative drugs either stimulate or lubricate the large intestine to promote bowel movement. These drugs are mainly indicated for constipation; their function is purgation. Through the function, impacted feces and fluid remaining in the intestines are discharged, pathogenic heat or cold is dispelled, and edema is relieved. The drugs are divided into three subgroups as follows.

Purging Drugs

Purging drugs are bitter and cold in property, and they function to reduce fire and promote bowel movement. They are indicated for retention of feces due to accumulation of excessive heat in the stomach and intestines. The most commonly used purging drugs are *Rhei* (root and rhizome), *Sennae* (leaf), and *Aloe* from medicinal plants have three kinds, *Rheum palmatum, Cassia angustifolia,* and *Aloe vera,* respectively.

Drugs Lubricating the Intestines

These drugs are mostly seeds of plants containing oil. They lubricate the intestines and move the stool, and indicated in constipation due to deficient body fluids in the aged person or patients with weakness of the body due to chronic diseases. The commonly used plants (*Cannabis* [fruit] and *Prunus* [seed]) for lubricating the intestines are *Cannabis sativa* and *Prunus japonica,* respectively.

Drugs Transforming Water

These drugs purge water. They cause the discharge of water, together with feces, from the body. The property of drugs transforming water is toxic. Overdoses or prolonged administration is harmful to human health. Some of them promote urination and are indicated in cases of edema in limbs, ascites, fullness in the chest, and asthma due to phlegm-dampness. The most commonly used drugs transforming water are *Genkwa* (flower), *Kansui* (root), *Euphorbiae* (root), and *Crotonis* (seed).

DRUGS EXPELLING WIND AND DAMPNESS

Drugs expelling wind and dampness do so mainly from skin, muscles, channels, and collaterals. These drugs work by relaxing tendons, promoting circulation of channels and collaterals, stopping pain, and strengthening joints and bones. The main indications are wind-damp obstruction pain, migrating pain, spasm of tendons, numbness of muscles, hemiplegia, soreness and pain in the lower back and knees, and flaccid lower limbs. The most commonly used drugs for expelling wind and dampness are *Angelica pubescens* (root), *Clematidis* (root), *Stephania tetrandra* (root), *Gentiana macrophylla* (root), *Siegesbeckiae* (herb), *Clerodentri trichotomi* (leaf), *Chaenomelis* (fruit), *Trachelospermi* (stalk), *Mori* (branch), *Taxilli* (branch), *Acanthopanacis* (bark), *Erythrinae* (bark), *Aristolochiae mollissimae* (herb), *Piperis futokadsurae* (stalk), *Homalomenae* (rhizome), and *Pini nodi* (wood).

AROMATIC DRUGS TRANSFORMING DAMPNESS

Aromatic drugs, which transform dampness, are mostly pungent, fragrant, warm, and dry. The functions are to promote Qi, transform dampness, and strengthen the spleen and stomach. The main indications of dampness are full sensation in the epigastric and abdominal regions, vomiting, sour regurgitation, poor appetite, lassitude, diarrhea, sweet taste in the mouth, and sticky and moist tongue coating. The most commonly used aromatic drugs for transforming dampness are *Atractylodis* (root), *Magnoliae officinalis* (bark), *Gastachis* (herb), *Eupatorii* (herb), *Amomi* (fruit), *Alpiniae katsumadai* (seed), and *Tsaoko* (fruit).

DRUGS BENEFITING URINATION AND DRAINING DAMPNESS

Drugs that benefit urination and drain dampness work by transforming accumulated dampness or fluid into urine. Some of them also clear heat and drain dampness. These drugs are sweet or tasteless, neutral or slightly cold, and are indicated in dysuria, edema, urinary tract diseases, phlegm-damp, jaundice, and eczema. These drugs should be used with caution in those with deficiency of yin and body fluids. The most commonly used drugs benefiting urination and draining dampness are *Poria, Poria cocos, Alismatis* (rhizome), *Coicis* (seed), *Plantaginis* (seed), *Aristolochiae seu Clematis* (stalk), *Medulla tetrapanacis* (pith), *Lysimachiae* (herb), *Lygodii* (spores), *Pyrrosiae* (leaf), *Dioscoreae hypoglaucae* (rhizome), *Artemisiae scopariae* (herb), *Kochiae* (fruit), *Exocarpium Benincasae, Lagenariae* (peel), *Phaseoli* (seed), *Polygoni avicularis* (herb), *Dianthi* (herb), and *Malyvae* (seed).

DRUGS WARMING THE INTERIOR

The interior cold syndromes can be caused by an invasion of exogenous pathogenic cold that leads to deficiency of "Yang-Qi" in the spleen and stomach, or by weakness of Yang-Qi, giving rise to excessive yin and cold in the interior. Drugs that warm the interior and expel cold are pungent and hot. The functions are to warm the spleen and stomach, to expel cold, and to stop pain in interior cold syndromes. The most commonly used drugs for warming the interior are *Aconiti lateralis praeparata* (root), *Zingiberis* (rhizome), *Cinnamomi* (bark), *Evodiae* (fruit), *Asari cum Radice* (herb), *Zanthoxyli* (peel), *Caryophylatae* (flower), *Alpinia officinarum* (rhizome), and *Foeniculi* (fruit).

DRUGS REGULATING QI

Qi means energy or bioenergy. Qi stagnation may be mainfested as stifling distension and pain, whereas Qi perversion presents as nausea, vomiting, belching, asthma, and cough. If Qi stagnation in the spleen and stomach impair the normal function of ascending and descending; if Qi stagnation in the lungs inhibits the function of descending; or if the liver is stagnated, the manifestations are hypochondriac pain, stifling sensations in the chest, hernia pain, nodules, distension and pain in the breasts, or irregular menstruation. The drugs regulating Qi are aromatic, pungent, and bitter. They are good for promoting the freeing up of Qi and are indicated in both Qi stagnation and Qi perversion. The most commonly used drugs regulating Qi are *Citri reticulatae* (peel), *Citrus reticulate* (peel), *Citri reticulatae viride* (peel), *Aurantii immaturus* (fruit), *Citri sarcodactylis* (fruit), *Citri* (fruit), *Aucklandiae seu vladimiriae* (root), *Cyperi* (rhizome), *Linderae* (root), *Aquilariae resinatum* (wood), *Toosedan* (fruit), *Allii macrostemi* (bulb), *Santati albi* (wood), *Calyx kaki, Rosae* (flower), and *Mume* (flower).

DRUGS RELIEVING FOOD STAGNATION

Drugs relieving food stagnation are indicated when distension and fullness in the epigastric and abdominal regions, belching, sour regurgitation, nausea, vomiting, abnormal bowel movement, or indigestion due to weakness of the spleen and stomach occurs. These drugs improve the appetite and digestion and remove stagnated food. They are usually used to treat dyspepsia or indigestion. The most commonly used drugs relieving food stagnation are *Crataegi* (fruit), *Hordei germinatus* (fruit), *Oryzae germinatus* (fruit), and *Raphani* (seed).

DRUGS EXPELLING PARASITES

These drugs are used to treat roundworm, tapeworm, pinworm, hookworm, and other intestinal parasites. Some of these drugs are toxic. Overdosage may cause side effects or toxicity. The most commonly used drugs expelling parasites are *Quisqualis* (fruit), *Meliae radicis* (bark), *Arecae* (seed), *Cucurbitae moschatae* (seed), *Agrimoniae* (bud), *Omphalia, Carpesii* (fruit), *Torreyae* (seed), and *Dryopteris crassirhizomae* (rhizome).

DRUGS STOPPING BLEEDING

Drugs that stop bleeding or hemorrhaging are used to treat cases of vomiting blood, epistaxis, cough with blood, bloody stool, bloody urine, uterine bleeding, or traumatic bleeding. The most commonly used drugs for stopping bleeding are *Cirsii japonica* (herb and root), *Cirsii* (herb), *Sanguisorbae* (root), *Imperatae* (rhizome), *Sophorae* (flower), *Agrimoniae* (herb), *Bletillae* (rhizome), *Trachycarpi carbonisatus* (leaf stalk), *Notoginseng* (root), *Rubiae* (root), *Typhae* (pollen), *Artemisiae argyi* (leaf), and *Nelumbinis rhizomatis* (node).

DRUGS REGULATING BLOOD CONDITIONS

These drugs are divided into two subgroups: hemostatics and invigorating drugs for improving blood circulation. Hemostatics are drugs for arresting bleeding. Invigorating drugs invigorate blood circulation and eliminate blood stasis, and are used to arrest bleeding, to invigorate blood circulation, and to eliminate blood stasis. The most commonly used hemostatics are *Agrimoniae* (herb), *Dioscoreae* (rhizome), *Bletillae* (rhizome), *Boehmeriae* (root), *Capsellae* (herb), *Imperatae* (rhizome), *Cirsii japonica* (herb and root), *Nelumbinis rhizomatis* (node), *Rumecis patientiae* (root), *Rubiae* (root), *Sanguisorbae* (root), *Cirsii* (herb), *Sophorae* (flower), *Sphorae immoturus* (flower), and *Sophorae* (fruit). The most commonly used drugs invigorating blood are *Lycopi* (herb), *Sparganii* (rhizome), *Typhae* (pollen), *Ligustici chuanxiong* (rhizome), *Corydalis* (rhizome), *Curcumae* (root), Cyathulae (root), *Leonuri* (herb), *Persicae* (seed), *Paeoniae rubra* (root), *Salviae miltiorrhizae* (root), *Carthami* (flower), *Sappan* (wood),

Siphonostegiae (herb), *Vaccaria* (seed), *Verbena* (herb), and *Zedoariae* (rhizome).

DRUGS TRANQUILIZING THE MIND

These drugs are used to treat deficient Qi of the heart, deficient blood in the heart, or flaring up of fire in the heart, which are manifested as restlessness, palpitations, anxiety, insomnia, dream-disturbed sleep, convulsions, epilepsy, and manic psychotic disorders. The most commonly used drugs for tranquilizing the mind are *Albiziae* (flower), *Biotae* (seed), *Polygoni multiflori* (stalk), *Ganoderma lucidum, Polygalae* (root), *Ziziphi spinosae* (seed), and *Valeriana* (rhizome).

DRUGS PACIFYING THE LIVER AND SUBDUING ENDOGENOUS WIND

Drugs that pacify the liver and subdue endogenous wind are used to stop tremors and subdue the yang. They are mainly indicated in tremors, convulsions, and spasms caused by the stirring of liver wind, and in dizziness or vertigo due to hyperactivity of liver yang. As endogenous wind can arise from extreme heat; hyperactivity of the liver yang or deficient yin and blood, corresponding herbs should be used in the prescriptions. The most commonly used plant medicines for pacifying the liver and subduing endogenous wind are *Uncaria* (branch), *Gastrodia* (rhizome), *Tribuli* (fruit), and *Senna* (seed).

DRUGS THAT OPEN ORIFICES

Drugs that open the orifices are aromatic substances. These drugs are called aromatic stimulants. They open the sense organs and restore consciousness. These drugs are used when heat attacks the pericardium or turbid phlegm mists the heart. Manifestations include loss of consciousness, delirium, epilepsy, convulsions, or coma from windstroke. The most commonly used two drugs for opening orifices are *Styrax* and *Acori graminei* (rhizome).

TONICS

Tonics are used to recover from deficiencies, or to strengthen or supplement the body's resistance against disease. These drugs have an invigorating action and reinforce vital function in the treatment of deficiencies of yin, yang, Qi, and Xue (blood), and therefore they are usually divided into four kinds of tonics: Qi, blood, yin, and yang tonics.

Drugs Replenishing Qi

Qi tonics are used to replenish vital energy, and are usually indicated in the treatment of Qi deficiency (functional activities) of the spleen and lung; they are also used in the treatment of blood deficiency together with blood tonics. The most commonly used Qi tonics are *Ginseng* (root), *Panacis quinquefolii* (root), *Codonopsis* (root), *Pseudostellariae* (root), *Astragali* (root), *Atractylodis macrocephalae* (rhizome), *Dioscoreae* (rhizome), *Dolichoris* (seed), *Glycyrrhizae* (root), and *Jujubae* (fruit).

Drugs Replenishing Yang

Yang tonics are used for reinforcing vital function (chiefly in the kidney). They are used in the treatment of insufficient vital function of the kidney marked by chilliness in the back and loins, impotence, chronic diarrhea that occurs daily before dawn, chronic ashmatic conditions, and so on. The most commonly used yang tonics are Morindae *officinalis* (root), *Cistanches* (herb), *Curculiginis* (rhizome), *Epimedii* (herb), *Eucommiae* (bark), *Dipsaci* (root), *Cibotii* (rhizome), *Drynariae* (rhizome), *Psoraleae* (fruit), *Alpiniae oxyphyllae* (fruit), *Cordiceps, Juglandis* (seed), *Cuscutae* (seed), *Allii tuberose* (seed), and *Cnidii* (fruit).

Drugs Replenishing Blood

Blood tonics nourish the blood. They are mainly used in the treatment of blood deficiency. Most of these drugs are also yin tonics used for replenishing vital essence. The most commonly used blood tonics are *Angelica sinensis* (root), *Rehmanniae praeparata* (root), *Polygoni multiflori* (root), *Paeoniae* (root), and *Arillus longan* (root).

Drugs Replenishing Yin

Yin tonics are used for replenishing the vital essence and fluid. They are indicated in the treatment of vital essence and fluid deficiency marked by thirst, tidal fever, heat sensation on the palms and soles, fidgeting, insomnia, red tongue, weak and rapid pulse, etc. The most commonly used yin tonics are *Glehniae* (root), *Ophiopogonis* (root), *Asparagi* (root), *Dendrobii* (herb), *Polygonati odorati* (rhizome), *Lilii* (bulb), *Lycii* (fruit), *Mori* (fruit), *Eciliptae* (herb), *Ligustri lucidi* (fruit), and *Seami* (seed).

ASTRINGENT DRUGS

Astringents are sour in nature. They are known to stop sweating, reduce diarrhea, control essence, hold in urine, and stop leukorrhea, bleeding, and coughing. They are indicated in cases with weakness from chronic disease or unconsolidated antipathogenic factors which lead to spontaneous perspiration, night sweats, chronic diarrhea, chronic dysentery, seminal emissions, nocturnal emissions, enuresis, frequent urination, chronic cough and asthma, uterine bleeding, leukorrhea, and prolapse of the uterus or rectum. They relieve symptoms by preventing the antipathogenic factor from weakening. The most commonly used astringent drugs are *Schisandrae* (fruit), *Mume* (flower), *Tritici levis* (fruit), *Ephedrae* (root), *Ailanthi* (bark), *Chebulae* (fruit), *Myristicae* (seed), *Papaveris* (peel), *Nelumbinis* (seed), *Corni* (fruit), and *Rosae laavigatae* (fruit).

NOTE

1. Geng, J.Y., Huang, W.Q., Ren, T.C., and Ma, X.F., *Medicinal herbs* (Beijing: New World Press, 1990), pp. 12-262; Liu, C.X. and Xiao, P.G., *An introduction to Chinese materia medica* (Beijing: Peking Union Medical College & Beijing Medical University Press, 1993), pp. 94-190; Liu, C.X. and Xiao, P.G., *Chinese medicinal plants* (Tianjin: Tianjin Institute of Pharmaceutical Research, 1995), pp. 1-90; Liu, C.X., Xiao, P.G., and Li, D.P., *Modern research and application of Chinese medicinal plants* (Hong Kong: Hong Kong Medical Publisher, 2000), pp. 1-100.

Chapter 4

Research and Development of New Drugs Originating from Chinese Medicinal Plants

Chang-Xiao Liu
Zohara Yaniv

INTRODUCTION

Almost all pharmacologically active compounds belong to the secondary metabolites of plants. In clinical trials, the relationships between the chemical structures of active substances and their pharmacological activities have been studied. It was found that the alkaloids with benzylisoquinoline type are of major importance and potentially could serve as the basis of new drugs for analgesic, sedative, antimicrobial, and cardiovascular ailments. The propane-type alkaloids, such as anisodamine and anisodine isolated from *Scopolia*, are attractive for their effect on microcirculatory systems. The diterpene-type alkaloids isolated from *Aconitum* and *Delphinium* provide a starting point in the quest for more effective anodyne, arrhythmia, and cardiotonic drugs. The alkaloids related to harringtonine were found to have an antileukemic action. In the phenolic compounds, several flavonoids, such as scutellaria, icariin, isorhanetin, and puerarin, and coumarins, such as daphnetin and daphnoretin, need to be considered for their cardiovascular effects. Within the terpenoids, many monoterpenes display an effect against respiratory problems and can perhaps be used in the development of more favorable remedial medicines. Several peroxide substances, such as quinhaosu and yingzhaosu A, hold promise as antimalarial agents.

About 140 new drugs have originated directly or indirectly from Chinese medicinal plants by means of modern scientific methods, confirming that Chinese medicinal plants are an important source of new drugs.

A surprisingly large amount of information on ethnopharmacological applications throughout China is available. For instance, over the past 40 years our two institutes (the Institute of Medicinal Plants, Chinese Academy of Medical Sciences, Beijing, China; and Tianjin Institute of Pharmaceutical Research, Tianjin, China) have collected 40,000 such pieces of information. On the basis of these data, some interdisciplinary systematization should be carried out. This kind of endeavor will certainly help predict the most promising botanical candidates for further investigation. Use of specialized software and dedicated databases may be the best means to this end.

Since 1949 more than 1,000 scientific reports on studies of Chinese medicinal plants have been published. These reports include botanical, chemical, pharmacological, and clinical studies of Chinese medicinal plants, and some have led to the development of new drugs. Chinese traditional and herbal drugs have gained importance in the international medical, biomedical, and pharmaceutical institutions as a potential source of valuable medicinal agents. [1]

PATHWAY OF RESEARCH AND DEVELOPMENT FROM CHINESE TRADITIONAL AND HERBAL DRUGS

Chinese traditional medicine is an integral part of Chinese civilization and is one of the major systems of medical practice in China today. Numerous prescriptions, folk drugs, and herbal drugs play an important role in maintaining people's health. They are also the resources for new drug research and development. The following pathways for the discovery and development of new drugs from Chinese traditional and herbal drugs should be reviewed and considered.

Research Conducted Under the Guidance of Experiences of Chinese Traditional Medicine

No records of the term *chronic myelocytic leukemia* can be found in the classic works on Chinese traditional medicine. However, based on the theories and methods of Chinese traditional medicines, diagnosis of this condition can be made and treated according to the symptoms involved. A complex prescription of "Danggui-Luhui" pills has been used for this purpose and was confirmed with a clinical trial on 22 patients. The prescription consists of the following: *Angelica sinensis* (root), *Aloe vera* (dried juice), *Gentiana scabra* (fruit), *Gardenia jasminoides* (root), *Scutellaria baicalensis* (root), *Phellodendron amurense* (stem bark), *Coptis chinensis* (rhizome), *Rheum palmatum* (root and rhizome), natural indigo (a product derived from the leaves of *Saphicacanthus cusia*), and *Aucklandia lappa* (root). Subsequent experimentation conducted with the animal model of leukemia 7212 (L7212) in mice revealed that only natural indigo among the ingredients of the prescription was effective, but it was associated with some side effects. In an effort to increase the therapeutic effect and reduce the side effects, further study of natural indigo has been made. Indirubin was identified as the active antitumor principle. The compound showed various inhibitory activities against Walker carcinoma in rats and Lewis lung cancer in mice, as well as mammary cancer (Ca 615) and L7212 in mice. Toxicological studies showed that when indirubin was given daily for one month at an oral dose of 100 to 500 $mg \cdot kg^{-1}$, no toxicological reactions, including the inhibition of bone marrow production, were observed. Another clinical trial was carried out in 314 cases of individuals with chronic meylocytic leukemia. Of these 314 patients, 82 achieved complete remission, 104 partial remission, and 87 beneficial effects; the total effective proportion was 87.3 percent. In the clinical observation, side effects such as abdominal pain and diarrhea sometimes occurred. Indirubin has been shown to have a similar therapeutic effect to Myleran in treatment of chronic myelocytic leukemia. It neither possesses serious side effects nor inhibits production of bone marrow.

It is well known that alpha-dichroine isolated from the Chinese traditional antimalarial drug *Dichroa febrifuga* has shown an antimalarial activity from 98 to 152 times as strong as quinine hydrochloride. After systematic investigation of another Chinese traditional antimalarial drug, *Artemisia annua,* it has been found that the neutral fraction of the ether extract showed marked malariacidal effects in *Plasmodium berghei* in mice and *P. inui* and *P. cynomolgi* in monkeys. Toxicological studies showed no obvious side effects in animals. A clinical study of 30 cases also gave satisfactory results. Further investigation led to the isolation of an active compound called qinghaosu. Clinical studies on different qinghaosu preparations were carried out involving 2,099 cases of malaria; among these 1,511 were *P. vivax* and 558 were *P. falciparum* malaria. All patients were clinically cured.

Exploration of Ancient Literatures of Chinese Traditional Medicine

In the course of treatment of hypertension among workers in a factory, it was found that blood pressure could be reduced to normal of suitable level. However, many of the patients still suffered from headache, dizziness, and a stiffness and soreness in the neck. In the ancient classic *Shanghanlun* (Treatise on febrile diseases), edited by Zhang Zhong-Jing, it is recorded that *gegen* decoction, a decoction of the root of *Pueraria lobata* as its chief ingredient, is recommended for the treatment of stiffness and soreness in the neck. As a part of the pursuit of this therapeutic lead, a series of clinical trials of preparation of *Pueraria lobata* gave satisfactory results. The active components of this plant were isolated; they are the isoflavones daidzein, daidzin, puerarin, and daidzein-4',7-diglucoside. Pharmacological studies revealed that the total isoflavone of the single compound puerarin was capable of increasing the cerebral and coronary blood flow, decreasing the oxygen consumption of the myocardium, increasing the blood oxygen supply, and depressing the production of lactic acid by the blood oxygen deficient heart muscle. These actions could partially explain the mechanism of the *Pueraria* preparation used for the relief of hypertensive diseases, angina pectoris, migraine, and sudden deafness. Daidzein has shown a papaverine-like spasmolytic action, and the clinical trial of this compound has demonstrated that it has an action similar to the total isoflavone preparation.

Exploration of the Effective and So-Called Secret Prescriptions

The first example to be offered in this category is the discovery of a new antiepileptic agent called antiepilepsirine. It possesses a wide anticonvulsive spectrum and a low toxicity. Piperine was found to be the active compound of an effective prescription for an antiepileptic, which consisted of pepper and turnip. Based on this interesting clue, a series of derivatives of piperine were synthesized. One of these derivatives was antiepilepsirine, which was found to possess a strong anticonvulsive action. After clinical trials, it was established as an antiepileptic agent having minimal side effects.

However, the scientists did not stop there. They prepared 100 piperine derivatives which were screened in animal models for central nervous system actions. Four of these derivatives, *N*-(*p*-chlorocinnamoyl) piperidine, *N*-(cinnamoyl) piperidine, *N*-cyclopentyl-*p*-chloro-cinnamoylamide, and *N*-(isopropyl-*p*-chlorocinnamoyl) amide were found to be more effective than antiepilepsirine against experimental animal epilepsy.

In 1970, secret prescriptions for antitaenia used by peasants in northeast China was explored. The method of the traditional treatment was to take it orally on an empty stomach, a powdered portion of *Agrimonia pilosa*. Five to six hours after administration of the drug not only the segments but also the scolex of the worm could be eliminated. Sixty-eight patients were then treated in hospital by a similar regimen of drug administration with effectiveness of higher than 98.5 percent. The active principle of the bud of *Agrimonia* was then isolated by Chinese chemists. Its structure was elucidated and named agrimophol. This compound was submitted to clinical study and its efficacy was fully substantiated. This drug enjoys the advantage of minor side effects and high anthelmintic effect.

THE TREATMENT PRINCIPLES OF CHINESE TRADITIONAL MEDICINE

Chinese traditional medicinal treatment is conducted under the guidance of systematic principles. Prescriptions are chosen in accordance with those same diagnosis and treatment principles. The most common treatment principles and their possible correlations with Western diagnoses are summarized in Table 4.1.

For instance, after an investigation of the so-called blood circulation stimulating drug *Ligusticum chuanxiong*, more than ten compounds were isolated from this plant, among which tetramethylpyrazine was found to be the active principle. It was present only in minute amounts in the plant. Pharmacological and clinical studies revealed that tetramethylpyrazine hydrochloride could improve microcirculation in the mesenteries of rabbits and could also dilate capillary vessels in vitro. Tetramethylpyrazine not only inhibited platelet aggregation caused by adenosine triphosphate (ADP) but also disaggregated the collected platelets. Clinically, it can be

TABLE 4.1. Treatment principles of Chinese traditional medicines and their possible correlations to Western diagnoses.

Treatment principles	Relation to Western diagnoses	Examples of treatment drugs
Mobilization of blood circulation and treatment of stasis	Cardiovascular diseases, thrombo angitis, obliterans, cerebral embolism	*Ligusticum chuanxiong, Carthamus tinctorius, Typha latifolia, Crataegus pinnatifida, Angelica sinensis, Paeonia lactiflora, Salvia miltiorrhiza*
Promotion of vigor *membranaceus* var. and stablization of vitality	General tonic for treatment of the weakness, probably with an immunostimulating action	*Astragalus mongholicus, Panax ginseng, Acanthopanaxesnti, Codonopsis pilosula orrhiza*
Treatment of fever and inflammation along with detoxicity	Antibacterial antivirotic, anticancer, possibly with an immunostimulating action	*Coptis chinensis, Scutellaria baicalensis,* natural indigo

used with satisfactory results for the treatment of occlusive cerebral blood vessel diseases, such as cerebral embolisms.

The Value of the Hints Emerging from the Phenomenon of Drugs or Plants Acting upon the Human Body

It has long been known that the use of crude cottonseed oil for salad dressing leads to infertility in young men. As a result of a series of clinical investigations that involved more than 8,000 healthy men receiving gossypol treatment for more than six months, it was found that the gossypol present in crude cottonseed oil is a reliable antifertility agent, and it is relatively safe to use, provided the dosage is kept at the antifertility level. If oral administration of gossypol is discontinued, fertility will gradually recover.

Wuweizi (*Schisandra chinensis,* fruit), a Chinese traditional drug commonly used as an astringent, was found in clinical use to exhibit therapeutic effects on certain types of hepatitis, particularly in lowering the elevated serum glutamic pyruvic transaminase (SGPT) level. This effect immediately piqued the interest of doctors as well as pharmacologists. In detailed chemical and pharmacological studies it was found that the active principles responsible for lowering elevated SGPT levels were a series of derivatives belonging to the dibenzo(a,c) cyclooctene series. Among the derivatives, schisandrin B, schizandrol B, and schisandrin C showed the best protective activity against liver damage induced by chemical substances.

Ling zhi cao (Ganoderma), a fungus, has been regarded as a panacea. Widely used in folk medicine for neurasthenia, chronic hepatitis, cardiovascular disease, and chronic ulcers of the digestive system, it is regarded as having a broad range of pharmacological actions. In clinical use, the water soluble preparation of *Ganoderma capense* has been found to have a satisfactory effect on progressive muscular dystrophy and atrophic myotonia. Since hyperaldolasemia is one of the biochemical indications of muscular dystrophy, the effect of *Ganoderma* on experimental aldolasemia caused by the administration of 2,4-dichlorophenoxy acetic acid was studied in mice. Uracil and uridine, among others, were found to be responsible for the specified actions. Clinical trials of uridine injection to patients with progressive muscular dystrophy indicated that the symptoms were improved to a certain degree.

VALUABLE CLINICAL RESULTS OF THE APPLICATION OF NATIONAL MINORITY DRUGS

Scopolia tangutica (Tangchom Nagbo) is a drug used by doctors of traditional Tibetan medicine in Qinghai, a western inland province. Overdosage of this medicine causes the symptom characteristic of atropine-like toxicity. Phytochemical investigation revealed that besides hyoscymine and scopolamine, it contained anisodamine and anisodine. Pharmacological studies indicated that the effect on the central nervous system of anisodamine is 6 to 20 times weaker than that of atropine, while the effect of anisodine on the peripheral nervous system is weaker than the effect of both atropine and scopolamine. Anisodamine has been used clinically for the treatment of septic shock from toxic bacillary dysentery, fulminant epidemic meningitis, and hemorrhagic enteritis. Anisodine is now used for the treatment of migraines and diseases of the fundus oculi caused by vascular spasm, organophosphorus poisoning, and acute paralysis caused by cerebral vascular accident; it is also being used as the chief component in Chinese traditional anesthesia.

ACTIVE COMPOUNDS ISOLATED FROM CHINESE MEDICINAL PLANTS

Of the 250 pharmacologically active compounds that have been reported, 100 or so have been subjected to systematic chemical, pharmacological, and clinical studies, and approximately 60 new drugs originating from Chinese medicinal plants or derivations of these have been identified.

The chemical structure classification of pharmacological active principles from Chinese medicinal plants is shown in Table 4.2. Alkaloids occupy the leading position in numbers (202), followed by phenolic compounds (177), and then terpenes (158).

Alkaloids

Ornithine Derived

The most widely used alkaloids in China are the tropans anisodamine (see Figure 4.1, part 1) and anisodine (see Figure 4.1, part 2), which have been isolated from the Tibetan plants *Scopolia tangutica* Maxim. and *Przewolskia tangutica* Maxim.[2] Both of these alkaloids affect the nervous system. Clinically, anisodamine is used for the treatment of septic shock. Anisodine is used for the treatment of migraine headaches and fungal dis-

TABLE 4.2. The pharmacological active principles from Chinese medicinal plants.

Chemical type	Number of compounds
Alkaloids	202
Terpenes	
Monoterpenes	27
Sesquiterpenes	32
Diterpenes	33
Triterpenes	43
Cardiac glucosides	23
Phenolic Compounds	
Quinones	25
Chromones	5
Flavonoids	34
Phenyl propanoids	17
Coumarins	27
Lignans	21
Others	48
Acids and miscellaneous	34
Total	571

eases. Baogongteng A (see Figure 4.1, part 3) is isolated from the stem of *Erycibe obtusifolia* Benth.[3] Clinically, it is used as a substitute for pilocarpine in the treatment of glaucoma.

Lysine Derived

Piperine (see Figure 4.1, part 4) has been discovered from *Piper nigrum* and *Raphanus sativus*.[4] It has strong anticonvulsive and marked sedative actions.

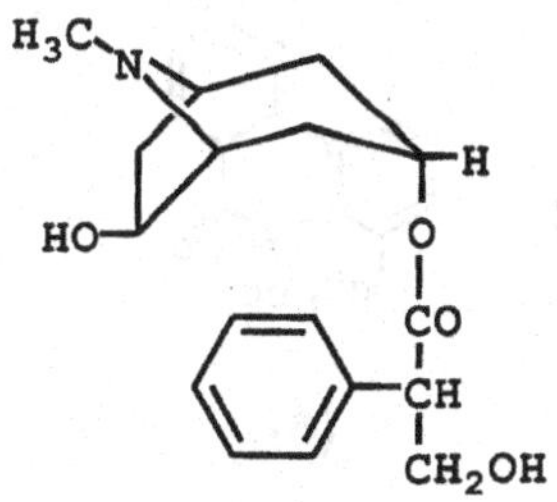

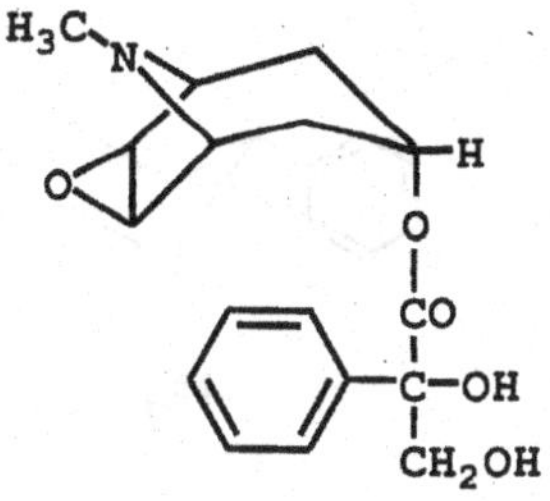

FIGURE 4.1. Illustrations of the chemical structures of anisodamine (1), anisodine (2), baogongteng A (3), and piperine (4).

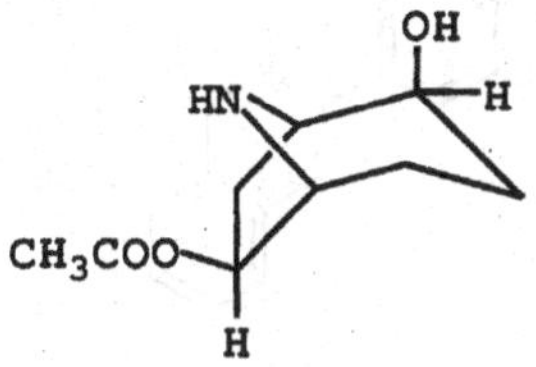

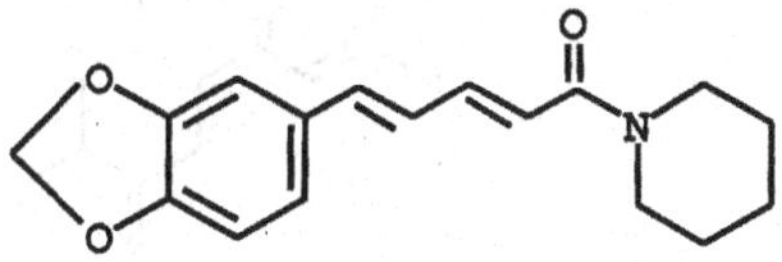

Nicotinic Acid Derived

The derivative dl-anabasine (see Figure 4.2, part 5) is isolated from *Alangium chinense* (Lour.) Harms.[5] It is well known as a muscle relaxant, and it exerts significant neuromuscular blocking effects.

Anthranilic Acid Derived

Zanthobungeanine (see Figure 4.2, part 6) isolated from the root of *Zanthoxylum bungeanum* Maxim, has a broad spectrum of antifungal activity. Febrifugine (β-dichroine) (see Figure 4.2, part 7) isolated from *Dichroa febrifuga* Lour, is an antimalarial constituent. This compound exhibits antimalarial activity and amebicidal effects.[6]

Phenylalanine and Tyrosine Derived

The derivative dl-demethylcoclaurine (higenamine) (see Figure 4.2, part 8) is isolated from the root of *Aconitum carmichaeli* Debx. It has an active

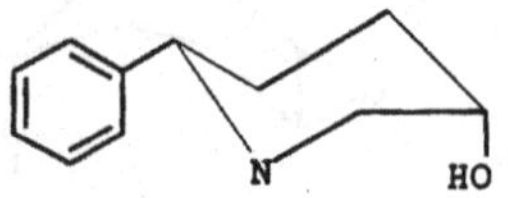

(5) 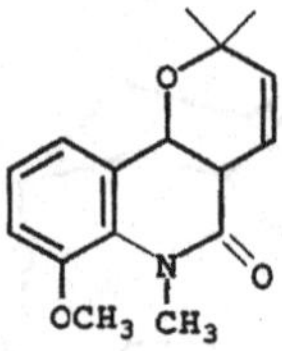 (6)

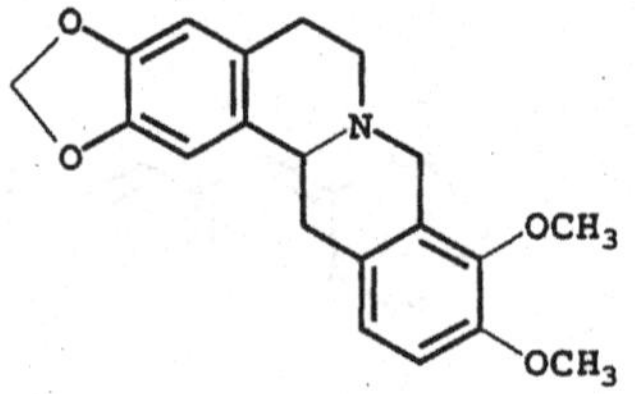

(7) 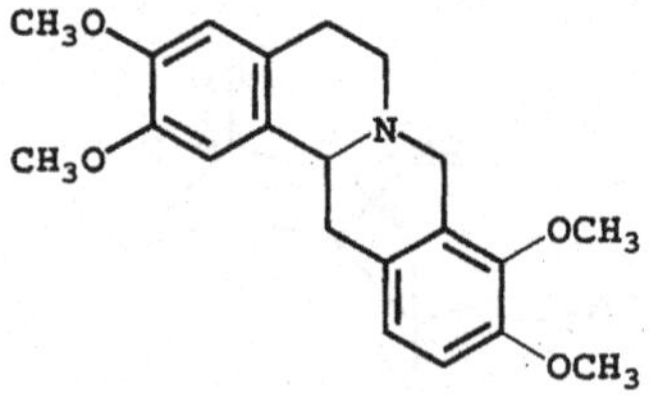 (8)

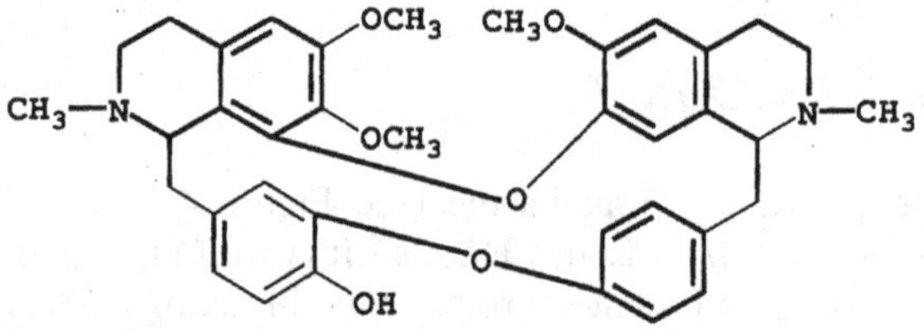

(9)

FIGURE 4.2. Illustrations of the chemical structures of dl-anabasine (5), zanthobungeanine (6), febrifugine (β-dichroine) (7), dl-demethylcoclaurine (higenamine) (8), and berbamine (9).

cardiac principle and has been demonstrated to be a partial agonist of the β-adrenergic receptor.[7] Berbamine (see Figure 4.2, part 9), a biobenzylisoquinoline alkaloid obtained from *Berberis poiretii* Schneid., has been applied to increase leukocyte content of blood.[8] Tetrahychoberberine (see Figure 4.3, part 10) and dl-tetrahydropalmatine (see Figure 4.3, part 11) isolated from *Stephania dielsiana* Y.C. Wu, possesses sedative, tranquiliz-

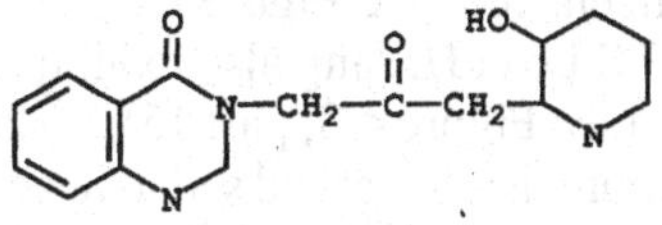

(10) 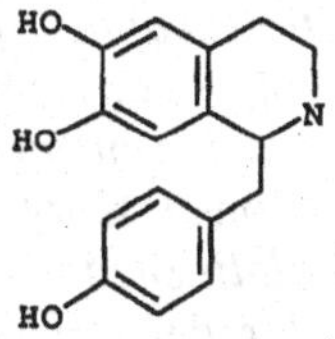 (11)

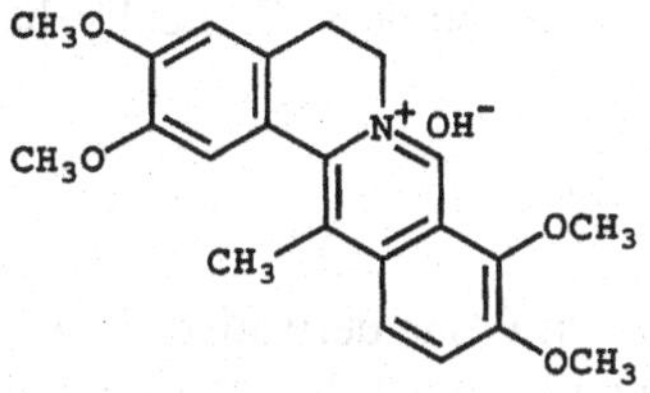

(12)

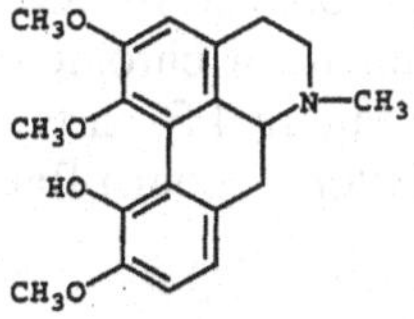

(13) 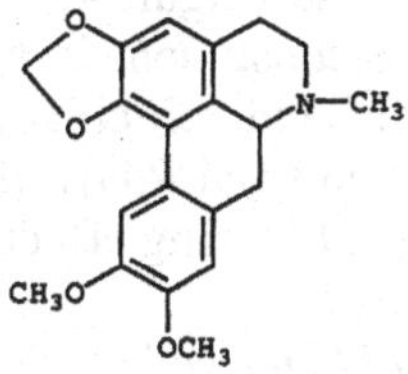 (14)

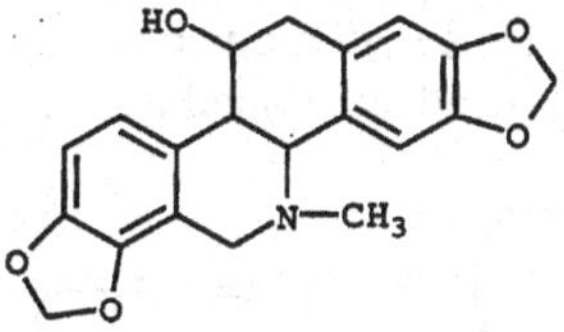

(15)

FIGURE 4.3. Illustrations of the chemical structures of tetrahydroberberine (10), dl-tetrahydropalmatine (11), dehydrocorydaline (12), d-isocorydine (13), l-dicentrine (14), and corynoline (15).

ing, and analgesic actions.[9] Dehydrocorydaline (see Figure 4.3, part 12), obtained from *Corydalis yanhusuo* W.T. Wang, improves the tolerance to monobaric and hypobaric hypoxia in mice. It is used for the treatment of coronary heart disease.[10] d-isocorydine (see Figure 4.3, part 13), isolated from *Dictylicapnes scaudens* (D.Don) Hutch, possesses marked analgesic and sedative activities; l-dicentrine (see Figure 4.3, part 14) isolated from *Stephania decentrinifera* H.S Lo et H.Yang also exhibits analgesic and sedative actions.[11] Corynoline (see Figure 4.3, part 15), isolated from the tuber of *Corydalis sheareri* S. Moore, has a marked sedative action in mice.[12] The four ester derivatives of cephalotaxine—harringtonine, homoharringtonine, isoharringtonine, and deoxyharringtonine (see Figure 4.4, parts 16, 17, 18, and 19 respectively)—demonstrated various degrees of inhibitory activities against animal tumors. Clinically, harringtonine is used for treating leukemia.[13]

Tryptophan Derived

A series of canthin-6-one derivatives have been isolated from *Picrasma quassioides* (D. Don) Bennet. It was shown that 4,5-dimethoxycanthin-6-one (see Figure 4.5, part 20) and 4-methyoxy-5-hydroxycanthin-6-one (see Figure 4.5, part 21) inhibited the growth of *Staphylococcus aureus.*[14] Indirubin (see Figure 4.5, part 22) obtained from *Indigofera tinctoria* L. is an active constituent in the treatment of chronic myelocytic leukemia.[15] Spegatrine and verticillatine (see Figure 4.5, parts 23 and 24 respectively) are isolated from *Rauvolfia verticillata* (Lour.) Baill and have exhibited a ganglianic blocking effect.[16]

Isoprenoid Derived

Lappaconitine (see Figure 4.5, part 25), isolated from the root of *Aconitum sinomontanum* NaKai, demonstrates an analgesic effect and local anesthetic activity.[17] It is used as a nonaddictive analgesic agent. Pingpeimine A

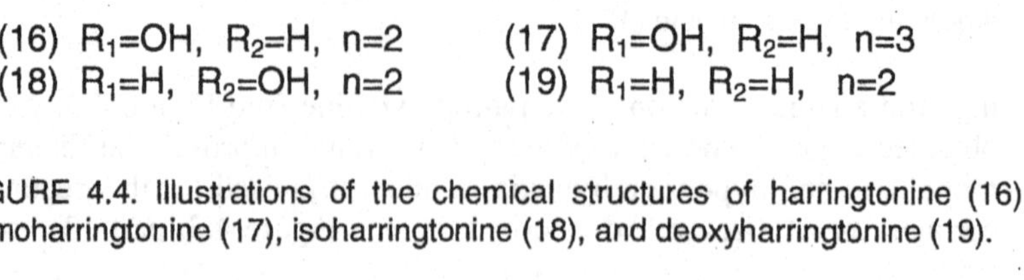

(16) R_1=OH, R_2=H, n=2 (17) R_1=OH, R_2=H, n=3
(18) R_1=H, R_2=OH, n=2 (19) R_1=H, R_2=H, n=2

FIGURE 4.4. Illustrations of the chemical structures of harringtonine (16), homoharringtonine (17), isoharringtonine (18), and deoxyharringtonine (19).

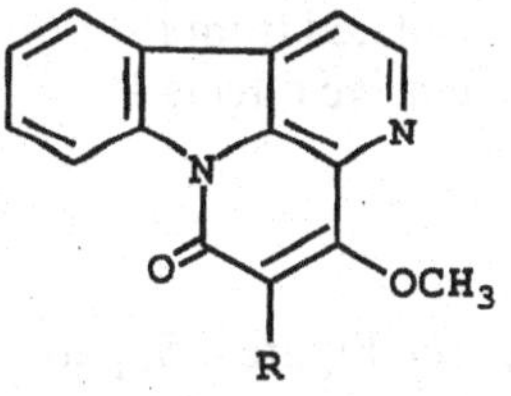

(20) R = $-OCH_3$ (21) R = $-OH$

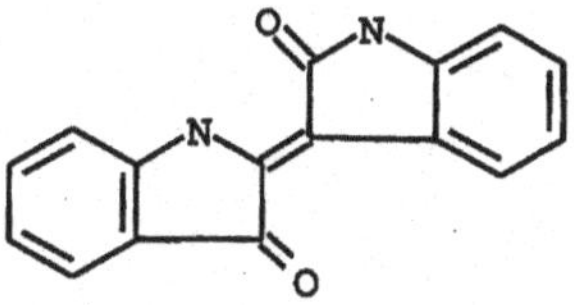

(22) 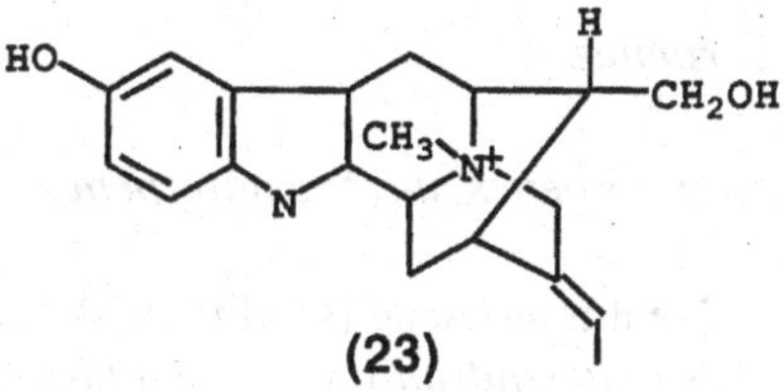 (23)

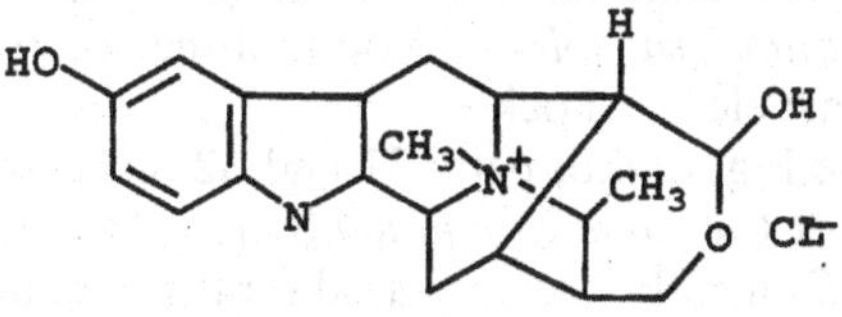

(24) 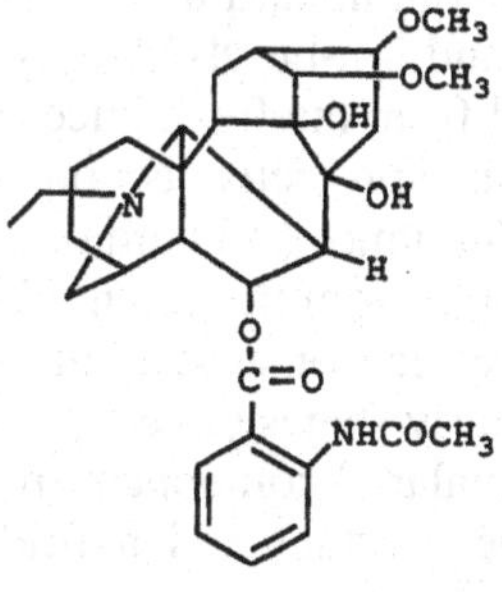 (25)

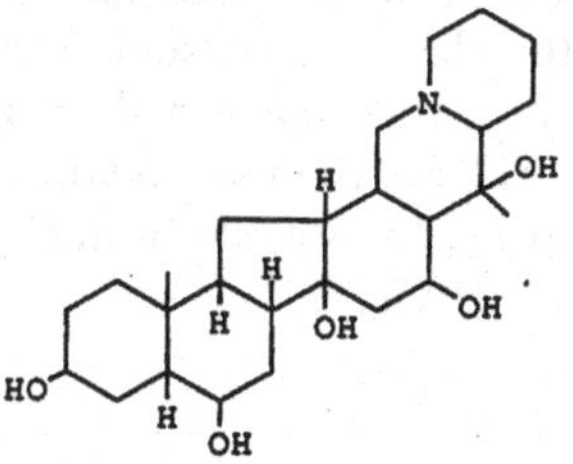

(26) 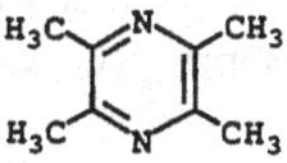 (27)

FIGURE 4.5. Illustrations of the chemical structures of 4,5-dimethoxycanthin-6-one (20), 4-methyoxy-5-hydroxycanthin-6-one (21), indirubin (22), spegatrine (23), verticillatine (24), lappaconitine (25), pingpeimine A (26), and tetramethylpyrazine (27).

(see Figure 4.5, part 26), isolated from *Fritillaria ussuriense* Maxim, exhibits expectorant and hypotensive effects.[18]

Miscellaneous Alkaloids

Tetramethylpyrazine (see Figure 4.5, part 27) is the active ingredient of *Ligusticum chuanxiang* Hort. Pharmacological and clinical studies indicate that the drug improves microcirculation. It is used for the treatment of occlusive cerebral blood vessel disease.[19]

Terpenes

Monoterpenes and Sesquiterpenes

Neofuranodiene (see Figure 4.6, part 28), isolated from the essential oil of *Rhododendron anthopogonides* Maxim, possesses expectorant activity and exhibits antitussive effect.[20] 3-n-butyl-phthalidl (see Figure 4.6, part 29) and 3-n-butyl-4,5-dihydro-phthalide (see Figure 4.6, part 30) are isolated from the fresh juice of *Apium graveolens* L. Both of the compounds produce protective action against electroshock.[21]

Curcumol and curdione (see Figure 4.6, parts 31 and 32 respectively), isolated from the essential oil of *Curcuma aromatica* Salisb., have inhibitory effects on animal tumors. Clinically, they are used for the treatment of the early stages of cervical cancer.[22] Qinghaosu (see Figure 4.6, part 33) is antimalarial component from *Artemisia annua* L. It is used against chloroquine-resistant and multidrug-resistant strains of *Plasmodicum falciparum*.[23]

Yingzhaosu (see Figure 4.6, part 34) has been extracted from the root of *Artabotrys hexapetallus* (L.f.) Bhandari and possesses antimalarial activity.[24] Tutin (see Figure 4.6, part 35), isolated from *Loranthus parasiticus* (L.) Merr., has been exhibited to be effective in the treatment of schizophrenia.[25] Pogostone (see Figure 4.6, part 36), isolated from *Pogostemon cablin* (Blanco) Benth., shows antifungal activity.[26]

Diterpenes

Rabdosin A (see Figure 4.7, part 37), isolated from *Rabdosia japonica* (Burm.f.) Hare, is significantly cytotoxic against tumor cells.[27] Macrocalin A (38) is a new dirpeniod isolated from *Rabdosia macrocalyx* (Dunn) Hara. It showed cytotoxic effects in vitro against cultures of HaLa cells. Rabdophyllin G (39) is isolated *Rabdosia macroohylla* (Migo) C. Y. Wu et H. W.

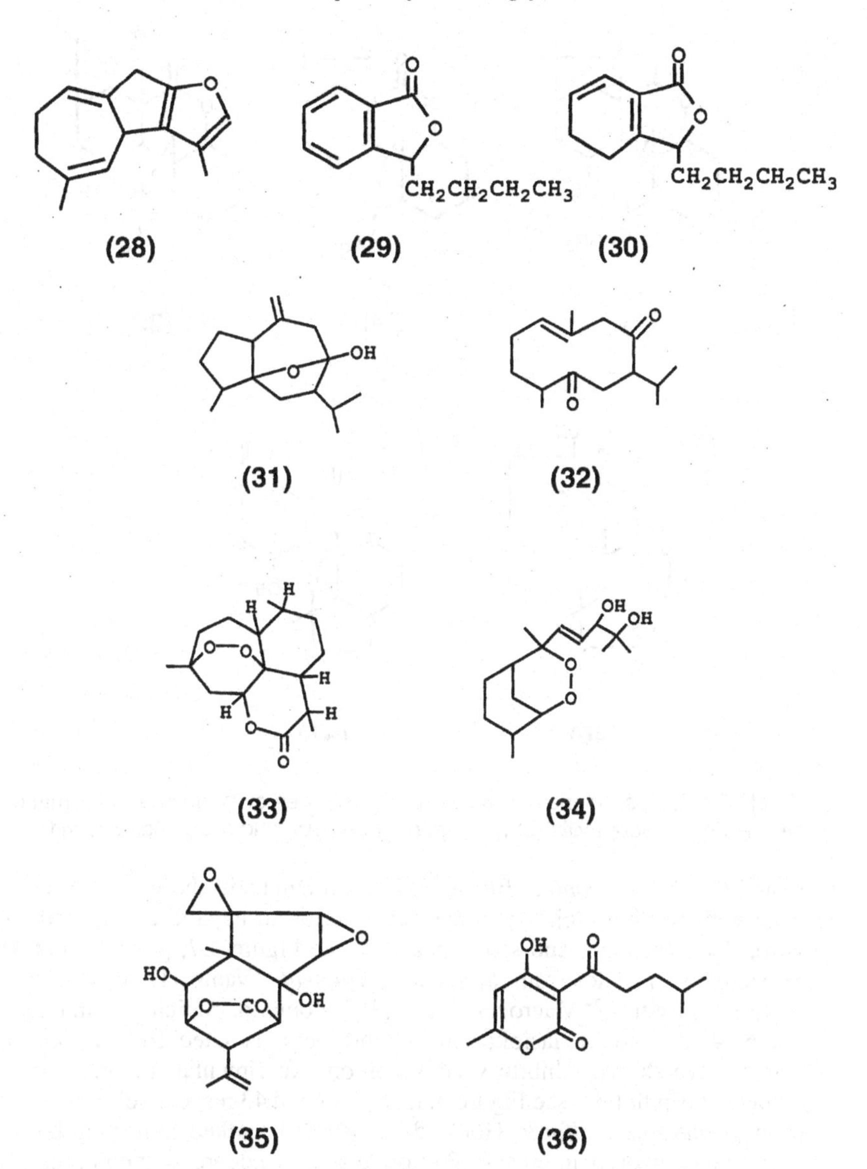

FIGURE 4.6. Illustrations of the chemical structures of neofuranodiene (28), 3-n-butyl-phthalidl (29), 3-n-butyl-4,5-dihydro-phthalide (30), curcumol (31), curdione (32), qinghaosu (33), yingzhaosu (34), tutin (35), and pogostone (36).

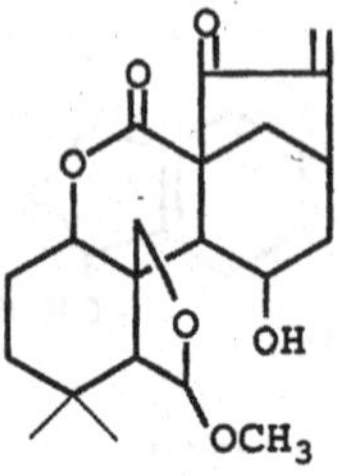

(37) 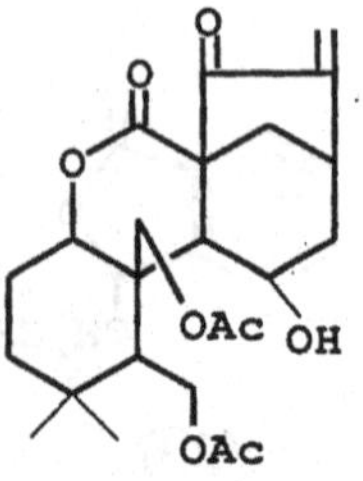 (38) 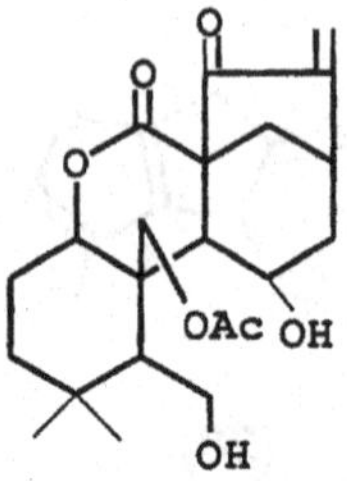 (39)

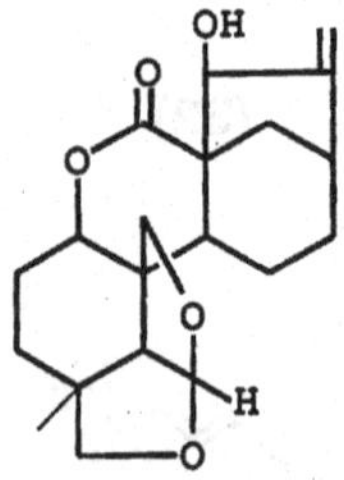

(40) 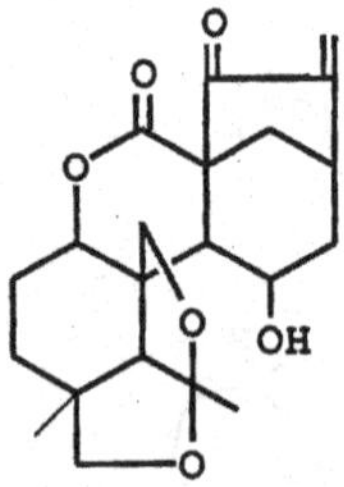 (41)

FIGURE 4.7. Illustrations of the chemical structures of rabdosin A (37), macrocalin A (38), rabdophyllin (39), sculponeata A (40), and sculponeata B (41).

Li and *Rabdosia japonica* (Burm. F.) Hare. It lengthens the survival time of mice with Ehrlich ascites and cytotoxic action to hepatic cancer cells in vitro. Sculponeata A and sculponeata B (see Figure 4.7, parts 40 and 41 respectively), isolated from *Rabdosia sculponeata* (Vaniot) Hare, have antibacterial activities.[28] Macrocalin A and B, ludongnin, ponicidine, amethystoidin A, lasiodonin, lasiokaurin and oridonin, isolated from *Rabdosia* plants, have shown inhibitory activities on experimental tumors.[29] Triptolide and tripdiolide (see Figure 4.8, parts 42 and 43 respectively), isolated from *Triptergium wilfordii* Hook, demonstrated marked tumor-inhibitory activity on leukemia in mice.[30] Triptolide acts as a depressant on humoral-mediated immunity. Brusatal, bruceine D, and bruceine E are isolated from *Brucea javanica* (L.) Mer., a Chinese drug with noted toxicity. They are active against sarcoma or malaria. Deoxyandrographolide, andrographolide, neoandrographolide, and deoxydidehyroandrographalide (see Figure 4.8,

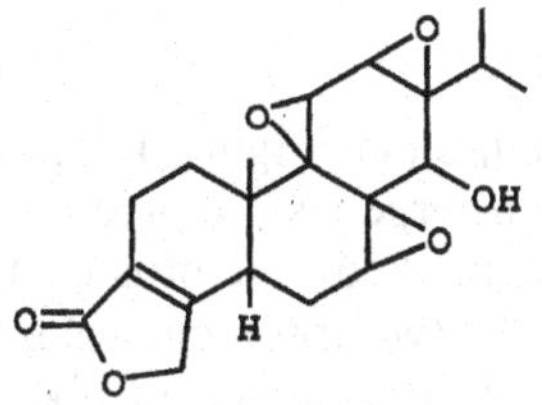

(42) 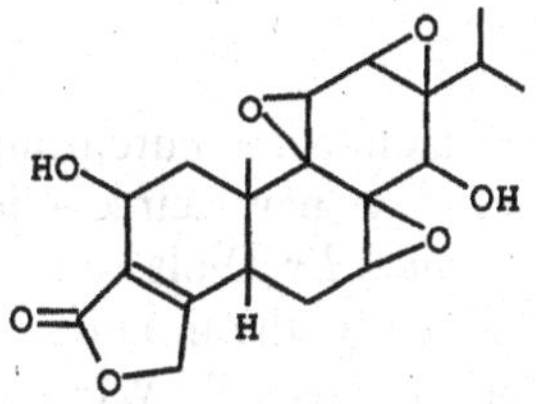 **(43)**

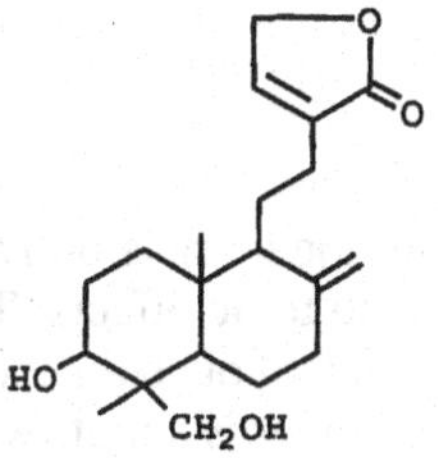

(44) 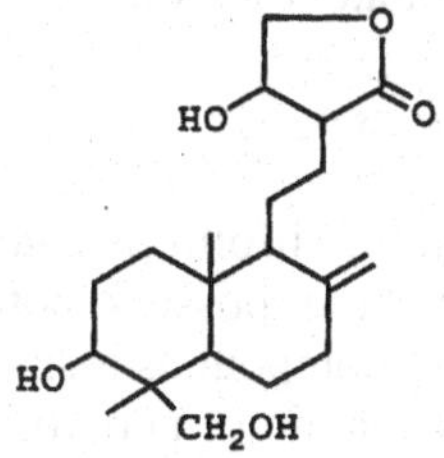 **(45)**

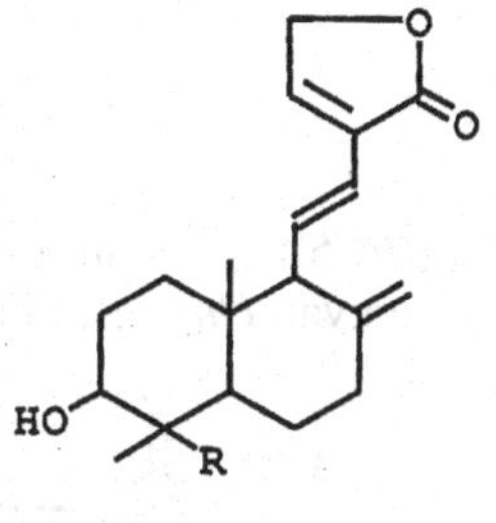

(46) R = $-CH_2OH$ **(47)** R = $-CH_2O$-Glu

FIGURE 4.8. Illustrations of the chemical structures of triptolide (42), tripdiolide (43), deoxyandrographolide (44), andrographolide (45), neoandrographolide (46), and deoxydidehyroandrographalide (47).

parts 44, 45, 46, and 47 respectively) are isolated from *Andrographis paniculata* (Burm.f.) Nees.[31] They possess various degrees of anti-inflammatory and antipyretic effects. Clinically, these compounds have been applied in the treatment of infectious diseases of gastrointestinal tract, respiratory organs, and urinary system.

Triterpenes

Curcurbitacin a and curcurbitacin b (see Figure 4.9, parts 48 and 49 respectively), two new tetracyclic triterpenes isolated from the tuber of *Hemsleya amabilis* Diels, have a marked and broad antibacterial spectrum.[32] Clinically, the mixture of the two triterpenes is used for the treatment of bacterial dysentery, bronchitis, tonsillitis, and tuberculosis. Astragolus saponin I, a triterpene saponin, is isolated from *Astragalus membranaceus* Bge. Esculentosides A, B, C, D, E, and F are also triterpene saponins isolated from *Phytolacca esculenta* Van Houtte. They have been considered to have anti-inflammatory action.[33]

Cardiac Glucosides

Peruvoside and neriifolin, isolated from the kernels of *Thevetia peruriana* (Pers.) K. Schum, possess inotropic and toxic effects. The maximum effect of these two compounds is similar to that of ouabain and strophanthin K, but the safety margin of peruvoside treatment is about twice that of the other three glycosides. Neriifolin has been clinically used for the treatment of heart failure.

Phenolic Compounds

Quinones

Irisquinone (see Figure 4.10, part 50) is an antitumor principle isolated from the seeds of *Iris pallasii* Fisch.var. *chinensis* Fisch.[34] The compound

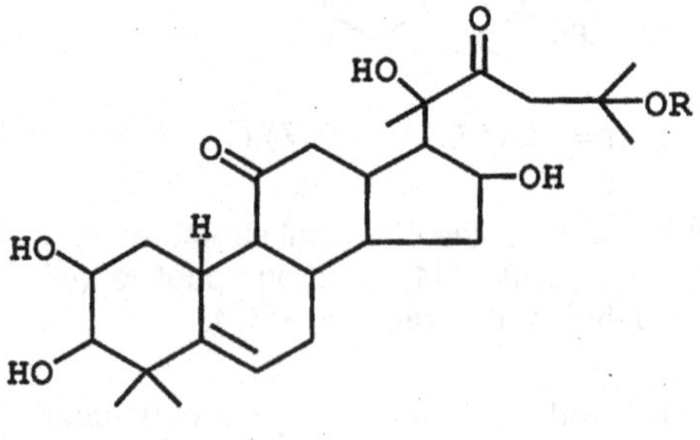

(48) R = $-COCH_3$ **(49)** R = –H

FIGURE 4.9. Illustrations of the chemical structures of curcurbitacin a (48) and curcurbitacin b (49).

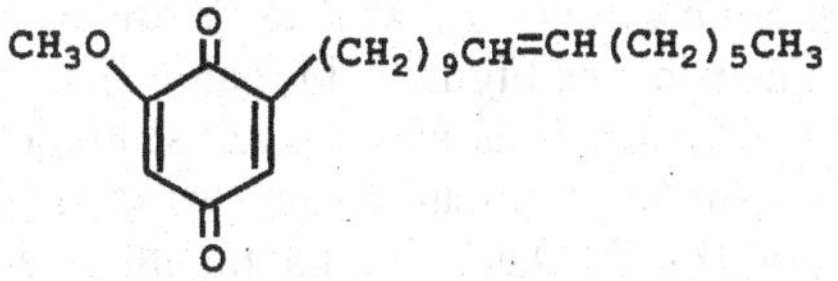

(50) 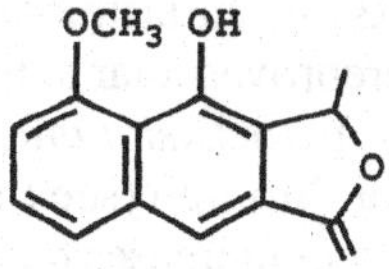 (51)

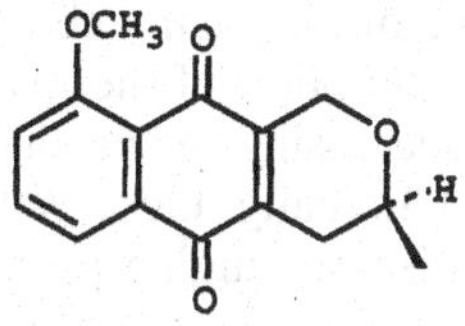

(52) 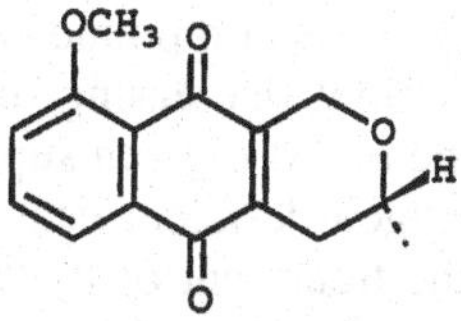 (53)

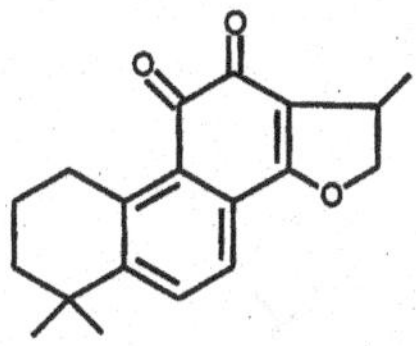

(54) 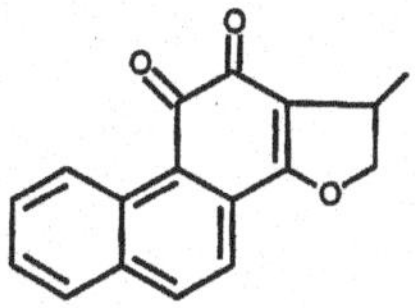 (55) 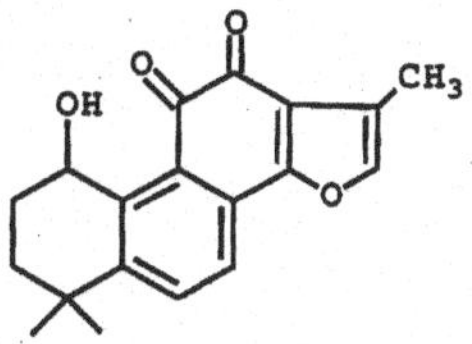 (56)

FIGURE 4.10. Illustrations of the chemical structures of irisquinone (50), eleutherol (51), elutherin (52), isoelutherin (53), cryptotanshinone (54), dihydrotanshinone (55), and hydroxytanshinone II-A (56).

has proven to be effective against U14 and lymph sarcoma, and has inhibited hepatic and Ehrlich ascite scancer. Eleutherol, elutherin, and isoelutherin (see Figure 4.10, parts 51, 52, and 53 respectively) are isolated from *Eleutherine americana* Merr. ex Heyne. [35] They can markedly increase the blood flow of atreriae coronaria. A series of phenanthrequinone compounds have been discovered in *Salvia miltiorrhiza* Bge., Tanshinone as well as cryptotanshinone, dihydrotanshinone I, and hydroxytanshinone II-A (see Figure 4.10, parts 54, 55, and 56 respectively) display bacteriostatic action. Sodium tanshinone II-A sulfonate exhibits marked cardiovascular activity.[36]

Flavonoids

Flavonoids are the most widely distributed natural products in medicinal plants. They have broad spectrum pharmacological activities, such as antitussive, antiasthmatic, antiallergic effects, as well as cardiovascular and cerebrovascular actions. Nevadensin (see Figure 4.11, part 57), identified in the herb *Lysionotus pauciflora* Maxim was active against tuberculosis.[37] Corylifolinin (see Figure 4.11, part 58) is isolated from *Psoralea corylifolia* L. and increases coronary blood flow.[38] Daidzein, daizin, and puerarin (see Figure 4.11, parts 59, 60, and 61 respectively) have been isolated from *Pueraria lobata* (Willd.) Ohwi.[39] They are isoflavone compounds. Pharmacologically, these compounds are capable of increasing cerebral and coronary blood flow, decreasing oxygen consumption of the myocardium, increasing the blood oxygen supply, and decreasing the production of lactic acid by oxygen-deficient heart muscle. Clinically, these isoflavones were used for the treatment of hypertensive disease, angina pectoris, migraine headaches, and sudden deafness.

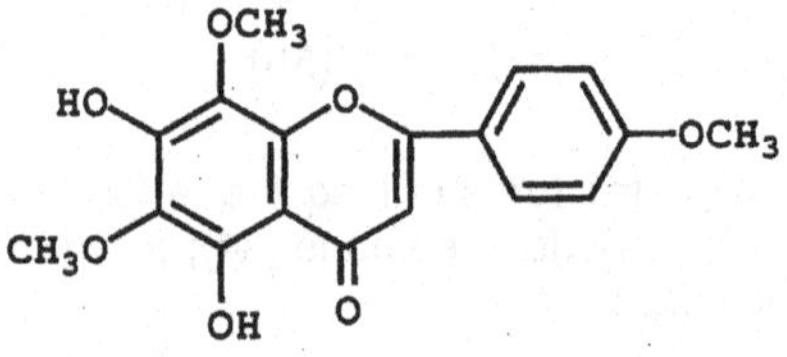

(57) 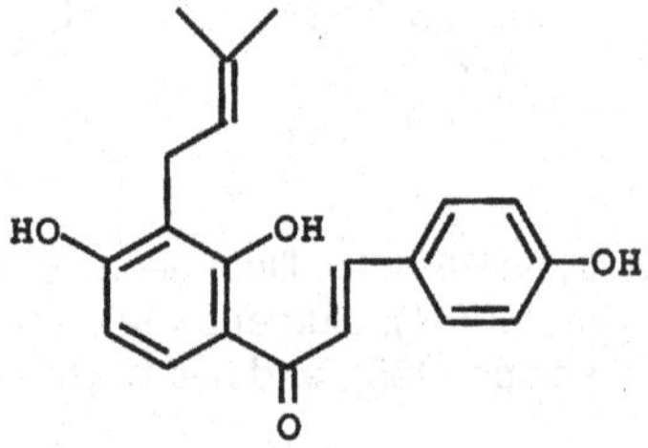 **(58)**

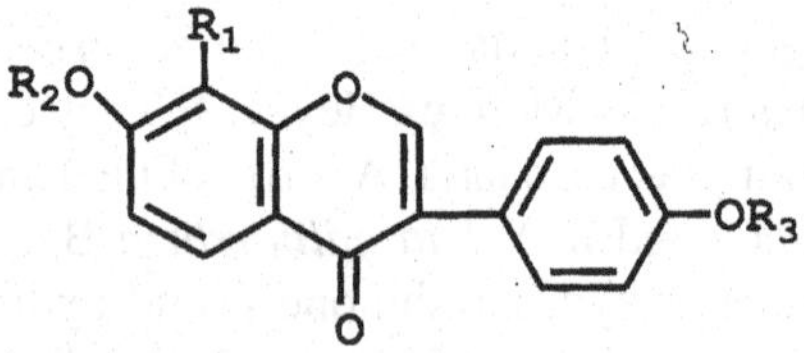

(59) R_1 = H, R_2 = H, R_3 = H
(60) R_1 = 1, R_2 = Glucopyranose, R_3 = H
(61) R_1 = Glucopyranose, R_2 = H, R_3 = H

FIGURE 4.11. Illustrations of the chemical structures of nevadensin (57), corylifolinin (58), daidzein (59), daizin (60), and puerarin (61).

Xanthones

The total xanthones from *Swertia mussotii* Franch. possess potent activity against acute liver injury caused by hypoxia.

Phenyl Propanoids

Ferulic acid and its sodium salt from *Angelica sinensis* (Oliv.) Diels. are some of the active ingredients of this plant which show inhibitory effect on rat platelet aggregation by ADP or collagen. Chlorogenic acid is isolated from *Cirsium segetum* Bge. The compound possesses inhibitory activity on bacteria in various degrees. Syringin is isolated from *Daphne giraldii* Nitshe and has been proved as a hemostatic both pharmacologically and clinically. Danshensuan B (see Figure 4.12, part 62) is from the water-soluble fraction of *Salvia miltiorrhiza* Bge.[40] This compound promotes fibrinolysis and increases coronary blood flow.

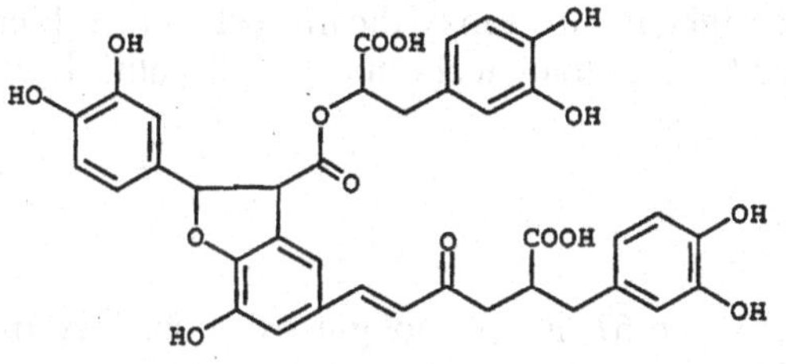

(62)

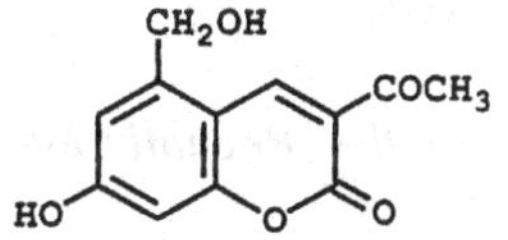

(63)

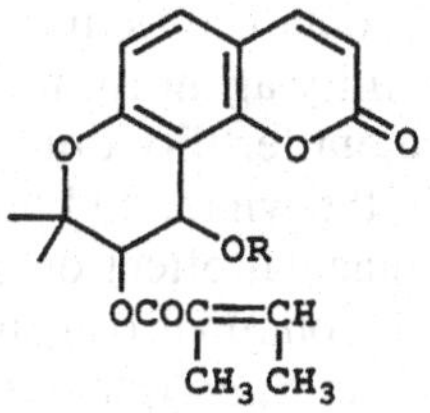

(64) R = CH_2CH_3
(65) R = $CH_2CH_2CH(CH_3)_2$

FIGURE 4.12. Illustrations of the chemical structures of danshensuan B (62), armillarisin A (63), (+)-praeruptorin A (64), and (+)-praeruptorin E (65).

Coumarins

Coumarin compounds are isolated from *Fraxinus rhynchophylla* Hance., *Daphne tangutica* Maxim., *Daphne koreana* Nakai, *Wikstroemia indica* C.A. Mey., *Erycibe obtusifolia* Benth, *Artemisia capillaris* Thunb., and *Sarcandra glabra* (Thunb.) Nakai etc. Armillarisin A (see Figure 4.12, part 63), a new coumarin compound, was isolated from the ethanol extracts of the culture medium of the fungus *Armillariella tabescens* (Scop. ex Fr.) Sing. It markedly increases bile secretion in rats and dogs.[41] The derivative (+)-praeruptorin A and E (see Figure 4.12, parts 64 and 65), isolated from *Peucedanum praeruptorum* Dunn, have shown pharmacological activities.[42] (+)-Praeruptorin A increases coronary blood flow in isolated guinea pig hearts and (+)-praeruptorin E increases the tolerance of mice to anoxia.

Lignans

A number of lignan compounds isolated from the kernels of genus *Schisandra* lower elevated serum glutamic pyruvic transaminase (SGPT) levels and help central nervous system (CNS) depression. All derivatives of dibebzo-cycloctene series isolated from the fruit of *Schisandra* were tested for lowering SGPT levels. Biphenyldimethoxy-dicarboxylate (see Figure 4.13, part 66) was developed for used to treat chronic virus hepatitis in clinics.

Other Phenolic Types

Gastrodin (see Figure 4.13, part 67) is a phenol glucoside obtained from *Gastrodia elata* Blume.[43] The pharmacological test indicated that the compound possesses active sedative and anticonvulsive properties. Clinically, gastrodin has been used for the treatment of neurasthenia, insomnia, and headache. Gossypol (see Figure 4.13, part 68), obtained from cottonseed oil *Gossypium hisrutum* L., is used as an antifertility agent for males and has been clinically demonstrated as an effective antifertility drug.[44] Agrimophol (see Figure 4.13, part 69), isolated from the winter bud of *Agrimonia pilosa* Ledeb., has a confirmed in vitro taeniacidal effect on Taenia solium.[45] This compound also displayed strong action in vitro against *Schistosama japonica*. Robustanol A (70) isolated from *Eucalypus robusta* Smith showed to have inhibitory activity against malaria in mice. Ardisonol I and II (71) and (72), respectively, obtained from *Ardisia japonica* (Horn.) Bume have specific activity against tuberculosis. As the active constituents of *Citrus aurantium* L., N-methyltyramine and synephrine (see Figure 4.14, parts

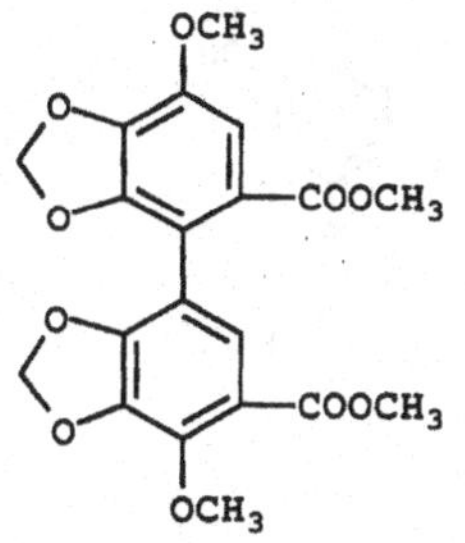

(66) 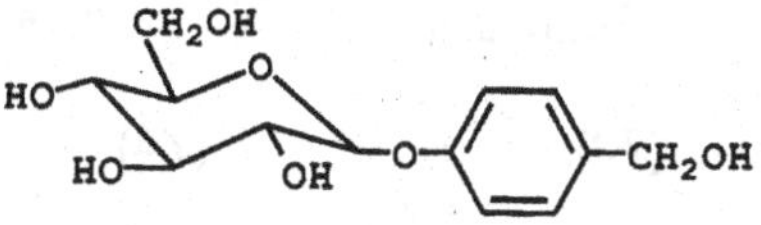 (67)

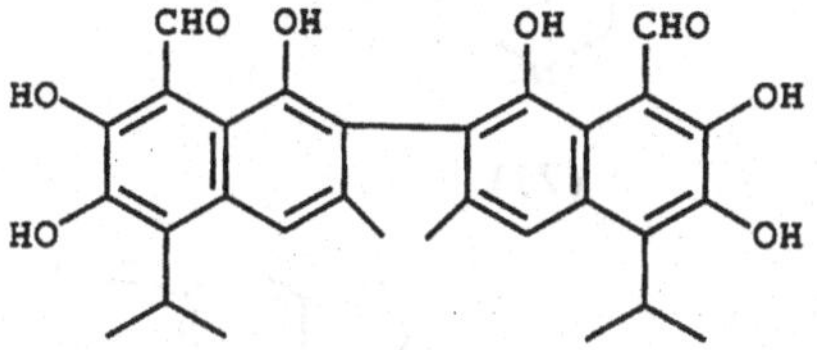

(68) 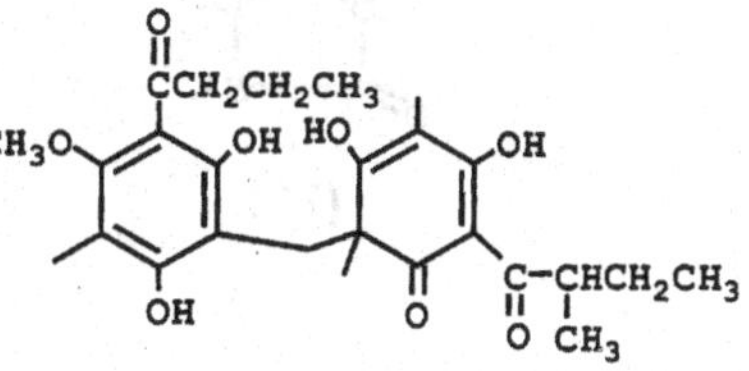 (69)

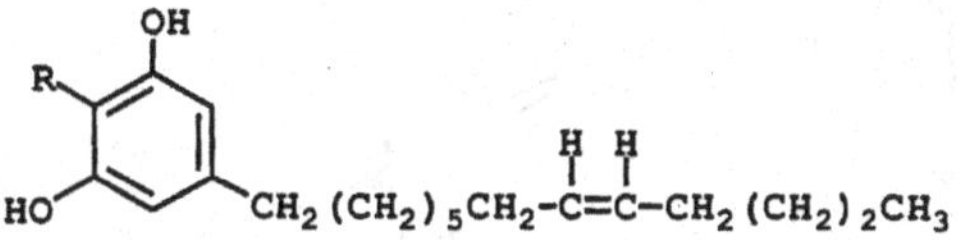

(70)

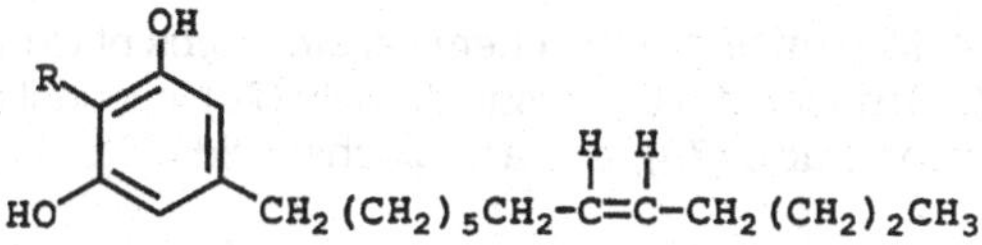

(71) R = $-CH_3$ (72) R = $-H$

FIGURE 4.13. Illustrations of the chemical structures of biphenyldimethoxy-dicarboxylate (66), gastrodin (67), gossypol (68), agrimophol (69), robustanol A (70), ardisonol I (71), and ardisonol II (72).

73 and 74, respectively 74) have been demonstrated to increase blood pressure.[46] Paeonol (see Figure 4.14, part 75), obtained from the root bark of *Paeonia suffruticosa* Andr., has bacterostatic, anti-inflammatory, and CNS depressive effects.[47] Clinically, 5 percent paeonol sulfonate displayed analgesic action.

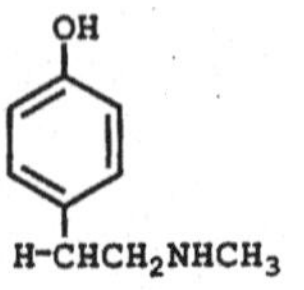

FIGURE 4.14. Illustrations of the chemical structures of N-methyltyramine (73), synephrine (74), paeonol (75), quisqualic acid (76), cucurbitine (77), danshensu (78), rubia naphaic acid (79), and aristolochic acid (80).

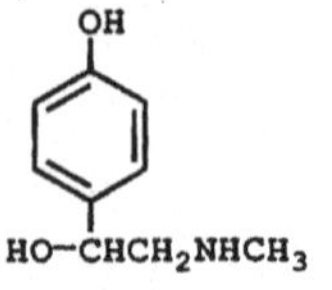

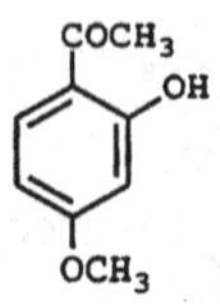

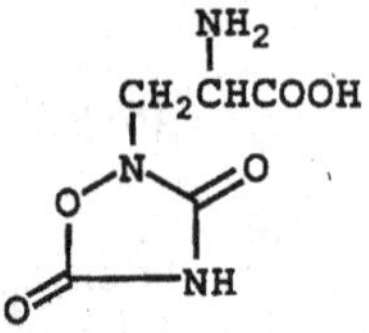

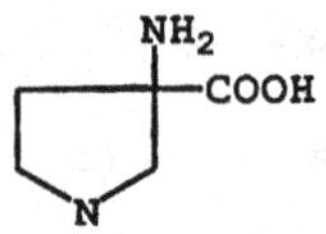

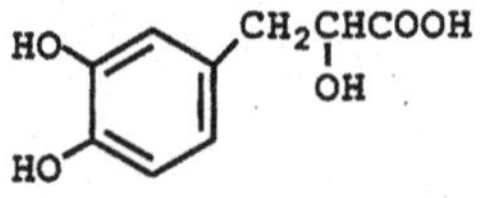

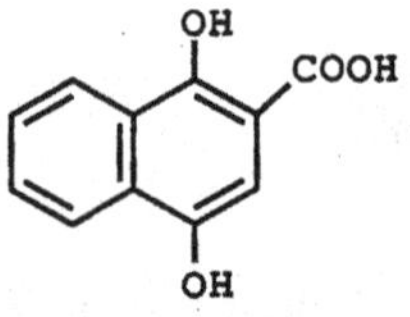

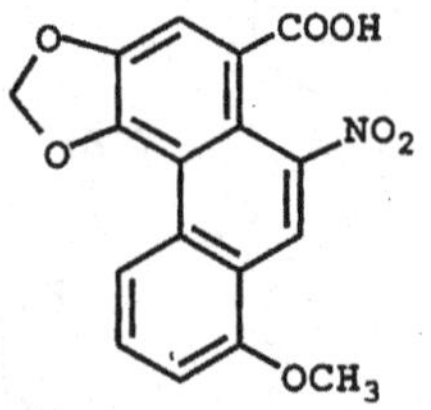

Acids

Quisqualic acid (see Figure 4.14, part 76) was isolated from *Quisquilis indica* L.[48] In clinical trials, potassium quisqualate was anthelmintic without any laxative side effects at dose of 0.125 g. Cucurbitine (see Figure 4.14, part 77) obtained from the pumpkin seeds of *Cucurbite moschata* Duch. posesses a protective activity against schistosomal infections.[49] Danshensu (see Figure 4.14, part 78) obtained from the water-soluble fraction of *Salvia miltirrhiza* Bge., is reported to dilate the coronary artery and markedly antagonize the constricting responses elicited by morphine and peopranolol.[50] Rubia naphaic acid (see Figure 4.14, part 79), isolated from *Rubia cordifolia* L., increases leucocyte numbers.[51] Aristolochic acid (see Figure 4.14, part 80), isolated from *Aristolochia mollissima* Hance, possesses stimulating phagocyte action.[52] It also displayed significant anti-inplantation and abortifacient effects.

Miscellaneous Compounds

Rorifone (see Figure 4.15, part 81), isolated from *Roripa montana* (Wall) Small, has demonstrated expectorant and antitussive activities.[53] The volatile oil of *Houttuynia cordata* Thunb. contains an antimicrobial principle, decanoyl-acetaldehyde (see Figure 4.15, part 82).[54] It also possesses immuno-stimulating action. Diallyl thiosulfonate (see Figure 4.15, part 83), isolated from *Allium sativum* L., has been clinically tested for use in the treatment of infections of the respiratory and digestive systems.[55]

TYPES OF PHARMACOLOGICAL ACTIVITIES

About 140 new drugs have originated directly or indirectly from Chinese medicinal plants by means of modern scientific methods. These new drugs are grouped under different pharmacological actions: 36 affect the cardiovascular system, 23 are anticancer drugs, 21 affect the nervous system, 15 affect the respiratory system, 18 are antimicrobials, 14 are for the digestive system, 9 are for eliminating parasites, 5 are for family planning, and 1 is an eye medication. The new drugs and their origins are summarized in Table 4.3. All of these drugs are further divided into four categories according to their characteristics:

A. Known compound but new usage
B. New compound and new usage
C. Semisynthesis, chemical modification of the active principles of medicinal plants
D. Total synthesis based on the active principle isolated from medicinal plants

Medical practice has taught us to understand that ethnopharmacological data is an important source of new drugs. Since medicinal plants have been used for centuries and tested by billions of people, there has been ample opportunity to find satisfactory medical agents and to solve problems of toxicity and side effects. Based on rough statistics, these 140 new drugs have been developed since 1949, of which about 80 originated directly or indirectly from medicinal plants. These data have shown that increasing emphasis on the use of medicinal plants in searching for new drugs in China is undoubtedly a correct strategy.

$CH_3\text{-}SO_2\text{-}CH_2\text{-}(CH_2)_7\text{-}CH_2CN$
(81)

$CH_3\text{-}(CH_2)_8\text{-}CO\text{-}CH_2\text{-}CHO$
(82)

$CH_2{=}CH\text{-}CH_2SO_2\text{-}S\text{-}CH_2\text{-}CH{=}CH_2$
(83)

FIGURE 4.15. Illustrations of the chemical structures of rorifone (81), decanoylacetaldehyde (82), and diallyl thiosulfonate (83).

TABLE 4.3. New drugs originating from Chinese traditional and herbal drugs.

Drug	Origin	Category
Drugs affecting the cardiovascular system		
Tetramethyl pyrazine	*Ligusticam chuanxing*	A, D
Glaberone	*Ilex pubescens* var. *glabra*	A
Ilexonin A	*Ilex pubescens*	
N-methyltyramine	*Citrus aurantium*	A, D
dl-muscone	*Moschus moschiferus*	A, D
Higenamine	*Aconitum japonicum*	A, D
Securinine	*Securinega suffruticosa*	A
Sodium tanshinone IIA	*Salvia miltiorrhiza*	B, C
Cyclovirobuxine D	*Buxus sinica*	B
Scutellarin	*Scutellaria baicalensis*	A, D
Papaverine	*Papaver somniferum*	A, D
Digitoxin	*Digitalis purpurea*	A
Digixin	*Digitalis purpurea*	A
Cedilanid	*Digitalis lanata*	A
Convallatoxin	*Convallaria majalis*	A
Peruvoside	*Thevetia perviana*	A
Lanatoside A, B	*Digitalis lanata*	A
Quinidine	*Cinchona succirubra*	A
Reserpine	*Rauvilfoa verticilata*	A
Rutin	*Sophora japonica*	A
Puerarin	*Pueraria lobata*	A
Rhynchophylline	*Uncaria rhynchophylla*	A
Daidzein	*Glycine max*	A
Dehydrocorydaline	*Corydalis yanhusuo*	A
Andromedoxin	*Rhododendron molle*	A
Cycleanine dimethobromide	*Cyclea tokinensis*	B, C

Drug	Origin	Category
Sodium ferulate	*Angelica sinensis*	A
Synephrine	*Citrus aurantium*	A, D
beta-sitosterol	plant oil	A
Caffeic acid	tea	A
Psoralen	*Psoralea corylifolia*	A
Xanthotoxin	*Pastinaca sativa*	A,D
Anethole	*Illicium verum*	A
Cepharanthine	*Stephania cepharantha*	A
Berbamine	*Berberis sargentiana*	A
Drugs showing anticancer activity		
Hainanolide	*Cephalotaxus hainanensis*	B
Taxol	*Taxus chienesis (Pilger) Rehd.*	A
Vinblastine	*Catharanthus roseus*	A
Vincristine	*Catharanthus roseus*	A
Camptothecine	*Camptotheca acuminata*	A
10-hydroxoycamptothecine	*Camptotheca acuminata*	A
Colchicine	*Colchicum autumnale*	A
Harringtonine	*Cephalotaxus fortunei*	A
Sophocarpine	*Sophora alopecuroides*	A
Monocrotaline	*Crotalia assamica*	A
Indirubin	*Baphicacanthus cusia*	A,D
Rubescensin	*Rabdosia rubescens*	A
Nitidine chloride	*Zanthoxylum nitidum*	A
Tetrandrine	*Stephania tetrandra*	A
Curcumol	*Curcuma aromatica*	A
Curdione	*Curcuma aromatica*	A
Aristolone	*Aristolochia debilis*	A
Matrine	*Sophora flavescens*	A
Drugs affecting the nervous system		
Morphine	*Papaver somniferum*	A
Corydalis B	*Corydalis yanhusuo*	A,D
Lobeline	*Lobelia inflata*	A
Caffeine	tea	A
Huperzine A	*Huperzin serraata*	B
Anisodamine	*Scopolia tangutica*	B,D
Anisodine	*Scopolia tangutica*	B,D
Scopolamine	*Scopolia japonica*	A,D
Gastrodin	*Gastrodia elata*	B,D
Antiepilepsirin	total synthesis	B

TABLE 4.3 *(continued)*

Drug	Origin	Category
Helicoside	*Aconitum sinomontanum*	A
Levodopa	*Mucuna cochinensis*	A
Daphnetin	*Daphne giraldii*	A
Paeonol	*Paeonia suffruticosa*	A
Rotudine	*Stephania rotunda*	A
	Stephania sinica	
Sinomenine	*Sinomenium acutum*	A
Tutin	*Coriaria sinica*	A
	Loranthus parasiticus	
	Dactylicapnos scandes	A
Metetrandriniodide	semisynthesis	A,C
Cissampelosine methiodide	*Cissampelos pareira*	A,C
Daurisoline	*Menispermum dauricum*	B
Curine methochloride	*Cyclea barbata*	A,C
Drugs affecting the respiratory system		
Codeine	*Papaver somniferum*	A
Rorifone	*Rorippa montana*	B
Bergenin	*Bergenia purpurascens*	A
Zhiketong	*Rhododendron anthopogonoides*	B,C
Protocatechuic acid	*Salvia plebeia*	A
Geraniol	*Cymbopogon citratus*	A
Farrerol	*Rhododendron farrerae*	A,D
	Rhododendron dauricum	
	Scopolia japonica	A,D
Mangiferin	*Mangifera indica*	A
Isomangiferin	*Pyrrosia sheareri*	A
Pinitol	*Lespedeza cuneata*	A
p-cymene	*Angelica acutiloba*	A
alpha-asarone	*Asarum europaeum*	A
Esculetindimethylether	total synthesis	A,D
Piperitone	*Mentha piperita*	A,D
Drugs having antimicrobial activity		
Andrographolide	*Andrographis paniculata*	A
Andrographolid sulfonic acid	total synthesis	B,D
Chuanhunin	total synthesis	B,D
Armillarisin A	*Armillariella tabescens*	B,D
Houttuynin bisisonicotinoyl hydrazone	total synthesis	B,D

Drug	Origin	Category
Allicin	*Allium sativum*	A,D
Novoallicin	*Allium sativum*	A,D
Berberine	*Berberis julianae*	A,D
Lysionotin	*Lysionotus pauciflorus*	A
Pyrolin	*Pyrola rotundifolia*	A,D
Chlorogenic acid	*Lonicera japonica*	A
Cryptotanshinone	*Salvia miltiorrhiza*	A
Dimethyoxycanthinone	*Picrasma quassioides*	A
Esculetin	*Fraxinus rhynchophylla*	A,D
Ardisinol I,II	*Ardisia japonica*	B,D
Tetrahydroxyflavanoldimer	*Fagopyrum cymosum*	B
Methyl hydroquinone	*Pyrola rotundifolia*	A
Drugs affecting the digestive system		
Carbenoxolone	semisynthesis	A,C
Schisantherin A	*Schisandra sphenanthera*	B
Schisandr B	*Schisandra sphenanthera*	B
Biphenyldicarboxylate bifenbate	*Lindera cammunis*	B
Cuscohygrinol ester	*Scopolia tangutica*	B,C
Sarmentoside	*Sedum sarmentosum*	B
Silymarin	*Silybum marianum*	A
Silybin phythalate sodium	semisynthesis	B,C
Swertiajaponin	*Swertia japonica*	A
Swetiamarin	*Swertia patens*	A
Curcubitacin B, E	*Curcumis melo*	A
Oleanlic acid	*Swertia mileensis*	A
	Ligustrum lucidum	
	Artemisia scopolia	A
Drugs having parasiticidal activity		
Arteannuin	*Artemisia annua*	B
Methylarteannuin	semisynthesis	B,C
Sodium arteannuin succinate	semisynthesis	B,C
Yingzhaosu A	*Artabotrys hexapetalus*	B
Toosendanin	*Melia toosendan*	B
Agrimophol	*Agrimonia pilosa*	B,D
Quisqualic acid	*Quisqualis indica*	B
Santonin	*Artemisia cina*	A
Quinine	*Cinchona ledgeriana*	A

TABLE 4.3 *(continued)*

Drug	Origin	Category
Drugs for family planning		
Yuanhuacin A	*Daphne genkwa*	B
Yuanhuacin B	*Daphne genkwa*	B
Gossypol	*Gossypium hirsutum*	A
Gossypolformate	semisynthesis	B,C
Trichosanthin	*Trichosanthes kirilowii*	B
Eye medicine		
Erycibe alkaloid II	*Erycibe obtusifolia*	B

A = known compound but new usage; B = new compound and new usage; C = semisynthesis, chemical modification of the active principles of medicinal plants; D = total synthesis based on the active principle isolated from medicinal plants.

For the purpose of promoting and mobilizing medicinal plant applications, modern scientific research is regarded as the most important measure. From a practical point of view, we look for medicinal plants that emerge from primary health care showing excellent clinical results. This is followed by multidisciplinary research including pharmacognostical, chemical, pharmacological, and clinical studies. After isolation and elucidation of the chemical structure of the active principles, we can further synthesize or modify the chemical structures in order to obtain more favorable therapeutic agents.

THE CHALLENGES OF MODERN RESEARCH AND THE DEVELOPMENT OF MEDICINAL PLANTS

Chinese traditional medicine played an important role in the health care needs of China and other countries of Asia for thousands of years. Recently, traditional Chinese and herbal drugs have gained popularity in international medical, biomedical, and pharmaceutical institutions as a potential source of valuable medicinal agents. This interest should be followed by modern research and development, and by exploring the full medical potential of resources of Chinese traditional and herb drugs, especially from medicinal plants.

The effectiveness of Chinese traditional and herbal drug products in the treatment of a variety of ailments and diseases has been established empirically over thousands of years of historical use. However, scientific data on their efficacy, pharmacological properties, and action mechanisms as well as on their active chemical constituents have so far been lacking. We recognize the importance of using established modern scientific methods and criteria to characterize these medicines so that their full medical potential can be harnessed and new insights into disease processes can be uncovered. In the research and development of Chinese traditional and herb drugs, we are tackling the many challenges faced in the scientific and technological development in innovative ways:

1. Establishing standardization for characterization of medicinal plants and products
2. Application of analytical methods to standardize methods
3. Analytical assays for safety evaluation; the use of medicinal plants through scientific verification of their therapeutic actions
4. Isolation of desired active fractions or compounds to solve potential souring problems for providing leads for new drug development
5. Use novel biosensors for studying macromolecular interactions and for rapid high-throughput screening
6. In order to provide safety, efficacy, and quality control of drug products, the international practices for research and development of new drugs, as well as the production of pharmaceuticals, are need.[56]

These international practices include good laboratory practice (GLP), good clinical practice (GCP), good agriculture practice (GAP), and good management practices (GMP).

Novel standardization methods will significantly raise the medical potential and quality of Chinese traditional and herbal drugs. Utilizing DNA fingerprinting techniques and developing cutting-edge gene chip-based applications will assure standardization for the quality of botanicals and products of Chinese traditional and herb drugs. Using modern molecular biology techniques, different species of commonly used Chinese traditional drugs are being studied to identify species-specific markers. Application of this information, different genera of these drugs as well as different families and species can be distinguished. For example, using a distinct cDNA sequence, information derived from *Fritillaria cirrhosae* and *Fritillaria thunbergii,* a silicon-based gene chip was manufactured which contained different oligonucleotide probes corresponding to the different species of *Fritillaria* plants. Thus, the gene chip was used to demonstrate its capability to correctly detect the sequence specific probes. The technology is useful in the development of gene chip technology as a standardization tool for other phytopharmaceutical materials. This standardization tool can also be applied to drug discovery research. This gene chip technology will be a part of research for biologically active fractions and compounds. We envisage that the successful development of Chinese traditional and herbal drugs will develop its own niche in today's fast-changing pharmaceutical industry.

CONCLUSION

The purpose of this chapter is to recall the experiences and achievements of research and development of new drugs from Chinese traditional and herbal drugs, and to discuss the challenges faced in modern research and development of medicinal plants. Chinese medicine and pharmacology are a great treasure-house, and efforts should be made to explore them and raise them to a high level. Reviewing the research and development of new drugs from Chinese traditional and herbal drugs, it is shown that developing natural drugs is an important pathway for new drug research. About 140 new drugs have been developed since 1949, of which about 80 originated directly or indirectly from medicinal plants. This background has shown that increasing emphasis on the use of medicinal plants in searching for new drugs in China is undoubtedly a correct strategy. Chinese traditional and herbal drugs have gained interest from the international medical, biomedical and pharmaceutical institutions as a potential source of valuable medicinal agents. This interest has to be followed by modern research and development in order to explore the full medical potential of resources of Chinese traditional and herbal drugs, especially from medicinal plants.

NOTES

1. Xiao, P.G. and Fu, S.L., Pharmacologically active substances of Chinese traditional and herbal medicines, in P.G. Xiaoi (ed.), *Herbs, spices, and medicinal plants* (Beijing: Institute of Medicinal Plant Development, 1986), pp. 49-103; Liu, C.X. and Xiao, P.G., *An introduction to Chinese materia medica* (Beijing: Peking Union Medical College And Beijing Medical University Press, 1993), pp. 246-283; Liu, C.X.., Xiao P.G., and Li, D.P., *Modern research and application of Chinese medicinal plants* (Hong Kong: Hong Kong Medical Publisher, 2000), pp. 1-111; Xiao, P.G., *General introduction of Chinese medicinal plant development* (Beijing: IMPLAD, 1988), pp. 1-162; Xiao, P.G., Some experience on the utilization of medicinal plants in China, WHO Interregional Meeting on the Standardization and Uses of Medicinal Plants, November 4-12, Tianjn, China, 1980; Xiao, P.G., Traditional experience of Chinese herb medicine, its application in drug research and new searching, in J. Beal and E. Reinhard (eds.), *Natural products as medicinal agents* (Stuttgart: Hippokrates Verlag, 1981), p. 351; Xiao, P.G., 1986, The role of traditional medicine in the primary health case system of China, *Economic and Medical Plant Research,* 4:17-26; Xiao, P.G., 1983, Recent developments on medicinal plants in China, *J Ethnopharmacology,* 7:95-100; Liu, C.X. and Xiao, P.G., *Chinese*

medicinal plants (Tianjin: Tianjin Institute of Pharmaceutical Research, 1992), pp. 1-497; Liu, C.X., *On Chinese traditional medicines* (Tianjin: Tianjin Institute of Pharmaceutical Research, 1988), pp. 1-180; Xiao, P.G. and Liu, C.X., Immunostimulants in traditional Chinese medicine, in H. Wagner (ed.), *Immunomodulatory agents from plants* (Basel: Birkhauser Verlag, 1999), pp. 325-356; Liu, C.X., 1987, Development of Chinese medicine based on pharmacology and therapeutics, *J Ethnopharmacology,* 11: 119-123; Liu, C.X., Ethnopharmacology, pharmacology and clinical application of medicinal plants in China, in C.X. Liu (ed.), *Studies on pharmacology and pharmacokinetics* (Hong Kong: Hong Kong Medical Publisher, 2000), pp. 56-134.

2. Department of Phytochemistry, Institute of Materia Medica, 1976, Chemical studies on anisodamine, *Acta Pharm Sin,* 34(1): 39-44; Department of Phytochemistry, Institute of Materia Medica, 1975, Comparison of the pharmacological actions of anisodine and scopolamine, *Nat Med J Chin,* 35(11): 795-798.

3. Yao, T.R. and Chen, Z.N., 1979, Chemical investigation of Chinese medicinal herbs: Baogongteng, I. The isolation and preliminary study of a new myotic constituent, Baogongteng A, *Acta Pharm Sin,* 14(12): 731-735.

4. Group of Pharmacology, Beijing Medical College, 1982, Anticonvulsive and sedative actions of piperine, *J Beijing Med College,* (4): 217-220.

5. Yang, Q.Z., Shu, H.D., and Lin, L.R., 1981, The blocking action of anabasin of neuromuscular junction, *Acta Pharm Sin,* 16(2): 84-88.

6. Zhang, T.M., Fong, T.C., and Lue, E.H., 1958, Studies of beta-dichroine, chloroguanide, cyclochloroquininde and baicalin on amebicides, *Acta Acad Med Wuhan,* (1): 11-16.

7. Zhou, Y.P., Fan, L.L., Zhang, L.Y., and Zeng, G.Y., 1978, The effect of higenamine on the cardiovascular system, *Nat Med J Chin,* 58(11): 664-669.

8. Liu, C.X., Liu, G.S., and Ji, X.J., 1979, Study on leukogenic effect of berbamine, *Chin Trad Herb Drugs,* 10(9): 36-37.

9. Kin, K.C., Wang, Y.E., and Hsu, B., 1964, Some neuripharmacological actions ofrotundine, *Acta Pharm Sin,* 11(11): 754-761.

10. Jiang, Y., Zhao, X.R., Wu, Q.X., Shui, H.L., Chen, W.P., Chang, S.Q., Zhao, S.Y., Tian, X.Y., Zhou, L.F., Guo, S.M., and Li, Y.J., 1982, Pharmacological actions of dehydrocorydaline on the cardiovascular system, *Acta Pharm Sin,* 17(1): 61-65.

11. Hu, B.Z., Xu, R.S., Chen, G.J., and Wu, S.X., 1979, The structure identification and pharmacological actions of l-dicentrine, *Chin Pharm Bull,* 14(3): 110-111.

12. Zhou, J.M., Yu, Z.Q., Cao, Y., Zhou, B.Z., and Liu, C.Y., 1981, Active principles in Shehanqi, *Chin Trad Herb Drugs,* 12(3): 99-107.

13. Wu, T.K., *Cephalotaxus fortunei* Hook. F., in Y.S. Wang (ed.), *The pharmacology and usage of traditional Chinese medicines* (Beijing: People's Health Publishers, 1983), p. 54.

14. Yang, J.S., Luo, S.R., Shen, X.L., and Li, Y.X., 1978, Chemical investigation of the alkaloids of Ku-Mu (*Picrasma quassioides* (D. Don) Benn.), *Acta Pharm Sin,* 14(3): 167-177.

15. Ji, X.J., Zhang, F.R., Lei, J.L., and Xu, Y.T., 1981, Studies on the antineoplstic action and toxicity of synthetic indirubin, *Acta Pharm Sin,* 16(2): 146-148.

16. Liu, G.T., We, G.T., Bao, T.T, and Song, Z.Y., 1980, The effect of *Ganodermas* on elevated serum aldolase levels induced by 2,4-dichlophenoxyacetic acid

in mice, *Acta Pharm Sin,* 15(3): 142-146; Lin, M., Yu, D.Q., Lin, X., Fu, F.Y., Zheng, Q.T., He, C.H., Bao, G.H., and Xu, C.F., 1985, Chemical studies on the quaternary alkaloids of *Rauvolfia verticillata* (Lour.) Baill. f. Rubwerocarpa H.T. Chang, Mss., *Acta Pharm Sin,* 2(3): 198-202.

17. Chen, C.H., 1984, Biological activities and medicinal potentialities of diterpene alkaloids, *Chin Trad Herb Drugs,* 15(4): 180-184.

18. Xu, D.M., Zhang, B., Li H.R., and Xu, M.L., 1982, Isolation and identification of alkaloids from *Fritillaria ussuruensis* Maxim., *Acta Pharm Sin,* 17(5): 355-359.

19. Yeng, H.W., Kong, Y.C., Lay, W.P., and Cheng, K.F., 1977, The structure and biological effect of leonurine, a uterotonic principle from Chinese drug I-mu-tsao, *Planta Med,* 31: 51-56; Yang, C.M., Hou, S.X., Zhang, Z.R., Wang B.N., and Wang, X.C., 2001, Recent advance in metabolic chemistry of traditional Chinese medicine, *Asian J Drug Metabol Pharmacokinet,* 1(1): 57-62; Liu, Y.M., Chen, H., and Zeng, F.D., 2001, The advance in studies on pharmacokinetics of Ligustrazin in China, *Asian J Drug Metabol Pharmacokinet,* 1(3): 217-220.

20. Lu, Y.C., 1980, Studies on the constituents of the essential oil of *Rhododendron tisnghaiense* Ching., *Acta Chim Sin,* 38: 241-249.

21. Yu, S.R. and You, S.Q., 1984, Anticonvulsant action of 3-n-butyphalide (Ag-1) and 3-n-butyl-4,5-dihydrophthalide (Ag-2), *Acta Pharm Sin,* 19: 566-570; Yu, S.R., You, S.Q., and Chen, H.Y., 1984, The pharmacological action of 3-n-butylphthalide (Ag-1), *Acta Pharm Sin,* 19: 486-490.

22. Yunnan Institute of Botany, 1975, Studies on d-8-acetoxycarvotanacetone, a new repellent, *Acta Botan Yunnan,* (1): 1-14; Luda Hospital, 1978, The clinical observation of Curcuma aromatica on cervical carcinoma, *New Med Pharm J,* (3): 109-111.

23. Qinghaosu Research Group, 1979, Antimalaria studies on qinghaosu, *Chin Med J,* 92: 811-816.

24. Liang, X.T., Yu, Q.X., Wu, W.L., and Deng, H.C., 1979, The structure of yingzhaosu A, *Acta Chim Sin,* 37: 231-240.

25. Yuan, D.J., 1979, Clinical observation on the effects of lactonin Coriae and tutin in the treatment of schizophrenia, *Chin J Neurol Psychiat,* 12(4): 196-200.

26. Caton Institute of drug Control, 1977, Isolation and structure of pogostone, an antifungal compnent from Chinese drug Kwang-ho-hsiang, *Pogostemon cablin* (Blanco) Benth., *Science Bull Sin,* 22: 318-320.

27. Li, J.C., Liu, C.J., An, X.Z., Wang, M.T., Zhao, T.Z., Yu, S.Z., Zhao, G.S., and Chen, K.F., 1982, Studies on the antitumor constituent of *Rabdosia japonica* (Burm. F.) Hara. I. The structure of rabdosin A and B, *Acta Pharm Sin,* 17: 682-687.

28. Wang, X.R., Wang, Z.Q., and Dong, J.G., 1982, A new diterpene from Huang-hua-xiang-cha-cai *(Robdosia sculponeata), Chin Trad Her Drugs,* 13: 491-492.

29. Cheng, P.Y., Lin, Y.L., and Xu, G.Y., 1984, New diterpeneoids of *Rabdosia macrocalyx,* the structure of macrocalin A and B, *Acta Pharm Sin,* 19: 593-598.

30. Zhang, T.M., Chen, Z.Y., and Lin, C., 1981, Antineoplastic action of triptolide and its effect on the immunological functions in mice, *Acta Pharmacol Sin,* 2: 128-131.

31. Deng, W.L., Nie, R.J., and Liu, J.Y., 1982, Pharmacological comparison of four kinds of andrographolides, *Chin Pharm Bull,* 17: 195-198.

32. Ma, J.Z., Zhao, Y.C., Yin, L., Han, D.W., and Ji, C.X., 1982, Studies on the effect of oleanolic acid on experimental liver injury, *Acta Pharm Sin,* 17: 93-97.

33. Deng, W.L. *Houttuynia cordata* Thunb, In T.S. Wang (ed.), *The pharmacology and usage of traditional Chinese drugs* (Beijing: People's Health Publishers, 1983), pp. 709-718.

34. Li, D.H., Hao, X.G., Zhang, S.K., Wang, S.X., Liu, R.Y., Ma, K.S., Jiang, H., Yu, S.P., and Guan, J.F., 1981, Antitumor action and toxicity of iriquinone, *Acta Pharmacol Sin,* 2: 131-134.

35. Chen, Z.X., Huang, B.S., Wang, C.R., Li, Y.H., and Ding, J.M., 1981, Studies on the active constituents of Hong-Cong (rhizome of *Eleutherine americana*), *Chin Trad Herb Drugs,* 12: 484.

36. Gao, Y.G., Song, Y.M., Yang, Y.Y., Liu, W.F., and Tang, J.X., 1979, Pharmacology of tanshinone, *Acta Pharm Sin,* 14: 75-82.

37. Xu, Y., Hu, Z.B., Feng, S.C., and Fan, G.J., 1979, Studies on the antituberculosis principles from *Lysionotus pauciflora* Maxim, I. Isolation and identification of nevadensin, *Acta Pharm Sin,* 14: 447-448.

38. Zhu, D.Y., Chen, Z.X., Zhou, B.N., Liu, J.S., Huang, B.S., Xie, Y.Y., and Zeng, G.F., 1979, Studies on chemical constituents of Bu-gu-zhi, the seeds of *Psoralea corylifolia* L., *Acta Pharm Sin,* 14: 605-611.

39. Tseng, K.Y., Chou, Y.P., Chang, L.Y., and Fan, L.L., 1974, Pharmacological studies on *Radix puerariae,* I. Its effects on tdog arterial pressure, vascular reactivity, cerebral and peripheral circulation, *Chin Med J,* 54: 265-270.

40. Chen, Z.X., Gu, W.H., Huan, H.Z., Yang, X.M., Sun, C.J., Chen, W.Z., Dong, Y.L., and Ma, H.L., 1981, Studies on the water soluble phenolic acid of *Salvia miltiorrhiza, Chin Pharm Bull,* 16: 536-537.

41. Sun, F.Z., Su, J.D., and Zhen, H., 1981, Studies on pharmacological activities and toxicities of armillarisin A, a new choleretic drug, *Acta Pharm Sin,* 16: 401-406.

42. Chen, Z.X., Huang, B.S., She, Q.L., and Zeng, G.F., 1979, The chemical constituents of bai-hua-qin-hu, the root of *Peucedanum praeruptorum* Dunn, four new coumarins, *Acta Pharm Sin,* 14: 486-489; Ye, J.S., Zhang, H.Q., and Yuan, C.Q., 1982, Isolation and identification of coumaria praeruption E from the root of Chinese drug *Peucedanum praeruptorum* Dunn, *Acta Pharm Sin,* 17: 431-434.

43. Deng, S.X. and Mo, Y.J., 1979, Pharmacological studies on *Gastrodia eleta* Bleme, I. The sedative and anticonvulsant effect of synthetic gastrodin and its genin, *Acta Botan Yunnan,* 1: 66-73.

44. Xiao, P.G., Traditional experience of Chinese medicine: Its application in drug research and new drug searching, in J. Beal and E. Rinhard (eds.), *Natural products as medicinal agents* (Stuttgart: Hippokrates Verlag, 1981), pp. 351-394.

45. You, J.Q., Le, W.J., and Mei, J.Y., 1982, The in vitro effect of agrimophol on *Schistosoma japonicum, Acta Pharm Sin,* 17: 663-666.

46. Hunan Institute of Pharmaceutical Industry, 1978, Pharmacological investigation of the pressure action of *Citrus aurantium, Science Bull Sin,* 23(1): 58-62.

47. Zheng, J.M., 1984, Clinical observation of the analgesic action of paeonol sulfanate injection, *Chin Trad Herb Drugs,* 15: 460.

48. Duan, Y.C., Li, C.H., and Chen, S.Y., 1957, A preliminary study of anthelmintic action of potassium quisquqalate, *Acta Pharm Sin,* 5: 87-91.

49. Shia, S.H., Shao, B.R., Ho, Y.S., and Yang, Y.C., 1962, Prophylatico-therapeutic studies of curcurbitine in *Schistosomiasis japonica* in mice, *Acta Pharm Sin,* 9: 327-332.

50. Dong, Z.T. and Jiang, W.D., 1982, Effects of danshensu on isolated swine coronary artery perfusion preparation, *Acta Pharm Sin,* 17: 226-228.

51. Xiao, P.G., Tong, Y.Y., Lou, S.R., Chen, B.Z., Wang, L.W., Shang, J.H., and Qian-cao, *Rubia cordifolia* L., in *Chinese materia medica* II (Beijing: People's Health Publishers, 1982), pp. 149-159.

52. Cheng, Z.L., Huang, B.S., Zhu, D.Y., and Yin, M.L., 1981, Studies on the active principles of *Aristolochia debilis,* II. 7-hydroxy aristolochic acid A and 7-methyloxy aristolochia acid A, *Acta Chim Sin,* 39: 237-242.

53. Xiao, 1981, Traditional experience of Chinese medicine.

54. Deng, 1983, *Houttuynia cordata* Thunb.

55. Shen, L.C., Wang, Y.Y., and Feng, L.H., 1983, Preparation of allicin microcapsule, *Chin Trad Herb Drugs,* 14: 161-164.

56. Liu, C.X. and Xiao, P.G., 2002, Recalling the research and development of new drugs from traditional Chinese medicine, *Asian J Drug Metabol Pharmacokinet,* 2(2): 133-126.

MEDICINAL PLANTS IN NATIVE CULTURES

Chapter 5

African Medicinal Plants

Ruth Kutalek
Armin Prinz

INTRODUCTION

In the early modern age, European travelers and explorers were obliged to study the local flora and fauna as well as food and plant medicines used by the local populations of newly discovered areas. At that time it was too expensive and too unsafe to transport everyday goods from their home countries. Due to these circumstances, scientists very soon became essential members of the discovery crews. This was especially the case in South America and south Asia. Europeans were able to penetrate the continents easily through the rivers and came into contact with local populations. Therefore as early as 1600, possibly even earlier, colonial doctors used such plants as the snakeroot—*Rauwolfia serpentina* or *Strychnos nux-vomica* from India, Ipecacuanha or *Cinchona* from South America—and incorporated them into the old European pharmacopoeias. Famous doctors such as Garcia da Orta (1490/1500-1570), a Portuguese of Jewish origin, or the Dutch Willem Piso (1611-1678) respectively published wonderful descriptions of the materia medica of Ceylon and Brazil.[1] However, no similar reports came from Africa; there the situation was completely different. The

Photographs in this chapter are courtesy of Armin Prinz.

continent south of the Sahara is a plateau; the rivers descend into cataracts and waterfalls toward the sea and were thus not traversable. In addition, malaria and other deadly tropical diseases made it difficult to penetrate the continent. Some of the most famous active plants relevant to modern medicine that are native to Africa were thus discovered relatively late.

In 1941 Paul Mangelsdorf, former director of the Harvard Botanical Museum, in the first lecture he gave to students on medicinal drugs, predicted that in 25 medicinal drugs of plant origin would be of little more than historical interest. At a symposium in 1968 on plants in the development of modern medicine, he made a follow-up comment to his 1941 statement:

> Twenty-seven years have now elapsed since I made this bold prediction, and it turns out that I could scarcely have been more wrong. Today there is probably more interest in drugs derived from plants than at any time in history.[2]

For his change of thinking with the discovery of penicillin (deriving from the mold fungus *Penicillium notatum*) and its "brilliant success," one plant was especially responsible: *Rauwolfia.*

The plants described in this chapter are some of the most important African folk medicines that are, were, or will probably be again of relevance to biomedicine. It is noteworthy that three of them—*Strophanthus* spp., *Catharanthus roseus* (L.) G. Don, and *Rauwolfia vomitoria* Afzel—belong to the family Apocynaceae; only one, *Physostigma venenosum* Balf., belongs to the family Fabaceae. It is also interesting to note that all four genera are or were used not only as medicines but also as hunting poisons in Africa.

STROPHANTHUS *SPP., APOCYNACEAE*

Strophanthus is surely one of the most interesting plants of Africa. It was the first effective and highly toxic plant discovered by the Europeans in Africa. *Strophanthus* is the widest spread, best documented, most traded, and most rapidly working poison. About 30 species are known in Africa. The most common ones used are *Strophanthus hispidus* A. P. De Candolle and *S. sarmentosus* A. P. De Candolle in West Africa and *S. kombe* Oliver in East Africa.

The discovery of *Strophanthus* started with the beginning of European colonization at the coasts of West Africa in the fifteenth century. European travelers and missionaries were often confronted with poisoned weapons—in attacks that many of them did not survive. Richard Hakluyt (1552?-1616), an English geographer who issued a number of accounts on voyages of discoveries, published a report, "The Voyage of M. George Fenner to Guinie, and the Islands of Cap Verde, in the year of 1566," in which it is described how the crew of the Mayflower landed near Cape Verde. After initial friendly contacts with the indigenous population, the situation seemed to deteriorate. Fenner writes:

> [A]nd presently the Negros laied handes to our men that were on the shore, and tooke three of them with great violence. . . . They [the English] got from them with their boates although many of them were hurt with their poysoned arrowes: and the poison is uncurable, if the arrow enter within the skin and drawe blood, and except the poison be presently suckt out, or the place where any man is hurt bee forthwith cut away, he dieth within foure dayes, and within three houres after they hurt or pricked, wheresoever it be, although but at the litle toe, yet it striketh up to the heart, and taketh away the stomacke, and causeth the partie marveilously to vomite, being able to brooke neither meat nor drinke. . . . Some of them [indigenous population] caried our men againe to the towne, and the rest shot at us with their poisoned arrowes, and hurt one of our men called Androwes in the smal of the leg, who being come aboord, (for all that our Surgeons could do) we thought he would have died. . . . Of our men that were hurt by the Negros arrowes, foure died, and one to save his life had his arme cut off. Androwes that was last of all hurt, lay lame not able to helpe himselfe: onely two recovered of their hurts.[3]

Even though this poison was legendary, it was relatively late until its plant source was documented scientifically. The Scottish doctor and adventurer Mungo Park in 1796 was the first to discover that the poison derived from a shrub:

> Poisoned arrows are used chiefly in war. The poison, which is said to be very deadly, is prepared from a shrub called *koona* (a species of *echites*), which is very common in the woods. The leaves of this shrub, when boiled with a small quantity of water, yield a thick black juice, into which the Negroes dip a cotton thread; this thread they fasten round the iron of the arrow in such a manner that it is almost impossible to extract the arrow, when it has sunk beyond the barbs, without leaving the iron point and the poisoned thread in the wound.[4]

Scottish botanist, doctor, and later British consul in Zanzibar, Sir John Kirk, who accompanied David Livingstone during his second journey to Africa (1858-1864), described the discovery of the plant:

> The source of the poison, namely *Strophantus kombe,* was first identified by me. I had long sought for it, but the natives invariably gave me some false plant, until one day at Chibisa's village, on the river Shire, I saw the "kombe," then new to me as an East African plant (I had known an allied or perhaps identical species at Sierra Leone 1858, where it is used as a poison). There climbing on a tall tree it was in pod, and I could get no one go up and pick specimens. On mounting the tree myself to reach the kombe pods, the natives, afraid that I might poison myself, if I handled the plant roughly or got the juice in an cut or in my mouth, warned me to be careful, and admitted that this was the "kombe" or poison plant. In this way the poison was identified, and I brought specimens home to Kew, where they were described.[5]

Kirk managed to send some examples of this plant, which was kept strictly secret by the indigenous population, to Edinburgh. In 1869 Thomas Fraser was able to isolate the glycoside responsible for the cardiac effectiveness and named it strophantine. It was he who introduced strophantine into therapy. Later he was also able to detect its chemical and pharmacological properties. Strophantine had been on the market since 1887 when the German company Merck produced it from the seeds of *Strophanthus kombe*. Anton Drasche was the first to do clinical tests with *Strophanthus* as a cardiac medicine outside of America and England.[6] Emil Pins, however, published his findings first.[7] In 1888 Rothziegel and Koralewski tested strophantine on 44 patients in the Viennese General Hospital. It was administered in the form of a powder but also through injections. Even though the seeds were obtained via wholesale trade from England, not a single voucher specimen with leaves, flowers, and seeds was available in Europe.[8] Albert Fraenkel introduced intravenous strophantine therapy in 1905-1906, a technique which was later declared as "one of the greatest advances in medicine in the 20th century."[9]

Louis Lewin describes the deadly effect of a poisoned arrow among the Bambara:

> Two marksmen were hit by these (Strophantus-arrows). The first one had the iron of the arrow 5 cm deep embedded in his right chest, between the third and fourth rib; it had pierced the pleura and had slightly grazed the lung. The arrowhead was immediately removed. Six minutes after the extraction and eleven minutes after he had been wounded, he died. That he clearly died of poisoning is drawn from the course of the second victim's disease. This person was hit in the right calf by an iron arrowhead. It was necessary to cut a large incision to

> pull it out. However, death occurred eight minutes after the extraction and fifteen minutes after he had been wounded. Both poisoned victims showed the same symptoms. About one minute after the extraction of the spearhead, both, who had been sitting, lay down slowly and uttered some inarticulate sounds. Their heads fell on their chests and a cold sweat quickly covered their bodies. Movements were slow and painful; breathing seemed to stop, it slowed down more and more. Their pulse was weak, difficult to feel. Suddenly the heart stopped beating. The dying went into a sudden spasm, their tongues were stretched out of their mouths, their eyeballs gazed upwards up and death immediately followed. All these symptoms quickly happened in succession. The injection of atropine and other things, of course, had to fail.[10]

In the 1970s Malawi was the main source of supply for seeds of *Strophanthus kombe,* which were then of commercial importance for medicinal purposes in Europe.[11] *Strophanthus* still remains a source of ouabain, a cardiac glycoside that is used as a heart tonic in emergencies because of its rapid onset and short duration of action.[12]

CATHARANTHUS ROSEUS *(L.) G. DON, APOCYNACEAE*

Antitumor medicines are quite common in folk medicine. Already in the papyrus Ebers (1550 B.C.) garlic is praised as effective against cancer.[13] The antimitotic activity of colchicine from *Colchicum autumnale* L., one of the oldest documented medicinal plants, was discovered when the effective mechanism of the substance on the acute gout attack was detected.[14] The Arabs used the plant long before Linnaeus named it after Colchis at the Black Sea. In the seventeenth century European medicine widely embraced it.[15] Meadow saffron was used both to treat gout as well as rheumatic complaints. These unclear indications and the high toxicity induced Nothnagel and Rossbach to advise against the application of colchicine even up to around 1900.[16] Nevertheless, until today it has remained the only specific therapy against acute gout attack.

The most significant discovery in the field of antitumor substances deriving from folk medicine is owed to a coincidence. At the end of the 1950s, two research groups independent of each other—the Collip Laboratories (University of Western Ontario) and the Lilly Research Laboratories—made a major finding in their search for antidiabetics. While studying the hypoglycemic properties of *Catharanthus roseus,* a small shrub indigenous to Madagascar and now widespread in all tropical regions of the world, the research workers lost many rats which had been given an extract of the leaves, due to *Pseudomonas* infection. Further investigations revealed that their death was due to a severe decline in the number of their lymphocytes. This discovery finally directed the researchers' attention toward the antitumor activities of *Catharanthus* alkaloids.[17] In the beginning, four indolalkaloids, which were found to inhibit cell growth, were discovered: vinblastine (vincaleucoblastine), vincristine (leurocristine), vinleurosine (leurosine), and vinrosidin (leurosidine).[18] Vincristine is now used in single-drug therapy for acute leukemia; in combination chemotherapy it is indicated for the treatment of Hodgkin's disease, non-Hodgkin's lymphoma, breast cancer, uterine and cervical cancer, small cell bronchial cancer, rhabdomyosarcoma, and various sarcomas. Vinblastine is indicated in the treatment of Hodgkin's disease, non-Hodgkin's lymphoma, advanced testicular cancer, breast and ovary epithelioma, Kaposi's sarcoma, choriocarcinomas, and in some cases of histiocytosis.[19] A semisynthetic derivative, vindesine, which can be prepared from vinblastine, is indicated in the treatment of acute lymphoblastic leukemia and refractory lymphomas; certain solid tumors are also indications: breast, esophagus, upper respiratory and digestive tract, and bronchopulmonary cancer. Another semisynthetic derivative, vinorelbine, is currently used in cases of metastatic breast cancer and bronchial cancer.[20]

The first report on *Catharanthus roseus* dates back more than 300 years. Etienne de Flacourt-Bizet in his book *Histoire de la Grande Isle Madagascar,* published in 1661, describes a plant that could well have been the famous Madagascar periwinkle:

> *Tongue,* herbe resseblante au *Saponaria,* qui a la fleur comme celle du Jassemin, l'une est blanche, l'autre est de couleur de pourpre, la racine est fort amere, de laquelle ils se servent contre le mal de cœur, & est bonne contre les poisons, elle approche du vincetoxin; ou asclepias, & ne vient pas plus haute. Celle qui a la fleur blanche a plus de vertu.[21]

Madagascar periwinkle is included in Madagascar's *Pharmacopoeia of Crude Drugs*.[22] The plant up until today is very common in the folk medicine of many tropical countries. In Guyana a decoction is drunk against "heart diseases."[23] Peckolt reports from Brazil that an infusion of the leaves is used against internal bleeding and scurvy, as a mouthwash against toothache, and for cleaning and healing chronic wounds.[24] In Togo a watery solution is used to treat dysmenorrhoea and, indeed, antibiotic and antiviral activities were found in *Catharanthus* alkaloids.[25] In Congo the stems, leaves, and roots of *C. roseus* are used against diabetes and diarrhea.[26] On the coast of Tanzania a decoction of the leaves is drunk for stomach problems.[27] In Brazil the plant is still used under the name *Boa noite* as an antidiabetic.[28] Also, in the Philippines, Jamaica, South Africa, India, and Australia, an infusion of the leaves is drunk against diabetes.[29] Other indications in African traditional medicine are against diarrhea, abdominal pain, hypertension, and as an antitussive, diuretic, purgative, and antihelmintic.[30].

It is assumed that plants used in folk medicine are a rich source of possible substances with antitumor effect. J. L. Hartwell enumerates more than 3,000 species that could be promising in this regard.[31] Many scientists have shown in their work that plants can actually be a good source for antineoplastic substances.[32] Systematic investigations of traditional medicinal plants in South Africa by Charlson brought the following results: Of 46 plants with different indications, 6 could be determined as having antitumor activity: *Raphionacme hirsuta, Cheilanthes contracta, Haemanthus natalensis, Urginea capitata, Brunsvigia radulosa,* and *Amaryllis belladonna.* Of these, two—namely *Raphionacme hirsuta* and *Cheilanthes contracta*— are also used traditionally against tumors.[33]

Catharanthus roseus is yet to be fully studied. New substances, such as the flavonol glycoside syringetin are still being discovered.[34] In a dual culture of *Trichoderma harzianum* and *Catharanthus roseus,* a novel antimicrobial compound, trichosetin, with a remarkable activity against gram-positive bacteria was produced.[35] The antidiabetic properties of the Madagascar periwinkle are also again the focus of research.[36] With new, rapid secondary-metabolite screening methods, the production of indole alkaloids by plant cells and tissues becomes more probable for the future.[37]

RAUWOLFIA VOMITORIA *AFZEL (APOCYNACEAE)*

The Indian snakeroot *Rauwolfia serpentina* (L.) Bent. ex Kurz was first described by Garcia da Orta (1490/1500-1570) as an important medicinal plant in South Asia. Da Orta, a Portuguese doctor of Jewish origin, was the senior consultant of the Portuguese-East Indian Company and had an understanding for Singhalese medicine in Ceylon. His book *Coloquios Dos Simples, e Drogas he Cousas Medicinaes da India,* published in 1563 in Goa, is the first book written by a European that describes a foreign traditional medical system. In imaginary dialogues between himself and a local practitioner, the medical system and materia medica of the peoples of Ceylon are presented. Under the title "Do Páo da Cobra" he reports how the local inhabitants discovered the use of snakeroot:

> On the wonderful island of Ceylon, which is rich in numerous and excellent fruits as well as small and large game, there are also plenty of snakes, named by the population spectacle snakes, in Latin we could name them *Regulus serpens*. Against them God gave this "snake wood." I know it helps against their bite. Because on this island there are animals, similar to ferrets, which are called *quil* (others say *qurpele*). They often fight with these snakes. When it (the ferret) knows or fears that it has to fight with it, it bites into a piece of this root, which is visible and rubs itself with the hand, which it has moistened in the juice (of the plant). This it does on the head, the body and on all parts, where, as it knows, the snake will bite. And it will fight until it has killed (the snake) by biting and scratching. And when it is not able to defeat her, or if it has no power any more, the *quil* or *quirpale* called animal rubs itself on the root and goes back to fight again with her, and thus it kills or defeats her at last. And this incident the Singhalese have observed. They have seen that this root and this wood is helpful against snake bites.[38]

This "snake wood" was introduced into European medicine only a few years later. It was used for the same indications as in Indian medicine: against intermittent fever, worms, snakebites, and delirious conditions. In the nineteenth century *Rauwolfia* disappeared from the European pharmacopoeias. The newly established "scientific medicine" started to remove from its materia medica "absurd" and "magical" elements. Its usefulness was not recognized due to the pathophysiological ignorance about hypertonia and ventricular tachycardia, today the most important indications for use of *Rauwolfia* alkaloids.

It was not a pharmacologist but an English colonial economist, George Watt, who at the turn of the century thought again about the possible therapeutic benefit of *Rauwolfia* in modern medicine:

> Altogether the popular beliefs with regard to this plant and the testimony of medical men in India who have practically tried it as a remedy for fevers, seem to indicate that it possesses strong and well-marked properties; it might, therefore, be advantageous to have a more complete analysis of its composition and more careful determinations of its action.[39]

More than fifty years had passed when, in 1952, the alkaloid reserpine with its sedative and hypotensive action was discovered. Finally *Rauwolfia* was again considered a highly effective drug. Other important alkaloids in

Rauwolfia are ajmalicine (a spasmolytic) and ajmaline (an antiarrhythmic).[40]

The triumphal advance of *Rauwolfia* alkaloids as components of important modern pharmaceuticals led to an enormous demand for appropriate raw drug material. Similar to the first boom of *Cinchona* bark approximately 200 years before, when Peru unsuccessfully tried to install an export prohibition for cinchona seedlings, India in 1953 tried to issue an export prohibition in order to attain a monopoly on *Rauwolfia* alkaloids.[41] It was also unsuccessful, because very soon a replacement was found in Africa. It even turned out that *Rauwolfia vomitoria* had twice the amount of reserpine of *R. serpentina*.

The search for an African alternative was simple. Practically all Central African people use *Rauwolfia* in their traditional medicine (see Photo 5.1). Its toxicological and therapeutic power is well known and highly regarded. Traditionally it is used against snakebites, fever, and nervous disorders. In Ghana and Nigeria herbalists use it as an emetic and a purgative. In the same regions children are treated with this plant for cerebral cramps, jaundice, and gastrointestinal disorders. A watery solution of the bark is administered against such parasites as lice and scabies.[42] In Mali the root is used to treat hemorrhoids and hepatomegaly.[43] Among the Pygmies of the Congo basin, *Rauwolfia* species are administered together with traditional ash salt against diarrhea and with red palm oil against elephantiasis of the legs.[44] Traditionally it was as well used against tetanus, which was also indicated in biomedicine in the late 1950s, but it had little success.[45] *Rauwolfia vomitoria* is used in traditional African medicine as a sedative for mentally ill people and against epilepsy. The Ubi on the Ivory Coast treat epilepsy in the following way: the roots of *Rauwolfia vomitoria* are pounded along with the leafs of *Tristemma incompletum* and *Alchornea cordifolia*. The extract obtained from squeezing the mixture is drunk or dribbled into the nose. Another very interesting combination is the mixture of the sap of *Parquetina nigrescens* (Afzelius) Bullock with the powdered root of *R. vomitoria* as an arrow poison in Central Africa. The cardiotoxicity of the glycosides of *P. nigrescens* is increased by the heart activities of the *Rauwolfia* alkaloids.[46]

By 1965 large collecting campaigns had reached the Nile-Congo watershed in Northeast Zaire. Due to the well-known feature of the plant, no special training was necessary for the people who gathered the plant. With the help of missionaries and adventurers, huge amounts of wild *Rauwolfia* roots were harvested and dried. People earned U.S. 32 cents per kilogram—the world market price at that time (1972) was up to U.S. $3.50.[47] In 1976, eleven years after the first campaigns were initiated, harvesting in the wild

PHOTO 5.1. Collected roots of *Rauwolfia vomitoria,* Azande area. Doruma, Northeast Zaire, September 1974.

was stopped—not because the therapeutic guidelines spoke against the use of reserpine in hypertonia (as was the case at the beginning of the 1980s) but because the yield of *Rauwolfia* roots in this region decreased enormously and further gathering was no longer profitable. For more than a decade people had been encouraged to harvest wild *Rauwolfia.* Whole families and communities had changed their survival strategies accordingly. They eventually had to move deeper and deeper into the bush to find these wild plants. One day, however, no one was interested in buying their yield. From one day to the next, people lost their economic base; great distress ensued. Families no longer had fields, storehouses, solid farmstead; they had given up everything to collect *Rauwolfia*. It took several years until the situation stabilized.

Rauwolfia is traditionally also used as abortifacient. In experiments, contractions of the uterus after administration of *Rauwolfia* were proven.[48] Also the Azande know *Rauwolfia*—or *gbutunga*—for these properties. In earlier times abortions were practiced exclusively for magicoreligious reasons. If an oracle stated that a pregnancy involved danger to the physical or spiritual well-being of the mother, family, or community, then an abortion was usually judged necessary.

Illegitimate children are happily accepted in Zande society. Indeed, women who have already given birth to a child are preferentially married in this population troubled by a high rate of sterility. Completely different, however, is the situation of young Azande girls in mission schools or government schools. These young women are prohibited from attending school in case of pregnancy. It is a sad irony that the Christian mission schools created the basis which made abortions a social necessity. The following is a letter by a 19-year-old woman, a high school pupil, to her former teacher and lover:

> My dear L., you know, I have always admitted my mistakes. I was pregnant. You knew that this could not have come from another boy. I hinted to you that I am pregnant. You kept silent for two days. I felt truly abandoned. What will I do with a child without a father? Thus I decided to undergo an abortion. I took *gbutunga*. You ask, how I got to know about this plant? In the Lyceum Ndolomo the girls always discuss about chaps, about sexuality, about abortions, about the way how the boys like it. I got this information from my friends. One of them recalled *gbutunga* as abortifacient. Many of us were interested. We asked her questions about its use. Because we girls are often victims of an unwanted pregnancy and the boys escape from their responsibility. She told us: "Friends, if you are pregnant and are at the beginning of your pregnancy, take the root of *gbutunga*. Peel it, put it into a cup and add water. Leave this solution to concentrate for several hours. When you think that the solution is well concentrated, drink it in the late evening. I am sure that you will be relieved of the developing child. In case of failure increase the dosage." I heard that and took my chance. In the evening of the 21st October 1982 I peeled the root of *gbutunga* and put it into water. The next evening, before going to bed I emptied the cup, without someone at home aware. After some hours I felt severe heat in my uterus. The labia became highly sensitive. I could close my legs only with difficulty. I had severe pain in my lower belly. I constantly had to change my position on the bed. This is pain-

> ful, my friend. I felt death approaching. I thought about everything. I said to myself that I have to wait patiently until I am ready to die like an animal in the wilderness. It became late. At around 4 or 5 o'clock in the morning I felt blood. I went out of the house. I withdrew into the bush. Almost one liter of blood ran out. After this had stopped, I bathed and fell into deep thoughts. Now I am happy. I am again able to go to school but the pain continues.[49]

Even though *Rauwolfia* alkaloids are not much used in medicine today except for ajmaline in the treatment of ventricular tachycardia, we are convinced that *Rauwolfia* will again play an important role in biomedicine, especially with improved developments to obtain indole alkaloids with biotechnological methods.[50]

PHYSOSTIGMA VENENOSUM *BALF. (FABACEAE)*

The famous toxicologist Louis Lewin describes the main active principle of *Physostigma venenosum* Balf., the physostigmine, as the most important substance used in medicine.[51] The seed of this West African liana, the so-called Calabar bean, was used as ordeal poison *esere* or *itunda* among the peoples of the Calabar coast from Sierra Leone to Congo. William Freeman Daniell first observed the use of the bean in a traditional judicial procedure in about 1840.[52]

Louis Lewin vividly illustrates the *esere* ordeal:

> The accused was dragged before the idol and had to consume a number of seeds which varied between twenty and one hundred; afterwards, in order to accelerate the absorption of the poison into the circulation, he had to walk to and fro—as the jailer in times past recommended to Socrates after drinking the mug of hemlock. If he vomits the poison while moving around, he is considered innocent. The consumption of a few seeds is considerable more dangerous than that of many, because the absorption into the circulation (blood and lymphatic vessels) in the former is so fast that gastric irritation can hardly develop. Therefore general illness can occur before vomiting takes place. Repeatedly European observers stated that if signs of a general poisoning appeared, the spectators would rush toward the poisoned person like wild animals and kill him.[53]

In sub-Saharan Africa it is very common to use oracles, especially before a court in the form of a poison ordeal. In former times opponents directly underwent a poison ordeal; today chicken or dogs are used instead. During these ceremonies, the poison is asked to make a decision and, depending on which animal dies and when, the ordeal will be interpreted either positively or negatively (see Photo 5.2).[54] Already in 1878 the English Consul Hopkins drew up an agreement treaty with the chiefs of the region that anybody who administered the *esere* bean or took the *esere* bean willfully should be punished.[55] Donald Simmons, however, states that even though Nigerian law forbade the use of the Calabar bean and the possession of the bean entailed fine and imprisonment, some Efik people kept one bean in their pocketbook or with a cache of coins in order to prevent witches from

PHOTO 5.2. Poison ordeal with *Physostigma venenosum*. *Journal de Voyage*, 231 (Paris: 1881), p. 353.

"drinking" the money. Simmons in the 1950s also gave one of the last accounts of the use of *esere:*

> The Efik believe the *esere* or Calabar bean *(Physostigma venenosum)* possesses the power of destroying witchcraft. An individual accused of witchcraft usually demanded his right to undergo the Calabar bean ordeal in order to establish his innocence. The suspect ate eight Calabar beans, and then drank an infusion of several ground Calabar beans and water. If the suspect possessed witchcraft, his mouth shook and mucus came from his nose, but if innocent of witchcraft, he lifted his right hand and then regurgitated. If the poison continued to affect the suspect after he established his innocence, the suspect received an infusion of the excrement of a person of the same sex mixed with water previously used to wash the external genitalia of a female. If the ordeal revealed guilt, the suspect died without aid, and the corpse was thrown into the forest.[56]

There are two explanations why the Calabar bean was used as an ordeal. One is that the judge was aware that boiling the bean would render the potion less toxic; eserine (physostigmine) degrades to the less toxic eseroline. The traditional judge would thus prepare a hot decoction for an accused whom he was convinced was not guilty and a cold infusion for the one he thought guilty. A second explanation is that a guilty person afraid of being sentenced to death would drink the potion hesitantly, which would result in the steady absorption of the poison and lead to death by asphyxia. The innocent person, convinced that the ordeal would be positive toward him, would drink the potion quickly, which resulted in gastric irritation and vomiting before the poison could be fully absorbed.[57]

Besides the Calabar bean, in Africa *Erythrophleum guineense, Adenium obesum, Datura metel, Calotropis procera,*[58] *Strychnos icaja,*[59] and various *Catharanthus* species are also used as ordeal poisons.[60] *Erythrophleum guineense* was examined especially intensively ethnopharmacologically.[61] The 30-year time span between the European discovery of the Calabar bean and the first therapeutic employment of physostigmine was extremely short in comparison to the 200-year time span of a similar working parasympathomimetic—the alkaloid pilocarpine from *Pilocarpus jaborandi.* The South American jaborandi had already been described in the sixteenth and seventeenth centuries by Soares, Piso, and Marcgrave. Pilocarpine was isolated 1875 and therapeutically used in cases of fever, stomatitis, enterocolitis, laryngitis, bronchitis, bronchiectasis, influenza, pneumonia, hydropericarditis, edema, psoriasis, poisoning, neurosis, and kidney diseases.[62]

Still a modern indication, Weber used pilocarpine against glaucoma for the first time in 1877.[63]

The circumstances that made identification of the plant so difficult were similar for many other poisons used by the indigenous peoples of Africa. People tried to keep the plant a secret in the presence of the white missionaries and colonial personnel. William Freeman Daniell was the first to bring Calabar seeds to Scotland. The seeds sprouted, the plants grew, but they had no blossoms—and an identification was therefore not possible. In the following years no blossoms could be obtained from Africa. The indigenous kings were able to keep the secret of the plant for centuries. Finally, Scottish missionaries were able to bring leaved branches, blossoms, and fruits to Edinburgh where John Hutton Balfour recognized it as a hitherto undescribed plant and founded a new genera, *Physostigma,* with its only species, *venenosum.*[64] After the first description of physostigmine by W. F. Daniell in 1846, its therapeutic effect was investigated in 1855 by Christison, who did an experiment on himself with a tiny amount of the bean:

> Having some doubts whether I had obtained the true ordeal-potion, as it tasted so like an eatable leguminous seed, I took one eighth of a seed or six grains, about an hour after a very scanty supper. During an hour that I passed in bed reading, I could observe no effect whatever, and next morning I could still observe none. I am now satisfied, however, that a certain pleasant feeling of slight numbness in the limbs, like that which precedes the sleep caused by opium or morphia, and which I remarked when awake for a minute twice or thrice during the night, must have been owing to the poison. On getting up in the morning I carefully chewed and swallowed twice as much, viz. the fourth of a seed, which originally weighed forty-eight grains. A slight giddiness, which occurred in fifteen minutes, was ascribed to the force of the imagination; and I proceeded to take a warm shower bath; which process, with the subsequent scrubbing, might take up five or six minutes more. The giddiness was then very decided, and was attended with the peculiar indescribable torpidity over the whole frame which attends the action of opium and Indian hemp in medicinal doses. Being now quite satisfied that I got hold of a very energetic poison, I took immediate means for getting rid of it, by swallowing the shaving water I had just been using, by which the stomach was effectually emptied. Nevertheless I presently became so giddy, weak, and faint, that I was glad to lie down supine in bed. The faintness continuing great, but without any uneasy feeling of alarm . . . I had, in fact, no uneasy feeling of any kind, no pain, no numbness, no prickling, not even any sense of suffering from the great faintness of the heart's action; and as for alarm,

> though conscious I had got more than I had counted on, I could also calculate, that, if six grains had no effect, twelve could not be deadly, when the stomach had been so well cleared out. . . . Next morning, after a sound sleep, I was quite well.[65]

Physostigmine was isolated for the first time in 1860 by Jobst and Hesse in Stuttgart, Germany.[66] Already in 1876 the substance was successfully used by Laqueur in case of glaucoma, the later main indication of this substance.[67] A detailed pharmacological investigation was made by Loewi* and Mansfeld;[68] Percy Julian succeeded in doing the first total synthesis of physostigmine in the 1930s.[69]

Today the liana sold for medicinal purposes can still be found in Central African markets.[70] In traditional medicine *Physostigma venenosum* is used against edemas, asthma in children, head lice, parasitic skin diseases, and stomach pain.[71]

The seed of *Physostigma venenosum* was formerly "officinal" used as medicine. Recently it was discovered that its alkaloid physostigmine or eserine can also be obtained by fermentation produced by a *Streptomyces*.[72] Eserine is an antidote to strychnine, nicotine, curare, and atropine.[73] It has been replaced by synthetic anticholinesterases (neostigmine and pyridostigmine). Their therapeutic indications today are myasthenia, postoperative intestinal or bladder atony, stubborn constipation, and postoperative recovery after curarization with a competitive neuromuscular blocking agent.[74] Physostigmine was also tested as an adjuvant treatment to overcome unwanted gastrointestinal side effects of morphine in mice[75] and in treating a carbon monoxide poisoning in a patient.[76]

A future potential use might lie in its ability to improve cognitive function, which could be useful in Alzheimer's disease and other cognitive disorders.[77] Rivastigmine, which was developed after the chemical structure of physostigmine, is licensed for use in the United Kingdom for the symptomatic treatment of mild to moderately severe Alzheimer's disease.[78]

FUTURE PROSPECTS

Several promising plants used in African traditional medicine might be of future relevance to modern medicine. *Securidaca longepedunculata* Fres. (Polygalaceae), for instance, is a legendary plant that is widespread in Africa and used as poison and in folk medicine. The Hausa in north Nigeria

---

*Otto Loewi was awarded the Nobel Prize for medicine in 1936. After the Nazi occupation of Austria in 1938 he was arrested and released only after he transferred his Nobel Prize money to a Nazi-controlled bank.

even call it "mother of medicines." The plant is known especially as an ordeal poison. Among the Gbaya in the southwest of the Central African Republic a decoction of the root bark is dribbled into the eye of the accused. If the individual becomes blind, he or she is proven guilty. If the eye only swells and does not lose its capability to see, the victim was wrongly accused and has to be amply compensated. The same ordeal is also done with a chicken as a kind of preselection. Preparations of the root are especially used against pain, various kinds of infections, snakebite, and for its psychoactive properties.[79] In Zimbabwe, it is used against headache, madness (the powdered root causes violent sneezing), pains, and rheumatism; in Kenya, against wounds.[80] The roots are also utilized as purgative, anthelmintic, against convulsions in children, and as a sedative.[81] The dried powdered roots were also put in stored grain to prevent insect pests, an effect the methyl salicylate seems largely responsible for.[82]

The alkaloid securinine has a stimulating effect on the CNS. It causes an increase of muscular tonus, intensive stimulation of respiration, increase of heart contraction, and elevated blood pressure. In high dosage it has a similar effect to strychnine and leads to lethal respiration paralysis.[83] The saponins are anti-inflammatory and keratoplastic and are used against psoriasis, ichthyosis, and eczema. The triterpenacids, especially presenegine, are promising against multiple sclerosis and to prevent host-versus-graft reaction in organ transplantation.

Other plant families appear very promising concerning their possible role in modern medicine, especially Euphorbiaceae. One plant belonging to this family is *Alchornea cordifolia* (Schumach. et Thonn.) Müll. Arg. It is common in tropical Africa and widely used in traditional medicine. The dried pulverized leaves are sprinkled on chancre lesions.[84] The fruits cure coughs and coated tongue; the pulverized dry leaves are applied to ulcers and yaws; and the twigs and leaves are used for tanning fishnets in Nigeria.[85] In Congo the young leaves of *Alchornea cordifolia* are used in a decoction against caries.[86] Neuwinger documents its use against dysentery, bronchitis, toothache, caries, yaws, ulcers, wounds, fever, gonorrhea, gum inflammation, eye infection, and other symptoms.[87] Photos 5.3 to 5.8 show the preparation and application of *Alchornea cordifolia* for a patient with an abscess.

It could be shown that among ten different plants used by Azande traditional healers against infections in the widest sense, extracts from the leaves of *vondio (Alchornea cordifolia)* have been found highly effective to inhibit the growth of *Staphylococcus aureus, Streptococcus faecalis,* and to a lesser degree of *Escherichia coli* and *Pseudomonas aeruginosa*. Azande traditionally use this plant against skin infections.[88]

PHOTO 5.3. Traditional treatment of an abscess with *Alchornea cordifolia* among the Azande, Zaire. The bark of the stem is scraped off.

PHOTO 5.4. The scraped bark is put into a funnel made of a leaf.

PHOTO 5.5. Water is added.

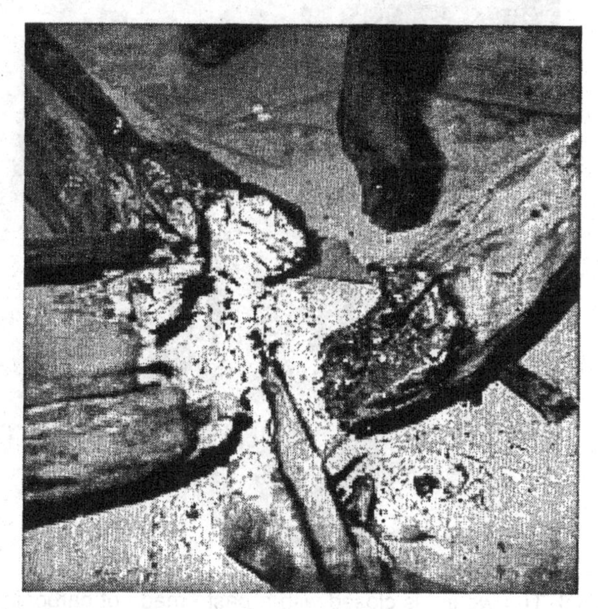

PHOTO 5.6. The preparation is placed on the wound.

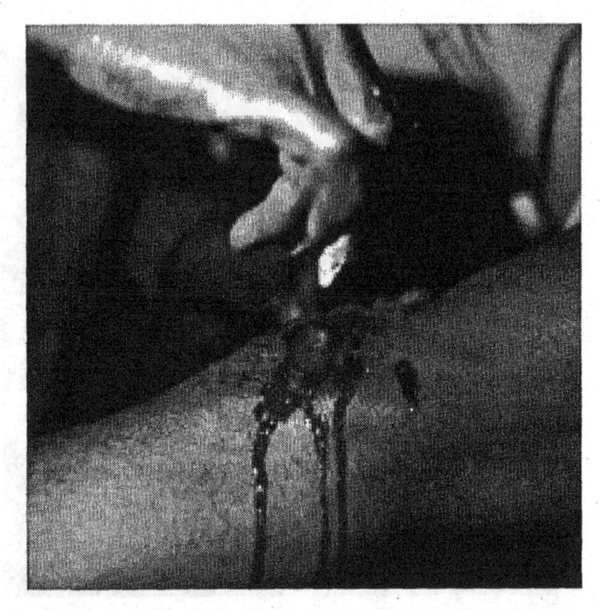

PHOTO 5.7. The decoction is dribbled into the abscess.

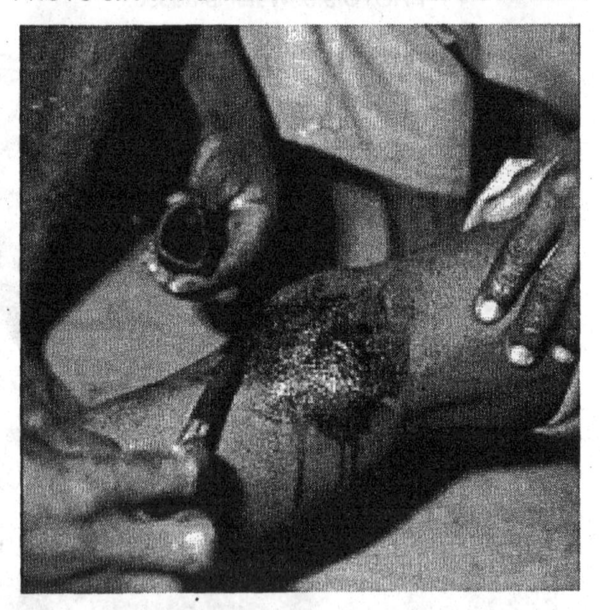

PHOTO 5.8. The wound is closed with a paste made of carbonized and powdered medicinal plants mixed with peanut oil.

Several studies at the University of Vienna proved that the extract of leaves and bark have a strong bacteriostatic effect which correlates with the tannins content, though tannins and flavonoids are not the only active principles.[89] *Alchornea cordifolia,* as well as other *Alchornea* species, contains alkaloids with a structure similar to alchorneine and guanidine derivates,[90] which are bacteriostatic toward pathogen gram-positive and gram-negative bacteria.[91] The content of these alkaloids is, however, very low. More recent studies showed that this plant has a very broad antimicrobial spectrum,[92] which is most significant against *Pseudomonas aeruginosa, Bacillus subtilis,* and *Escherichia coli.*[93] The plant has also been found to possess antiplasmodial activity, due to the active constituent ellagic acid, as well as trypanocidal and antiamoebic activity.[94]

Another most interesting medicinal plant is the *tongbiloli* of the Azande in Northeast Zaire. It has not been determined botanically yet because blossoms are not available. It is a rare liana that grows in the dense forests along the rivers, where it is extremely difficult to obtain its blossoms. Zande healers use this poison as a kind of diagnostic oracle to find out whether a sick person can be cured (see Photos 5.9 and 5.10). A water infusion of the bark or roots is administered into eyes, nose, ears, and mouth. The positive reac-

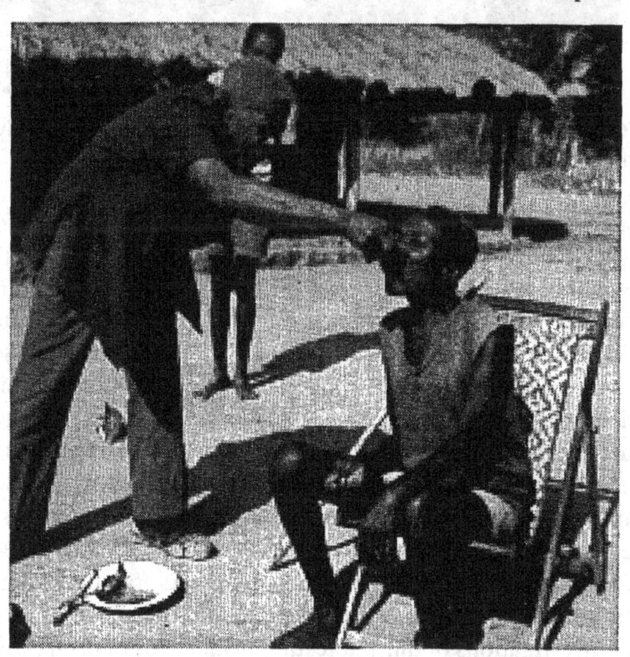

PHOTO 5.9. The diagnostic oracle *tongbiloli* is administered orally.

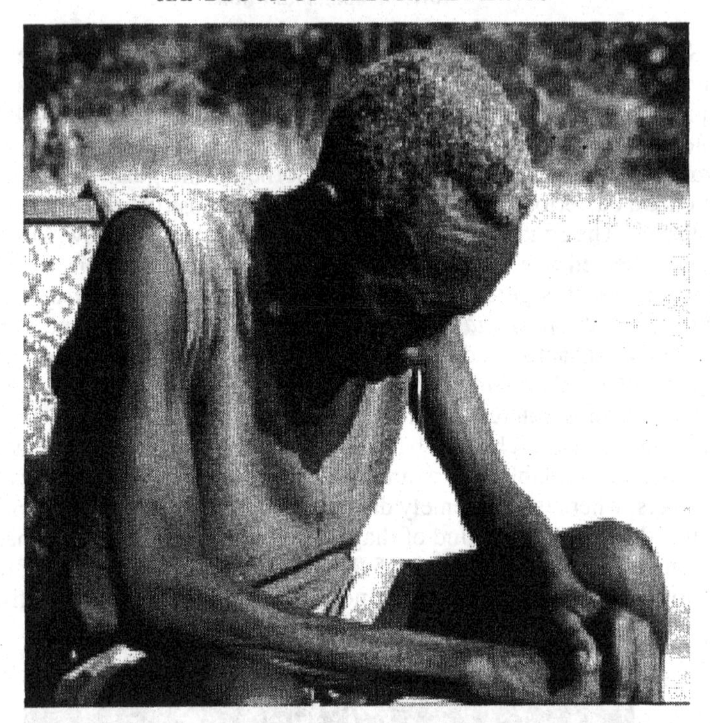

PHOTO 5.10. The positive reaction to tongbiloli is a yellowish, slightly bloody secretion that oozes out of the patient's eyes, mouth, and nose.

tion, which shows that a treatment will be successful, follows few minutes later. A yellowish, slightly bloody secretion oozes out of the patient's eyes, mouth, and nose. Characteristically the eyes are closed in a cramp.[95]

Later the patient often vomits and his or her general condition is disturbed for many hours. Under the name *gbanga* the plant is also used as a poison ordeal, allegedly often ending in death. In its first pharmacognostic investigations, no alkaloids could be found that would suggest a parasympatomimetic action similar to physostigmine. Toxicological investigations confirmed the symptoms observed in the field and revealed that this plant must indeed be extremely poisonous. In laboratory animals the poison was lethal after 72 hours, with various symptoms such as bleeding in the lungs, coagulation disorders, centrally triggered cramps, and gangrene of the extremities. Currently, however, no responsible content can be found. Further investigations of this interesting plant have not been possible due to the difficult political situation in northeast Congo.

NOTES

1. Da Orta, G., *Coloquios dos simples e drogas e cousas mediçinais da India, e assi dalgumas frutas achadas nella, onde se tratam algumas cousas tocants a medicina pratica, e outras cousas boas pera saber, compostos pello doutor Garcia d'Orta, fisico del-rey nosso senhor, vistos pello muyto reverendo senhor, o liçençiado Aleixo Dias Falcam, desenbargador da Casa da Supricaçam, inquisidor nestas partes* (Goa: Joannes de Endem, 1563); Pisonis, G. (Willem Piso), De Indiae Utriusque Re Naturali et Medica, libri quatuordecim. Quorum contenta pagina sequens exhibet. Amsterdam: Elsevier, 1658.

2. Mangelsdorf, P.C., Introduction, in T. Swain (ed.), *Plants in the development of modern medicine* (Cambridge: Harvard University Press, 1972), pp. XI ff.

3. Hakluyt, R., *The principal navigations, voyages, traffiques and discoveries of the English nation* (Edinburgh: E. and G. Goldsmid, 1889), pp. 178 ff.

4. Park, M., *Travels in the interior of Africa* (Edinburgh: Adam and Charles Black, 1867), p. 261.

5. Cited in Neuwinger, H.D., *Afrikanische Arzneipflanzen und Jagdgifte. Chemie, Pharmakologie, Toxikologie* (Second edition) (Stuttgart: Wissenschaftliche Verlagsgesellschaft, 1998), p. 156.

6. Rothziegel, A. and Koralewski, R., 1888, Ueber Strophanthus und Strophanthin, *Wiener Medizinische Blätter,* 6: 16-24.

7. Pins, E., Ueber die Wirkung der Strophantus-Samen im Allgemeinen und deren Anwendung bei Herz- und Nierenkrankheiten, *Therapeutische Monatshefte* (Berlin: N.N., 1887).

8. Rothziegel and Koralewski, 1888, Ueber Strophanthus und Strophanthin.

9. Hedinger, M., 1957, Albert Fraenkel und seine intravenöse Strophanthintherapie, *Münchner Medizinische Wochenschrift,* 99(9): 307.

10. Lewin, L., 1984, *Die Pfeilgifte* (Hildesheim: Gerstenberg) [Reprographic reprint from Leipzig, 1923], pp. 202 ff.

11. Williamson, J., *Useful plants of Malawi* (Zomba, Malawi: The Government Printer, 1972), pp. 113 ff.

12. Bruneton, J., *Pharmacognosy* (Second edition) (Paris: Intercept, 1999), p. 745.

13. Charlson, A.J., 1980, Antineoplastic constituents of some southern African plants, *Journal of Ethnopharmacology,* 2: 323-335.

14. Zoellner, N., Urikosurika und Urikostatika: Insulin und Antibiabetika: Pharmakotherpaie von Stoffwechselkrankheiten, in W. Forth, D. Henschler, W. Rummel (eds.), *Allgemeine und spezielle Pharmakologie und Toxikologie* (Mannheim, Wien, Zürich: B.I. Wissenschaftsverlag, 1983), pp. 325-344.

15. Claus, E.P., *Pharmacognosy* (Philadelphia: Lae and Febiger, 1956).

16. Nothnagel, H. and Rossbach, M.J., *Handbuch der Arzneimittellehre* (Berlin: Verlag August Hirschwald, 1894), pp. 762 ff.

17. Oliver-Bever, B., *Medicinal plants in tropical West-Africa* (Cambridge: University Press, 1986), pp. 250 ff.

18. Creasey, W.A., Biochemistry of dimeric *Catharanthus* alkaloids, in W.I. Taylor and N.R. Farnsworth (eds.), *The* Catharanthus *alkaloids* (New York: Marcel Decker, 1975).

19. Bruneton, 1999, *Pharmacognosy,* pp. 1016 ff.

20. Ibid., pp. 1019 ff.

21. de Flacourt-Bizet, E., *Histoire de la Grande isle Madagascar. Avec une relation de ce qui s'est passe es annees 1655, 1656 et 1657 non encor veue par la premiere Impression* (Paris: Pierre l'Amy, 1661), p. 130.

22. Sofowora, A., *Medicinal plants and traditional medicine in Africa* (New York: John Wiley and Sons, 1982), p. 86.

23. Mihalik, G.J., 1978-1979, Guyanese ethnomedical botany: A folk pharmacopeia, *Ethnomedizin,* 5(1/2): 83-96.

24. Peckolt, T., 1910, Heil-und Nutzpflanzen Brasiliens, *Ber. Deut. Pharm. Ges.,* 20: 36-58.

25. Adjanohoun, E., et al., *Médecine traditionelle et pharmacopée: Contribution aux études ethnobotaniques et floristiques au Togo* (Paris: Agence de Coopération Culturelle et Technique, 1986), p. 51; Svoboda, G.H. and Blake, D.A., Phytochemistry and pharmacology of *C. roseus,* in W.I. Taylor and N.R. Farnsworth (eds.), *The* Cataranthus *alkaloids* (New York: Marcel Decker, 1975); Farnsworth, N.R., Svoboda, G.H., and Blomster, R.N., 1968, Antiviral activity of selected *Catharanthus* alkaloids, *Journal of Pharmaceutical Sciences,* 57: 2174-2175.

26. Adjanohoun, E., Adjanohoun, E.J., Ahyi, A.M.R., Aké Assi, L., Akpagana, K., Chibon, P., El-Hadji, A., Eyme, J., Garba, M., Gassita, J.-N., et al., *Médecine traditionelle et pharmacopée: Contribution aux études ethnobotaniques et floristiques en République Populaire du Congo* (Paris: Agence de Coopération Culturelle et Technique, 1988), p. 95.

27. Heine, B. and Legère, K. *Swahili plants: An ethnobotanical survey* (Köln: Rüdiger Köpper Verlag, 1995), p. 240.

28. de Mello, J.F., 1980, Plants in traditional medicine in Brazil, *Journal of Ethnopharmacology,* 2: 49-55.

29. Oliver-Bever, B., 1980, Oral hypoglycaemic plants in West Africa, *Journal of Ethnopharmacology,* 2(2): 119-127.

30. Fortin, D., Modou, L., and Maynart, G., *Plantes medicinales du Sahel* (Montreal, Dakar: CECI, ENDA, 1989), p. 103.

31. Cited in Arcamone, F., Cassinelli, G., and Casazza, A.M., 1980, New antitumor drugs from plants, *Journal of Ethnopharmacology,* 2: 149-160.

32. Barclay, A.S. and Perdue, R.E., 1976, Distribution of anticancer activity in higher plants, *Cancer Treat. Rep.,* 60: 1081-1113; Cordell, G.A. and Farnsworth, N.R., 1976, A review of selected potential anticancer plant principles, *Heterocycles,* 4: 393-426; Perdue, R.E., Abbott, R.J., and Hartwell, J.L., 1970, Screening plants for antitumour activity, II. A comparison of two methods of sampling herbaceous plants, *Lloydia,* 33: 1-6; Spuijt, R.W. and Perdue, R.E., 1976, Plant folklore: A tool of predicting sources of antitumor activity? *Cancer Treat. Rep.,* 60: 979-985.

33. Charlson, 1980, Antineoplastic constituents of some southern African plants.

34. Brun, G., Dijoux, M.-G., David, B., and Mariotte, A.-M., 1999, A new flavonol glycoside from *Catharanthus roseus, Phytochemistry,* 50: 167-169.

35. Marfori, E.C., Kajiyama, S., Fukusaki, E., and Kobayashi, A., 2002, Trichosetin, a novel tetramic acid antibiotic produced in dual culture of *Trichoderma harzianum* and *Catharanthus roseus* callus, *Zeitschrift für Naturforschung* 57(5-6): 465-470.

36. Chattopadhyay, R.R., 1999, A comparative evaluation of some blood sugar lowering agents of plant origin, *Journal of Ethnopharmacology,* 67: 367-372.

37. Thikhomiroff, C. and Jolicoeur, M., 2002, Screening of *Catharanthus roseus* secondary metabolites by high-performance liquid chromatography, *Journal of Chromatography A*, 955: 87-93.

38. Rieppel, F.W., 1956, Zur Frühgeschichte der Rauwolfia, *Sudhoffs Archiv*, 40(3): 231-239 (translation by Armin Prinz).

39. Watt, G., *A dictionary of the economic products of India,* Volume 6, Part 1 (London, Kalkutta: Allen and Government Printing, 1892), pp. 398 ff.

40. Bruneton, 1999, *Pharmacognosy,* p. 1026.

41. Kerstein, G., 1957, Zur Geschichte der *Rauwolfia, Planta Medica,* 5(5/6): 131-134.

42. Oliver-Bever, 1986, *Medicinal plants in tropical West-Africa,* p. 36; Adjanohoun, et al., 1988, p. 99.

43. Adjanohoun, E.J., Ake Assi, L., Floret, J.J., Guinko, S., Koumare, M., Ahyi, A.M.R., and Raynal, J., *Médecine traditionelle et pharmacopée. Contribution aux études ethnobotaniques et floristiques au Mali* (Paris: Agence de Coopération Culturelle et Technique, 1985), p. 29.

44. Motte, E., *Les plantes chez les pygmées Aka et les Monzombo de la Lobaye (Centafrique)* (Paris: Société d'études linguistiques et anthropologiques de France, 1982), p. 351.

45. Schadewaldt, H., 1958, Zur Geschichte der Rauwolfia, *Veröffentl. Internat. Ges. Gesch. Pharm.*, 13: 139-155.

46. Neuwinger, 1998, *Afrikanische Arzneipflanzen und Jagdgifte. Chemie, Pharmakologie, Toxikologie,* pp. 125 ff.

47. Prinz, A., Das Ernährungswesen der Azande Nordost-Zai res. Ein Beitrag zum Problem des Bevölkerungsrückganges auf der Nil-Kongo Wasserscheide University of Vienna, doctoral dissertation, 1976, p. 176.

48. Chopra, R.N., *Indigenous drugs of India: Their medical and economic aspects* (Calcutta: Art Press, 1933), p. 373; Schneider, J.A., The pharmacology of *Rauwolfia*, in R.E. Woodson, H.W. Younken, E. Schlittler, and J.A. Schneider (eds.), Rauwolfia: *Botany, pharmacognosy, chemistry and pharmacology* (Boston, Toronto: Little, Brown and Co., 1957), pp. 109-143.

49. Original letter is in possession of the authors of this chapter.

50. Klyushnichenko, V.E., Yakimov, S.A., Tuzova, T.P., Syagailo, Ya,.V., Kuzovkina, I.N., Wulfson, A.N., and Miroshnikov, A.I., 1995, Determination of indole alkaloids from *R. serpentina* and *R. vomitoria* by high-performance liquid chromatography and high-performance thin-layer chromatography, *Journal of Chromatography A*, 704: 357-362.

51. Lewin, L., 1929, Gottesurteile durch Gifte und andere Verfahren, *Beiträge zur Giftkunde,* 2: 10.

52. Holmstedt, B., The ordeal bean of Old Calabar: The pageant of *Physostigma venenosum* in medicine in T. Swain (ed.), *Plants in the development of modern medicine* (Cambridge: Harvard University Press, 1972), p. 311.

53. Lewin, 1929, Gottesurteile durch Gifte und andere Verfahren, p. 10 (translation by Armin Prinz).

54. Prinz, A., 1978, Azande (Äquatorialafrika,, Nordost-Zaire) Gift-Orakel Film E 2380 des IWF, Göttingen 1978. Publikation von A. Prinz, *Publ. Wiss. Film, sekt. Ethnol.*, Ser. 8, Nr. 19E, 2380.

55. Holmstedt, 1972, The ordeal bean of Old Calabar, p. 317.

56. Simmons, D.C., 1956, Efik divination, ordeals, and omens, *Southwestern Journal of Anthropology,* 12(2): 225.

57. Sofowora, 1982, *Medicinal plants and traditional medicine in Africa,* pp. 68 ff.

58. Kerharo, J., 1968, Les plants magiques dans la pharmacopée sénégalaise traditionelle, *Etnoiatria,* 2: 3-6.

59. Prinz, A., 2003, Ethnomedical Filmwork among the Azande—Part 1, *Viennese Ethnomedical Newsletter,* 5(2): 10-16.

60. Neuwinger, H.D., *African traditional medicine* (Stuttgart: Medpharm, 2000).

61. Cronlund, A. and Oguakwa, J.U., 1975, Alkaloids from the bark Erythrophleum couminga, *Acta Pharm. Suec.,* 12: 467-478.

62. Holmsted, B., Wassén, H.S., and Schultes, R.E., 1979, Jaborandi: An interdisciplinary appraisal, *Journal of Ethnopharmacology,* 1: 3-21.

63. Arlt, F.R.v., 1912, Eine neue Methode der Glaukom Behandlung mit Pilocarpin und Dionin-Merck, *Wochenschrift Ther. Hyg. Aug.,* 15: 20-21.

64. Neuwinger, 1998, *Afrikanische Arzneipflanzen und Jagdgifte. Chemie, Pharmakologie, Toxikologie,* p. 712.

65. Holmstedt, 1972, The ordeal bean of Old Calabar, pp. 323 ff.

66. Anonymous, 1864, Neues Alkaloid von der Calabar Bohne, *Wiener Medizinische Wochenschrift,* 14, Sp. 692.

67. Arlt, 1912, Eine neue Methode der Glaukom Behandlung mit Pilocarpin und Dionin-Merck.

68. Loewi, O. and Mansfeld, G., 1910, Über den Wirkungsmodus des Physostigmins, *Archiv für experimentelle Pathologie und Pharmakologie,* 62: 180-185.

69. American Chemical Society, 2003, Synthesis of physostigmine, available online at <http://www.chemistry.org/portal/a/c/s/1/feature_ent.html?id=2094bceece1311d5f2944fd8fe800100>.

70. Adjanohoun, E., Adjanohoun, E.J., Ahi, A.M.R., Aké Assi, L., Baniakina, J., Chibon, P., Cusset, G., Doulou, V., Enzanza, A., Eymé, J., et al., 1988, *Médecine traditionelle et pharmacopée,* p. 293.

71. Neuwinger, 2000, *African traditional medicine,* pp. 398 ff.

72. Bruneton, J., 1999, *Pharmacognosy,* p. 975.

73. Oliver-Bever, 1986, *Medicinal plants in tropical West-Africa,* p. 42.

74. Bruneton, 1999, *Pharmacognosy,* p. 977.

75. Patil, C.S. and Kulkarni, S.K., 2000, Effect of physostigmine and cisapride on morphine-induced delayed gastric transit in mice, *Indian Journal of Pharmacology,* 32: 321-323.

76. El-Dawlatyl, A.A., Bakhamees H., Al-Majid, S., and Al-Saddique A., 1997, Physostigmine reversal of carbon monoxide coma, *Annals of Saudi Medicine,* 17(3): 334-336.

77. Thal, L.J., Ferguson, J.M., Mintzer, J., Raskin, A., and Targum, S.D., 1999, A 24-week randomized trial of controlled-release physostigmine in patients with Alzheimer's disease, *Neurology,* 52(6): 1146; van Dyck, C.H., Newhouse, P., Falk, W.E., Mattes, J.A., 2000, Extended-release physostigmine in Alzheimer disease, *Archives of General Psychiatry,* 57: 157-164.

78. Howes, M.-J.R., Perry, N.S.L., and Houghton, P.J., 2003, Plants with traditional uses and activities, relevant to the management of Alzheimer's disease and other cognitive disorders, *Phytotherapy Research,* 17: 1-18.

79. Neuwinger, 1998, *Afrikanische Arzneipflanzen und Jagdgifte. Chemie, Pharmakologie, Toxikologie,* pp. 760 ff.

80. Nyazema, N.Z., Medicinal plants of wide use in Zimbabwe, in A.J.M. Leeuwenberg (ed.), *Medicinal and poisonous plants of the tropics* (Wageningen: Pudoc, 1987), p. 40; Kokwaro, J.O., Some common African herbal remedies for skin disease: With special reference to Kenya, in A.J.M. Leeuwenberg (ed.), *Medicinal and poisonous plants of the tropics* (Wageningen: Pudoc, 1987), pp. 44-69.

81. Oliver-Bever, 1986, *Medicinal plants in tropical West-Africa,* p. 108.

82. Jayasekara, T.K., Stevenson, P.C., Belmain, S.R., Farman D.I., and Hall, D.R., 2002, Identification of methyl salicylate as the prinipal volatile component in the methanol extract of root bark of *Securidaca longepedunculata* Fers., *Journal of Mass Spectrometry,* 37: 577-580.

83. Neuwinger, 1998, *Afrikanische Arzneipflanzen und Jagdgifte. Chemie, Pharmakologie, Toxikologie,* pp. 764 ff.

84. Ibid., p. 261.

85. FAO, *Some medicinal forest plants of Africa and Latin America* (Rome: FAO Forestry Paper 67, 1986), pp. 15 ff.

86. Adjanohoun, E., Adjanohoun, E.J., Ahyi, A.M.R., Aké Assi, L., Baniakina, J., Chibon, P., Cusset, G., Doulou, V., Enzanza, A., Eymé, J., et al., 1988, *Médecine traditionelle et pharmacopée,* p. 189.

87. Neuwinger, 2000, *African traditional medicine,* pp. 29 ff.

88. Prinz, A., Wewalka, G., Stanek, G., Kraus, G., 1987, Neue Ergebnisse zur antimikrobiellen Wirksamkeit traditioneller Heilpflanzen in Zentralafrika, *Mitteilungen der Österreichischen Gesellschaft für Tropenmedizin und Parasitologie,* 9: 45-49; Prinz, A., 1999, Ethnopharmacologic research on poisonous and medicinal plants from the Azande, Central Africa, *Viennese Ethnomedicine Newsletter,* 2(1): 6-11.

89. Kögl, A., *Zur Kenntnis der antimikrobiellen Wirkung von "Vondio"—* Alchornea cordifolia *(Euphorbiaceae)* (Wien: Diplomarbeit zur Erlangung des akademischen Grades Magister der Pharmazie, 1989); Kern, E., *Zur Kenntnis der antimikrobiellen Wirkstoffe von Alchornea cordifolia* (Wien: Diplomarbeit zur Erlangung des akademischen Grades Magister der Pharmazie, 1990); Rauch, M., *Zur Kenntnis der Alkaloide von* Alchornea cordifolia *(Euphorbiaceae)* (Wien: Diplomarbeit zur Erlangung des akademischen Grades Magister der Pharmazie, 1991); Müller, B., *Zur Auffindung der antimikrobiellen Wirkstoffgruppen von* Alchornea cordifolia *(Euphorb.)* (Wien: Diplomarbeit zur Erlangung des akademischen Grades Magister der Pharmazie, 1992).

90. Rauch, 1991, *Zur Kenntnis der Alkaloide von* Alchornea cordifolia *(Euphorbiaceae)*; Seeliger, A., *Beiträge zur Kenntnis der Alkaloide von* Alchornea cordifolia *(Euphorbiaceae)* (Wien: Diplomarbeit zur Erlangung des akademischen Grades Magister der Pharmazie, 1991).

91. Bahr, K., *Zur Kenntnis der Alkaloide von* Alchornea cordifolia *(Euphorbiaceae)* (Wien: Diplomarbeit zur Erlangung des akademischen Grades Magister der Pharmazie, 1993).

92. Ajali, U., 2000, Anibacterial activity of *Alchornea cordifolia* stem bark, *Fitoterapia,* 71: 436-438; Okeke, I.N., Ogundaini, A.O., Ogungbamila, F.O., and

Lamikanra, A., 1999, Antimicrobial spectrum of *Alchornea cordifolia* leaf extract, *Phytotherapy Research,* 13: 67-69.

93. Ebi, G.C., 2001, Antimicrobial activities of *Alchornea cordifolia, Fitoterapia,* 72: 69-72.

94. Mustofa, Valentin, A., Benoit-Vical, F., Pélissier, Y., Koné-Bamba, D., and Mallié, M., 2000, Antiplasmodial activity of plant extracts used in west African traditional medicine, *Journal of Ethnopharmacology,* 73: 145-151; Banzouzi, J.-T., Prado, R., Menan, H., Valentin, A., Roumestan, C., Mallie, M., Pelissier, Y., and Blache, Y., 2002, In vitro antiplasmodial activity of extracts of *Alchornea cordifolia* and identification of an active constituent: Ellagic acid, *Journal of Ethnopharmacology,* 81: 399-401; Adewunmi, C.O., Agbedahunsi, J.M., Adebajo, A.C., Aladesanmi, A.J., Murphy, N., and Wando, J., 2001, Ethno-veterinary medicine: Screening of Nigerian medicinal plants for trypanocidal properties, *Journal of Ethnopharmacology,* 77: 19-24; Tona, L., Kambu, K., Ngimbi, N., Cimanga, K., and Vlietinck, A.J., 1998, Antiamboebic and phytochemical screening of some Congolese medicinal plants, *Journal of Ethnophamracology,* 61: 57-65.

95. Prinz, A., 1984, Ergebnisse pharmakologischer Untersuchungen von Gift- und Heilpflanzen aus Zentralafrika, *Mitteilungen der Österreichischen Gesellschaft für Tropenmedizin und Parasitologie,* 6: 157-165.

Chapter 6

Dilemmas and Solutions: Native American Plant Knowledge

Daniel E. Moerman

A MOTHER'S DILEMMA

Imagine this scene: 2000 years ago, a Native American woman is tending to a sick child who is hot, feverish, and unhappy. She finds a plant which she makes into a tea; she feeds it to the child. She also washes him with it—his head, chest, back, and legs—and she massages some into her own hair and face as well. The child calms down, cools off, and goes to sleep. In a few days, he seems much better.

A year later, the child of a friend is similarly sick: hot, feverish, unhappy. The friend asks, "Can you get for me some leaves from that plant you used last year when your son was sick?" These people have no botanists or botanical gardens, no floras, no field guides to medicinal plants, no herbaria, no professional pharmacists or pharmacologists—not so much as a spiral notebook. The only thing this mother has to go on is what she can remember about that plant from a year ago. Suppose the plant were a rare annual forb, an undistinguished little thing with no particularly visible flowers, no distinctive scent or foliage, which grows for a few weeks in the spring, or maybe the summer, or—some years—maybe not at all. Its habitat is tiny; its range is small. It is, of course, unlikely that this mother would have ever picked this plant in the first place for treating her child; but even if she had, it is unlikely that she would have found it the second time. Even if she did find it, it is unlikely that she would be able to teach many younger mothers where to find this nondescript little weed. Although it is possible, it is unlikely that this plant would become part of the medical botany of this, or any other, community.

Special thanks to my good friend Eva Huseby-Darvas who helped me understand the Hungarian uses of *Achillea*. And thanks to my colleague George Estabrook who read the manuscript very closely and made many helpful suggestions. For Joe and Martha.

To be of value in such a situation, a plant has to be recognizable, distinctive, and available. If a hike into a mountain fastness is required, the child might be beyond help by the time that the herbalist gets back.

A PLANT'S DILEMMA

Why do plants produce medically valuable chemicals? To simplify a very complex problem, plants generally serve as useful drugs because of the biologically active secondary chemicals they produce. These chemicals may act as toxic deterrents to browsing by mammals or insects (like morphine from poppies or nicotine from tobacco),[1] or defenses against various pathogens (like the salicylates found in willows and birches),[2] or as herbicides against competing plants (like phenolic compounds such as chlorogenic acid from sunflowers)[3]. All of these actions are in place in the interest of the plant producing them. Such biologically active secondary chemicals are responsible for the healing action of these plants when humans use them as medicines.

In particular, plants that attract pollinators or browsers by producing attractive flowers or fruits to facilitate their reproduction or dispersal often evolve protection to minimize the damage such visitors may cause.[4] As a result, for human beings, it is very often the case that foods and drugs come from the same plant.[5] It is possible that these attracting functions in part connect these dimensions of selectivity with the ones mentioned earlier, i.e., the visibility which allowed our historical mother to find them a second time.

Successful perennials, especially trees, will have been able to protect themselves against the most withering insect attack over a period of centuries; they seem more likely to contain chemicals of interest to human beings than would annual forbs which, with life cycles measured in weeks, may have turned themselves into seeds again before they are found by plant-eating insects. The characteristics of plants that enable people to readily find and remember them also make plants more perceptible and available to insects and other browsers. Such plants, said by plant ecologists to have high levels of apparency, are more likely to have evolved significant numbers of biologically active secondary chemicals potentially valuable as medicines.

Recently, a similar argument has been made for weeds, adventive plants growing primarily in disturbed soils:[6] In both Chiapas, Mexico, and in North America north of the Rio Grande, many ubiquitous weeds can be shown for several reasons to be particularly richly endowed with bioactive secondary chemicals. Ecological phytochemistry is a study in its infancy, involving people from a variety of established disciplines. Students of botany, plant ecology, allelopathy (plant herbicides), and plant insecticides all tend to operate independently and seem to know little of one another's work. Inderjit's definitive book on plant herbicides, for example—filled with discussion of standard medicinal plants—never once hints that the chemicals discussed have any uses other than as agrochemicals which may facilitate various tillage systems.[7]

TWO DILEMMAS YIELD A SOLUTION

I suggest that these two arguments—the mother's memory and the apparency problem—overlap. The characteristics making a plant recognizable to nonliterate nonspecialists are likely to overlap with the characteristics that will increase the chance that plants contain substantial quantities of biologically active chemicals. Medicinal plants will tend to be more abundant than are plants not used medicinally; they will be perennials, not annuals; they will be more widespread than others; they will be long-lived (in particular, they will be trees); and they will tend to be visible and showy, with distinctive flowers, leaves, foliage, or smell—the kinds of plants that we put in flower gardens and that tend to attract animals that collect nectar (hummingbirds, honeybees, butterflies). Such plants will be easier to find, to commit to memory, to teach about; they are also likely to have more medicinally valuable chemicals than their rarer, annual counterparts.

Two large databases are available to test these ideas. One on medicinal plants has been collected by this author.[8] The other on the flora of North America is the result of the work of John Kartesz and Christopher Meacham.[9*]

Native American peoples were very selective in their use of plants. While the numbers are large—2,703 taxa (species, subspecies, varieties)

---

*Unless otherwise specified, all counts of plants or medicinal plants come from two sources. For the overall flora of North America, data are derived from Kartesz, J. T. and Meacham, C. A., *Synthesis of the North American Flora,* Version 1.0 (Chapel Hill, NC: North Carolina Botanical Garden, 1999). The data used here are those for plants found in North America north of the Rio Grande, including the United States, Canada, and Greenland. Kartesz and Meacham count 28,014 taxa in that region. All data on plants used by Native Americans are derived from Moerman, D.E., *Native American Ethnobotany* (Portland, OR: Timber Press, 1998), to the extent that the items in the medicinal plant database can be cross-matched with the flora. This means that a small number of species identified in the ethnobotanical database are not considered here, including some vascular plants with incorrect scientific names and a number of nonvascular plants, e.g., liverworts, mosses, and mushrooms. The ethnobotanical data are also freely available on the World Wide Web at <http://www.herb.umd.umich.edu/cgi-bin/herb>.

were used as drugs—the percentage is not. They used just under 10 percent of the flora for medicinal purposes. Which 10 percent?

I cannot directly test the proposition of medicinal plants that are common as I know of no data on the frequency of plant appearance. But I can test the opposite proposition: Medicinal plants tend *not* to be rare and endangered; in particular, they tend not to be on the endangered species list. Only two of the 659 taxa listed on the U.S. endangered species list were used medicinally by Native American people;* if all were random, we would expect about 10 percent of them to be so used, or about 65 of them. This difference is highly statistically significant ($\chi^2 = 73.4$, $p = 0.0000$).

I can also directly test the proposition that drug plants are more widespread than others. The Kartesz database includes information on the distribution of plants across all 50 states, plus the District of Columbia, plus 15 Canadian provinces and territories; this makes a total of 66 "states" in this region. The average North American plant occurs in about seven states. Those used medicinally occur in 19 states, while those not so used occur in 6 ($t = 64$, $p = 0.0000$).

Friends and colleagues (even botanists) often express surprise when they find that the average (seven states) is so low. It seems to us that the plants we know are more widespread than that, and they probably are. But we do not *know* a representative sample of plants. There are, for example, 14 plant species that occur only in the Canadian Yukon Territory and 59 species that occur only in Greenland; few readers of this chapter are ever likely to encounter them. In this—our tendency to know widespread species—I think we are much like our prehistoric ancestors. Thirty-nine percent of nonmedicinal plants, but only 7 percent of medicinal plants, occur in only one state, whereas 6 percent of medicinal plants but only 0.6 percent of nonmedicinal plants, occur in 50 or more of 66 states. Medicinal plants tend to be widely distributed.

North America has four times as many perennials as annuals. But it also has six times as many medicinal perennial as medicinal annuals ($\chi^2 = 59.1$, $p = 0.0000$). Medicinal plants tend to be perennials.

Also in North America, 16.5 percent of trees are used as medicinals, while only 9.7 percent of other plants are used as medicinals ($\chi^2 = 102.5$, $p = 0.0000$). Medicinal plants tend to be trees.

*The two endangered medicinal species are both endemic to Hawaii (that is, they grow nowhere else). Hawaii's native fauna is particularly endangered; 274 of 659 endangered species are found in Hawaii.

Twenty-three percent of North American weeds are used as medicinals, while only 7 percent of nonweedy species are used as medicinals ($\chi^2 = 795$, $p = 0.0000$). Medicinal plants tend to be weeds.

Using somewhat less precise data and working only to the level of the genus, it is also possible to show that medicinal plants are far more likely to appear in gardens, or in a list of plants of virtue for the garden compiled in *Ortho's Guide to Successful Gardening.*[10]* My interpretation of this—based on helpful discussion with my gardener wife—is that we put plants in gardens because they have distinctive appearance: they have showy flowers, distinctive foliage, unusual scent, or the like. More than half of the 411 genera on the garden list are used medicinally by Native Americans, whereas only one-quarter of the remaining 2,534 genera are so used ($\chi^2 = 100.7$, $p = 0.0000$). Medicinal plants are distinctive and showy.

Addressing this same issue in a somewhat different way, Kartesz characterized a number of taxa that commonly attract butterflies, honeybees, or hummingbirds. Combining the three categories, 14 percent of drug plants attract these seekers of nectar, whereas only 4 percent of the nondrug plants do ($\chi^2 = 498.6$, $p = 0.0000$). These animals are commonly found on sweet-smelling plants with large or vivid flowers and ample nectaries, or with high apparency. Medicinal plants tend to be attractors. Some complex interactions are occurring here; these are not discrete categories. For example, attractor plants are more often found in gardens than are nonattractors. And weedy species of any sort are more likely to be used medicinally than nonweedy ones.

About 9 percent of all plants in North America are listed as "weeds." Of these, 26 percent are used medicinally by native peoples. By contrast, only 8 percent of nonweedy species are used medicinally. The same relationship holds—a larger percentage of weeds are selected as medicines—for every other category of plants, by habit and character, shown in Figure 6.1. For example, weedy trees, such as sassafras *(Sassafras albidum),* smooth sumac *(Rhus glabra),* prickly ash *(Zanthoxylum americanum),* and American elm *(Ulmus americana),* are much more likely to be selected for use as medicines than are nonweedy trees. Likewise, weedy plants that attract hummingbirds (e.g., *Monarda*) are more likely to be used medicinally than are nonweedy attractors (e.g., *Phlox paniculata,* a nonweedy hummingbird and

*This somewhat unorthodox choice was made for two reasons. I could not locate anywhere a more formal listing of horticultural plants from North America, and this choice has the virtue of "independence." The authors of the book had no interest at all in, or knowledge of, Native American plant use. They made no reference at all to any cross-cultural data. Yet their list of garden plants significantly overlaps the list of medicinal plants.

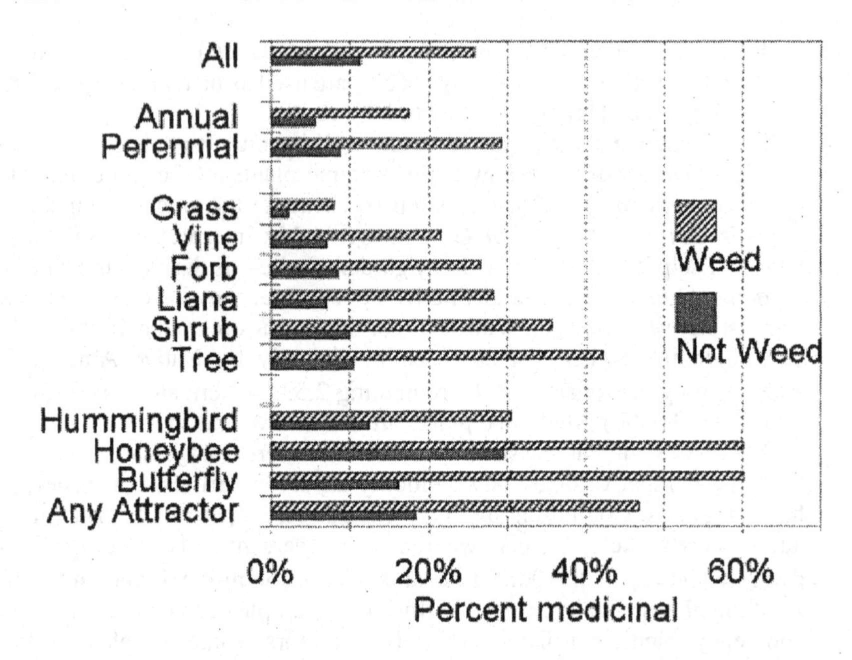

FIGURE 6.1. Medicinal use of weeds versus nonweeds.

butterfly attractor found in 40 states, with no Native American medicinal uses).

Therefore, medicinal plants tend to be

- highly selectively chosen,
- relatively abundant or at least not rare and endangered,
- relatively widespread,
- perennials,
- trees,
- weeds,
- showy, distinctive, and commonly found in ornamental gardens, and/or
- attractive to animals who seek nectar.

They tend to be large, long-lived, widespread plants with high apparency. These are precisely the ones we would predict to have high levels of biologically active compounds that can be utilized as medicines.

CASE EXAMPLE: YARROW (ACHILLEA MILLEFOLIUM)

The most widely used medicinal plant in North America is yarrow, *Achillea millefolium*. My database of Native American medicinal plant use includes mention of 355 uses of several varieties of the species by 77 different tribes, with 258 uses attributed to *Achillea millefolium* var. *millefolium.*[11] (*Acorus calamus,* sweetflag, is in second place.) These uses vary substantially. The plant is used internally and externally. It is used to treat pain in several different forms including headache, body aches, chest pain, and other sorts of pain; it is also used to treat fevers. It is used as an antidiarrheal (usually in a formulation using the leaves) and for other gastrointestinal distress.* The classic use of the plant, discovery of which the ancient Greeks attributed to Achilles (after whom the plant is named), is to treat wounds; in this way, Native Americans used it on cuts, wounds, bruises, strains or sprains, sore throats, burns, and arthritic joints.

It is in many ways the definitive medicinal plant, bearing many of the qualities just noted. It is often extremely common; although I have not counted them, I have seen fields with thousands of individual plants in the summer in Michigan. The plant is among the most widespread of species. It is found in all 50 U.S. states plus all Canadian provinces and territories, as well as Greenland. It has a circumpolar distribution and is found all across Europe, Asia, and North Africa. In Hungary, for example, it is known as *cickafark* ("kitten tail," perhaps referring to the form of the leaf) and is used for internal ailments as well as for burns and wounds.[12] It is an important weed in New Zealand, although it seems not to grow on Groote Eylandt, in Australia.[13] In China, the dried herb was used to treat bleeding, wounds, snakebites, and sores.[14] In the New World, it was an important medicinal plant for the pre-Columbian Mexicans,[15] and it is found at least as far south as Argentina (it grows on the Falkland Islands). Yarrow was one form of pollen found in the famous site of Shanidar IV, a Neanderthal burial in a cave on Mount Carmel, in Iraq, dating approximately 60,000 years ago interpreted as part of a bouquet of medicinal plants placed in a grave,[16] and it is an important contemporary European medicine, subject of a German

---

*It may be worth noting that while several Native American groups (Chehalis, Iroquois, Skagit, Snohomish, Thompson, and Shoshoni) use yarrow as an antidiarrheal, several other groups are reported to use it as a cathartic (the Makah chew the leaves "to clean one out"). Nancy Turner has reported that the Okanagan-Colville use an infusion of the roots to treat diarrhea, and use a decoction of the roots along with the leaves of scarlet gilia, *Ipomopsis aggregata*, as a laxative or cathartic (see Turner, N.J., Bouchard, R., and Kennedy, D.I.D., Ethnobotany of the Okanagan-Colville Indians of British Columbia and Washington, British Columbia Provincial Museum, Occasional Paper 21, 1980).

Commission E monograph.[17] It could be the single most widely distributed medicinal plant, perhaps tied with the dandelion.

Yarrow is a perennial that propagates from a small rhizome. It is very weedy and often recognized by weed scientists in their compendia.[18] The wild plant is easily recognized, as it is a plant of some beauty. It has also been adapted to ornamental gardens, where it takes a prominent place (see Photo 6.1) often showing lovely shades ranging from gold to red. (Photographs in this chapter courtesy of Denial E. Moerman, © 2004.) Although

PHOTO 6.1. *Achillea* and *Asclepias.*

not attractive to hummingbirds, it is common to see it host various bees and other insects.

Achillea seems to be used medicinally nearly everywhere it grows. There may be a few exceptions: Brian Compton (personal communication, 2002) has advised me that no evidence shows that it is used by Tsimshian Indians of the coast of British Columbia; others report that the Tsimshian use it as a gargle for sore throat.[19] Given this near universal use, I was puzzled to notice that I had no reports of use of yarrow by the Hopi Indians of New Mexico. Yarrow is used by many neighbors of the Hopi: the Apache, Navajo, Ute, Papago, and Paiute, and by the Zuni. I knew, for example, that the Zuni had a ceremonial society called the Yayat which, among other things, included dramatic conjuring with fire which members handled in many ways; one authority says they "juggled fire like confetti," and others say they ate it.[20] But before they did, they washed themselves with the cooling juice of the chewed blossoms and roots of yarrow, *Achillea millefolium,* to protect themselves from burns. So, I approached officials of the Hopi tribe and asked about it. Some time later I was contacted by a young Hopi man who consults with the tribe on ethnobotanical matters. He informed me that the Yayat society was made up of *both* the Hopi and Zuni, which I had not known, and that, indeed, his great-grandfather had been a member of the society (Loma'omvaya, personal communication, 1999). So, this was in all likelihood such an important plant that knowledge of its use was kept to small circles of initiates.

Achillea millefolium is a veritable pharmacy of secondary chemicals. It produces hundreds of them, of a broad variety of types. In one review, Chandler and his colleagues report that "over 120 compounds" have been found in yarrow;[21] a more recent list includes 243 different chemicals (Jim Duke, personal communication, 2002). The plant contains a complex blue volatile oil containing dozens of substances (including azulene, which gives it the blue color). It contains dozens of sesquiterpene lactones, including achillin, millefin, and achillicin; it also contains a long list of flavonoids, nearly 30 alkaloids, sterols, triterpenols, plus acids, coumarins, proteins, resins, tannins, and a number of minerals (including selenium, cobalt, and chromium).

Such a huge range of chemicals creates a number of dilemmas in attempting to understand how (or even if) the plant does what people want it to do. For example, the alkaloid achilleine has been shown to be an active hemostatic agent[22] and may account for the plant's widespread use on wounds; but yarrow also contains coumarin, which facilitates bleeding.[23] Whatever the situation may be, it is not simple. But no doubt exists that this plant is richly endowed with chemicals and that it is biologically active in a number of ways. Indeed, in many ways—rational and otherwise—it is, I suggest, the paradigmatic medicinal plant: a highly bioactive, extremely widespread weed.

Achillea and *Asclepias* often grow near each other in midwestern American fields, meadows, and roadsides. The two are named after the Greek heros Achilles and Asclepias. Each was tutored by the wisest of the ancients, Chiron the Centaur, the immortal son of Chronos and Philyra. Asclepias, the father of medicine, was the son of Apollo and Coronis, yet was raised by Chiron, who taught him the arts of medicine and the chase. Chiron was, in addition, Achilles' grandfather, from whom the youth learned both medicine and the art of war. Achilles fought with Telephus, king of the Mysians, wounding him in the thigh. Years later, Telephus, the only one who knew the correct way to Troy, was still suffering from Achilles' wound. Achilles met Telephus, and scraped into his wound some copper rust from the very sword with which he had wounded his erstwhile enemy. Atop the wound, he lay the leaves of yarrow whose virtues as a vulnerary he had discovered. Telephus, healed, directed the fleet to Troy, and the great war began in earnest. Such at least is the testimony of Pliny and Apollodorus, Hyginus, and Homer. It may be worth adding that two species of the plant named after Chiron the Centaur, *Centauria,* an invasive weed, were used medicinally in many ways by Native Americans.

Photo 6.2 is of a flower garden designed to be pleasing to the senses; it is distinctly not a "medicinal plant garden." Yet most of the plants in the image are well-known medicinal species. At the front center are several forms of *Achillea.* They are surrounded by other medicinals. To the right is the purple coneflower, *Echinacea purpurea;* several species of *Echinacea* had more than 100 uses by Native American peoples primarily as analgesics, to dress burns, and as a treatment for toothache. To the left is a clump of bee balm, *Monarda fistulosa;* a number of species of *Monarda* were used in more than 150 ways, as analgesics, for coughs, colds, and fevers. It was also used as a spice, particularly for sausage, and as a perfume. A moment before I took this photograph, a hummingbird was feeding in this patch of bee balm. In the upper left is a lily; native peoples used four different kinds of lilies medicinally to treat coughs, skin problems, and as an antidiarrheal. It was also widely used as a food. At the upper right is another variety of *Achillea.* Just to the left of this yarrow is a small clump of mallow, *Malva* species. Four kinds of *Malva,* all introduced nonnative species, were utilized by native peoples; *Malva neglecta,* the common mallow, was a particular favorite of the Iroquois who used it to treat skin problems, broken bones, and as a love medicine. Edward Voss, in his definitive *Michigan Flora,* characterizes common mallow as a Eurasian weed found throughout Michigan on "roadsides, railroads, filled land, dumps, gravel pits, lawns, gardens and

PHOTO 6.2. *Achillea* in the garden.

waste places generally."[24] It is not clear to me if he thinks lawns and gardens are "waste places," but this is a fine description of the habitats preferred by a "weed"; *Malva neglecta* is found in 58 of the 66 North American "states" I listed earlier.

Elsewhere in this garden, at various times of the year, other medicinal species might be found including poppies *(Papaver somniferum),* fleabane *(Erigeron),* St. John's wort *(Hypericum),* meadow rue *(Thalictrum),* marigolds *(Tagetes),* and, inadvertently, the occasional dandelion *(Taraxacum).*

FINAL WORDS

Across the northern temperate zone, people of a thousand ethnic, cultural, and historical traditions shared a wealth of useful medicinal plants. It can be argued that the roots of this rich system of knowledge date back in evolutionary history to before we were human at all, back to our primate ancestors.[25] As I have already noted, there is plausible evidence of medicinal plant use by Neanderthals in the Middle Paleolithic period, some of which were plants highly favored by Native American people. We can be certain that human beings across space and time have a powerful shared heritage of the knowledge of useful medicinal plants.

NOTES

1. Arnason, J.T., Philogene, B.J.R., and Morand, P., *Insecticides of plants* (Washington, DC: American Chemical Society, 1989).

2. Walton, J.D., Biochemical plant pathology, in P.M. Dey and J.B. Harborne (eds.), *Plant pathology* (San Diego: Academic Press, 1997), pp. 487-502.

3. Inderjit, Dakshini, K.M., and Einhellig, F.A., *Alleopathy: Organisms, processes, and applications* (Washington, DC: 1995).

4. Blumenthal, M. (senior editor), *The complete German Commission E monographs: Therapeutic guide to herbal medicines* (Austin, TX: American Botanical Council; Newton, MA: Intgrative Communications, 1998), pp. 233-234.

5. Moerman, D.E., 1996, An analysis of the food plants and drug plants of native North America, *Journal of Ethnopharmacology,* 52(1): 1-22.

6. Stepp, J.R. and Moerman, D.E., 2001, The importance of weeds in ethnopharmacology, *J Ethnopharmacol,* 75(1): 19-23.

7. Inderjit Dakshini and Einhellig, 1995, *Alleopathy.*

8. Moerman, D.E. *Native American ethnobotany* (Portland, OR: Timber Press, 1998).

9. Kartesz, J.T. and Meacham, C.A., *Synthesis of the North American flora,* Version 1.0 (Chapel Hill: North Carolina Botanical Garden, 1999).

10. Kolos, E. and Kolosné Pathe, E., *Hazai Gyógnynöve nyeink* (Domestic medicinal plants) (Second edition) (Budapest: Müvelt Nép, 1956).

11. Moerman, 1998, *Native American ethnobotany.*

12. Kolos and Kolosné Pathe, 1956, *Hazai Gyógnynöve nyeink.*

13. Levitt, D., *Plants and people: Aboriginal uses of plants on Groote Eylandt* (Canberra: Australian Institute of Aboriginal Studies, 1981).

14. Duke, J.A. and Ayensu, E.S., *Medicinal plants of China* (Algonac, MI: Reference Publications, 1985), pp. 41-46.

15. Heinrich, M., Ethnobotany of Mexican Compositae: An analysis of historical and modent sources, in D.J.N. Hind (ed.), *Proceedings of the International Compositae Conference, Kew, 1994* (Kew, UK: Royal Botanic Gardens, 1996), pp. 475-503.

16. Solecki, R.S., 1975, Shanidar IV: A Neanderthal flower burial in northern Iraq, *Science,* 190: 880-881; Leroi-Gourhan, A., 1975, The flowers found with Shanidar IV: A Neanderthal burial in northern Iraq, *Science,* 190: 562-564; Leroi-Gourhan, A., 1998, Shanidar et ses fleurs, *Paléorient,* 24(2): 79-88.

17. Stubbendieck, J., Friisoe, G.Y., and Bolick, M.R., *Weeds of Nebraska and the Great Plains* (Lincoln, NE: Nebraska Department of Agriculture, Bureau of Plant Industry, 1994).

18. See for example Stubbendieck, Friisoe, and Bolick, 1994, *Weeds of Nebraska and the Great Plains*; Uva, R.H., Neal, J.C., and DiTomaso, J.M., *Weeds of the Northeast* (Ithaca, NY: Cornell University Press, 1997).

19. Pojar, J. and MacKinnon, A., *Plants of costal British Columbia, including Washington, Oregon, and Alaska* (Edmonton: Lone Pine Publishing, 1994), p. 279.

20. Chandler, R.F., Hooper, S.N., and Harvey, M.J., 1982, Ethnobotany and phytochemistry of yarrow, *Achillea millefolium,* Compositae, *Economic Botany,* 36(2): 203-223.

21. Chandler, Hooper, and Harvey, 1982, *Ethnobotany and phytochemistry.*

22. Miller, F.M. and Chow, L.M., 1954, Alkaloids of *Achillea millefolium* L., I. Isolation and characterization of achilleine, *Journal of the American Pharmaceutical Association,* 76: 1353-1354.

23. Huffman, M.A., 2001, Self-medicative behavior in the African great apes: An evolutionary perspective into the origins of human traditional medicine, *BioScience,* 51(8): 651-661.

24. Voss, E.G., *Michigan flora II: Dicots* (Ann Arbor, MI: Cranbrook Institute of Science and University of Michigan Herbarium, 1985).

25. Huffman, 2001, Self-medicative behavior in the African great apes.

Chapter 7

Ethnobotany and Ethnomedicine of the Amazonian Indians

Sir Ghillean T. Prance

INTRODUCTION

Ethnobotany, the study of plant uses by indigenous and local peoples, has contributed many leads for medicines because the use of plants to heal is an integral part of the culture of these peoples. I have had the opportunity to visit many different indigenous tribes of the Amazon region and have found that each group has certain basic requirements that they have been able to fulfill by using the diversity of plant species that surrounds them. Each tribe has arrow poisons, fish poisons, narcotics, stimulants, and a large array of medicines to treat the common diseases and infestations of parasites from which they suffer. A striking feature of studying Amazonian tribes is that each one has a different array of plant uses and to get a complete picture it is necessary to study many different tribes. A book by Richard Schultes, a pioneer of Amazonian ethnobotany, and chemist Robert Raffauf listed 1,500 different medicinal plants from only the northwest corner of the Amazon.[1] However, today much of this tribe-based information is being lost because of the acculturation, extinction, and in some cases development of indigenous people. This ongoing change makes it all the more urgent to document this useful information before it is lost; we hope to preserve it also for the tribal peoples themselves. I have chosen to present here a few of the chemically interesting plants that I have encountered in several tribes, and I have grouped them under the different functions for which they are used. The primary use is not necessarily medicinal but for arrow poisons, narcotics, and stimulants. This fact indicates interesting chemistry, and some of these substances, such as the arrow poison curare, have given rise to therapeutic agents. However, I begin with a few more general aspects

This research was carried out while I held the McBryde chair at the National Tropical Botanical Garden, Kauai, Hawaii. I thank Brenda Carey for typing the manuscript.

of medicinal ethnobotany from experiences with the Yanomami, a tribe that I have visited extensively.

THE YANOMAMI

The Yanomami live in the hilly forest region on the frontier between Brazil and Venezuela in the northwest part of Amazonia and in Orinocia (see Photo 7.1). Until recently they were relatively undisturbed by Western culture, but in the 1980s their territory was invaded during a massive gold rush. The 50,000 miners that entered the region both came into conflict with the Yanomami and brought with them many diseases, such as malaria and venereal diseases. Early studies of the Yanomami drew much attention to their shamanism, use of narcotics, and intertribal disputes, but they paid little or no attention to the Yanomami's use of medicinal plants.[2] In fact, some studies concluded that their medicine was entirely shamanistic magic. This misconception was corrected by the studies of Milliken and Albert who demonstrated that the Yanomami group which they studied had a substantial pharmacopoeia of at least 113 different species of plants and fungi.[3] The study introduced two important new features to Amazonian ethnobotany. First, Milliken and Albert disclosed only the scientific identifications of plant use that had already appeared in the literature, in order to protect the

PHOTO 7.1. Yanomami Indian beside stream, Auaris, Roraima, Brazil.

intellectual property rights of the Yanomami. This commendable practice drew considerable surprise and criticism from editors of journals and others. This practice must be encouraged in future work, which in any case must now be carried out with in-place consent agreements that protect the rights of any group being studied. Second, the purpose of their research was also to study the plants used to combat malaria by other tribes and the Luso-Brazilian population of the Roraima state of Brazil.[4] This study was undertaken in order to find effective plants to treat malaria and to introduce them to the Yanomami who did not have remedies to combat this recently introduced disease. A total of 99 species in 82 genera were collected, and these plants are being screened in laboratories in Brazil and at Kew for antimalarial activity and toxicity. Several have shown strong antimalarial effects, indicating the value of ethnobotanical studies in the search for medicines. The innovative aspect of this study is that it was carried out to aid the Yanomami rather than to exploit the intellectual property of the Amazon people for use in the developed world. This type of study, for the benefit of indigenous people, sets an example for future work. I hope that in future ethnobotanical studies we will be always thinking of how we can benefit local people rather than concentrating only on the benefits to the developed world. This is particularly important for studies in ethnomedicine. Now that the Convention on Biological Diversity is firmly in place, equitable sharing of the benefits gained from the knowledge of indigenous peoples is more likely to occur.

Another important aspect of the Yanomami is their diet. Prior to the recent introduction of many new diseases by gold miners, the Yanomami were a healthy people, probably due to their varied diet. In addition to their staple crop of bananas and the normal products of hunting, the Yanomami eat a large number of other forest products such as fruit, insects (beetle and termite larvae and ants), and fungi.[5] The Yanomami are the only Amazonian people known to make extensive use of fungi in their diet. We have now identified more than 30 species of fungi eaten by the Yanomami.[6] Most of the fungal species used are wood-rotting polypores that grow on the dead logs that remain in the fields. They are therefore inadvertently cultivating an important second crop. The use of these fungi makes a considerable contribution to their diet and consequently to their good health.

ARROW POISONS

The Yanomami use an arrow poison that is different from any other tribe's, made from the bark sap of *Virola theiodora* (Spruce ex Benth.) Warb, a species of Myristicaceae (see Photo 7.2). Bark is stripped from a

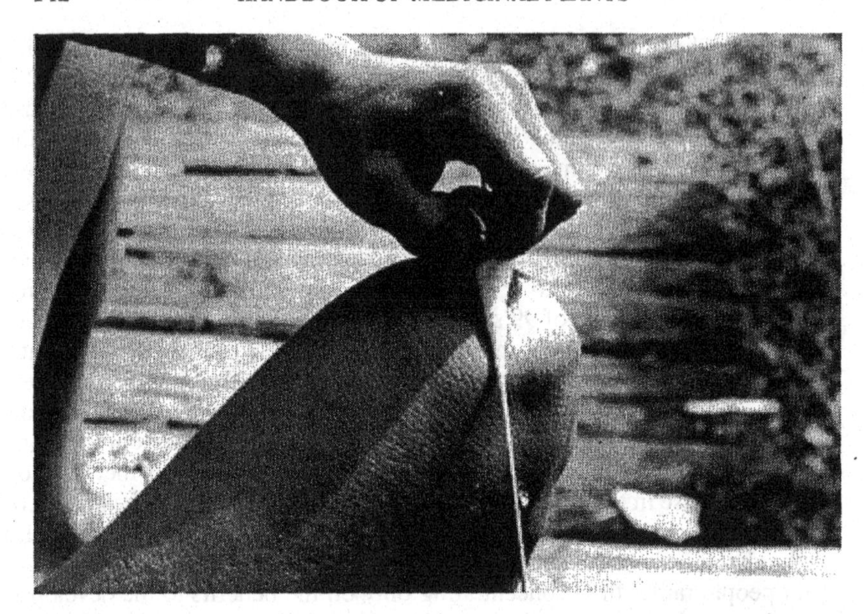

PHOTO 7.2. Yanomami Indian putting flight of cotton onto a poisoned blowgun dart.

tree and is placed over a fire, which is made nearby. The heat from the fire causes a copious quantity of red sap to ooze from the inner bark. The sap is collected into a utensil that is placed on the fire to boil down to a sticky consistency. Arrows and blowgun darts are then dipped in the preparations and allowed to dry.[7] The *Virola* sap contains a variety of harmaline alkaloids (tryptamines), and the arrow poison is a slow-acting one. Game that has been shot with coated arrows often has to be followed for a considerable distance before it succumbs to the effect of this poison. Further details of the chemistry of *Virola theiodora* are given in the next section, on narcotics, since this species is also the source of the Yanomami's snuff.

The Makú tribe inhabits the upper Rio Negro region of Brazil and extends into Colombia. This tribe has created a most effective poison for their blowgun darts, made from the untreated latex of *Naucleopsis mellobarretoi* (Standl.) C.C. Berg, a species of the fig family, Moraceae. The latex is collected by making a small nick in the trunk and gathering it into a folded leaf. The darts are then dipped directly into the latex with no further treatment. The active substances in this poison are cardiac glycosides.[8] The latex has a mixture of cardenolide glycosides including α-antiarin. Poison actually scraped from a Makú blowgun dart contained about 4.4 percent of a cardiac glycoside mixture. Shrestha, Kopp, and Bisset gave a good summary of the use of Moraceous dart poisons in South America and showed that species of both *Naucleopsis* and *Maquira* are used in the Chocó of Colombia, in Ecuador, and by the Tikuna Indians of Brazil.[9]

The most well-known and widely used Amazonian arrow poison is curare, a complex mixture of species of *Strychnos* (Loganiaceae) and/or Menispermaceae. Each tribe has a slightly different recipe and different plant admixtures for their curare. As an example I have studied the specific mixture used by the Jamamadí Indians, an Arawak tribe in the Purus River basin of Central Amazonian Brazil (see Photo 7.3). The Jamamadí use cu-

PHOTO 7.3. Jamamadí Indian preparing a blowgun dart poisoned with curare.

rare for blowgun darts and arrow tips. It is made from a mixture of the bark of at least seven species from five plant families (see Table 7.1).[10] The bark from the stem or trunk of all seven species is used, with the *Strychnos* and *Curarea* being used in greater quantity than the other species. More *Strychnos* than *Curarea* is used. The barks are boiled in water until a sticky liquid residue remains. This is then used to coat darts and arrow tips. The wet darts are passed through a fire to dry the poison. *Strychnos solimoesana* was first reported as an arrow poison by Krukoff for the Cauichana Indians from the Rio Tocantins, and it was studied chemically by Marini-Bettolo and colleagues, who reported the presence of no less than 40 alkaloids in the stem bark.[11] *Strychnos solimoesana* is said to be one of the most powerful and effective paralyzing curares. Together with the many alkaloids of *Curarea toxicofera* the Jamamadí arrow poison is an extremely complex chemical mixture.[12]

The Jamamadí make their arrow points from a species of bamboo (*Bambusa* sp.) that they call *hado*. Before being dipped in the poison the carefully shaped arrow points are cured in the smoke of the annonaceous ingredient of the poison, *Duguetia asterotricha*. The arrows or dart tips are held in the smoke of *Duguetia* bark, and the Indians insist that the poison is more effective after this treatment.

NARCOTICS AND STIMULANTS

Every tribe that I have visited uses some sort of narcotic. This is very much part of Amerindian culture, and they have been extremely successful and creative in finding substances to use for their ceremonial life.

TABLE 7.1. The constituents of the Jamamadí curare arrow poison.

Jamamadí name	Scientific name	Family
Ihá	*Strychnos solimoesana* Krukoff	Loganiaceae
Bicafa	*Curarea toxicofera* (Wedd.) Barneby & Krukoff	Menispermaceae
Bicafa	*Abuta splendida* Krukoff & Moldenke	Menispermaceae
Unknown	*Guarea carinata* Ducke	Meliaceae
Barafa	*Guarea grandifolia* C. DC	Meliaceae
Unknown	*Picrolemma sprucei* Benth.	Simaroubaceae
Boa	*Duguetia asterotricha* Diels	Annonaceae

The Yanomami's principal narcotic is from *Virola theiodora,* the source of their arrow poison. After use for coating arrow tips, the sticky sap of *Virola* is then boiled dry and pulverized. This powder is usually mixed with the dried leaves of *Justicia pectoralis* Jacq. (Acanthaceae), which are rich in coumarins and give a pleasant odor to the snuff. Coumarins were long ago banned from food products because of their carcinogenic properties. The snuff resulting from the mixture of *Virola* and *Justicia* is used by the shamans before treating a patient with their magic. It is also taken by adult males in various ceremonies such as that of the final day ritual after the death of a member of the village. It is also taken following certain hunting parties. The snuff is administered through small blowpipes. One person blows the snuff hard into the nostrils of another to introduce a sufficient quantity to be effective. The effect is rapid and powerful because of the hallucinatory tryptamines in the snuff. *Virola* resin contains approximately 8 percent 5-methoxy-*N, N*-dimethyltryptamine, and lesser amounts of *N, N*-dimethyltryptamine.

Although *Virola*-based snuffs are commonest among the Yanomami, some groups, especially in the southwestern part of their territory, use *Anadenanthera peregrina* (L.) Speg.[13] The disclike seeds of this tree in the Legume family are pulverized with alkaline ash to produce a snuff which they call *yopo.* This snuff is rich in β carbolines and bufotenin, a substance that also occurs in the poison-arrow frogs used by tribes of the Colombian Chocó.

These strongly hallucinogenic snuffs are used strictly for ceremonial purposes, but the Yanomami also grow tobacco. The leaves are softened by crushing and then sprinkled with ash to create an alkaline environment. The leaves are then rolled into a quid which is placed between the lower lip and teeth to allow the stimulating juices of the tobacco to gradually enter the digestive system. Tobacco quids are used daily by all Yanomami males.

Various lowland tribes of western Amazonia get their stimulation from the lowland variety of coca, *Erythroxylum coca* L. var. *ipadu,* the cocaine plant. The Makú use this species. They cultivate the shrubs in their fields and harvest large quantities of leaves. These are put into large flat pans to toast until they are crisp and dry. At the same time, green banana leaves are burned on a fire and their ash is collected. The dried coca leaves and banana leaf ash are then placed in an extremely deep, long hollow mortar where they are ground to a fine powder with a hard wooden pestle. Considerable ceremony is attached to the pulverizing, and a rhythm is often beaten with the pestle while other members of the tribe chant. The wooden mortar makes a loud drumlike noise when the pestle is knocked against the side. The different rhythms that the tribe members beat inform the rest of the tribe how the preparation is progressing. When a fine powder is produced, it is ready for use. The powder is mixed with cassava flour and forms part of the daily diet of the Makú. The powder is not unpleasant to eat and tastes of dry leaves.

I have also observed the use of coca by the Bora Indians of Peru, who use the ash of the leaves of *Pourouma cecropiaefolia* Mart. rather than bananas. The powder is placed between the lower teeth and gums and worked with the tongue into a packed mass that slowly dissolves.[14]

The stimulant of the Jamamadí Indians is a tobacco-based snuff (*Nicotiana tabacum* L.) mixed with the bark ash of a wild species of cacau (*Theobroma subincanum* Mart.). Tobacco leaves are dried beside a fire or on top of a convex metal bowl where they are rolled with a wooden rolling pin to squeeze out the juices. A fire of the cacau bark is made and the ash gathered. The dried tobacco leaves and the cacau ash are mixed in about equal quantities in a mortar and pestle made of an empty Brazil nut pyxidium and a hard bit of wood. This produces a fine powder which the Jamamadí call *shinã*. It is in common use among all members of the tribe, and they prefer freshly made snuff which they prepare each day.

The *shinã* snuff is administered by sucking it into the nostrils through a small pipe made from a hollow leg bone of a monkey. One person will hold out the snuff on a leaf while the other draws it into his or her nostrils (see Photo 7.4). Even small children take this snuff, which has an intoxicating effect on the users. They appear inebriated and talk of light-headedness. This snuff is a powerful intoxicant, but it is not a hallucinogenic. The Jamamadí were insistent that the snuff is ineffective without the *Theobroma* bark ash, and said that they never take pure tobacco snuff. This use of tobacco-cacau ash snuff is taken by many different tribes in the area between the Rio Purus and the Rio Juruá. I have seen identical processes with the Dení and the Jarawara tribes which, like the Jamamadí, are of Arawak origin.

Another Arawak tribe of the Purus River Basin is the Paumarí. They have a different and more powerful snuff which they call *koribó*. This is made from a mixture of tobacco and *Tanaecium nocturnum* (Barb. Rodr.) Bur. & K. Schum (Bignoniaceae). The leaves of *Tanaecium* are pungent and have an almondlike smell when crushed. These leaves are shredded and then roasted until they are dry. The crisp dry leaves are then ground to a powder with a mortar and pestle and then passed through a cloth to ensure that the powder is fine in texture (see Photo 7.5). The resultant powder is mixed with tobacco snuff prepared by the same method (see Photo 7.6).

The Paumarí snuff is used only on special occasions or by the shamans before the treatment of a patient. It is mainly used in the rites of passage that are performed for the protection of children. Such rituals are frequent and

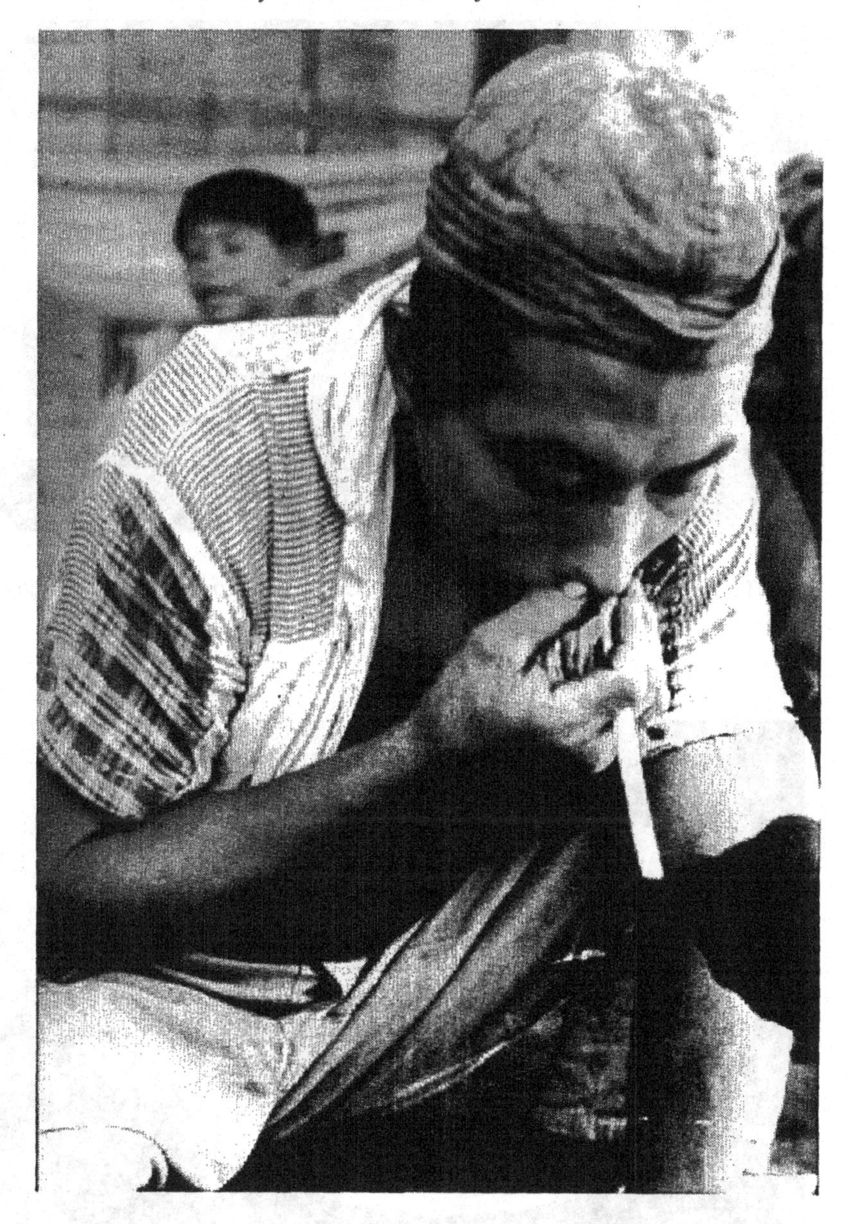

PHOTO 7.4. Jamamadí Indian inhaling tobacco-based snuff through a hollow bone.

PHOTO 7.5. Paumarí snuff made from *Tanaecium nocturnum* is sifted through a cloth.

PHOTO 7.6. Paumarí snuff after sifting through a cloth, ready for use.

are performed so that children may begin to eat new foods. They are accompanied by sacred songs, and rituals take place for each kind of animal food. A child may not begin to eat the food of any animal until he or she has been through the ceremony correlating to that animal. The ceremony is performed by the men of the tribe for their children. The men take *koribó,* and the ritual includes imitation of the animal that the child may now include in his or her diet. The snuff is also taken by tribesmen at the puberty rights for girls. The snuff is inhaled through a hollow leg bone of a water bird. Shamans take the snuff more frequently since most days they have someone to treat. Under the influence of *koribó* they suck violently on the patient to extract the illness. After a while, they run into the forest and retch until they vomit. They then return to the patient and display some object such as a grasshopper or a bone, which they claim to have drawn out of the patient.

Paumarí women do not usually take the snuff, but during festivals they take *koribó* in a tea made from the root bark of the *Tanaecium*. About two tablespoons of fresh root bark are brewed in water and drunk. This produces drowsiness and an inability to concentrate, and reduces awareness. The effect of drinking half a cup of this brew is slight but apparent.

We found several shamans who were paralyzed from the waist down. This was almost certainly due to their daily intake of *koribó*. Grajales Diaz reported an extremely high concentration of hydrogen cyanide in the leaves of *Tanaecium nocturnum.*[15] However, the toasting and preparation of the snuff probably removes much of the cyanide, leaving intact other intoxicating compounds as well as ones that damage the nervous system. This is the only case in which I observed adverse effects from compounds taken by tribal people. Further details of the effects of *koribó* are given by Prance.[16]

In western Amazonia the most widely used hallucinogenic plant is the vine *Banisteriopsis* (Malpighiaceae), the source of *ayahuasca,* a hallucinogenic beverage. This is used by many tribes such as the Cachinahua in Acre State of Brazil, but it is also in common use by local people both by spiritists and by local *curanderos* or medicine men. The leaves of *Banisteriopsis* are usually mixed with those of a species of *Psychotria, P. viridis* Ruiz & Pavón in the Rubiaceae. This has been extensively researched elsewhere, so I do no more than mention it here.[17]

FISH POISONS

Almost all Amazonian tribes use plant material to stun fish in small rivers and streams. In a review of the use of fish poisons, researcher Acevedo-Rodriguez listed 935 species of plants employed as fish poisons.[18] The active substances of many fish poisons are either rotenones or saponins, which interfere with the gill membrane of the fish. The poisons are strewn into streams and the asphyxiated fish rise to the surface and are easily gathered up. This is a drastic way of fishing since even the smallest fish are poisoned. Many tribes are aware of this and only poison a particular stream occasionally to avoid depleting their supply.

All of the tribes mentioned in this chapter have fish poisons. The Yanomami most commonly use the stems of the legume vine *Lonchocarpus utilis* A.C. Smith, but they also use *Clibadium* in the Asteraceae. The Jamamadí mainly use another legume vine *Derris latifolia* H.B.K. The tribe with the greatest variety of fish poisons that I have visited is the Makú (see Photo 7.7). The five different fish poisons that I collected are used in different sized streams and at different times of the year. Two of these species are the ones used by the Yanomami (see Table 7.2). When the Makú use the leaves of either species of Euphorbiaceae (*Euphorbia cotinifolia* or *Phyllanthus brasiliensis*), they build a small rack of logs over a stream and place baskets of leaves on it. They then beat the leaves with sticks to release the plant juices into the stream. At the same time some of the women stir up the stream bed above the site of the poison to make the water muddy. Other women and children are placed downstream of this activity and gather up

PHOTO 7.7. Makú Indians beating the leaves of *Euphorbia cotinifolia* into a stream to stun fish.

TABLE 7.2. Fish-poison plants used by the Makú.

Family	Species	Part used	Poison
Asteraceae	*Clibadium sylvestre* (Aubl.) Baill.	Leaves	Ichthyothereol
Caryocaraceae	*Caryocar glabrum* (Aubl.) Pers.	Pericarp and fruit	Saponins
Euphorbiaceae	*Euphorbia cotinifolia* L.	Leaves	Triterpenes
Euphorbiaceae	*Phyllanthus brasiliensis* (Aubl.) Poir.	Leaves	Triterpenes
Fabaceae	*Lonchocarpus utilis* A.C. Smith	Stems	Rotenone

the fish that come to the surface. After this is done, they spend the rest of the evening enjoying and celebrating an enormous fish feast.

CONCLUSION

Given in this chapter are just a few examples of the enormous diversity of chemically active substances that have been discovered by the indigenous population of the lowland Amazon region. I have focused more on poisons and narcotics than medicines because these are the aspects that I have studied the most. However, the pharmacopeia of each of these tribes and of many more is also extremely diverse. Several recent studies have shown that ethnobotanical work is a good way to search for chemically active substances.[19] Although the use of a plant in local medicine frequently indicates the presence of biologically active substances, when analyzed in the laboratory they are often found to have only a small effect. One of the biggest differences between indigenous use of medicines and Western medicine is the mixture of substances used by native people, rather than a pure extract. By not analyzing these actual mixtures and isolating each plant species individually, we are missing a lot of the chemical reactions that take place in local medicinal brews. There is still a great need for chemical studies in the field.

Finally, the information carried by local and indigenous people is fast disappearing through their acculturation. There has never been a greater urgency for research on ethnomedicine. The results of 10,000 years of experimentation with Amazonian plants by the native people is rapidly being lost. What is needed is much more than the sort of traditional ethnobotanical studies reported here; the need is for studies that use field laboratories to analyze the actual mixtures used by local people.

NOTES

1. Schultes, R.E. and Raffauf, R.F., *The healing forest: Medicinal and toxic plants of the Northwest Amazon* (Portland, OR: Dioscorides Press, 1990).

2. Chagnon, N.A., *Yanomamö: The fierce people* (New York: Holt, Rinehart and Winston, 1968); Lizot, J., *Les Yanömami centraux* (Paris: Cahiers de l'homme, 1984); editions de L'Ecole des Hautes Etudes en Sciences Sociales; Schultes, R.E., 1954, A new narcotic from the Northwest Amazon, *Botanical Museum Leaflets,* 9: 241-260.

3. Milliken, W. and Albert B., 1996, The use of medicinal plants by the Yanomami Indians of Brazil, *Economic Botany,* 50: 10-25.

4. Milliken, W., 1997, Traditional anti-malarial medicine in Roraima, Brazil, *Economic Botany,* 51: 212-237.

5. Smole, W.J., *The Yanomama Indians in a cultural geography* (Austin, TX: University of Texas Press, 1976).

6. Fidalgo, O. and Prance, G.T., 1976, The ethnomycology of the Sanama Indians, *Mycologia,* 68: 201-210; Prance, G.T., 1984, The use of edible fungi by Amazonian Indians, *Adv. Economic Botany,* 1: 127-139.

7. Schultes, R.E. and Holmstedt, B., 1968, The vegetal ingredients of the myrsticaceous snuffs of the northwest Amazon, *Rhodora,* 70: 113-156; Prance, G.T., 1970, Notes on the use of plant hallucinogens in Amazonian Brazil, *Economic Botany,* 24: 62-68.

8. Bisset, N.G. and Hylands, P.J., 1977, Cardiotonic glycosids from the latex of Naucleopsis mello-barretoi, a dart poison plant from north-west Brazil, *Economic Botany,* 31: 307-311; Shrestha, T., Kopp, B., and Bisset, N.G., 1992, The Moraceae-based dart poisons of South America: Cardiac glycosides of *Macquira* and *Naucleopsis* species, *Journal of Ethnopharmacology,* 37: 129-143.

9. Shrestha, Kopp, and Bisset, 1992, The Moraceae-based dart poisons of South America.

10. Prance, G.T., 1973, An ethnobotanical comparison of four tribes of Amazonian Indians, *Acta Amazonica,* 2: 7-27.

11. Krukoff, B.A., 1965, Supplementary notes on the American species of Strychnos VII, *Memoirs of the New York Botanical Garden* 12: 42-43; Marini-Bettolo, G.B., 1957, Nota 8. Gli alcaloidi dell *Strychnos solimoesana* Krukoff, *Rend Istituto Super Sanita,* 20: 342-357.

12. Krukoff, B.A. and Barneby, R.C., 1970, Supplementary notes on American Menispermaceae VI, *Memoirs of the New York Botanical Garden,* 20: 1-80.

13. Schultes, R.E., 1963, Hallucinogenic plants of the New World, *The Harvard Review,* 1: 18-32; Schultes, 1954, A new narcotic.

14. Schultes, R.E., 1957, A new method of coca preparation in the Colombian Amazon, *Botanical Museum Leaflets,* 17: 241-246.

15. Grajales Diaz, J.C., Estudo fitoquimico de plantas Colombians, in IX Convencion Nacional Quimicos Farmaceuticas, Medellin, 1967. Mimeographed in August.

16. Prance, G.T., 1978, The poisons and narcotics of the Dení, Paumarí, Jamamadí and Jarawara Indians of the Purus river region, *Revista Brasileira Botânica,* 1: 71-82; Prance, G.T., 1999, The poisons and narcotics of the Amazonian Indians, *Journal of the Royal College of Physicians, London,* 33: 368-376.

17. Prance, 1999, The poisons and narcotics of the Amazonian Indians; Schultes, R.E., 1957, The identity of the Malpighiaceous narcotics of South America, *Botanical Museum Leaflets,* 18: 1-56; Der Maderosian, A., Pinkley, H.V., and Dobbins, M.F., 1968, Native use and occurrence of *N, N*-dimethyltryptamine in the leaves of *Banisteriopsis rusbyana, American Journal of Pharmacology,* 140: 137-147; Pinkley, A.V., 1969, Plant admixtures to *ayahuasca,* the South American hallucinogenic drink, *Lloydia* 32: 305-314; Lamb, F.B., *Wizard of the upper Amazon* (Boston: Houghton Mifflin, 1974).

18. Acevedo-Rodriguez, P., 1990, The occurrence of piscicides and stupefactants in the plant kingdom, *Advances in Economic Botany,* 8: 1-23.

19. Balick, M.J., Ethnobotany and the identification of therapeutic agents from the rainforest, in *Ciba Foundation Symposium,* 154 (Chichester: Wiley, 1990), pp. 22-39; Chapuis, J.C., Sordat, B., Hostettman, K., 1988, Screening for cytotoxic activity of plants used in traditional medicine, *Journal of Ethnopharmacology,* 23: 273-284.

Chapter 8

Healers and Physicians in Ancient and Medieval Mediterranean Cultures

Alain Touwaide

TRADITIONAL HISTORIOGRAPHY

According to a traditional picture, the field of therapeutics in ancient and medieval Mediterranean cultures was characterized by both extraordinary scientific achievement in antiquity and then by a real odyssey around the Mediterranean Basin and an alternating fortune.[1] During antiquity, Hippocrates (460-between 377? B.C.), Dioscorides (1st cent. A.D.) and Galen (129-200? A.D.) truly created the discipline. This early body of knowledge constituted the reference for the later ages and cultures, which progressively impoverished it. While Hippocrates laid down the principles to be followed in any treatment (*Primum non nocere,* for example), Dioscorides inventoried in his work *De Materia Medica* (On therapeutic substances) all of the natural products of vegetable, mineral, and animal nature used at his time for therapeutic purposes that were known to him.[2] He not only described them (particularly the plants) but also mentioned their therapeutic properties and listed the diseases for the treatment of which they were prescribed. He also might have included visual representations of plants, following the example of the physician Crateuas (first century B.C.) who is credited with the production of the first illustrated herbal in antiquity.[3] As for Galen, he created in his treatise *De Simplicium Medicamentorum Temperamentis et Facultatibus* (On the mixtures and properties of simple medicines) a system aimed at measuring the intensity of the therapeutic properties of medicines.[4] This was particularly useful in the preparation of compound medicines (that is, medicines associating more than one active substance) to correctly dose their different components and to balance their actions.

Diffused through the whole Mediterranean Basin from the Near East to Spain and from Gaul to North Africa, the achievements of classical antiquity were quickly followed by a decline.[5] In Byzantium, Dioscorides' encyclopedia was transformed into an alphabetical dictionary as early as the fourth century A.D., if not earlier, and before the sixth century it become a short handbook containing only some of the medicinal plants of the original version.[6] Further on, therapeutics became more open to faith healing, magic, and superstition—losing if not all then at least a large part of the ancient heritage, particularly in the popular manuals of therapeutics, the so-called *iatrosophia*.[7] Plant representations were submitted to the same erosion: the realistic pictures illustrating Dioscorides' text on the model of Kratevas' herbal became more and more schematic and no longer enabled the recognition of plants in the field.[8]

In the West the legacy of antiquity was very soon reduced to poor manuals principally aimed at practical use. Among the many anonymous works that proliferated, one could quote the so-called *Medicina Plinii* (late third century or fourth century A.D.) based on the *Natural History* of Pliny (23/24-79 A.D.), the great Roman encyclopedist, and also including popular prescriptions and magic.[9] The translation of Dioscorides' *De Materia Medica* into Latin, made perhaps during the sixth century A.D. in North Africa or Southern Europe, did not have much impact, if any.[10] Textual decline was completed by iconic schematism: plant representations not only became oversimplified but also included imaginary elements transforming plants in chimeric aberrations.[11]

In the Islamic world, the Greek heritage was translated into Arabic from the ninth century onward.[12] *De Materia Medica* by Dioscorides in particular became the basis of an intense scientific work that very quickly led to new original treatises, such as those by al-Kindî (d. ca. 873 A.D.), al-Birûnî (973-after 1050 A.D.), ibn-Sînâ (980-1037 A.D.), better known as Avicenna, al-Ghâfiqi (d. 1165 A.D.), and ibn-al Baytâr (ca. 1190-1248 A.D.).[13] These scientists enlarged Dioscorides' range of natural products by systematically inventorying the plants of the Mediterranean Basin and beyond, and developed the methods to be used in the production of compound medicines, following Galen's example. While al-Kindî created a mathematical algorithm, Avicenna developed a speculative theory of philosophical nature.[14]

Again, a decline followed, and scientific activity continued its clockwise movement around the Mediterranean Basin.[15] In Salerno (southern Italy), Constantine the African (d. after 1087) translated into Latin many Arabic medical treatises, launching a vast enterprise of assimilation of Arabic science in the West.[16] However, mistakes in the translations, an overemphasis on compound medicines, and the speculation on their properties quickly led therapeutics to an impasse. This was all the more so because, at that time, works were reproduced by hand from one copy to another and from one generation to another; over time, mistakes were cumulatively added and eventually annihilated the original messages of the texts.

It was the merit of the Renaissance to directly rediscover the ancient heritage from Greek manuscripts, especially after the fall of Constantinople in 1453 and the arrival of Byzantine scholars and scientists to the West. The printing press created around the same time by Gutenberg allowed reproducing texts in an invariable form and in a higher number of copies, thus contributing to the revival of ancient science. The full text of Dioscorides' *De Materia Medica* was printed in Greek in Venice as early as 1499 and no less than five times during the sixteenth century, giving rise to an unprecedented quantity of commentaries and translations, the most famous of which was by Italian Pietro-Andrea Mattioli (1501-1577), first published in 1544.[17] The so-called "German fathers of botany," Otto Brunfels (1489-1534), Leonhart Fuchs (1501-1566), and Hieronymus Bock (1498-1554), along with Valerius Cordus (1515-1544), created premodern botany and botanical illustrations.[18] While Dioscorides' botany and plant representations became rapidly obsolete, the pharmaceutical knowledge contained in his work, instead, constituted the basis of therapeutics until André-Laurent Lavoisier (1743-1794) formulated the theory of chemistry, making it possible to extract the active principles from vegetable therapeutic substances.[19]

However historically founded it might seem, such a reconstruction is misleading. It results from a historiographical paradigm focusing on the so-called great scientists, their works and contributions, and is almost exclusively interested in the authors of classical antiquity credited with the highest degree of achievement. The historiography resulting from these premises necessarily leads us to consider postclassical history as a decline interrupted by attempts to recover the grandeur of antiquity and restore its scientific systems. As a consequence, it negates any other contributions and even eliminates the possibility of bringing any to light—not to speak of the limitations resulting from focusing on the written record limited to the works by educated physicians.

A renewed study of ancient and medieval therapeutic lore using the concepts and methods of ethnobotany and ethnopharmacology, particularly the analytical approach developed by Daniel Moerman, allows reconstituting a different picture: less schematic, more detailed, dynamic, not static, and thus probably more realistic.[20] In this chapter I highlight this new approach by studying three aspects of ancient and medieval therapeutics:

1. The range of materia medica
2. The transformation of the ancient body of knowledge over time
3. Plant representations

To this end, I apply the methods just mentioned to two works, the *Corpus Hippocraticum* and Dioscorides' *De Materia Medica*. In analyzing the former, I will show that knowledge was gradually constituted over time rather than resulted from the action of an individual (or a group of individuals) and, in studying a phase of the latter's history, I will bring to light some of the principles underlying its transformation over time. As for plant representations, I will question the theory according to which they were realistic at the origin and were further transformed into schematic and oversimplified drawings.

In so doing, I reintroduce the thickness of reality and the fluidity of history into the picture of ancient therapeutics, transforming it from the flat representation of traditional historiography into a three-dimensional image susceptible to transformations over time.

TWO MAIN WORKS

The *Corpus Hippocraticum* is a collection of more than 60 treatises once thought to be written by Hippocrates himself and now dated from the fifth century B.C. to the second A.D. and attributed to a wide range of authors, including Hippocrates.[21] Whoever their author may be, these treatises deal with very different topics—from anthropology to gynecology and traumatology, including cardiology, embryology, and also the method of medicine and its status in ancient science and society.[22]

De Materia Medica by Dioscorides contains more than 1,000 chapters, each of them devoted to a natural substance used for therapeutic purpose.[23] All chapters contain three major kinds of data:[24]

- the description or the preparation of the product dealt with, unless it is well known. For vegetable materia medica, this part contains a description of the plants from which drugs were taken;
- its major therapeutic property (or properties); and
- the full list of its therapeutic indications.

According to contemporary bibliography, *De Materia Medica* was divided into five books, each of which corresponded to one or more kinds of materia medica:[25]

Book 1: Perfumed plants, perfumes and trees
Book 2: Animals and plants
Book 3: Plants
Book 4: Plants
Book 5: Liquids and minerals

It has not been established, however, that this division is original. All books contain approximately the same number of lines, a fact suggesting that these books were tomes, that is, units determined by the techniques of book production (at that time, papyrus rolls). If so, *De Materia Medica* was a sequence of 1,000-plus chapters ordered according to the cultural values attributed to the materia medica, from iris in the first chapter to the black used to make ink in the very last, that is, from polychromy to the absence of color (in the Greek chromatic palette, black was not a color but the absence of color).[26]

THE RANGE OF MATERIA MEDICA

Case stories and therapeutic prescriptions in the *Corpus Hippocraticum* contain 3,100-plus mentions of vegetable products used for medical purposes and coming from 380 different plants.[27] At first glance, these numbers could suggest (as has been often the case) that Hippocratic physicians mastered a wide range of therapeutic substances and that ancient medicine reached a high degree of development.[28] However, 44 plants (about 12 percent of the 380 species mentioned in the *Corpus*) produce more than 1,500 of the 3,100-plus previous mentions (nearly 50 percent). This suggests that the range of vegetable substances used as pharmaceuticals by Hippocratic physicians was not as extended as the raw numbers might seem to indicate. Furthermore, the 44 plant species form two groups of plants and substances according to their origin: exotic and native. Interestingly, exotic substances—that is, substances coming from regions at the edge of the world, themselves characterized by their proximity to the sun according to ancient cosmology and geography—have the highest number of occurrences: myrrh appears 87 times, cumin 72, incense 45, and silphium 45. This fact throws a new light on the origin of Hippocratic medicine and, in our case, therapeutics. Traditionally considered as a creation of Greek culture, if not one of its highest achievements, it probably included practices imported from the East, that is, Egypt, Persia, and the Near East.[29] Although such an origin contradicts the image that Greeks tried to give of themselves, it is probably more realistic. It could result either from an Oriental basis of Greek culture or later importations through trade over time, particularly during the period of the so-called Oriental century (750-650 B.C.).[30]

Among the most used 44 plants in the *Corpus Hippocraticum,* those native to the Mediterranean Basin are common; they are listed in columns according to their total number of mentions, in decreasing order, in Table 8.1. This list shows that Hippocratic materia medica was simple, could be found everywhere in the Greek world (that is, in the Mediterranean biota), was

TABLE 8.1. The native plants of the Greek world among the most used vegetable substances in the *Corpus Hippocraticum*.[a]

Plant	Mentions	Plant	Mentions	Plant	Mentions	Plant	Mentions
hellebore	63	pomegranate	37	olive tree	26	earth almond	22
garlic	49	elder tree	35	peony	26	love-in-a-mist	21
French mercury	47	myrtle	35	fennel	26	stavesacre	19
parsley	46	squirting cucumber	33	wheat	26	bramble	19
leek	46	sage	31	lentil	25	millet	19
flax	45	cypress	31	daphne	25	scammony	19
anis	43	barley	30	*lôtos*	25	chaste tree	18
white beet	43	rue	30	bottle-gourd	25	sesam	18
cauliflower	41	laurel	29	mint	24	onions	17
origanum	40	fig	29	cyclamen	23	coriander	14

Source: Touwaide, Alain, 1997-1998, Bibliographie historique de la botanique: Les identifications de plantes médicinales, *Lettre Jean Palerne,* 30: 2-22; 31: 2-65.

[a]The identification of plants mentioned in ancient texts is a critical issue. We have adopted here the common English name used in current literature.

easy to procure, and was not subject to deep changes according to the seasons. This contradicts a hypothesis according to which Hippocratic physicians (who were itinerant professionals) carried in a bag the therapeutic substances they needed to treat their patients. Judging from this list, they were more likely to select their materia medica from among the resources of the place where the patients were living. None of the substances listed was rare, difficult to find, or expensive. The ancient therapeutic arsenal included simple people living in the countryside and using the natural resources at their disposal.

Furthermore—and this is probably more important—Table 8.1 throws a very new light on the way the range of drugs used by Hippocratic physicians and, more generally, by ancient physicians was constituted, in other words, how the therapeutic properties of plants were discovered.

Ancient literature—be it scientific or not—contains several accounts on this point that can be divided into two major categories: mythological tales and naturalistic stories. According to mythological tales, the curative properties of plants were revealed to humankind by different gods or mythological creatures especially interested in medicine. Betony is probably the most significant example since it gave rise to an illustration: Apollo cuts the plants and gives it to a man who was surely a physician or, probably more exactly, a lay healer who became a medicine man precisely thanks to this revelation. Naturalistic stories do not report a revelation but instead usually show a discovery, often involving animals. Stating the effects of some plants on the game, shepherds and stock persons applied the same plants to themselves or to other people, obtaining the effect naturally provoked on the game.

Table 8.1 suggests a different reconstruction. Its plants indeed present typical characteristics: Some have flowers and fruits with notable color, shape, or seasonality. This is the case of pomegranate, with its scarlet flower and ruddy applelike fruit, or myrtle, with its delicate white flowers open during the winter and sharply contrasting with its evergreen foliage. Many were—and still are—consumed as foodstuffs, with different categories: vegetables (onion, garlic, and leek, for example), cereals (barley and wheat), condiments (origanum, sage, coriander), or fruits (pomegranate and fig). Others, instead, were—and still are—typical of the Mediterranean biota. Pomegranate, cypress, laurel, fig tree, and olive tree contribute to create the maquis, garigue, and thicket of the Mediterranean area, with the perfume they emanate at the end of a hot summer day. Parsley, fennel, and anise, instead, punctuate open spaces with their high stems and their umbels. Squirting cucumber, finally, and to mention just some characteristics, could not escape attention with its fruit that explodes its seeds outward to some distance when it is disturbed! All of these plants were linked in one way or another with human activity, or were remarkable enough to have been noticed, tasted, and further sought after for their action on human physiology.[31] If so, Hippocratic physicians did not discover the therapeutic properties of these plants, but just reproduced previous uses, subsumed them, and included them into a theoretical frame, provided that they did not shape their theories on the basis of the action of the plants on human physiology. They thus appear as a professional group who absorbed, organized, systematized, and also took control of a knowledge that was already traditional at their time.

In Dioscorides' *De Materia Medica,* the total number of plants is much higher than in the *Corpus Hippocraticum:* more than 630, which is approximately twice the number of the *Corpus*. Again, numbers should not be misleading and suggest that the range of materia medica dramatically increased further to a systematic exploration and exploitation of the natural resources provided by environment. In fact, as Dioscorides himself stated in the preface of the work, he aimed to provide readers with an exhaustive inventory of the natural substances used at that time for medicinal purposes.[32] In the work, indeed, he mentions substances with very few uses (sometimes no more than one) and even products lacking medicinal properties, but linked in one way or another to a real drug. *De Materia Medica* is an encyclopedia rather than a manual of pharmaceutical therapy. As such, it reflects the cultural and intellectual climax of an apolitical moment also echoed in the *Natural History* by Pliny, probably contemporary to Dioscorides: the Roman Empire was at its zenith, searching for unusual and rare products from its remotest parts and beyond, and trading them all across the Mediterranean Sea.[33] *De Materia Medica* does not describe the actual practice of a group of physicians in a determined place at a determined moment, but encompasses all substances known at one moment in a determined political, social, and economic structure. To be used, it necessarily required readjustments according to local needs and resources.

TRANSFORMATIONS OF TEXTS OVER TIME

If availability was a major factor in the selection of plants used for medicinal purposes, as the *Corpus Hippocraticum* suggests, usefulness was another. This is particularly clear in the rearrangements of Dioscorides' *De Materia Medica*.

According to a historical reconstruction widely diffused in contemporary literature, the chapters of *De Materia Medica* were arranged in alphabetical order of their heading (that is, substance names) sometime between the second and the fourth century, for this way of sequencing data was more and more diffused in late antique science and culture.[34] On the basis of this alphabetical version, a more reduced version was further constituted, which contained only some of the chapters devoted to vegetable substances: 390 out of the more than 630 of the original text. This is the so-called *Alphabetical Herbal*. Its creation supposedly evidences progressive scientific entropy, that is, a gradual loss in the information bequeathed by antiquity to the later ages due to the gradual rise of medieval obscurantism.

Again, the application of ethnobotanical and ethnopharmacological methods to historical data leads to different conclusions, which probably correspond more closely to historical reality. Dioscorides' full text contains a total number of 5,300-plus prescriptions that can be divided into 41 categories, according to the diseases for the treatment of which they were prescribed: from skin diseases to gynecological troubles of any kind, including alopecia, gastric problems, alimentary diseases, and toothaches.[35] If we sum up the total number of occurrences of each disease and if we classify the diseases according to the descending order of their numeric importance, we obtain Table 8.2, where the name of the different categories of diseases is followed by the number of occurrences of that disease and its percentage in Dioscorides' treatise (only the first ten items are mentioned). The identification of the diseases mentioned in ancient texts is particularly problematic.[36] We have adopted here a classification on the basis of a *philological* reading of the texts. If we do a similar work on the herbal supposedly resulting from a double process of transformation of Dioscorides' full text, we obtain Table 8.3 (data are presented according to the same principles as in Table 8.2).

In comparing Tables 8.2 and 8.3, we need not to focus on the total number of occurrences: 5,375 in Dioscorides' full text and 2,233 in the *Herbal*.

TABLE 8.2. Diseases in Dioscorides' *De Materia Medica*.

Affected systems, organs, or functions, diseases, or cause of the diseases	**Number of occurrences**	**Percentage**
Skin, nails, and mucous membrane	626	11.64
Gastrointestinal system	596	11.08
Venoms and poisons	481	8.94
Gynecology, obstetrics	377	7.01
Urinary tract	348	6.47
Respiratory system	327	6.08
Eyes, sight	316	5.87
Bones and articulations	231	4.29
Wounds and ulcers	223	4.14
Humors	140	2.60

Their differences result from the fact that the *Herbal* contains only a limited number of chapters from Dioscorides' full text. The important number to be taken into consideration here is the percentage of each category.

A comparison of the two tables reveals first that the list of disease categories is identical in the two tables, apart from the last item: humors in Table 8.2 and inflammations in Table 8.3. We shall return later to the differences. To start our analysis, we need to focus on the similarities, that is, the fact that the sequence of diseases (or affected organ, part of the body, etc.) is almost identical in both cases and that the respective percentage of each category of disease is similar. This global similarity suggests that the transformation of Dioscorides' full text into the *Herbal* did not result either from an arbitrary selection or from a mere elimination of some chapters of *De Materia Medica*. Instead, the creation of the *Herbal* relied on a preliminary work aimed at evaluating—in fact, quantifying—the uses of therapeutic agents. To this end, a comparison was made between *De Materia Medica* and actual practice, and only the substances that were required in the daily exercise of therapeutic treatment were conserved. Hence the reduction of the number of chapters and occurrences of diseases, as well as the parallelism between *De materia medica* and the *Herbal*. This work of quantifying is an epidemiological analysis, that is, a statistical approach to diseases allowing to evaluate the needs for therapeutic agents.

The factor of reduction of the number of diseases from *De materia medica* to the *Herbal* oscillates between 1.3 minimum to 10 maximum, with a medium of 2.4. For example, for the diseases of skin, nails, and mucous membrane, the total number of occurrences in *De materia medica* is 626 and in the *Herbal* 256, a reduction by 2.4. A lower factor of reduction

TABLE 8.3. Diseases mentioned in the *Alphabetical Herbal*.

Affected systems, organs, or functions, diseases, or cause of the diseases	Number of occurrences	Percentage
Skin, nails, and mucous membrane	256	11.46
Gastrointestinal system	252	11.28
Venoms and poisons	182	8.15
Gynecology, obstetrics	184	8.24
Urinary tract	162	7.25
Respiratory system	132	5.91
Bones and articulations	128	5.73
Eyes, sight	102	4.56
Wounds and ulcers	90	4.03
Inflammations	67	3.00

indicates a higher need of therapeutic substances for a determined kind of diseases and, instead, a higher factor reveals a reduced need.

Our reconstitution of the process underlying the creation of the *Herbal* is confirmed *a contrario* by the analysis of the chapters of Dioscorides' full text not included in the *Herbal*. The three substances of *De Materia Medica* used to treat such rare a disease as nyctalopia are eliminated in the *Herbal*, as is also the only one recommended to treat bites by humans, crocodile bites, and bites by nonvenomous spiders. For the following diseases the number of therapeutic agents kept in the *Herbal* is strongly reduced (between parenthesis I respectively give the total number of drugs mentioned in *De materia medica* and in the *Herbal*): *opisthotonos* (9/1), iatrogenic pathologies (120/23), hysteric suffocation (27/8). Similarly, the many mentions of bites and stings by snakes, scorpions, fishes, and other venomous animals in *De materia medica* are strongly reduced in the *Herbal:* vipers (25/14), cobra (4/1), venomous spiders (24/14), and scorpions (40/17), for example. This is also the case for rabies, which was then considered an envenomation by a rabid dog (34/9). In fact, the high number of mentions of venomous animals in *De Materia Medica* resulted from the fact that, during the period first century B.C. to first to second century A.D., toxicology was used as a heuristic tool for a better understanding of the action of medicines on human physiology. Their elimination or strong reduction in the *Herbal* evidences a readjustment of texts aimed at putting them in correspondence with practice.

Apart from these, the main differences between *De Materia Medica* and the *Herbal* are shown in Table 8.4. While in three cases there is a reduction (humors: 2.60 percent to 2.07 percent; iatrogenic diseases: 2.23 percent to 1.03 percent; head pathologies: 1.28 to 0.62 percent), in all others an increase occurs. The reduction affecting the diseases due to humors probably results from a transformation of medical theory: humors were no longer considered as the main component of human physiology. The reduction of iatrogenic diseases, instead, might indicate that collateral effects of medicines were paid better attention, whereas the reduction of head pathologies cannot be explained by any clear factor. The categories for which there is a higher percentage of occurrences (inflammations of all kinds, swellings and dropsy, and pathologies of the hepatic system and the spleen) are interesting. They seem to evidence a higher concern for internal diseases that could be linked to a better clinical observation and a deeper understanding of pathological processes.

TABLE 8.4. Major differences in the diseases mentioned in *De Materia Medica* and in the *Alphabetical Herbal*.

	Dioscorides' full text			***Alphabetical Herbal***		
Affected systems, organs, or functions, diseases, or cause of the diseases	**No. of occurr.**	**Percentage**	**Position in the table**	**No. of occurr.**	**Percentage**	**Position in the table**
Humors	140	2.60	10	46	2.06	14
Inflammations	138	2.56	11	67	3.00	10
Iatrogenic diseases	120	2.23	14	23	1.03	23
Swellings, dropsy	96	1.78	17	54	2.41	12
Hepatic system	92	1.71	18	56	2.50	11
Spleen	76	1.41	21	46	2.06	15
Head	69	1.28	22	14	0.62	28

PLANT REPRESENTATIONS

As for plant representations, they are considered to date back to Kratevas, supposedly the personal physician of the king of Pontos Mithridates VI Eupator (132-63 B.C.), better known for his interest in compound medicines.[37] Kratevas' herbal is not conserved, but some of its pictures are thought to be included in the set of plant representations contained in a copy of the *Alphabetical Herbal* produced from Dioscorides' *De Materia Medica.* This copy is a manuscript now conserved in the National Library of Austria and, for this reason, is commonly designated as the *Vienna Dioscorides* (*Vindobonensis Medicus Graecus* 1).[38]

This manuscript is truly exceptional: dating back to ca. 513 A.D., it is among a few of the most ancient books of Byzantium that are currently conserved. Produced in Constantinople, the capital of the empire, as a gift for a princess, it has an imperial size (380 × 330 mm) and contains no less than 383 plant representations, most of which cover the full surface of the page. Some of the pages containing these pictures also include short textual extracts preceded by the name Kratevas, hence the conclusion that the pictures corresponding to these fragments could come from Kratevas' *Herbal.* Since these figures and others in the same manuscript are characterized by an effort toward realism, Kratevas' pictures (in fact, their supposed models) have been considered to be realistic.[39] This feature is all the more characteristic because other illustrated copies of *De Materia Medica* offer more schematic drawings, including the manuscript that is said to contain the original version of the text.

On the basis of these elements, three main conclusions have been reached that are generally adopted in contemporary literature:[40]

1. Dioscorides did not necessarily illustrate the original version of his work, even though in a miniature of the Vienna manuscript he is represented writing *De Materia Medica* while an artist depicts a plant (mandrake);
2. Plant representations in the manuscripts of *De Materia Medica* are additions gradually introduced into the manuscripts and coming from different sources—among others, Kratevas' *Herbal*;
3. Originally realistic, plant representations were progressively transformed by the artists who reproduced them from one copy to another; they thus became more and more schematic according to a process of deformation similar to that affecting the text.

A closer comparison of all pictures of all extant illustrated copies of *De Materia Medica* in Greek sheds a new light on this question. The *Vienna Dioscorides* is not the only illustrated copy but is a member of a group of three closely related manuscripts identified by the place where they are currently conserved: the so-called *Naples Dioscorides* and *New York Dioscorides* (manuscripts *graecus* 1 of the National Library in Naples, Italy[41] and M 652 of the Pierpont Morgan Library in New York[42]). The Naples copy probably dates back to the first half of the seventh century A.D. and might have been produced in Rome. The New York copy is from the tenth century and was made in Constantinople. A systematic comparison of their plant representations evidences a closer affinity between the Naples and the New York copies than between these manuscripts and the Vienna version. Yet the Naples and the New York manuscripts do not depend on each other (they were not copied from each other) and thus are copies of the same model or, if this is not the case, of different models that eventually proceeded from a unique original. Their common features can thus be attributed to this ultimate original. They share a similar schematism, as opposed to the realism of the Vienna manuscript. As a consequence, it has to be admitted that the realism of the Vienna copy is typical of this manuscript, of its model or of one among its ancestors, without coming from the common original of the group constituted by the *Vienna,* the *Naples,* and the *New York Dioscorides.* It could be admitted, however, that the Vienna realism is the original form of plant representations and that, instead, the Naples and New York schematism comes from a manuscript common to these two copies. This possibility is invalidated, however, by further elements.

If the pictures of the Vienna, Naples, and New York manuscripts group are compared to those of the other manuscripts of *De Materia Medica,* some have an equivalent in the most ancient copy currently conserved of the full text, a manuscript now conserved in the National Library of France in Paris, manuscript *graecus* 2179, possibly of the ninth century.[43] Yet the similar pictures have a more schematic aspect that recalls the Naples and New York manuscripts rather than the Vienna one. Furthermore, figures in the Paris manuscript closely correspond to the Dioscorides' text, in other words, they seem to be a kind of pictorial translation. Rather than exact reproductions of botanical reality, these representations are in fact visualizations of the text. Hence their schematism that did not result from an inaptitude of artists to render nature, but probably was voluntary and intended to visually express the analytical description made in the text by means of words. Plant representations thus result from the addition of single bits of botanical data assembled together, rather than being a synthetic and global view of plants.

Plant representations adhering so closely to the text could result from the same project of the text itself and thus could have been conceived as part of *De materia medica.* In other words, they could very well date back to Dioscorides.

CONCLUSION

The application of ethnobotanical and ethnopharmacological concepts and methods to ancient and medieval drug lore allows us to reconstitute a multidimensional, dynamic, and probably realistic picture of the creation and development of botanical and medical knowledge among the cultures that flourished around the Mediterranean Sea.

As a result, the primary sources dealing with the topic should not be considered any longer to be works authored by single individuals, but rather as the written record of a practice gradually constituted over time and resulting from a pragmatic and empirical approach to botanical environment. Many of the plants used as medicinal agents were foodstuffs; their range was not necessarily extended, and their inclusion into the alimentary and medicinal arsenal of human groups was probably due to their striking characteristics, such as shape and color, to sensory perception and physiological effects, such as sweetness, bitterness, perfume, balsamic odor, and strong cathartic action, for example. Before being recorded in the written works of historical physicians from Hippocrates onward, knowledge thus was constituted over millennia by trial and error. In other words, historical texts are the final phase of a long development that can be correlated with repeated clinical trials. The body of information they contain thus expresses actual facts verified by repeated experiments, first orally transmitted among medicine men and women and only recently (so to speak) recorded by physicians and expressed by them according to their contemporary medical systems.

Once this body of knowledge was constituted, it passed from one generation to another (be it before they were recorded in written works or after) and was constantly adapted to possible changing needs. This work of adaptation, far from being the result of entropy, witnesses a quantified approach to epidemiology and hence a corresponding adaptation of the needs for medicines in the groups using previous knowledge. Healers constantly updated their therapeutic arsenal, thus controlling the health conditions of the members of the group and of the groups themselves, and continuously rethinking their needs for pharmaceuticals. At the same time, educated physicians reexpressed previous knowledge according to contemporary theoretical systems so as to be able to explain experimentally gained facts by means of changing epistemic tools and systems, and to integrate traditional knowledge into formal science.

NOTES

1. Ernst Meyer, *Geschichte der Botanik*, 4 volumes (Königsberg: Gebrüder Bornträger, 1854-1857; Reprint: Amsterdam: A. Asher and Co., 1965).

2. Jacques Jouanna, *Hippocrate* (Paris: Fayard, 1992; English translation: *Hippocrates*, Baltimore, London: The Johns Hopkins University Press, 1999); Max Wellmann, 1903, Dioskurides 12, *Realencyclopädie der classischen Altertumswissenschaft*, 5(1): 1131-1142; John Riddle, *Dioscorides on pharmacy and medicine*, History of Science Series, 3 (Austin, TX: University of Texas Press, 1985); Max Wellmann, *Pedanii Dioscuridis Anazarbei, De materia medica libri quinque*, 3 vols. (Berlin: Weidmann, 1906-1914; reprint: 1958); Tess Anne Osbaldeston and Robert P.A. Wood, *Dioscorides De materia medica: Being an herbal with many other medicinal materials.* Written in Greek in the first century of the common era. A new indexed version in modern English (Johannesburg: IBIDIS Press, 2000).

3. Max Wellmann, *Kratevas* (Abhandlungen der Königlichen Gesellschaft der Wissenschaften zu Göttingen, Philologish-historische Klasse, N.F. 2, 1) (Göttingen: Königliche Gesellschaft der Wissenschaften, 1897); Minta Collins, *Medieval herbals: The illustrative traditions* (London: The British Library, and Toronto: University of Toronto Press, 2000).

4. John Scarborough, Galen of Pergamon, in Ard W. Briggs (ed.), *Dictionary of literary biography*, volume 176: *Ancient Greek Authors* (Detroit, Washington DC, London: Gale Research, 1997), pp. 156-170; Luis García Ballester, Galen's medical works in the context of his biography, in Jon Arrizabalaga, Montserrat Cabré, Lluís Cifuentes, and Fernando Salmón (eds.), *Galen and Galenism: Theory and medical practice from antiquity to the European Renaissance* (Aldershot, Burlington: Ashgate, 2002), no. I; Karl Gottlob Kühn, *Clauddi Galeni Opera omnia*, Volumes 11 and 12 (Leipzig: K. Knobloch, 1826); Alain Touwaide, La thérapeutique médicamenteuse de Dioscoride à Galien: Du "pharmaco-centrisme" au "médico-centrisme," in Armelle Debru (ed.), *Galen on pharmacology: Philosophy, history and medicine* Proceedings of the Fifth International Galen Colloquium, Lille, March 16-18, 1995 (Studies in Ancient Medicine, 16) (Leiden, New York, Köln: Brill, 1997), pp. 255-292.

5. Alain Touwaide, Le *Traité de matière médicale* de Dioscoride en Italie depuis la fin de l'Empire romain jusqu'aux débuts de l'école de Salerne, in Antje Krug (ed.), *From Epidaurus to Salerno* Symposium held at the European University Centre for Cultural Heritage, Ravello April 1990 (PACT 34) (Rixensart: PACT Belgium, 1994), pp. 175-305.

6. Wellman, 1903, Disokurides 12; Riddle, 1985, *Dioscorides on pharmacy and medicine.*

7. Iwan Bloch, Byzantinische Medizin, in Theodor Puschmann (ed.), *Handbuch der Geschichte der Medizin* (Jena: Gustav Fischer, 1902), pp. 492-568; Herbert Hunger, *Die hochsprachliche profane Literatur der Byzantiner*, Volume 2 (Handbuch der Altertumswissenschaft, 12. Abteilung, 5. Teil) (Münich: C.H. Beck: 1978), pp. 285-320; Gotthard Strohmaier, La ricezione e la tradizione: La medicina

nel mondo bizantino e arabo, in Mirko D. Grmek (ed.), *Storia del pensiero medico occidentale,* Volume 1 (Rome, Bari: Laterza, 1993), pp. 167-215.

8. Collins, 2000, *Medieval herbals.*

9. Alf Önnerfors, *Plinii Secundi Iunioris qui feruntur De medicina libri tres* (Corpus Medicorum Latinorum, 3) (Berlin: Academia Scientiarum, 1964); Klaus Sallmann, Plinius, in Hubert Cancik and Helmuth Schneider (eds.), *Der Neue Pauly. Enzyklopädie der Antike,* Voume. 9 (Stuttgart, Weimar: J.B. Metzler, 2000), cols. 1135-1141.

10. John M. Riddle, Dioscorides, in Ferdinand Edward Cranz and Paul Oskar Kristeller (eds.), *Catalogus translationum et commentariorum: Mediaeval and Renaissance Latin translations and commentaries—Annotated lists and guides,* Volume 4 (Washington DC: Catholic University of America Press, 1980), pp. 1-143.

11. Collins, 2000, *Medieval herbals.*

12. Dimitri Gutas, *Greek thought, Arabic culture: The Graeco-Arabic translation movement in Baghdad and Early cAbbâsid society (2nd-4th/8th-10th centuries)* (London, New York: Routledge, 1998).

13. Alain Touwaide, L'intégration de la pharmacologie grecque dans le monde arabe. Une vue d'ensemble, in Luciana R. Angeletti and Alain Touwaide (eds.), *Medieval Arabic medicine,* Volume 2 (Rome: Delfino Editore, 1995), pp. 159-189; Martin Levey, *The medical formulary or Aqrâbâdhîn of al-Kindî: Translated with a study of its materia medica* (Madison, London: University of Wisconsin Press, 1966); Muhammad ibn Ahmad Bîrûnî, *Book on pharmacy and materia medica,* Hakim Mohammed Said (ed. and trans) (Pakistan Series of Central Asian Studies, 2) (Karachi: Hamdard National Foundation, 1973); Avicenna, *The canon of medicine,* adapted by Laleh Bakhtiar (Chicago: Great Books of the Islamic World: 1999); Max Meyerhof and Georgy P. Sobhy, *The abridged version of "The book of simple drugs" of Ahmad Ibn Muhammad Al-Ghâfiqî by Gregorius Abu'l-Farag (Barhebraeus)* (edited from the only two known manuscripts with an English translation, commentary and indices (The Egyptian University, The Faculty of Medicine, Publication no. 4) 4 volumes (Cairo: Al-Ettemad, 1932; and Cairo: Government Press, Bûlâq, 1937-1940); Ibn al-Baytâr, *Traité des simples: Traduction par Lucien Leclerc,* 3 volumes (Paris: Imprimerie Nationale, 1877-1883).

14. Alain Touwaide, 1996, Theoretical concepts and problems of Greek pharmacology in Greek Arabic medicine: Reception and re-elaboration, *Forum,* 6(1): 21-39.

15. Danielle Jacquart, The influence of Arabic medicine in the medieval West, in Roshdi Rashed (ed.), *Encyclopedia of the history of Arabic science,* Volume 2 (London, New York: Routledge, 1996), pp. 963-983.

16. Charles Burnett and Danielle Jacquart (eds.), *Constantine the African and cAlî ibn al-cAbbâs al-Magûsî: The Pantegni and related texts* (Studies in ancient medicine, 10) (Leiden, New York: Brill, 1994).

17. Sara Ferri (ed.), *Pietro Andrea Mattioli, Siena 1501-Trento 1578. La vita e le opere, con l'identificazione delle piante* (Perugia: Quattroemme, 1997); Riddle, 1980, Dioscorides.

18. Agnes Arber, *Herbals, their origin and evolution: A chapter in the history of botany, 1470-1670* (Cambridge: University Press, 1912; third edition, 1986).

19. Gunther Stille, *Der Weg der Arznei. Von der Materia Medica zur Pharmakologie. Der Weg von Arzneimittelforschung und Arzneitherapie* (Karlsruhe: G. Braun, 1994).

20. See Chapter 6 in this book.

21. Jouanna, 1999, *Hippocrates.*

22. Geoffrrey E.R. Lloyd (ed.), *Hippocratic writings,* introduction by Lloyd, translated by John Chadwidk, William N. Mann, Iain M. Lonie, and Edward T. Withington (London: Penguin Books, 1950); Hippocrates, *Works,* 8 volumes (Loeb Classical Library, volumes 147-150, 472-473, 477, 482), (Cambridge MA: Harvard University Press, 1923-1995).

23. Wellman, 1958, *Pedanii Dioscuridis Anazarbei;* Osbaldeston and Wood, 2000, *Dioscorides de materia medica.*

24. Alain Touwaide, La botanique entre science et culture au Ier siècle de notre ère, in Georg Wöhrle (ed.), *Geschichte der Mathemathisch- und Naturwissenschaft in der Antike,* Volume 1: *Biologie* (Stuttgart: Franz Steiner Verlag, 1999), pp. 219-252.

25. Wellman, 1958, *Pedanii Dioscuridis Anazarbei.*

26. Towaide, 1999, La botanique entre science et culture.

27. Giovanni Aliotta, Daniele Piomelli, Antonino Pollio, and Alain Touwaide, *Le piante medicinali nel Corpus Hippocraticum* (Milan: Guerini, forthcoming).

28. Jerry Stannard (1961). Hippocratic pharmacology, *Bulletin of the History of Medicine,* 35: 497-518; John Scarborough, Theoretical assumptions in Hippocratic pharmacology, in François Lasserre and Philippe Mudry (eds.), *Formes de pensée dans la Collection hippocratique: Actes du IVe Colloque international hippocratique,* Lausanne, September 21-26, 1981 (Université de Lausanne, *Publications de la Faculté des Lettres,* XXVI) (Geneva: Droz, 1983), pp. 307-325; Monique Moisan, *Lexique du vocabulaire botanique d'Hippocrate* (Université Laval, Laboratoire de recherches hippocratiques, *Documents,* 7) (Québec: Université Laval, 1990).

29. Renate Gerner, *Flora des pharaonischen Ägypten* (*Deutsches Archäologisches Institut,* Abteilung Kairo, *Sonderschrift* 14) (Mainz: Philipp von Zabern, 1985); Nathalie Baum, *Arbres et arbustes de l'Egypte ancienne: La liste de la tombe thébaine d'Ineni (no. 81)* (Orientalia Lovaniensia Analecta, 31) (Leuven: Universiteit, Departement Oriëntalistiek, 1988); Renate Germer, *Die Heilpflanzen der Ägypter* (Düsseldorf, Zürich: Artemis und Winkler, 2002).

30. Walter Burkert, *Die orientalisierende Epoche in der griechischen Religion und Literatur* (Sitzungsberichte der Heidelberger Akademie der Wissenschaften, Philosophisch-historische Klasse, 1) (Heidelberg: Carl Winter, 1984). English translation: *The orientalizing revolution: Near Eastern influence on Greek culture in the early archaic age* (Revealing antiquity, 5) (Cambridge MA: Harvard University Press, 1995).

31. Jonathan Ott, 1998, The Delphic bee: Bees and toxic honeys as pointers to psychoactive and other medicinal plants, *Economic Botany,* 52(3): 260-266.

32. John Scarborough and Vivian Nutton, 1982, The "Preface" of Dioscorides' "Materia Medica": Introduction, translation and commentary. *Transactions and Studies of the College of Physicians of Philadelphia,* 5(4): 187-227.

33. James I. Miller, *The spice trade of the Roman Empire, 29 B.C. to A.D. 641* (Oxford: Clarendon Press, 1969); Manfred Raschke, New studies in Roman com-

merce with the East, in Hildegard Temporini and Wolfgang Haase (eds.), *Aufstieg und Niedergang der römischer Welt.* II, Principat, Volume 9/2 (Berlin, New York: De Gruyter, 1978), pp. 604-1361.

34. Wellman, 1903, Diokurides; Riddle, 1985, *Dioscorides on pharmacy and medicine;* Riddle, 1980, Dioscorides.

35. Mirko D. Grmek, *Les maladies à l'aube de la civiliation occidentale. Recherches sur la réalité pathologique dans le monde grec histoique, archaïque et classique* (Bibliothèque historique Payot) (Paris: Payot, 1983). English translation: *Diseases in the ancient Greek world* (Baltimore, London: The Johns Hopkins University Press, 1989).

36. Ibid.

37. Wellmann, 1897, *Kratevas;* Collins, 2000, *Medieval herbals;* Gilbert Watson, *Theriac and mithridatum: A study in therapeutics* (Wellcome Historical Medical Library, Publications, New Series, vol. 9) (London: Wellcome Historical Medical Library, 1966).

38. Paul Buberl, *Die Byzantinischn Handschriften,* 1. Der Wiener Dioskurides und die Wiener Genesis (Leipzig, Hiersemann, 1937); Hans Gerstinger, *Dioscurides. Codex Vindobonensis med. gr. 1 der Österreichischer Nationalbibliothek. Kommentarband zu der Faksimileausgabe* (Codices Selecti Phototypice Impressi, XII*) (Graz: Akademische Druck- u. Verlagsanstalt, 1979); *Der Wiener Dioskurides. Codex medicus graecus 1 der Österreichische Nationalbibliothek,* 2 volumes (Glanzlichter der Buchkunst, 8) (Graz: Akademische Druck- u. Verlagsanstalt, 1998); Leslie Brubaker, The Vienna Dioskorides and Anicia Juliana, in Antony Littlewood, Henry Maguire, and Joachim Wolschke-Bulmahn (eds.), *Byzantine garden culture* (Washington DC: Dumbarton Oaks Research Library and Collection, 2002), pp. 189-214.

39. Collins, 2000, *Medieval herbals.*

40. Ibid.

41. Salvatore Lilla, Guglielmo Cavallo, Giulia Orofino, and Carlo Bertelli, *Dioskurides, Codex Neapolitanus. Napoli, Biblioteca nazionale, Ms. ex Vindob. gr. 1. Vollständige Faksimile-Ausgabe in Originalformat* (Codices selecti phototypice impressi, Volume 88; Codices mirabiles, no.2) (Graz: Akademische Druck- u. Verlagsanstalt; Rome: Salerno editrice, 1988); Mauro Ciancaspro, Guglielmo Cavallo, and Alain Touwaide, *Dioscurides De materia medica. Codex Neapolitanus graecus 1 of the National Library of Naples.* Facsimile reproduction of the manuscript with introductory texts (Athens: Militos Editions, 1999).

42. *Pedanii Dioscuridis Anazarbei De materia medica libri VII. Accedunt Nicandri et Eutecnii opuscula medica. Codex Constantinopolitanus saeculo X. exaratus et picturis illustratus olim Manuelis Eugenici Caroli Rinuccini Florentini, Thomae Phillipps Angli nunc inter thesauros Pierpont Morgan Bibliothecae asservatus,* 2 volumes (Paris: s.n., 1935); Ann Van Buren, *De Materia Medica* of Dioscorides, in Gary Vikan (ed.), *Illuminated Greek manuscripts from American collections: An exhibition in honor of Kurt Weitzmann* (Princeton: The Art Museum, 1973), pp. 66-69.

43. Brigitte Mondrain, Dioscoride: *De materia medica,* in *Byzance: L'art byzantin dans les collections publiques françaises.*

Chapter 9

Remedies from the Bush: Traditional Medicine Among the Australian Aborigines

Emanuela Appetiti

INTRODUCTION

Prior to the British colonization in 1788, Aborigines were the sole inhabitants and owners of the huge territory now called Australia. Their adaptation to the environment was total: hunter-gatherers, they moved according to the seasons within their tribal country; recognizing plants, animals, and minerals; using them as food and medicine; and knowing the location of water holes (billabongs).

Contacts with foreign people, such as the Macassan seafarers coming from Indonesia, were regular among the inhabitants of the northern coastal areas, but not among those living on the interior of the continent. The harsh environment prevented contact with other peoples and protected the Aborigines from contagious illnesses over the centuries. This ended when the British came. The whites brought with them syphilis, tuberculosis, measles, and smallpox. Later on the development of alcoholism, drug addiction, and diabetes became common as well. The Aborigines had no natural protection for, and bush medicine had no effect on, these new infirmities.

In their traditional life, Aborigines had to cope with such respiratory ailments as coughs and colds, and such complaints as eye and intestinal disorders (dysentery, diarrhea, and constipation), headache, rheumatism, skin eruptions due to scabies or to other infectious diseases, wounds, burns, cuts, snakebites, and insect bites. Most serious and sudden illnesses almost inevi-

This research has been possible thanks to the resources of the libraries of the National Museum of Natural History, Smithsonian Institution, particularly the library of the botany section. I am indebted to Dan Nicolson, distinguished botanist and wonderful friend, who read the draft of this work and greatly helped me in verifying the accurateness of the identifications of plants as well as the abbreviation of the author's names. Any remaining errors are of course my own.

tably ended in death, causing great concern within the group, who attributed it to sorcery and evil spirits—especially if the victim was a young man or a child.

The Aboriginal concept of illness is complex beyond the scope of this chapter. Suffice to say that Aborigines traditionally recognize two main categories: illnesses due to understandable causes, which are normally treated with medicinal plants or other products found in nature, and illnesses due to unknown causes. The latter are generally ascribed to sorcery and require the help and skills of the medicine man or woman, who will apply complex treatments involving magical rituals.[1] This second type of illness is discussed here. On the other hand, plants play a great role in ceremonies and some of special totemic value are highly prized and used in secret or sacred rites. These ritual aspects and the spiritual implications will not be presented in this work.

THE DREAMTIME

To better understand Aboriginal culture, and the special relationship with the land and the environment, we must first understand the meaning of Dreamtime. Dreamtime (an English word for *alchera, alcheringa,* or *tjukurrpa,* in some Aboriginal languages) is the core of Aboriginal culture and gives an explanation to the origin of the universe and to any phenomenon. It refers to the period when the creators shaped the Australian territory and all it contains. During this Mythic Era, the ancestors traveled across the vast and empty continent and, in so doing, shaped the land and created rivers, the sky, plants, animals, and people. They established the law and taught people how to make utensils and to hunt and gather food, before resting in the sky, in caves, in billabongs, and in other secret-sacred places in which they still reside.

Most of these creators are tied to a particular geographical area, as well as to the Aboriginal people living in that area, while the others are common throughout Australia, for example the great and powerful Rainbow Serpent, who is believed to rest in water holes. Thousands of stories and songs (Dreamings) are told, sung, enacted, and danced during ceremonies, so as to re-create and perpetuate the Dreamtime. Dreamings not only tell about the creation of the sun, the moon and all the animals and plants, but also contain codes of behavior and teach daily life. They provide an explanation to facts otherwise mysterious: why the kangaroo stands upright or why there is day and night, for example. Finally, they teach obligations and responsibilities, first and foremost the care of the country.

Myths are taught by the elders to the younger generations at different stages and through different ceremonies. While some are told to everybody (children included), others are secret and known only by initiated men, others only by women. Nature and humankind, land and people, are so deeply interlinked that Aboriginal people used to say—and still say—that they belong to the land, and not the contrary.

PLANTS AND PEOPLE

Nearly all adult Aborigines, but especially women, are competent to treat illnesses of the first type, drawing on an extensive knowledge of the therapeutic properties of materia medica, above all plants. Aborigines generally treat the same disease in more than one way, using different plants, this depending mainly upon plant availability in their territory: the wider the choice, the better the chance of a cure. But some plants are used for almost all complaints, for example, *Eremophila* spp. These plants are so much valued that in the past they were dried and stored for future applications, something fairly unusual among seminomadic people.

As noted by Peter Latz, many Aboriginal people in Central Australia still take advantage of traditional plant remedies, and at least 70 plant species are or were used for medicinal purposes.[2] However, less than 10 percent are taken internally, due to two main factors. First, it is extremely difficult to know the precise quantity to be administered, above all when the active principles are not known and they can vary from plant to plant. Secondly, seminomadic people did not have all the necessary facilities handy and, before the introduction of the billycan, boiling water was not so easy and practical. Aborigines used to put very hot stones in water or, more frequently, they left crushed parts of the herbs to soak in water during the night. All this explains why most of the plants were and still are applied externally either as a wash, as an ointment, or as a smoke medicine, for which approximation of doses or mistakes in their administration do not alter the final result.

The Aborigines of central Australia faced droughts, which considerably reduced the range of plants. For them, the knowledge of location of permanent water holes, edible plants, and other resources was critical. The Aborigines living in the northern tropical areas have always experienced a different situation, since the lush vegetation, especially in the tropical rain forest, provides abundant resources. Here, Aborigines can rely on the many species of plants, including orchids, along with a variety of fauna and abundant resources from the sea, as excellent sources for food and medicine.

We have mentioned smoke medicine, a particular method of administration widespread all over Aboriginal Australia. It consists in fumigating the person to be cured: leaves or bark of a plant are burned and the patient lies close to the fire to be able to inhale or to be enveloped by the smoke. This kind of administration is used for several complaints: cough, diarrhea, and fever, for example, and also after a woman gives birth. In this case, mother and child are both "smoked" (as they say) by burning the leaves of *Eremophila longifolia,* which purifies them and gives strength and vigor to the infant. This method also stops the mother's bleeding and increases her milk supply.[3]

According to ethnobotanist Dulcie Levitt, the Aborigines of Groote Eylandt (in the Gulf of Carpentaria, in Northern Territory) are well aware that flies can be dangerous in case of wounds, and also that smoke can drive them away. She gave a detailed description on how the Aborigines proceeded when visiting an injured friend:[4]

> They lit a smoky fire with the smoke blowing toward the injured man, then walked in the smoke as they approached him to make sure that they did not carry flies on their bodies. If they found themselves getting out of range of the smoke they lit another fire, and another, until the last fire was near the patient with the smoke blowing over him. This kept flies away while the wound was being treated. (p. 67)

Sarcostemma australe is employed among the Pitjantjatjara of the Central Desert for smoke treatment: the sufferer lies over the smoke produced to alleviate pain.[5] They also use the common hopbush *(Dodonaea viscosa)*: the patient squats over the smoke, lying on one side then the other for relief of the pain. Finally, initiation rites always include smoking the boys, who look forward to it, since this signifies the end of a particularly grueling stage of the ceremony, which includes circumcision and, in certain areas, the extremely painful subincision.[6] Among the preferred plants is *Flagellaria indica,* while the leaves of *Syzygium suborbiculare* are heated and placed over the wound to prevent swelling.[7]

TREATMENT OF COMMON AILMENTS

Intestinal and Digestive Diseases

Diarrhea and dysentery are very common diseases among the Aborigines, and several treatments are used. For example, the Aborigines of Queensland generally administer the juice of *Grewia retusifolia* or parts of an orchid, *Cymbidium canaliculatum,* valued for its astringent properties.[8] In Victoria, Aborigines make a paste by mixing eucalyptus gum and water,[9] while those of Groote Eylandt use the rather bitter interior of *Pandanus spiralis*, which they eat raw.[10] Warlpiri people of the Central Desert smoke babies afflicted by diarrhea, burning the leaves of several species of *Acacia*.[11] The Wardaman people around Katherine (Northern Territory) eat a mixture made of the fruit of *Grewia retusifolia* and the fruit of *Terminalia platyphylla* to treat digestive disorders and diarrhea.[12] Constipation sufferers, conversely, are given the fruits of *Tamarindus indica*, whose laxative properties are widely known.[13]

Toothache

Toothache is treated with mouthwashes and dental pastes. For mouthwashes, infusions are obtained from various plants, depending, once again, upon geographical area and availability. Aborigines of New South Wales and Queensland utilize the bark and wood of *Alphitonia excelsa*,[14] while those of the Northern Territory prefer *Buchanania obovata*, even though the liquid obtained from its bark must be administered with great care and must not be swallowed due to its toxicity.[15] The gum of *Eucalyptus dichromophloia* and *E. opaca*, mixed with water, is also used in the Northern Territory.[16] Dental pastes, applied to the aching tooth may be made from the bark, roots, gum, or fruit of different plants, such as *Denhamia obscura*, *Pemphis acidula*, and *Dodonaea viscosa* in Northern Territory, *Eucalyptus* in the state of Victoria, *Flagellaria indica* in Arnhem Land, and *Petalostigma pubescens* in Northern Territory, Queensland, and New South Wales.[17]

Eye Complaints

As an eyewash for sore eyes and to prevent eye infections, the Aborigines prepare a decoction by steeping in water or in mother's milk (much valued for its gentler action) the woody stems of *Grewia retusifolia*, *Sterculia quadrifida*, *Persoonia falcata*, or *Brachychiton diversifolius*.[18] The resulting liquid is filtered and applied as an eyewash. The juice of *Scaevola taccada* (among Tiwi Islanders) or of *Spinifex longifolius* (in the Northern Territory) is used as an antiseptic directly instilled into the eyes, as eyedrops.[19] The juice of *Spinifex* proves effective also to treat infected ears.[20] The Aborigines of the Central Desert prepare an infusion from the bark of the bloodwood *(Eucalyptus opaca)* or of the ghost gum *(Eucalyptus papuana)*.[21] In the latter case, the gum is mixed with water and applied. This is also a strong and well-known antiseptic among Aborigines all over Australia; they apply it on cuts and wounds for a more rapid healing.[22]

Bites and Stings

Snakebites and insect bites, along with other similar minor wounds, are common among peoples who live in close contact with nature and are exposed to the hazards associated with hunting and gathering.

The Yolngu of Arnhem Land (Northern Territory) deal with snakebites in a very peculiar way. As they say, "Never run away after a snakebite!"[23] For them, indeed, it is important to capture the snake, which will be released only if the victim recovers. Otherwise it will be killed. Scorpion stings, as well as cup moth caterpillar stings, are treated by immediately crushing the animal on the wound. Groote Eylanders treat snakebites with the leaves of *Ipomoea pes-caprae,* applying them to the affected part, after cutting open the bite.[24] The Aborigines of Queensland treat the pain caused by the sting of the deadly jellyfish with *Crinum pedunculatum*.[25] The crushed leaves are applied to the affected area and the bulb is used as a powerful antiseptic. *Alocasia macrorrhizos* and *Capparis lasiantha* prove to be a strong antidote for the same complaint, as well as relieving insect bites of all kinds.[26]

Respiratory Ailments

In Central Australia, *Pterocaulon serrulatum* and *Stemodia viscosa* are used to treat colds, the former being preferred.[27] Its leaves are inserted in the nasal septum or, after being mixed with fat, are used as a rubbing ointment. For the same illness, the Arrernte people of the Central Desert particularly value *Senna planitiicola*.[28] They crush its leaves and immerse them in hot water; the steam is inhaled. In Northern Territory and Queensland, the leaves of *Melaleuca leucadendron* are inhaled to relieve nasal congestion and reduce cough.[29] The Aborigines of Arnhem Land, too, use *Melaleuca* (at least four species: *M. acacioides, M. cajuputi, M. dealbata,* and *M. viridiflora*) to treat chest congestion and general respiratory ailments.[30]

Finally, *Cymbopogon ambiguus* is an important medicinal plant whose distribution is always well known by local people. The scent from the torn leaves is inhaled to ease chest congestion. Central Desert Aborigines soak the crushed leaves in water and drink the juice. This is one of the few Australian plants occasionally taken internally.[31] The plant is also much prized to treat scabies and sores.[32]

Wounds, Cuts, and Burns

The *Ficus macrophylla,* a giant rain forest tree diffused on the New South Wales and Queensland coasts, produces a milky latex used by the Aborigines of these areas to treat minor wounds.[33] The Aborigines of Yirrkala, in Northern Territory, favor the leaves and bark of *Grewia retusifolia* for the same problem.[34] The corky bark of *Hakea divaricata,* a corkwood tree, is burned to ash and applied to sores resulting from burns by the Anangu people of South Australia.[35] Sometimes it is also applied to the mothers' nipples during breast-feeding.[36]

WOMEN'S MEDICINE

Women's medicine is so vast that it deserves a book on its own. To stimulate milk flow, women of Groote Eylandt apply the heated leaves of the *Buchanania obovata* (already mentioned as a treatment for toothache) to the breast, while women in New South Wales use the leaves of *Pittosporum phylliraeoides.*[37] The seeds of *Leichhardtia australis* are ground and pounded by Aboriginal women in the Geraldton area (Western Australia) and used as an oral contraceptive.[38]

Generally, products of this kind have drastic effects, causing permanent rather than temporal sterility. For example, women who do not wish to have any more children drink the liquid obtained by steeping the peelings of the fruit of *Petalostigma pubescens,* following a ritual. They must drink it in a small coolamon (a wooden container) with their eyes shut, bury the coolamon immediately after drinking it, and do all of this without telling anyone about it to ensure a good result. They will feel very sick for a certain time, but they never conceive again.[39] For the same purpose, the bark of *Denhamia obscura,* mentioned for use as a dental paste, and that of *Erythrophleum chlorostachys* are also useful.[40] In both cases, administration is by fumigation: the women stand facing into the wind, enveloped by the cloud of smoke from the burning bark. This practice is also believed to stop lactation.[41] Finally, *Mentha australis* and *Scleria polycarpa* are both credited with abortifacient properties.[42]

TWO REMARKABLE PLANTS

All of the above are specific remedies that are delimited to particular geographical areas. I mentioned at the beginning that the ability to rely on plants that are more or less always obtainable and suitable for more than one disease is of great help, when the need for a remedy occurs. *Acacia* spp. and *Eremophila* spp. are both extremely prized for their medicinal properties throughout Aboriginal Australia, and they are very popular because of their widespread availability and their great range of uses.

Acacia spp. (more than 700 species in Australia) is one of the most remarkable plants in Aboriginal life. Its bark, leaves, and roots are employed to treat complaints as different as diarrhea, laryngitis, venereal diseases, cuts and abrasions, skin irritation, and even the pain of childbirth, to name but a few.[43] *Acacia dictyophleba, A. kempeana, A. lysiphloia,* and *A. salicina* are particularly valuable as smoke medicines for the mother and newborn baby. The bark of several *Acacia* species is also used for making bandages.[44]

Acacia aneura (mulga) is a major source of food and medicine, especially for the Aborigines of Northern Territory. They benefit from its healing properties in various ways. For colds, leaflets and twigs are picked and immediately boiled in water. The brown liquid is applied as a wash several times during the day. Headaches caused by colds can be alleviated by heating young leaves and twigs on hot ashes or hot stones, and placing them over the aching area. As a smoke medicine for postnatal care, mulga leaves and small pieces of termite mound are layered over hot coals; the mother and her child recline on and sleep over this on a layer of branches in order to deeply inhale the steam.[45] Several wattles are also widely used to prepare decoctions to treat venereal diseases, sores, and scabies; to soothe aching joints; and to relieve congestion.[46] Finally, wood from various *Acacia* species is particularly valued to make boomerangs, spears, axe handles, digging and music sticks, shields, and many other implements.[47]

As for *Eremophila* spp., *E. alternifolia* is a much-valued medicinal plant, especially in the dry interior, and is employed for general sickness.[48] Its leaves are cut, reduced to a paste, and used as a rubbing medicine for the head, or mixed with grass and secured to the head. The patient will sleep all night with this poultice firmly tied to him or her, waking up in the morning without any pain. Generally speaking, all the *Eremophila* species have medicinal properties as a wash for scabies or to smoke babies. According to Peter Latz, *E. longifolia* (commonly known as emu bush) can be considered as "the most sacred, mystical or magical of the Central Australian plants."[49] It is used as smoke medicine, eye wash (preparing a decoction from the crushed leaves), or body wash to treat skin ailments and headaches. For its purifying properties, it also plays an important role in such ceremonies as the initiation rites.

OTHER USES OF MEDICINAL PLANTS

Aborigines use herbs not only as food and medicine but also for pleasure. The Aborigines of Queensland use the leaves of *Acacia salicina* as a mild psychotropic drug: they burn its leaves and smoke the ash to obtain a state of inebriation.[50] For a stronger effect, which often results in a prolonged sleep, they mix and chew this ash with leaves of the prized *Duboisia hopwoodii* (better known as pituri), which contains powerful alkaloids.[51] The Aborigines of Central and Western Desert areas use their local *Duboisia* only to hunt emus and other game, because this species is much more toxic than the one native to Queensland.[52] It contains, in fact, nor-nicotine, an alkaloid four times stronger than nicotine. The poisonous leaves are dried, crushed, and placed in water holes to stupefy emus; once drugged in this way, the animal is easily caught. Aborigines never touch or handle this plant near children and, whenever they poison a water hole with *Duboisia,* they leave a branch of the plant to make possible visitors aware of the danger. With the same purpose they also use *Prostanthera striatiflora*.[53]

Though employed by white Australians in a generic way, without any distinction between the *Nicotiana* or *Duboisia* species, the term *pituri* is a Queensland word (from the upper Mulligan River area). The more common words among Central Desert people to refer to their own pituri are mingulpa or ingulpa, meaning *Nicotiana* species (especially *N. excelsior* and *N. gossei*).[54] This wild tobacco is a much-valued and sought-after plant: dried leaves, powdered and mixed with the ashes of such plants as *Eucalyptus camaldulensis* (bark) or *Grevillea stenobotrya* (leaves), are chewed for their narcotic effect. The quid is normally chewed for a certain time, then it is passed around the group until it returns to the original owner, who normally places it behind the ear till further use. In the past, pituri had enormous value to the Aborigines and it was the main good in trade and exchange market in central Australia. As can be expected, it played a leading role in initiation rites.

Plants are also effective to drive mosquitos away (for example, *Eucalyptus citriodora, Pterigeron odorus, Santalum lanceolatum*) and are appreciated for their cosmetic properties (a decoction from the fruit of *Petalostigma pubescens* is applied to make skin shiny).[55]

Plants can be of help as weight-loss medicine: Tiwi Islanders poke the young and sharp points of blady grass (*Mnesithea rottboellioides* or *Imperata cylindrica*) into the backs of the knee joints of overweight people to help make them lose weight.[56] Finally, we cannot forget to mention an unusual way of using the plants among the Tiwi people of Bathurst and Melville Islands as a direct reaction to the assimilation policy. Welfare officers used to take children born from mixed parents (called at that time half-castes) away from their Aboriginal families in an attempt to integrate them into the wider Australian society. In order to hide their babies and children, Tiwi mothers rubbed their lighter skin with the very black soot of the burned bark of *Alstonia actinophylla* or *Terminalia grandiflora*.[57]

NONPLANT MATERIA MEDICA

Aboriginal people utilize also other natural elements as medicines. Among Groote Eylanders, white clay is widely consumed as a food remedy for diarrhea, even though it may cause constipation when taken in excessive quantities.[58] Tiwi people eat the cooked flesh of the black and little-red flying foxes, as well as wakatapa, the small mangrove worm, to treat asthma and chest congestion caused by cold and flu.[59] For the same sickness, Wardaman people eat the green-yellow abdomen of ganbowun, the green ant *(Oecophylla smaragdina)*.[60]

Other natural resources were or are still used alone or more often as excipients to amalgamate the drugs: mother's milk, human blood, human and animal urine, ash, mud, termite mound, ocher, and fat. The antiseptic properties of human urine are well known and are mainly exploited for treating cuts and abrasions to the feet and hands.[61] Urine is splashed directly onto the injured area, sometimes in conjunction with the flower of *Cycas media* in the Northern Territory. This practice is particularly common as a treatment for stab wounds. Tiwi Islanders use urine to treat snakebites: they urinate on the affected part, then cut it open and tie off above the cut with a string made from *Pandanus* leaves.[62] The Aborigines of Kimberley (northwestern Australia) use crocodile urine to treat sores, laceration, and bites. Central Desert Aborigines utilize the fat of goanna (iguana or *Varanus*), which is added to parts of plants to make ointments. They also appreciate goanna oil to relieve muscle and joint pain.

CONCLUSION

Despite the fact that Aborigines have adopted Western medicines in their pharmacopoeia and go to the hospital when necessary, they still regard traditional medicine with great respect. Efficacy of bush medicine does not result only from the chemical components but also, if not above all, from the strong connection plants have with the land. The simple consumption of bush food already helps maintain a good spiritual and physical well-being. Common afflictions such as cold, rheumatism, headaches, muscular pains, skin diseases, and general infections all find cures from the bush or can at least be relieved by natural means. Also, the alimentary and medical qualities of the natural resources employed by Aborigines are enhanced by specific and complex rituals aimed at securing the reproduction and increase of plants and game as well.

Even today Aborigines are reluctant to cultivate and produce bush medicine, justifying their unwillingness with the fact that only spontaneous plants and herbs found in nature bear therapeutic properties. Furthermore, they firmly believe that only plants growing on their tribal land—to which both plants and people belong—should be used. Their local flora are the only plants credited with healing principles.[63]

By the time missionaries, doctors, anthropologists, and botanists took some real interest in systematically studying the Aboriginal plant lore, much of the traditional knowledge had been lost.[64] This was especially the case in the areas of New South Wales, Victoria, South Australia, and Queensland, where colonies were first established and cattle and pastoral stations soon became operative, forcing Aboriginal people to move further into the outback.[65] On the contrary, indigenous people of central Australia, Top End, Kimberley, and Cape York—all areas where contact with whites took place later—better kept their traditional knowledge alive and passed it orally through the generations. This is all the more true for the Aborigines living nowadays in remote outstations (in fact, their homeland) where they again follow ancestral therapeutic methods. Elders instruct younger people about plant uses as in the past and, wishing to preserve and share their wisdom with the wider community, they collaborate with ethnobotanists, anthropologists, linguists, and other scientists to collect and record in written format all the oral tradition. However, they have decided not to reveal specific uses of plants in initiation rites and other secret and sacred ceremonies.[66]

APPENDIX: LIST OF PLANTS CITED

Acacia aneura F. Muell. ex Benth., mulga
Acacia dictyophleba F. Muell., desert wattle
Acacia farnesiana (L.) Willd., mimosa bush, mimosa wattle
Acacia kempeana F. Muell., witchetty bush
Acacia lysiphloia F. Muell., turpentine wattle
Acacia salicina Lindl., cooba, willow wattle
Alocasia macrorrhizos (L.) G. Don, cunjevoi
Alphitonia excelsa (Fenzl) Benth., red ash, soap tree
Alstonia actinophylla (A. Cunn.) K. Schum., milkwood
Brachychiton diversifolius R. Br., kurrajong

Buchanania obovata Engl., green plum
Capparis lasiantha R. Br. ex DC., balgarda, split jack
Crinum pedunculatum R. Br., swamp lily
Cycas media R. Br., cycad, zamia
Cymbidium canaliculatum R. Br., native arrowroot, tree orchid
Cymbopogon ambiguus (Steud.)A. Camus, lemon-scented grass
Denhamia obscura (A. Rich.) Meissn. ex Walp., emu tucker
Dodonaea viscosa Jacq., hopbush
Duboisia hopwoodii (F. Muell.) F. Muell., pituri
Eremophila alternifolia R. Br., fuchsia bush, native honeysuckle
Eremophila longifolia (R. Br.) F. Muell., emu bush
Erythrophleum chlorostachys (F. Muell.) Baillon, ironwood
Eucalyptus camaldulensis Dehnh., river red gum
Eucalyptus citriodora Hook, lemon-scented gum
Eucalyptus dichromophloia F. Muell., bloodwood
Eucalyptus opaca D. Carr & S.G.M. Carr, desert bloodwood
Eucalyptus papuana F. Muell., ghost gum
Ficus macrophylla Pers., Moreton Bay fig
Flagellaria indica L., supplejack, gadji
Grevillea stenobotrya F. Muell., sandhill grevillea
Grewia retusifolia Kurz, emu berry
Hakea divaricata L. Johnson, fork-leafed corkwood
Imperata cylindrica (L.) P. Beauv., blady grass
Ipomoea pes-caprae (L.) R. Br., beach morning glory, goatfoot
Leichhardtia australis R. Br., native pear
Melaleuca acacioides F. Muell., small-leaved paperbark
Melaleuca cajuputi Powell, paperbark, cajeput
Melaleuca dealbata S.T. Blake, smelly paperbark, soapy tea-tree
Melaleuca leucadendron (L.) L., weeping paperbark
Melaleuca viridiflora Sol. ex Gaertner, forest tea-tree
Mentha australis R. Br., river mint
Mnesithea rottboellioides (R. Br.) de Koning & Sosef, blady grass
Nicotiana excelsior (J. Black) J. Black, wild tobacco, mingulpa
Nicotiana gossei Domin, wild tobacco, mingulpa
Pandanus spiralis R. Br., screw palm
Pemphis acidula J.R. Forster & G. Forster, pemphis
Persoonia falcata R. Br., milky plum, geebung
Petalostigma pubescens Domin, quinine bush, emu tucker
Pittosporum phylliraeoides DC., native willow, butterbush
Prostanthera striatiflora F. Muell., striped mint bush
Pterigeron odorus (F. Muell.) Benth., smelly bush
Pterocaulon serrulatum (Montr.) Guillaumin, apple bush, ragwort
Santalum lanceolatum R. Br., quandong, sandalwood, plumbush
Sarcostemma australe R. Br., caustic vine
Scaevola taccada (Gaertn.) Roxb., pipe bush, lettuce tree
Scleria polycarpa Boeck., nutrush
Senna planitiicola (Domin) Randell, arsenic bush, yellow pea
Spinifex longifolius R. Br., spinifex

Stemodia viscosa Roxb., sticky blue rod, jirrpirinypa
Sterculia quadrifida R. Br., peanut tree
Syzygium suborbiculare (Benth.) T.G. Hartley & L.M. Perry, red bush apple
Tamarindus indica L., tamarind
Terminalia grandiflora Benth., nut tree
Terminalia platyphylla F. Muell., wild plum

NOTES

1. Elkin, A.P., *Aboriginal men of high degree* (St. Lucia, Queensland: University of Queensland Press, 1977); Cawte, J., *Healers of Arnhem Land* (Sydney: University of NSW Press, 1996).

2. Latz, P., *Bushfires and bushtucker: Aboriginal plant use in Central Australia* (Alice Springs, NT: IAD Press, 1995).

3. Ibid.

4. Levitt, D., *Plants and people: Aboriginal uses of plants on Groote Eylandt* (Canberra: Australian Institute of Aboriginal Studies, 1981).

5. Latz, 1995. *Bushfires and bushtucker.*

6. Ibid.

7. Levitt, 1981, *Plants and people.*

8. Cribb, A.B., and Cribb, J.W., *Wild medicine in Australia* (Sydney: Fontana/Collins, 1981).

9. Isaacs, J., *Bush food: Aboriginal food and herbal medicine* (Sydney: Lansdowne Publishing Pty Ltd., 1987).

10. Levitt, 1981, *Plants and people.*

11. Latz, 1995. *Bushfires and bushtucker.*

12. Raymond, E., Blutja, J., Gingina, L., Raymond, M., Raymond, O., Raymond, L., Brown, J., Morgan, Q., Jackson, D., Smith, N., et al., Wardarman ethnobiology: Aboriginal plant and animal knowledge from the Flora River and south-west Katherine region, north Australia, *Northern Territory Botanical Bulletin,* No. 25 (Darwin: Parks and Wildlife Commission of the Northern Territory, 1999).

13. Isaacs, 1987, *Aboriginal food and herbal medicine.*

14. Ibid.

15. Ibid; Barr, A. and Aboriginal communities of the Northern Territory of Australia, *Traditional bush medicines: An Aboriginal pharmacopoeia* (Richmond, Victoria: Greenhouse Publ., 1988); Wightman, G. and Smith, N.M., Ethnobotany, vegetation and floristics of Milingimbi, northern Australia, *Northern Territory Botanical Bulletin,* No. 6 (Darwin: Conservation Commission of the Northern Territory, 1989).

16. Isaacs, 1987, *Aboriginal food and herbal medicine.*

17. Ibid.

18. Levitt, 1981, *Plants and people;* Isaacs, 1987, *Aboriginal food and herbal medicine.*

19. Puruntatameri, J., Puruntatameri, R., Puruntatameri, A., Burak, L., Tipuamantymirri, Tipakalippa, M., Puruntatameri, J., Puruntatameri, P., Pupangamirri, J.B., Kerinaiua, R., et al., Tiwi plants and animals: Aboriginal flora and fauna knowledge from Bathurst and Melville Islands, northern Australia, *Northern Terri-*

tory Botanical Bulletin, No. 24 (Darwin: Parks and Wildlife Commission of the Northern Territory, 2001); Levitt, 1981, *Plants and people;* Isaacs, 1987, *Aboriginal food and herbal medicine.*

20. Levitt, 1981, *Plants and people.*

21. Latz, 1995, *Bushfires and bushtucker.*

22. Lindsay, B.Y., Waliwararra, K., Miljat, F., Kuwarda, H., Pirak, R., Muyung, A., Pambany, E., Marruridgi, J., Marrfurra, P., and Wightman, G., Malak Malak and Matngala plants and animals: Aboriginal flora and fauna knowledge from the Daly River area, northern Australia, *Northern Territory Botanical Bulletin,* No. 26 (Darwin: Parks and Wildlife Commission of the Northern Territory, 2001); Wightman, G., Dixon, D., Williams, L., and Dalywaters, I., Mudburra ethnobotany: Aboriginal plant use from Kulumindini (Elliott), *Northern Territory Botanical Bulletin,* No. 14 (Darwin: Conservation Commission of the Northern Territory, 1992).

23. Cawte, 1996, *Healers of Arnherm.*

24. Levitt, 1981, *Plants and people.*

25. Cribb and Cribb, 1981, *Wild medicine in Australia.*

26. Ibid.

27. Lats, 1995, *Bushfires and bushtucker.*

28. Ibid.

29. Smith, N.M., and Wightman, G.M., Ethnobotanical notes from Belyuen, Northern Territory, Australia, *Northern Territory Botanical Bulletin,* No. 10 (Darwin: Conservation Commission of the Northern Territory, 1990); Blake, N.M., Wightman, G., and Williams, L., Iwaidjia ethnobotany: Aboriginal plant knowledge from Gurig National Park, Northern Australia, *Northern Territory Botanical Bulletin,* No. 23 (Darwin: Parks and Wildlife Commission of the Northern Territory, 1998).

30. Wightman, G., Jackson, D., and Williams, L., Alawa ethnobotany: Aboriginal plant use from Minyerri, northern Australia, *Northern Territory Botanical Bulletin,* No. 11 (Darwin: Conservation Commission of the Northern Territory, 1991); Wightman, G., Roberts, J.G., and Williams, L., Mangarrayi ethnobotany: Aboriginal plant use from the Elsey area, northern Australia, *Northern Territory Botanical Bulletin,* No. 15 (Darwin: Conservation Commission of the Northern Territory, 1992); Yunupingu, B., Yunupingu-Marika, L., Marika, D., Marika, B., Marika, B., Marika, R., and Wightman, G., Rirratjingu ethnobotany: Aboriginal plant use from Yirrkala, Arnhem Land, Australia, *Northern Territory Botanical Bulletin,* No. 21 (Darwin: Parks and Wildlife Commission of the Northern Territory, 1995); Blake, Wightman, and Williams, 1998, Iwaidjia ethnobotany.

31. Latz, 1995, *Bushfires and bushtucker.*

32. Barr and Aboriginal Communities of the Northern Territory of Australia, 1988, *Traditional bush medicines.*

33. Cribb and Cribb, 1981, *Wild medicine in Australia.*

34. Yunuping et al., 1995, Rirrajingu ethnobotany.

35. Latz, 1995, *Bushfires and bushtucker.*

36. Ibid.

37. Levitt, 1981, *Plants and people;* Cribb and Cribb, 1981, *Wild medicine in Australia;* Isaacs, 1987, *Bush food.*

38. Cribb and Cribb, 1981, *Wild medicine in Australia.*

39. Levitt, 1981, *Plants and people.*

40. Isaacs, 1987, *Bush food.*
41. Ibid.
42. Cribb and Cribb, 1981, *Wild medicine in Australia.*
43. Lassak, E.V., and McCarthy, T., *Australian medicinal plants* (Kew, Victoria: Reed, 1997; first published by Methuen, 1983); Marrfurra, P., Akanburru, M., Wawul, M., Kumunerrin, T., Adya, H., Kamarrama, K., Kanintyanyu, M., Waya, T., Kannyi, M., Wightman, G., et al., Ngan'gikurunggurr and Ngan'giwumirri ethnobotany: Aboriginal plant use from the Daly River area, northern Australia, *Northern Territory Botanical Bulletin,* No. 22 (Darwin: Conservation Commission of the Northern Territory, 1995); Wightman, G., Kalabidi, G.J., Dodd, T.N.N., Frith, R.N.D., Jiwijwij, M.N., Oscar, J.N.N., Wave Hill, R.J.W., Holt, S., Limbunya, J.J., and Wadrill V.N., Gurindji ethnobotany: Aboriginal plant use from Daguragu, northern Australia, *Northern Territory Botanical Bulletin,* No. 18 (Darwin: Conservation Commission of the Northern Territory, 1994); Wightman et al., 1992, Mudburra ethnobotany.
44. Latz, 1995, *Bushfires and bushtucker.*
45. Barr and Aboriginal Communities of the Northern Territory of Australia, 1988, *Traditional bush medicines.*
46. Lassak and McCarthy, 1983, *Australian medicinal plants.*
47. Latz, 1995, *Bushfires and bushtucker.*
48. Ibid.
49. Ibid.
50. Isaacs, 1987, *Bush food.*
51. Peterson, N. (1979). Aboriginal uses of Australian Solanaceae, in J.G. Hawkes, R.N. Lester, and A.D. Skelding (eds.), *The biology and taxonomy of the Solanaceae* (London: Academic Press, 1979), pp. 171-188; Watson, P., *This precious foliage: A study of the Aboriginal psycho-active drug pituri* (Sydney: University of Sydney, 1983), Oceania monograph no. 26; Webb, L.J., 1969, The use of plant medicines and poisons by Australian Aborigines, *Mankind* 7(2): 137-146; Cribb and Cribb, 1981, *Wild medicine in Australia;* Isaacs, 1987, *Bush food.*
52. Latz, 1995, *Bushfires and bushtucker.*
53. Ibid.
54. Ibid.
55. Cribb and Cribb, 1981, *Wild medicine in Australia;* Isaacs, 1987, *Bush food.*
56. Puruntatameri et al., 2001, Tiwi plants and animals.
57. Ibid.
58. Levitt, 1981, *Plants and people;* Barr and Aboriginal communities of the Northern Territory of Australia, 1988, *Traditional bush medicines.*
59. Puruntatameri et al., 2001, Tiwi plants and animals.
60. Raymond et al., 1999, Wardarman ethnobiology.
61. Levitt, 1981, *Plants and people.*
62. Puruntatameri et al., 2001, Tiwi plants and animals.
63. Smith, N., Wididburu, B., Harrington, R.N., and Wightman, G., Ngarinyman ethnobotany: Aboriginal plant use from the Victoria River area, northern Australia, *Northern Territory Botanical Bulletin,* No. 16 (Darwin: Conservation Commission of the Northern Territory, 1993).
64. Appetiti, E., Barlow A., and Hill, M., 2000, Ethnobotanical research in Australia: A history—Proceedings of the 1st International Symposium on Ethnobotany

and Ethnomedicine, Costa Rica, September 14-18, 1999 (CD-ROM, San Jose, Costa Rica).

65. Appetiti, E., 1999, Black and white Australia, 1770-1970: A history of dispossession, *Revista de Indias,* LIX(217): 837-856.

66. Raymond et al., 1999, Wardarman ethnobiology; Barr and Aboriginal communities of the Northern Territory of Australia, 1988, *Traditional bush medicines;* Wightman and Smith, 1989, Ethnobotany; Puruntatameri et al., 2001, Tiwi plants and animals; Lindsay et al., 2001, Malak malak; Wightman et al., 1992, Ludburra ethnobotany; Smith and Wightman, 1990, Ethnobotanical notes from Belyuen; Blake, Wightman, and Williams, 1998, Iwaidjia ehthnobotany; Wightman, Jackson, and Williams, 1991, Alawa ethnobotany; Wightman, Roberts, and Williams, 1992, Mangarray ethnobotany; Yunupingu et al., 1995, Rirratjingu ethnobotany; Marfurra et al., 1995, Ngan'gikurunggurr; Wightman et al., 1994, Gurindji ethnobotany; Smith et al., 1993, Ngarinyman ethnobotany.

PART III: TECHNOLOGIES IN MEDICINAL PLANT RESEARCH

Chapter 10

Production and Breeding of Medicinal Plants

Éva Németh

SHORT HISTORY OF MEDICINAL PLANT PRODUCTION

Production of medicinal plants is a complex concept, compared to other horticultural crops. For medicinal and aromatic plants, production includes both the collection in the wild and the cultivation in agrarian systems. The majority of medicinal plant species—70 to 90 percent—even today are wild harvested.[1]

Nevertheless, cultivation also dates back to ancient ages. In developing countries of Africa, Asia, and South America, where medicinal plants have long been and still are used, they provided the only tool of curing, disinfecting, and aromatizing. In those parts of the world, they had been cultivated in small gardens for centuries as well as harvested in the wild. Chinese medicine would not exist without medicinal plants, many of which have been cultivated since ancient times.

In Europe and surrounding areas, cultivation of these species has also been documented. In addition to cereals and vegetables, medicinal plants were cultivated too in the fertile soils near the Nile. Remains of *Aloe* and *Sinapis* species, *Artemisia absinthium,* and *Cannabis indica* have been found in ancient tombs.[2] In the fourth and fifth century B.C., specific cultures of *Inula helenium, Pimpinella anisum, Cnicus benedictus, Coriandrum sativum,* and *Thymus serpillum* have been mentioned.[3] One of the oldest cultivated medicinal plants of the world seems to be poppy *(Papaver somniferum),* according to written papers dating from 2700 B.C. Cultivation of mints has a tradition of about 3,000 years as well. A wide variety of medicinal plants is also mentioned in the Bible, many of them cultivated in gardens.

During the ancient Greek and Roman periods, a wide range of medicinal plants was used. Aromatic plants were essential in foods as condiments and preservatives. Some of the curing species and the majority of the spices were obtained by regular cultivation.

In the early Middle Ages, gardens of monasteries became popular places for cultivation where rich collections of aromatic and medicinal species could be found. Via the Christian churches, these plants spread north from Italy. Even today, the majority (62 percent) of the cultivated species in Europe originates from Mediterranean areas.[4] Some well-known examples are *Lavandula angustifolia, Salvia officinalis, Foeniculum vulgare,* and *Coriandrum sativum.*

Later, gardens of pharmacies became new cultivation and production sites of medicinal plants. These pharmacies cultivated the most important medicinal species for their own use and processing as well as organized the collection of wild-growing ones.[5] According to Heeger, regular teaching of medicinal plant production started in the fifteenth century.[6] The sixteenth and seventeenth centuries can be characterized as widening the scientific background of pharmacognosy, the development of pharmacobotany, and the establishment of medical-botanical gardens which contributed not only to the more efficient utilization but also to the regular production of medicinal plants. Cultivation of fennel, lemon balm, marshmallow, marjoram, and others was already practiced in Germany at that time.[7] However, we know very little about the specific methods of cultivation of any single species. Existing descriptions deal mainly with their processing, utilization, and therapeutic effects.

The rapid development of chemistry and the synthesis of new drugs in the nineteenth century resulted in a temporary decline in the interest in raw plant production for medical application. The majority of European countries enjoyed a new boom of medicinal plant utilization and production during World War I. An increasing demand for plant drugs occurred due to the lack of other medicines, spices, and teas. The need to establish a home supply of these articles resulted in a considerable development of the medicinal and aromatic plant branch. During these times, several new commercial firms and research institutions dealing with the scientific aspects of medicinal plants were founded. They contributed significantly to the further development of agricultural and processing technologies as well as the perfection of methodology of chemical investigation and analytical control. Some of these institutions were well known and prosperous; some were so successful that they are still in business (e.g., Research Institute of Medicinal Plants in Budakalász, Hungary; Federal Scientific Research Institute of Medicinal Plants: VILAR in the Soviet Union; Bayer Institute for Plant Growing and Protection, Munich, Gemany).

The 1930s were characterized by impressive achievements in medicinal plant production. For example, licensing the procedure of extracting morphine from the dry capsules of poppy (J. Kabbay, unpublished); working out the technology of parasitic production of ergot on rye (N. Bekesi, unpublished), or beginning to breed, describe, and register selected varieties of medicinal plant species (Leipzig, Germany) contributed to the long-term prosperity of this horticultural branch.

The number of cultivated species increased. In larger countries, special districts were selected for growing special groups of plants. In Russia, already in the 1920s *Hyoscyamus niger, Datura stramonium, Althaea officinalis, Valeriana officinalis,* and many other species were regularly processed, altogether up to a yearly quantity of 500 to 600 tons.[8] In Germany, the production area from 1934 to 1940 increased from 820 to 8,300 hectares respectively, with caraway, mustard, marjoram, peppermint, fennel, coriander, and thyme being the main products.[9] In other countries, such as Poland and Hungary, several hundred hectares were regularly occupied by medicinal and aromatic plant cultures: mints, lavender, poppy, caraway, valerian, and dill.[10]

The last and perhaps the longest increase of the cultivation happened in the past 15 to 20 years. In this period, even countries with no medicinal plant production tradition started to cultivate, e.g., Finland and Ireland. The main reasons for this last boom can be suggested as follows:

- There is a new, general increasing interest for products of natural origin. It includes not only medicines but also plant-based cosmetics, natural dyes, food preservatives, and seasonings. Pank predicted the doubling of the market of medicinal and aromatic plants in Europe during the next ten years and a tenfold increase in North America.[11]
- An important trend is occurring in the economy of developed countries to support the local production not only of major agricultural products such as cereals, fodder, and meat but also of specialized minor crops.
- Requirement of high-quality products and quality control becomes more and more essential in agricultural production. Assuring stable quality and the desired documentation necessitates elaboration of standard production technologies and selected plant varieties.
- Since a considerable proportion of medicinal plant species occurs in rare abundance, low densities, or endangered habitats, protection of these taxa or habitants imposes restrictive measurements against their harvesting. Elaboration of cultivation techniques may assure a proper supply of the market.
- For the reduction of regular overproduction of major agricultural crops, stimulation and support of alternative nonfood crops becomes a

usual tool in market regulation. In most countries medicinal plants fit into this category.

- Species, which traditionally have been called medicinal and aromatic plants, enjoy several new areas of utilization, e.g., sources of various extracts, aromas, renewable raw materials, dietary supplements, antioxidants, and veterinary medicines which widen their market possibilities.

THE PRESENT SITUATION AND CHARACTERISTICS OF MEDICINAL PLANT CULTIVATION IN EUROPE

Countries of the European Union have been and remain the biggest market for medicinal plant products, although their own production has considerably increased too.[12] Their import of drugs exceeds exportation (see Figure 10.1). The local production is focused on high-quality or special cultures. Some examples are depicted in Table 10.1. However, it should be remembered that the comparison of the most significant crops of different countries faces a severe difficulty because of the often diverse interpretation of the term *medicinal plant.* While in some countries even parsley, horseradish, or oil pumpkin are included in this category, in other countries the statistics exclude similar items. Cultivation areas are changing from coun-

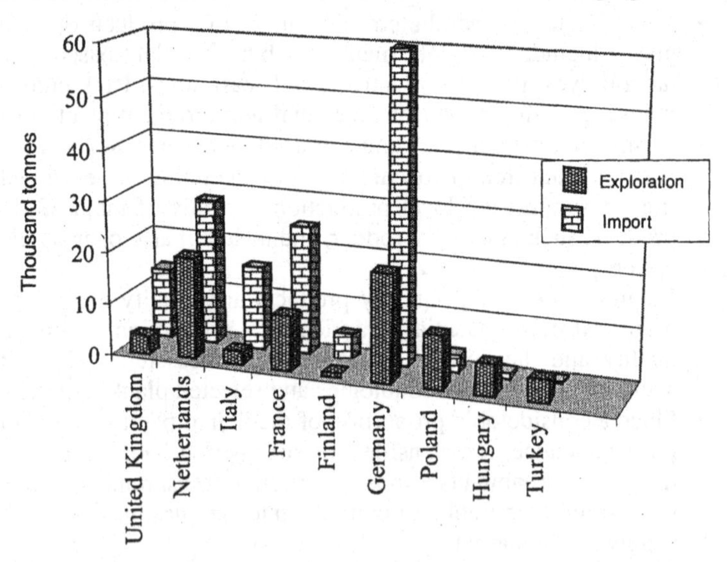

FIGURE 10.1. Import and export of medicinal plant drugs of some European countries.

TABLE 10.1. Characteristic cultures and areas of medicinal plant production in selected European countries.

Country	Area (ha)	Main species
Austria	4,300	Pumpkin, poppy, caraway, milk thistle, peppermint
Belgium	~300	Valerian, angelica, lovage
Bulgaria	~800	Rose, mint
Finland	1,900	Caraway, mustard
France	25,000	Lavender, poppy, muscat sage, fennel, ginkgo, hyssop, plantain
Germany	5,700	Camomile, marjoram, caraway, fennel, lemon balm
Greece	4,000	Crocus, oregano, peppermint
Hungary	35,000	Mustard, poppy, fennel, caraway, marjoram, mints, basil, dill
Italy	2,300	Peppermint, chamomile, iris, orange
Netherlands	2,500	Caraway, digitalis, evening primrose, poppy
Poland	~20,000	Valerian, caraway, milk thistle, chamomile
Romania	17,000	Peppermint, coriander, poppy, fennel
Spain	19,000	Thyme, anis, crocus, lavender, mint, fennel
United Kingdom	2,000	Mustard, evening primrose, comfrey

Source: Pank, F., 1998, Arznei und Gewürzpflanzenmarkt in der EU, *Z. Arzn. Gew. pfl.*, 3(2): 77-81; Schwierz, A., Kupke, J., and Niefind, B., *Arznei-und Gewürzpflanzen in Osteuropa* (Bonn: Zentrale Markt-und Preisberichtstelle GmbH, 2000), p. 95; Lange, D., 2002, Osteuropa-Mehr Exporte als Importe, *Marktbericht Arznei- und Gewürzpflanzen,* 4(2): 8.

try to country but also from year to year. Larger areas are found in France and Spain, while in the other countries cultivated areas remain at less than 5,000 hectares.[13]

In several Middle and East European countries the production of medicinal drug plants has a longer tradition and the cultivated areas reach a multiple of the above mentioned (see Table 10.1). These countries, especially France, Bulgaria, Hungary, and Poland, are the most significant European suppliers of the present EU member states. Higher quantities are imported to Europe from India, Chile, and Argentina.[14] However, the main source of the exported items, besides cultivation, remains harvesting from the wild even today.

Outside Europe, Asia and Africa have the most significant producer countries. In the United States, larger areas are occupied by peppermint and cornmint fields.[15] Production of *Echinacea, Hypericum, Salvia sclarea,* and *Panax ginseng* is also well known. In Argentina, a rather new but rapidly developing sector of medicinal plant production can be observed with a focus on chamomile. India is one of the largest producers—about 3,500 indigenous species are of medicinal value.[16] Although 80 percent of the drugs comes from collection, cultivation of *Rauvolfia, Cassia, Psyllium, Catharanthus,* and *Mentha* species, among others, are of great economic significance. Beside the well-known producer China, other Asian countries have started to establish their own production of medicinal plants, for example, Korea. In the Near East, cultivation in Israel and Egypt are the best known. In the first place, spices and aromatic plants are cultivated for commercial purposes: caraway, chamomile, marjoram, coriander, cumin, sage, and eucalyptus. In Australia, poppy is a main product for the pharmaceutical industry. Recently, the cultivation and processing of tea tree became extremely popular. The technology of cultivation depends basically on the species. However, three main groups of agrotechnologies can be distinguished:

1. *Large-scale technologies:* Crops produced on fields of 10 to 100 hectares. Mechanization of sawing, harvesting, and postharvest processing can be solved by machines and equipment similar to cereal production. Chemical weed control is well developed; cultivation does not need much manpower. These are efficient and widespread technologies for fennel, coriander, poppy, and milk-thistle. Individual differences among cultures exist especially in nutrition, plant protection, method, and time of harvest. However, sawing, spraying, cutting, and postharvest processing may be carried out by machines commonly used in cereal production (Photo 10.1). In some countries, special, intensive, industry-like large-scale technologies have been developed for the production of plant raw material. The cultivation of *Taxus brevifolia* or *Ginkgo biloba* for pharmaceutical industry and the growth of *Salvia sclarea* as raw material sclareol for perfume is proceeding in the United States as fully mechanized and as organized as a factory.[17] These plantations assure the basis for high-quality processed products. Thus, these technologies are product-oriented closed systems, including breeding activity, maintenance of varieties, development of postharvest handling, and optimization of yields of active agents.

2. *Smaller-scale technologies requiring higher amount of labor:* Critical steps in this respect are propagation, weed control, or harvesting. Fields rarely exceed 2 to 5 hectares, although in some regions special situations may appear. To a smaller extent they are also found in gardens. Typical species under this type of cultivation are lovage, lemon balm, marjoram, and valerian (see Photos 10.2 and 10.3).
3. *Special cultivation technologies:* These should be applied in situations in which at least one of the steps can scarcely be carried out by usual equipment (see Photo 10.4). The size of the cultivated area depends on ecological circumstances and market situations. Examples are chamomile, for which well-adapted harvesting equipment seems to be necessary. For peppermint propagation, special planting machines are required. In case of borretsch, PVC foils are laid in the rows for easier and more effective harvest. In production of saffron, propagation is carried out by bulbs, and processing includes cutting out the style; ginseng requires constant shading.

PHOTO 10.1. Cultivation of *Carum carvi* under large-scale conditions.

PHOTO 10.2. Cultivation of *Rosmarinus officinalis.*

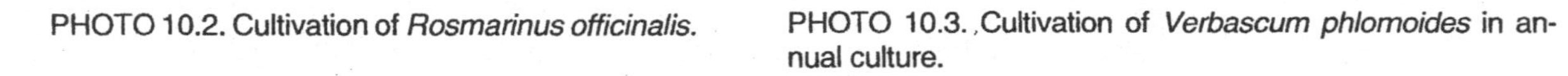

PHOTO 10.3. Cultivation of *Verbascum phlomoides* in annual culture.

PHOTO 10.4. Cultivation of *Borago officinalis.*

Beside these general features, the applied agrotechnology often depends on the purpose of the production. According to this, deviations in cultivation of the same species may happen. Thus, if peppermint is produced for its leaves, the stand is harvested before blossoming. However, if the purpose of the cultivation is essential oil, harvesting at full flowering stage is optimal. Similarly, dill, marketed as fresh spice, should be cut very early after shooting, and cutting can be repeated three to four times. When producing seeds, harvest is made by combine harvester at full ripening.

Modifications in cultivation technologies are also determined by the natural circumstances. Annual caraway is sown in Israel or Egypt in November and can be harvested April to May. Among temperate climatic conditions (e.g., Hungary, Netherlands) it cannot be sown earlier than end of March and harvesting usually takes place in August to September. In case of fennel, the climate basically influences the duration of the culture: in Italy it is a perennial (three to five years) culture, in the north (Germany) economic yields are rarely achieved far more than a single year because of its sensitivity to frost.

According to these experiences, organization and economy of the cultivation of medicinal plants shows some practical characteristics compared with other horticultural crops. Although considerable differences are found among medicinal plant species and growing years, the value of the end product per unit area is usually eight- to tenfold higher than in cereals.[18] To achieve this, producers should consider the following organizational steps and economical aspects.

- In Europe, the area of agricultural firms dedicated primarily to medicinal plants ideally reaches at least 25 to 30 hectares. In order of crop rotation, vegetables should be grown on fallow lands. Profitable production on smaller surfaces is only realized as an activity done in addition to other basic ones.
- In medicinal plant production, a basic principle is to avoid putting all your eggs in one basket. The market continuously faces big fluctuations and sudden changes of demand and prices of different drugs. As drugs are not products of vital importance, overproduction may easily occur. Producers should therefore consider to cultivate three to six different crops simultaneously. In this case a proper income of the firm can be assured even under uncertain market situations. From a practical point of view, the cultivated crops should be species that require similar agrotechnology and equipment.
- Considering the uncertainty of the drug market, producers are advised to procure purchase contracts in advance.[19] Even if the agreed-upon price is low, in low market demand it is the only possibility of obtaining a sufficient income. Contract-based production often has the advantage of providing the consumer with propagation material and technology information or consultation.
- The final step in the cultivation of medicinal and aromatic plants should include drug production and at least some drying. Without drying at proper temperatures and carrying out a minimal level of primary processing (cleaning, cutting, chopping, sifting, etc.) drugs of acceptable quality cannot be produced. At the same time, postharvest processing, especially drying, may increase production costs considerably. Often cooperation among farmers enables purchasing and a better utilization of processing machines and equipment.[20]
- Currently, the lack of knowledge on technology and tools (selecting variety, registering herbicides, using specialized machines) are the major problems in cultivation. In several countries, progress may be observed among those farmers with state support.

- In the 1980s, the requirement and possibility of applying quality control systems attracted considerable attention.[21] At that time, the Working Group of the ISHS Section MAP began to deal with recommendations and guidelines for Good Agricultural Practice (GAP). After a first summary document by Pank,[22] on August 5, 1998, the final European version of the guidelines was finished.[23] It consists of all critical points, including choosing the proper site of cultivation, propagation material, fertilization, irrigation, plant protection, and harvest and postharvest handling till storage (see Box 10.1). The system operates with full documentation and auditing. On this basis, efforts are being made by the World Health Organization (WHO) for regulation and elaboration of new guidelines for the whole production process of drugs, including collection from wild habitats. At the same time, with the WHO Guideline on Good Sourcing Practice (GSP) of medicinal plants, another proposal by the Working Party of Herbal Products of EMEA is also under discussion. The international interest for this topic shows the importance of building up a proper system for quality assurance of medicinal plant drugs.

Today, ecological farming in the production of medicinal and aromatic plants is becoming more and more popular. It seems to satisfy a well-defined, relatively tight demand of a certain groups of the population. While the general interest for natural products may increase the market, the considerably small quantities of consumption—compared to other foodstuff products—decreases the need for ecological products. Among the cultivated species, culinary herbs are the most important ones. The number of cultivated species may reach 50. The most widely grown are basil, marjoram, sage, caraway, fennel, peppermint, chamomile, lemon balm, nettle, coriander, and thyme.[24] Today, the largest territories (500 ha) in ecological cultivation of medicinal plants are registered in Germany, which provides about 1 to 2 percent of the country's total drug production.[25] Regular cultivation is known also in Italy, Finland, Norway, Austria, Switzerland, and to a small extent in each developed country. The principles of the cultivation are equivalent to the general guidelines of ecological farming. Special difficulties may arise due to the abundance of weeds (perennial species or shedding ones), lack of propagation material, and tolerant varieties. Methods such as growing crops in raised beds may be useful for production of roots *(Levisticum officinale, Valeriana officinalis)*; mulch cover for preserving humidity in delicate crops e.g., mints. In plant protection operations medicinal and aromatic plants are applied with preference as protective media. Water extract of horstail + comfrey is used as a principal agent for general strengthening of the plant and stimulation of soil life. Additional species of

BOX 10.1. Main aspects of Guidelines for Good Agricultural Practice (GAP).

General part: The aim of GAP is to ensure that raw plant material meets the demand of the consumer and as such upholds standards of the highest quality. Most important aspects include hygiene, caretaking, and documentation.
Propagation material: Identity and origin are properly identified; quality meets the standards.
Soil and fertilization: Soil is free of pesticide and heavy metal contaminants; fertilizer is at optimal dosages.
Irrigation: Water free is of contaminants; irrigation is kept to the minimal necessary level.
Plant protection: Pesticides and herbicides are minimized to the necessary level; staff is trained; applications are documented.
Harvesting: Under the best conditions of weather and plant stand, equipment is clean and avoids damage of plants and mixing with strange plants or contaminants.
Primary processing: Building and equipment is clean and well serviced; drying method crop-specific; avoiding direct sunlight and drying on the ground; personnel is trained.
Packaging: Clean and dry, nontoxic packaging is used.
Storage and transport: It is in dry, well-aired place, with protection against pests.
Equipment: The equipment is regularly cleaned and serviced, preferably nonwooden.
Personnel and facilities: Conform with EU guidelines on Food Hygiene of Codex Alimentarius and GMP; personnel are healthy, clean, and protected from toxic and allergenic materials.
Documentation: Labeling, batch documentation of all processes and procedures which could bear impact on the quality of the product, audits documented in the audit report, agreement between producer and buyer.

Source: Franz, C., 1999, Quality assurance systems and total quality management of medicinal plant production, *ICMAP News*, 3(6): 4-9.

ecological farming widely used are *Plantago* spp., *Allium* spp., *Betula pendula*. Species containing essential oils (sage, tansy, thyme, yarrow) are reportedly effective primarily against insects.

A special type of cultivation of medicinal plant species is the so-called quasicultivation. This means is a method of sustainable utilization of natural populations by intervention in propagation and harvesting. The natural population is enriched by seed sowing or vegetative propagation, and the harvest is controlled and scheduled to avoid exhausting of the stand. No further treatments are applied. It has a special significance in case of those species which are difficult to cultivate via classical methods because of physiological, ecological, or economical backgrounds. Recent efforts and results for a sustainable cultivation of *Arnica montana* at original alpine sites, of *Taxus wallichiana* in the Himalayas in India, or of *Harpagophytum procumbens* in the desert area of Namibia.[26] In the future, the role of this type of culture may increase. It represents an acceptable area of ecological production.

THE ROLE AND SIGNIFICANCE OF HARVESTING FROM WILD HABITATS

The importance of natural habitats in providing raw material for plant-based medicines and drugs cannot be underestimated. As for the world market, the majority of drugs still come from the wild. However, in individual countries the mode of collection may depend on different ecological development and the traditions of the state. Collection beyond the level of self-supply is hardly known among the majority of states within the EU—except in France, Spain, Portugal, Italy and Greece, where collection of medicinal plants is under state control.[27] On the other hand, in some eastern European countries, the total production of drugs is based on wild growing plants, e.g., in Albania, Bulgaria, and Turkey. In Spain, 102 of 195 items are of wild origin; in France not more than 3 of 59; in Bulgaria, around 85 percent of the total production is collected.

In Hungary and Poland, collection takes place at a significant level, which could assure 100 to 120 kinds of drugs and would be hard to replace. The most popular drugs are among the ones obtained by harvesting from the wild: yarrow, horsetail, nettle, rose hips, lime, and a large proportion of St. John's wort, chamomile, hawthorn, dandelion, and others originates from natural vegetation. Basically, the collectors are private individuals, especially villagers, often women, children, or people without any other official engagement. At best, collection and buying is organized by merchant firms or processing factories. In other systems, private enterprises deal with the purchase of medicinal plants and turnover is regulated by the market.

To provide drugs of optimal quality, knowledge, consideration, and keeping the following is necessary:

- Collectors should have the basic knowledge on plant species and their main features for realizing differences between similar or related ones. In addition, they should be aware of the status of protection and existence of toxic materials in the plants. Handling with protected species is regulated by law in each country. Poisonous species should be harvested separately and with special care.

- Characteristics of the growing area may have a great influence on drug quality. Polluted areas (such as neighborhoods, dunghills, fields treated by pesticides) are not suitable for collection. In environmentally protected areas, regulations should be followed.
- The optimal period for harvesting in the wild depends on the species and the maximum level of accumulation of active agents.
- Collection should be carried out using the utmost care of the plant population and habitat in order to assure sustainable use of them.

Ecological production is spreading also in the case of wild growing plants. Natural habitats—first of all, the protected ones—are often optimal sites for providing certified biological products. In natural reserve areas, human disturbance as well as pollution are minimal.

Since 1980, a continuous shift of the proportion of collection/cultivation can be observed for the benefit of cultivation. Introduction of medicinal plants into agriculture is a more and more common feature in developed countries and can be traced back to the following reasons.[28]

The growing quantitative demand for the raw plants cannot be assured by collection of wild populations. Demand is suddenly growing as new active materials and new uses become evident. New products require standard amounts of the same drug through the year. Characteristic examples are *Oenothera* species, introduced in the 1980s, or *Solidago virgaurea* L., being under introduction today.[29] As one of the latest cases, the enormous increase in demand—above 6,000 tons per year—of *Hyperici herba* in the 1990s required an intensive introduction of St. John's wort.[30]

- Wild growing stocks are hardly appropriate to assure standard quality, especially concerning chemical constituents. Application of quality assurance systems is more and more a prerequisite for optimal marketing or for any marketing at all. Cultivation under controlled conditions and utilization of certified propagation material may create the basis for the desired quality. Some current problems may be solved in the future in this way. Thus, cultivation of only special varieties of chamomile can provide a special spectrum of active agents (chamazulene, bisaboloids, flavonoids, etc.) for different purposes of utilization—although the quantity of the flowers from nature might be enough. Elaboration of cultivation technology of *Equisetum arvense* may provide a clean drug of proper quality without pollination and mixing with the poisonous *E. palustre* as often happens in the wild. By controlling cultivation methods and by choosing the proper growing site, nitrate pollution of *Chelidonium majus* or cadmium accumulation in *Hypericum perforatum* may be avoided.

- More and more plant populations are becoming extinct due to general industrialization, pollution, and even exhaustive overcollecting. Natural stands are not able to regenerate and serve as sources of raw plants, species become endangered. A passive way of preservation is the often used for official protection of species or habitats. An active method of preservation and at the same time of sustainable collection is cultivation. According to Lange, "European countries should promote cultivation of threatened medicinal and aromatic plant species, since cultivation is a very effective mean of reducing or eliminating pressure on their wild populations."[31] As a well-known example, *Arnica montana* L. was recently successfully introduced, even under high pH conditions of the soil, by developing a resistant cultivar and by using an optimal seedling-raising method.[32]
- The controlled quality of drugs from developing countries and high shipping costs result in the decrease of importation of drugs, which may support the domestic production even of species which are not indigenous in a given area. Thus, in the past twenty years the European cultivation of North American *Echinacea* species has been widely undertaken.[33] Similarly, *Ginkgo biloba* L. has been introduced from the Far East into other regions, and about 6,000 tons of leaves are produced in France and the United States in addition to the amount produced in China, its native habitat.[34]
- In several countries, failing social conditions may limit collection activities in the future. In southeastern Europe, reduction in growing spaces and high labor costs have reduced drug production. A well-planned and coordinated cultivation is desired to supply the in-demand drugs. Thus even hawthorn, elder, or linden may be introduced in Europe in the future.[35]

The number of cultivated species is continuously growing, although several drugs are regularly obtained from both cultivated and wild stock (e.g., *Althaea* [root], *Achillea* [flower], *Sambuci* [fruit]). However, even if a considerable progress is made in the introduction of several species into agriculture, the significance of collection will not diminish in the near future. A significant increase of cultivated areas faces some problems: Cultivation technology is quite difficult for several species; the drug is required only in small quantities; collection from the wild is much cheaper, yet some theories state that collected material is of higher quality.[36] In developing countries, collection of medicinal plants from the wild represents an important tool in earning money.

CHARACTERISTICS OF BREEDING MEDICINAL PLANTS

Quality control of medicinal plant drugs, phytopharmaceuticals, and their other products, even from the beginning of cultivation, is a prerequisite for marketing. Although several agrotechnical factors have a considerable influence on the quality of harvested products, the improved genetic material must be the basis for the cultivation.

In European countries, a remarkable progress in breeding activity and the supply of varieties of medicinal plants is being noticed. However, in France, the Netherlands, Italy, and Germany, the activity is mainly focused on a limited number of species (e.g., *Salvia* spp., *Origanum* spp., *Carum carvi,* and *Chamomilla recutita*). According to Franz, the low number of selected varieties among medicinal plants can be traced back to the minor economical significance of these cultures compared to cereals or fodder, as well as to insufficient regulation.[37]

Financial support from a processing perspective—from state projects or from the cooperation of agricultural producers—can encourage a more active breeding. Without support, breeding cannot be prosperous because of the relatively small surface area of these cultures.

Breeding goals can be defined as follows:

- Reaching positive economic yield of active materials: For that goal, the genetic material must have an optimal biomass production as well as a proper level of active agents. Enhancement of biomass formation of the overground parts had been one of the focal points during the development of the new Hungarian marjoram (*Majorana hortensis* Mönch.) variety Magyar. Selection for higher drug yield was carried out by examining plant height and width, plant habit, proportion of leaf and stem mass, leaf size (surface), and mass ratio of crumbled: whole drug.[38] Selection for higher biomass production from populations of great intraspecific variability could be enhanced by vegetative propagation.
- Enhancement of the accumulation level of the desired active components: An example of this is our own intensive selection work which has been carried out recently with *Hyssopus officinalis* L. to reach a high essential-oil content of the drug (above 1 percent d.w.). In some special cases, the decrease of accumulation level or elimination of certain components in the drug may come into the limelight of breeding. In case of *Salvia officinalis* L. the aim is the development of a chemotype with low level of β-thujone in the essential oil. Develop-

ment of poppy cultivars of low (under 0.01 percent) alkaloid content provides the possibility of producing legal poppy for seeds as foodstuff.[39]

- Appropriate drug quality with appropriate sensory characteristics: It is even more important in the case of raw plants for alimentary and cosmetic industry (spices, condiments, aromas). While medicinal plants as pharmaceuticals should be characterized by a stable ingredients' spectrum defined by chemical analysis, products of the mentioned uses have to face sensory analysis. For species of multiple uses, new cultivars desirably should show a stable and distinguishable sensory profile.
- Improvement of agrotechnical parameters, such as effective harvest, processing, and economic yields of appropriate quality: It is sometimes connected with culture duration (annual-perennial), growth intensity (earliness), competition ability (weed tolerance), propagation specialities (closed fruits, uniform germination), or plant morphology (size of organs, growth habit, bush form).
- Resistance against pests or insects: It is extremely important to have such varieties, because cultivation is becoming more and more intensive.

Special attention is paid to determine distinction between the genetically fixed chemism and its modifications. The most important influencing factors in evaluating chemical features include the phenological phase, the organic differentiation of plants, weather conditions, effects of year or age of plant individuals, extraction, and analytical methods. Aim of the breeding activity should be the combination and fixation of the genetic potential of special advantageous characteristics, dividing them from phenotypic modifications. Some examples of recently bred varieties are listed in Table 10.2.

Breeding methods for establishment of cultivars of medicinal and aromatic plants are rather uniform. Based on the genetic variability of existing populations, simple individual or mass selection may assure successful work. Searching through this natural diversity, chemical differences especially seem to be a basic element in prosperous breeding. Selection may result relatively shortly in the desired genotype. Good examples for these achievements are cultivars of *Mentha spicata, Papaver somniferum,* and *Melaleuca alternifolia.*[40]

Other methods, such as hybridization, mutation, polyploidization, and genetic transformations are seldom used today in breeding medicinal and aromatic plant species (see also Chapter 12 in this book). Crossing trials are

TABLE 10.2. Examples of registered varieties of medicinal plants from recent years.

Species	Variety	Origin	Special characteristics
Achillea collina	*Alba*	Slovakia,1992	High proazulen content
Arnica montana	*Arbo*	Germany, 1998	Tolerant for soil, high yield
Carum carvi	*Karzo*	Netherland,1994	Annual, high carvone content
Chamomilla recutita	*Robumille*	Germany, 2002	Rich in α-bisabolol
Coriandrum sativum	*Jantar*	Ukraine, 1988	Rich in essential oil
Foeniculum vulgare	*Magnafena*	Germany, 1993	Mainly for annual culture
Majorana hortensis	*Mara*	Austria, 1999	High *cis*-sabinene hydrate
Mentha piperita	*Mexian*	Hungary, 1997	Large leaves, rost-tolerance
Papaver somniferum	*Tebona*	Hungary, 2000	High tebaine content
Petasites hybridus	*Petzell*	Switzerland, 2001	Low pyrrolizidine alkaloid cont.

more often applied either as genetic investigation or for combining different species.

Accumulation of information about the genetic background and regulation of each species would accelerate efficacy of breeding.

NOTES

1. Lange, D., *Europe's medicinal and aromatic plant—Their use, trade and conservation* (Cambridge, UK: Traffic International, 1998), p. 77.

2. Kerekes, J., *Production of medicinal plants (Gyógynövénytermesztés)* (Budapest: Mezögazdasági Kiadó, 1969), pp. 10-20.

3. Heeger, E.F., *Handbuch de Arznei- und Gewürzpflanzenbaues* (Berlin: Deutscher Bauernverlag, 1956), pp. 19-42.

4. Bernáth, J., 2001, Ecological diversity of Hungarian medicinal and aromatic plant flora and its regional consequences, *Int. J. Hort. Sci.*, 7(2): 10-19.

5. Zemlinszkij, E., *Medicinal plants (Lekarstvennije rastenija* SSSR) (Moskva: Moskovskaja obsestva ispitatelej narodi, 1949), pp. 5-9.
6. Heeger, 1956, *Handbuch de Arznei.*
7. Ibid.
8. Zemlinszkij, 1949, *Medicinal plants.*
9. Heeger, 1956, *Handbuch de Arznei.*
10. Halmai, J., 1967, A review on the Hungarian medicinal plant branch, *Herba Hung.*, 1(1): 5-11; Ruminska, A., *Medicinal plants (Rosliny lecznicze)* (Warszawa: Panstwow Wydawnictvo Naukove, 1973), p. 325.
11. Pank, F., 1998, Arznei- und Gewürzpflanzenmarkt in der EU, Z. *Arzn. Gew.pfl.*, 3(2): 77-81.
12. Ibid.
13. Verlet, N. and Leclercq, G., The production of aromatic and medicinal plants in the European Union. *Proceedings of the Internat. Symp. on the Conservation of Medicinal Plants in Trade in Europe* (Kew, England, 1998), pp. 121-126.
14. Lange, D., *Untersuchungen zum Heilpflanzenhandel in Deutschland* (Bonn: Bundesamt für Naturschutz, 1996), p. 130.
15. Lawrence, B.M., The spearmint and peppermint industry of North America, *Third International Conference on Aromatic and Medicinal Plants,* Nyons, France, 1991, pp. 59-90.
16. Chatterjee, S.K., 2001, Cultivation of medicinal and aromatic plants in India—A commercial approach, *Acta Hort,* 576: 191-202.
17. Piesch, R.F and Wheeler, N.C., 1993, Intensive cultivation of *Taxus* species for the production of taxol, *Acta Horticult.*, 306: 219-227; Lawrence, B.M., Production of clary sage oil and sclareol in North America, *Fourth International Conference on Aromatic and Medicinal Plants,* Nyons, France, 1994, p. 41; Schmid, W., Anbau von *Ginkgo biloba Kurzfassung der Referate und Poster des 11. winterseminars zu Fragen der Arznei- und Gewürzpflanzenproduktion,* 2001, Sauplanta EV., Bernburg, pp. 32-33.
18. Kasbohm, A., 1999, Arznei- und Gewürzpflanzenanbau- nur ein Nischenmarkt?, ZMP, *Marktbericht Arznei- und Gewürzpflanzen.*, 1(1): 2.
19. Ibid.
20. Dachler, M. and Pelzmann, H., *Arznei- und Gewürzpflanzen, Anbau- Ernte-Aufbereitung* (Klosterneuburg: Österreichischer Agrarverlag, 1999), pp. 25-35.
21. Jens, P., 1987, Guide to good agricultural practice for herbs (EHIA), *Newsletter of Med. and Arom. Plants, Budakalász,* 1(1): 72-77; Franz, C., 1989, Report on the meeting of the sub-commission GAP, *Newsletter of Med. and Arom. Plants, Budakalasz,* 3(1): 14-24.
22. Pank, F., 1991, Richtlinien für den integrierten Anbau von Arznei- und Gewürzpflanzen, *Drogenreport,* 2(3): 45-60.
23. Franz, C., 1999, Quality assurance systems and total quality management of medicinal plant production, *ICMAP News,* 3(6): 4-9.
24. Dehe, M., 1999, Marktchancen ökologisch erzeugter Arznei- und Gewürzpflanzen in Deutschland, Z. *Arzn. Gew.pfl.*, 4(1): 44-50.
25. Lück, L., 1996, Der Anbau von Heil- und Gewürzpflanzen im ökologischen Landbau in Deutschland, *Drogenreport,* 9(16): 32-33.
26. Ellenbeger, A., Assuming responsibility for a protected plant: *Arnica montana, Proceedings of the Internat. Symp. on the Conservation of Medicinal Plants in*

Trade in Europe (Kew, England, 1998), pp. 127-130; Schmidt, M., Improvement of pharmaceutical drug quality: A cultivation project for *Harpagophytum procumbens* in Namibia, *Proceedings of the Internat. Symp. on the Conservation of medicinal plants in trade in Europe* (Kew, England, 1998), pp. 140-146; Phillips, D., Sustainable harvesting of Himalaya yews, *Proceedings of the Internat. Symp. on the Conservation of medicinal plants in trade in Europe* (Kew, England, 1998), pp. 147-154.

27. Verlet and Leclercq, 1998, The production of aromatic and medicinal plants.

28. Németh, É., 2001, Beweggründe und Ergebnisse der Inkulturnahme wildwachsender Arzneipflanzen im ladwirtschaftlichen Anbau (Motivations and results of the introduction of wild growing medicinal plants into the agriculture), *Drogenreport,* 14(25): 3-8.

29. Anonymous, *Domestication of the evening primrose,* Information sheet of the Scotia Pharmaceuticals Ltd. Woodridge Meadows, Guilford, GUI, 1BA. England, 1993, p. 6; Bohr, C. and Plescher, A., 1997, Empfelungen für den Anbau von echter Goldrute *(Solidago virgaurea)* in Thüringen, *Drogenreport,* 10(18): 27-34.

30. Franke, R., 1999, Neuentstehung von Kulturpflanzen heute, *Z. Arzn. Gew. pfl.,* 4(1): 24-38.

31. Lange, 1998, *Europe's medicinal and aromatic plants,* p. 77.

32. Bomme, U., 1999, Anbau und Züchtung von *Arnica montana* L., *Z. Arzn. Gew. pfl.,* 4(4): 202-203.

33. Franke, R., Schenk, R., Schaser, J., and Nagell, A., 1997, Einfluss von Anbaumassnahmen auf Ertrag und Inhaltstoffe von Echinacea pallida, *Z. Arzn. Gew. pfl.,* 4(4): 173-185.

34. Shan-An, H., Shi, S., and Wan-Gou, L., 1999, Leaf cultivation of Ginkgo for medicinal use, Proceedings of WOCMAPII, *Acta Horticult.,* 502: 143-151.

35. Bohr, C., 1997, Inkulturnahme von bisher aus Wildsammlungen stammenden Wirkstoffpflanzen, *Drogenreport,* 10(17): 37-39.

36. Lange, 1998, *Europe's medicinal and aromatic plants.*

37. Franz, C., 1996, Züchtungsforschung und Züchtung an Arznei- und Gewürzpflanzen, *Z. Arzn. Gew. pfl.,* 1(1): 30-38.

38. Németh, É., 2001, Goals and results in improvement of biological background of medicinal plant production, *Internat. J. of Horticult. Sci.,* 7(2): 20-27.

39. Németh, É., Bernáth, J., Sztefanov, A., and Petheö, F., 2002, New results of poppy (*Papaver somniferum* L.) breeding for low alkaloid content in Hungary, *Acta Horticult.,* 576: 151-158.

40. Bernath, J., 2001, Strategies and recent achievements in selection of medicinal and aromatic plants, *Acta Horticult.,* 576: 115-120.

Chapter 11

Biological Screening of Medicinal Plants

Ming-Wei Wang

INTRODUCTION

Medicinal plants have been applied to human health care for thousands of years. Drug discovery in ancient times was largely by chance and based on clinical practices. As understanding of therapeutic benefits deepens and demands for medicinal plants increase, serendipitous discovery evolves into an active search for new medicines. Modern drug screening uses laboratory animals as testing subjects. Scientists are looking for lead compounds with specific structures and pharmacological effects usually from natural sources. Many drugs presently prescribed by physicians are either directly isolated from plants or artificially modified versions of natural products. The tremendous progress made in life sciences has resulted in defining many physiological processes and their detailed mechanisms of drug action. This leads to the establishment of various molecular and cellular bioassays in conjunction with high-throughput screening (HTS) methods. The HTS system utilizes knowledge and techniques of a variety of disciplines such as medicinal chemistry, molecular and cell biology, biochemistry, pharmacology, computer science, automation engineering, informatics, and mathematics. It decreases the amount of testing compound required such that only microgram quantities are needed. This is advantageous and permits certain natural products that are difficult to isolate and purify, and allows compounds that are difficult to synthesize to be assayed. It also reduces screening costs and enhances screening capability significantly—numerous compounds can be tested multiple times in different screening models, thereby increasing the probability of discovering new leads and drug candidates.

Thanks Mai Huang for her technical assistance, and Dale E. Mais for his valuable comments in the preparation of this manuscript.

In the long history of humankind, combating disease has been an important aspect in the interactions between humans and natural environment, affecting the very existence and propagation of human beings. Today, although science and technology are very advanced, many diseases still trouble us and even threaten our lives. During the process of understanding and treating diseases, humans have discovered a variety of plants with therapeutic value. Many of these medicinal plants have been used for thousands of years by a significant fraction of the population and are still applied to health care, either alone or in combination with modern medicines. Indeed, it is estimated that about 25 percent of the drugs prescribed worldwide come from plants and 60 percent of antitumor/anti-infectious drugs already on the market or under clinical trial are of natural origin. For example, compounds isolated from plants such as quinine (from *Cinchona ledgeriana*), maytansine (from *Maytenus diversifolius*), camptothecine (from *Camptotheca acuminata*), berberine (from *Coptis chinensis*), andrographolide (from *Andrographis paniculata*), silybin (from *Silybum marianum*), schisandrin (from *Schisandra chinensis*), reserpine (from *Rauvolfia verticillata*), and trichosanthin (from *Trichosanthes kirilowii*) have been used as antimalarial, antitumor, antibacterial, antiviral, antihypertensive and abortifacient agents, respectively.

Clinical experience suggests that some plant medicinals show poor efficacy and severe side effects, while individual active constituents separated from medicinal plants often demonstrate the opposite, i.e., high efficacy and low toxicity. One example is artemisinin (or arteannuin) extracted from the non-root part of *Artemisia annua*. This natural product has been widely used in traditional Chinese medicine (TCM) to treat malaria with a proven efficacy and safety history. Certain diseases (e.g., kala-azar) that are not curable with modern therapeutics may be successfully treated with folk medicines. The effectiveness of medicinal plants in mitigating chronic diseases has elicited long-lasting inspiration of pharmaceutical scientists to search for new directions in drug discovery and development. The transition from fortuitous discovery to systemic screening through validation in experimental models has been taking place since the 1930s. High-throughput screening technologies developed in recent years have significantly improved the speed, scale, and quality of this process. This chapter briefly reviews the history of finding new drugs from medicinal plants as well as novel technologies and approaches established recently for drug screening. According to the methodology used in medicinal plant research, the historical course is divided into three periods which are described next.

DRUG DISCOVERY IN ANCIENT TIMES

In primitive societies, certain materials were found to alleviate pain or sickness. With accumulated experience and knowledge, the relationship between these materials and certain diseases or symptoms was gradually established. As the understanding of therapeutic benefit deepens and demands for such materials increase, passive discovery evolves into active search for new medicines.

Plants have supplied virtually all ancient cultures with food, clothing, shelter, and medicines. It is estimated that approximately 1 percent of the roughly 300,000 different species of higher plants that exist have a history of food use. By contrast, 10 to 15 percent have extensive documentation for use in traditional medicine.

The story of the ancient Chinese doctor, Shennong, testing hundreds of plants may be regarded as the best example of humanity's active search for medicines and the earliest record of drug screening. Through thousands of years of clinical practice and constant screening and evaluation, people have discovered a variety of plant resources possessing medicinal value. In China, from the end of the Spring and Autumn Period (770-476 B.C.) through the Warring States (472-221 B.C.) to the Qin dynasty (221-207 B.C.) and the early Han dynasty (206 B.C.-220 A.D.), medicinal plants were not only specially recorded and described in several medical monographs but also mentioned in historical literatures such as *Lu Shi Chun Qiu* (239 A.D.), *Huai Nan Zi* (139 A.D.), and *Shan Hai Jing* (the Warring States).

In ancient China, medicines were called *Bencao* (Chinese materia medica). The oldest herbal medicine book, *Shennong Bencao,* was written late in the Han dynasty. It recorded 365 types of herbs, including 252 plants, 67 animal parts, and 46 minerals. Some related medical indications were also described, such as *Ephedra* (herb) for asthma, *Rheum* (root and rhizome) for thyroid enlargement and *Artemisia annua* (herb) for malaria. Furthermore, the author divided these herbs into three different categories in accordance with their properties, applications, and toxicities. In *Handbook of Prescriptions for Emergency* (ca. 333 A.D.), the method to treat malaria was described: "A handful of *Herba Artemisiae Annuae* is added to two liters of water and pounded into juice to take"; in Tang dynasty (618-907 A.D.), *The Newly-Revised Materia Medica* recorded a plant extraction approach by pounding natural indigo into pieces, then "immersing in water for a whole night, evaporating and drying in air." The final product is in powder form called *qingdai* for treatment of bacterial infections. Since then, the speed of herbal use greatly accelerated and the application range constantly expanded. From Eastern Han dynasty (25-220 A.D.) to the end of

Song dynasty (960-1279 A.D.), for about one thousand years, both the varieties and sources of medicinal plants were tremendously increased. The classical herbal pharmacopoeias in the period included *Prescription in Jade Box* written by Hong Ge, *Essential Prescriptions Worth a Thousand Gold* written by Simiao Sun, who at that time was regarded as the "drug king," and *The Historically Classified Materia Medica* (which recorded 1,558 kinds of medicines in Song dynasty). During the next millennium, from the late Ming dynasty (1368-1644 A.D.) to the Qing dynasty (1644-1911 A.D.), pharmacist Shizhen Li (Ming dynasty), through a long period of medical practice, herbal collection, field investigation, experience verification, and consultation on historical references, summarized his knowledge in a book titled *Compendium of Materia Medica* (1578) with detailed descriptions of 1,195 kinds of medicinal plants. The contents of this book were disseminated all over the world.

In Western countries, herbal medicine has a long history of use as well. Hippocrates (460 B.C.-377 B.C.), the father of ancient Greek medicine, paid great attention to the therapeutic value of diets and used *Hordei alga* (fruit), *Codii cylindrici* and *Veratri nigri* (root and rhizome), to treat certain ailments. In the fourth century B.C., Diocles of Carystius of Greece, a student of Aristotle, put together a list of plants, along with their uses, titled *Rhizotomika*. Galen (~129 A.D.-200 A.D.), the famous Roman physician and pharmacist, once composed a series of books describing various therapeutic methods and herbal medicines. He also classified many herbs based on botanical categories and invented opiate and a number of other pharmaceutical preparations. Indeed, many simple herbal extracts are still called "galenicals." Later on, some well-known herbal works appeared, including *Liber de Proprietatibus Rerum* written by an English Monk, Bartholomew Glanvil, in the fifteenth century.

Although drug discovery in ancient times was largely by chance and based on practices in humans, it provides a precious legacy and knowledge that benefit us even today. In fact, botanical medicine changed when the advancement of chemical techniques in the beginning of the ninteenth century made possible the isolation of chemical constituents from plants. In 1805 German pharmacist Serturner extracted morphine from opium, and in the 1820s a French pharmacist isolated several alkaloids from plants, including quinine and caffeine. In 1893, aspirin was synthesized. In 1906, Paul Ehrlich developed his theory of "magic bullets," the idea of a selective drug that would home in on its target while leaving intact the surrounding physical environment. Single entity chemicals, which are more consistent and easier to quantify, have been judged more specific in their therapeutic focus than medicinal plants.

RANDOM DRUG SCREENING USING ANIMAL MODELS

Modern drug screening (general screening and combination screening) uses laboratory animals as testing subjects. It applies different kinds of techniques and methods, such as mechanical equipment, electronic and optical instruments, behavioral observation and mathematical modeling. Scientists are looking for lead compounds with specific structures and pharmacological effects often from natural sources (extracts, fractions, or pure compounds). Based on initial discoveries, further modification and optimization of lead structures via medicinal chemistry and biological screening are carried out in order to develop new drugs with therapeutic value. This process has been constantly updated, improved, and widely used for nearly a full century. In fact, many drugs presently prescribed by physicians are either directly isolated from plants or artificially modified versions of natural products.

One good example is the discovery of paclitaxel. In 1856, Lucas first extracted paclitaxel, an alkaline powder—taxine—from leaves of *Taxus baccata* in Europe, but its pharmacological property remained unknown for about 100 years. In 1958, the National Cancer Institute (NCI) organized a group of pharmacologists to screen antitumor activities from the extracts of 35,000 natural products. In 1963, Wall and Wani, who participated in this NCI-sponsored program, succeeded in isolating crude paclitaxel from the bark and chip of *Taxus brevifolia* indigenous to ancient forests of the Pacific Northwest. Using in vitro screening assays, they discovered potent inhibitory activities of this cured extract on a number of cultured cancer cell lines. Further purification work was carried out and prolonged for several years due to the extremely low content of the active component in pacific yew trees. Until 1971, with the help of McPhail (Duke University), they finally identified the structure of the lead compound, a complex polyoxygenated diterpene, and named it Taxol. In 1979, Horwitz and her colleagues at Albert Einstein Medical College discovered the unique anticancer mechanism of Taxol through animal experiments. Taxol binds to tubulin, thereby stabilizing the microtubule and blocking its disassembly. As a result, tumor cell mitosis is stopped at G_2/M phase leading to cell death. This novel mode of action is believed to be a prototype of a new class of anticancer drugs. From 1983 to 1992, a ten-year multicenter clinical trial was conducted and the results revealed that Taxol's response rate in advanced ovarian carcinoma (failed in the first-line chemotherapy) was 56 percent with a cure rate of 15 percent; the response rate in metastatic breast cancer was 30 percent with a cure rate of 10 percent. In addition, Taxol showed good efficacies in non-small-cell lung cancer, prostate cancer, upper gastrointestinal cancer, and leukemia. In 1992, the U.S. Food and Drug Administration (FDA) approved Taxol as a new drug to treat advanced ovarian cancer with a trade name of paclitaxel (see Figure 11.1).[1]

The TCM combination prescription *Angelica sinensis* (root) and *Aloe pill* has been used to treat chronic granulocyte leukemia (CGL) in China for some time. It includes medicinal plants such as *Angelica sinensis* (root), Aloe, Gentiana (root), Gardenia (fruit), *Scutellaria* (root), *Phellodendron* (bark), *Coptidis* (rhizome), *Isatidis* (leaf), and *Aucklandia* (root). Using the mouse L7212 leukemia model, scientists found that the only active component contained in this prescription is natural indigo, which has certain side effects. With joint efforts made by medicinal chemists and pharmacologists, an active compound was identified as indirubin (see Figure 11.2), which showed different degrees of inhibitory effects on rat Walker tumor, mouse Lewis lung tumor, C615 breast cancer, and L7212 leukemia. It did

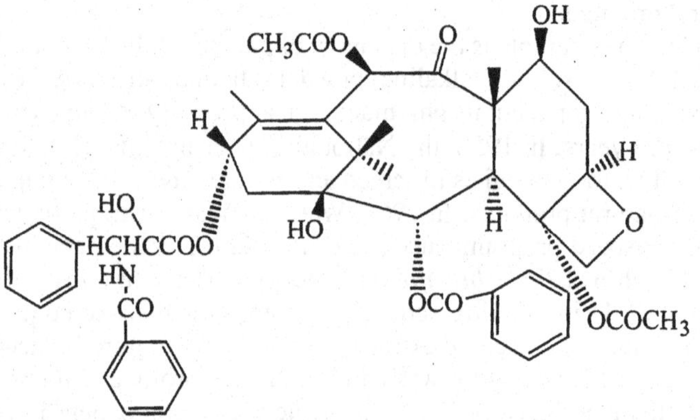

FIGURE 11.1. The chemical structure of paclitaxel.

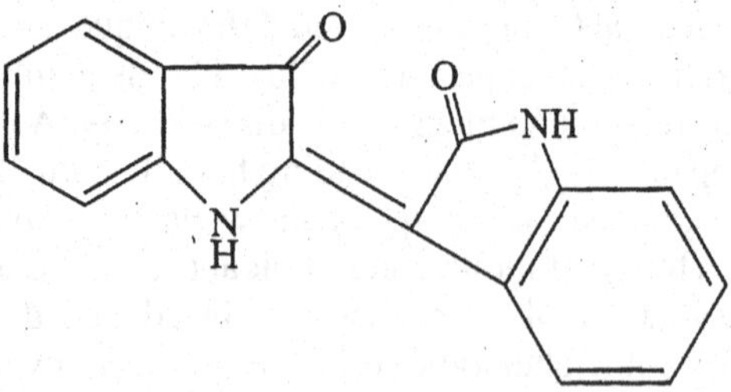

FIGURE 11.2. The chemical structure of the compound indirubin.

not exhibit observable side effects and bone marrow suppression at a daily oral dose range between 100 and 500 mg/kg. The response rate in treating CGL was 87.3 percent. The synthetic version of this drug has been approved for marketing.

Pure compounds isolated from plants are normally restricted by available resources or have very low content in crude extracts. Hence, chemical synthesis becomes the usual way for large-scale production following structure determination. However, structure-activity relationship (SAR) studies are often required to overcome poor efficacy, side effects, or toxicity and to circumvent the fact that some drugs are too complex to be produced by organic syntheses. Using the natural compound as a lead, active scaffolds and groups can be identified for chemical modification, thereby synthesizing a series of new analogs guided by SAR studies, with an ultimate goal of developing more new drugs. For example, a series of pharmacological experiments demonstrated that the neutral fraction derived from ethanol extract of *Artemisia annua* L. had a significant inhibitory effect on *Plasmodium falciparum* in both the Camp strain (chloroquine sensitive) and the Smith strain (chloroquine resistant). Toxicology studies proved that not only did the ethanol extract have no major liabilities in animals but clinical trial involving 30 cases also yielded satisfactory results. Further research work revealed that the active compound is an endoperoxide sesquiterpene lactone called artemisinin (see A in Figure 11.3), which was first purified in the 1970s. Its structure was determined in 1979 to be 3,6,9-trimethyl-octahydro-3,12-epoxy-pyranol[4,3-j]-1,2-benzodioxepin-10-one by X-ray diffraction analysis. It has a unique structure and lacks the nitrogen-containing heterocyclic ring, which is common in most antimalaria drugs.[2] Through numerous clinical studies in chemistry, pharmacology, toxicology, artemisinin was proven to be highly effective in treating malaria including multidrug-resistant strains. The isolation and characterization of Artemisinin from *Artemisia annua* L. was considered one of the most important discoveries in contemporary herbal research.[3] Although artemisinin has good efficacy against malaria with properties such as low toxicity and rapid onset, it has two major shortcomings: insolubility in water and oil and poor oral absorption, resulting in difficulties with formulation. To overcome these shortcomings, in vivo metabolic and structural modification studies were carried out. All the metabolites isolated from urine had no antimalaria activity due to the loss of the endoperoxide bridge. This suggests that the endoperoxide bridge is the active group responsible for its biological activity (see Figure 11.3). Thus, scientists semisynthesized a series of derivatives of artemisinin which were tested in malaria animal models. More than 20 derivatives exhibited a better efficacy than artemisinin; some examples follow.

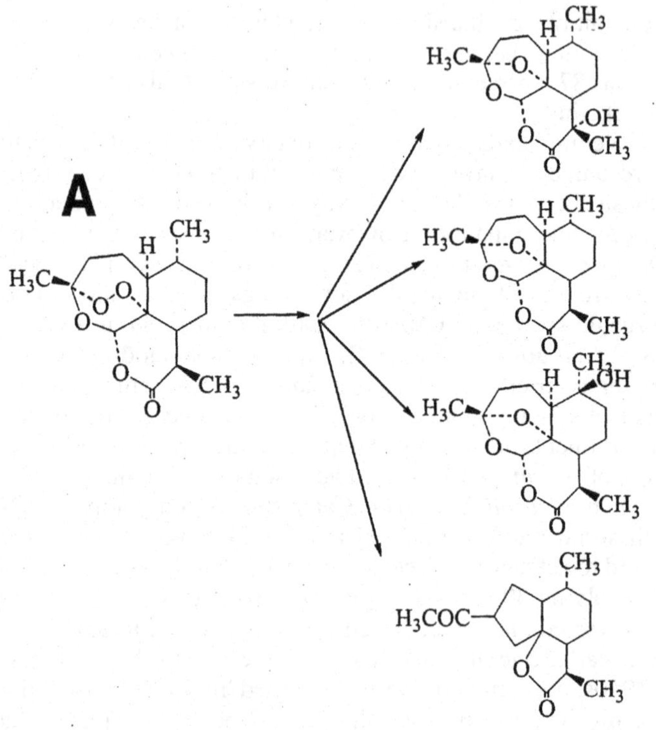

A

FIGURE 11.3. Artemisinin and its metabolites in vivo.

Deoxyartemisinin was successfully prepared from artemisinin in one step using $NaBH_4$ and BF_3Et_20 in tetrahydrofunan (THF). It showed an eightfold increase in bioactivity in vitro against chloroquine-resistant malaria as compared to artemisinin.

α-*Artelinic acid* is a potent, stable, and water-soluble compound synthesized from artemisinin with blood schizonticidal activity against *Plasmodium falciparum*.

Artemether is a methyl ether derivative of artemisinin found to be more active than artemisinin. This compound is reduced with sodium borohydride to produce dihydroartemisinin as a mixture of epimers. The mixture is treated with methanol and an acid catalyst to provide artemether that is effective in treating cerebral malaria by intramuscular injection. Artemether is the β-anomer of the ethyl ether of dihydroartemisinin. It was synthesized originally in China by Li and colleagues in 1981.[4] A. Brossi and co-workers synthesized artemether by borohydride reduction of artemisinin to hydroartemisinin, etherification in boron trifluoride ethyrate, and finally separation of anomers.[5] The compound has been formulated as a solution in sesame oil for intramuscular injection at a concentration of 160 mg/mL for preclinical and clinical studies. Artemether is virtually insoluble in water, but it is soluble in 50 to 100 percent ethanol with water and a variety of oils. It has been proven to be stable in accelerated stability studies at 50°C for several months. It is very efficacious against drug-sensitive strains of *Plasmodium falciparum* in vitro as well as in vivo in different malaria animal models. Presently, artemether is registered in 27 countries including France, Thailand, Pakistan, Sudan, and Niger.[6]

Huperzine A (see Figure 11.4) is a new alkaloid separated from *Huperzia serrata,* one of the most commonly used Chinese herbs for the treatment of contusion, strain, and swelling. Pharmacological studies indicated that it has a highly selective and long-lasting inhibitory effect on acetylcholinesterase (AchE) in the brain, and is capable of enhancing learning ability and memories in rat and mouse experimental models.[7] It is known that AchE inhibitors can be used to treat Alzheimer's disease (AD). Several clinical reports suggested that huperzine A facilitates cholinergic neurotransmission by increasing the concentration of acetylcholine in the central nervous system. Its activity is about 100 times greater than that of tetrahydroaminoacridine (THA, tacrine), a drug widely used against AD. Moreover, the toxicity is lower than several other AchE inhibitors.[8] Huperzine A is presently marketed in China as a new therapeutic agent to treat AD. This discovery has attracted significant interest among AD researchers around the world. Using huperzine A as a lead compound, a large number of analogs or derivatives were synthesized to find an AchE inhibitor with higher selectivity and lower toxicity. Little progress has been made to date relative to better

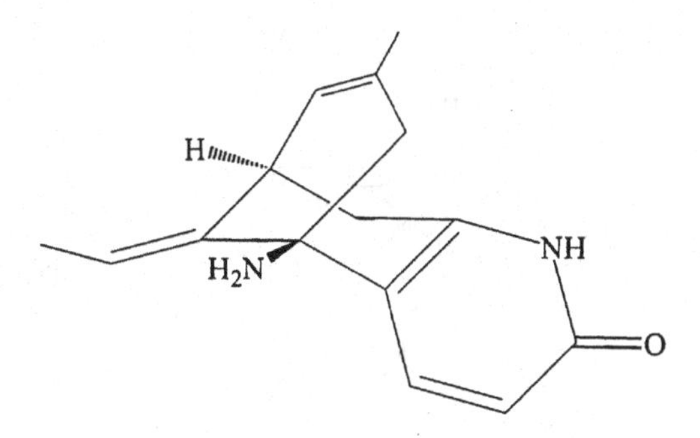

FIGURE 11.4. Chemical structure of huperzine A.

analogs, but among the 30 or so derivatives synthesized, HA-1 (see Figure 11.5) and ZT-1 (see Figure 11.6) demonstrated a 1.5-fold increase in selectivity and a sixfold decrease in toxicity, with an equivalent bioactivity to huperzine A. Relevant patent application (ZT-1) has subsequently been filed.[9]

Mylabris phalerata Palls is used in TCM for the treatment of goiter. Its principal active compound is cantharidin (see Figure 11.7), which can inhibit experimental liver cancer and certain sarcomas in mice. Clinical data showed a good efficacy in treating primary liver cancer without affecting the peripheral white blood cell counts.[10] Recent studies on hepatocellular carcinoma cells (Hep 3B) revealed that cantharidin is an acute cytotoxic agent.[11] The major side effects of this compound relate to the urinary and digestive systems. Bioassay-guided structural modification studies delivered two less toxic analogs, norcantharidin and *N*-hydroxycantharidin, and both of them have been used in the treatment of primary liver cancer.[12]

Ginkgo biloba has long been used in herbal practices and is officially listed in the Chinese pharmacopoeia. It is prescribed for treatment of cerebro-

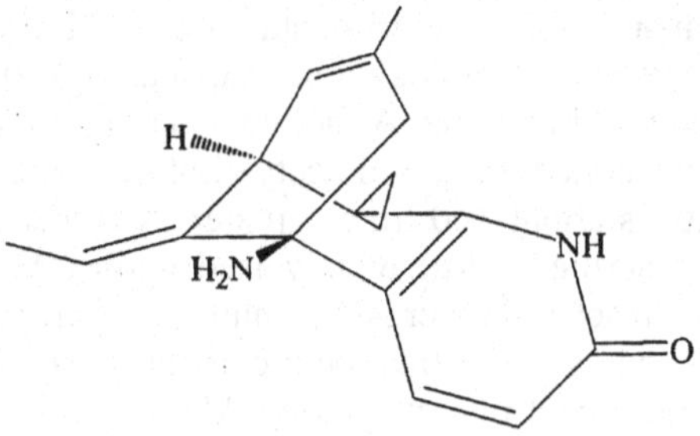

FIGURE 11.5. Chemical structure of HA-1.

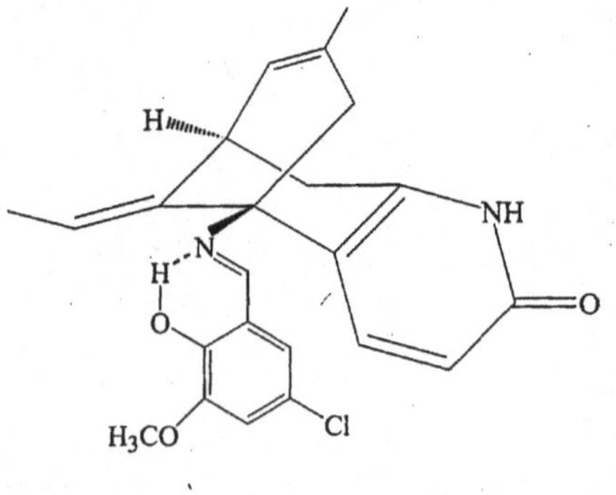

FIGURE 11.6. Chemical structure of ZT-1.

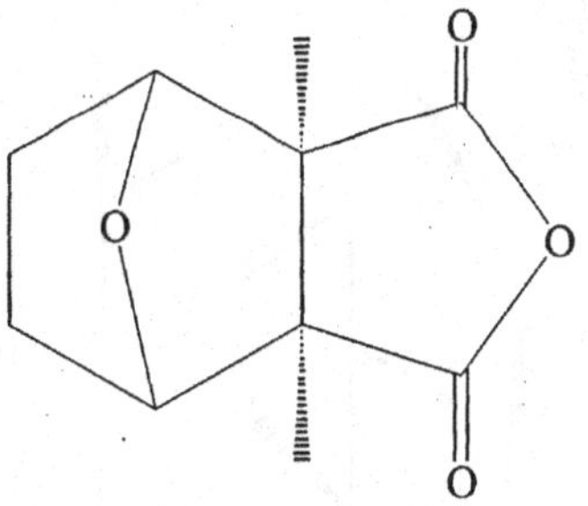

FIGURE 11.7. Chemical structure of cantharidin.

vascular and cardiovascular disorders, neurosensory-related problems, disturbances in vigilance, short-term memory loss, and other cognitive dysfunctions associated with aging and senility. Ginkgolides, isolated from the root bark and leaves of *Ginkgo biloba,* possess important pharmacological properties. Structural investigations on ginkgolides revealed that they are unique cage molecules, representing diterpene lactones incorporating a tertiary butyl group and six five-membered rings including Ginkgolides A, B, C, M, and J (see Figure 11.8).[13] Ginkgolides specifically inhibit platelet-activating factor from binding to its receptor, thereby preventing platelet aggregation. This group of natural compounds has a long history of safe use in humans, lack of toxicity, and total resistance to metabolism. Among them, ginkgolide B is the most bioactive.[14] In 1998, comparative molecular field analysis (CoMFA), a three-dimensional quantitative SAR (3D-QSAR) study, was conducted to understand the correlation between the physicochemical properties and the in vitro bioactivities of ginkgolide analogues. Based on the results of CoMFA, scientists designed some new compounds, three of which demonstrated a two- to fourfold increase in potency compared with ginkgolides.[15]

Although random compound screening in animal models is still a useful approach to discover new drugs, the disadvantages are obvious. It requires a large amount of compound; its sensitivity is low; and it is very time-consuming. In an average of 5,000 to 10,000 compounds, one might be able to identify four to five good leads. Since the amount of active constituents present in medicinal plants is usually very low, it is impractical, in most cases, to supply sufficient quantities of pure natural compounds for animal experimentation.

The tremendous progress made in molecular biology has resulted in defining many physiological processes and their detailed mechanisms of drug action. This advancement in life sciences leads to the establishment of vari-

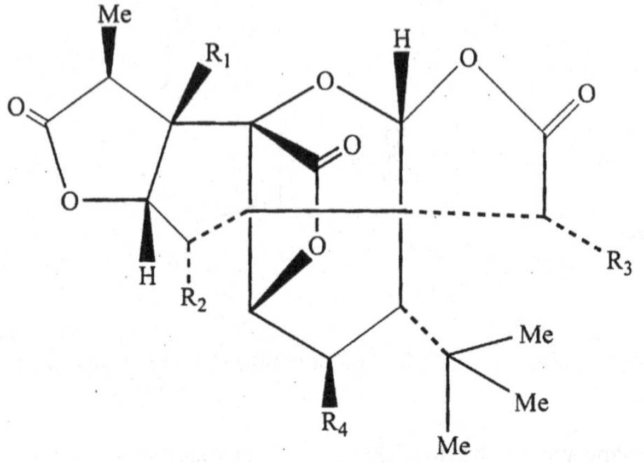

A:	R1=OH	R2=H	R3=OH	R4=H
B:	R1=OH	R2=OH	R3=OH	R4=H
C:	R1=OH	R2=OH	R3=OH	R4=OH
D:	R1=H	R2=OH	R3=H	R4=OH
E:	R1=OH	R2=H	R3=OH	R4=OH

FIGURE 11.8. Chemical structures of ginkgolides.

ous molecular and cellular bioassays in conjunction with high-throughput screening (HTS) methods.

HIGH-THROUGHPUT SCREENING TECHNOLOGIES

The discovery of pharmaceutical agents with novel structures and potential therapeutic activities is a complex process. It usually begins with intensive studies on the relationship between physiology and disease manifestation, followed by identification of relevant genes or proteins as therapeutic targets. Recent advances in life sciences provide exciting opportunities for setting up rapid and large-scale screening assays that use trace quantities of compounds. Coupled with this progress is the development of combinatorial chemistry, where large and structurally diverse chemical libraries can be generated at an unprecedented rate using different techniques including parallel synthesis. Innovations in computer applications, automation technologies, and software design have made it possible to screen hundreds of thousands of compounds within a short period time, thereby accelerating the pace of identifying active molecules or "hits" that can be further developed to potential drug "leads" with desired therapeutic activity. Hence, the phrase *high-throughput screening* is widely used to describe this new approach in drug discovery.

High-throughput screening is a relatively new approach, developed in the 1980s. It is based on the theoretical principles of drug discovery through application of multidisciplinary knowledge and integration of various advanced technologies, thereby forming a rapid, effective, and large-scale drug screening system. The contemporary state-of-the-art HTS techniques are highly automated and computerized to handle sample preparation, assay procedures, and subsequent processing of large volumes of data. Each of these steps requires careful optimization in order to operate efficiently. The HTS system utilizes knowledge and techniques of a variety of disciplines such as medicinal chemistry, molecular and cell biology, biochemistry, pharmacology, computer science, automation engineering, informatics, and mathematics.[16] Major components of HTS may include the following aspects.

Molecular and Cellular Screening Assays

With the application of technology platforms based on molecular and cell biology, many new therapeutic targets, such as genes, receptors, and enzymes, have been identified and validated for drug discovery in recent years. Bioassays developed using such an approach could facilitate our understanding of the relationship between active compounds and target molecules, underlining mechanisms of actions and potential therapeutic value. Most molecular and cellular level bioassays are either in or can be optimized to HTS format. HTS decreases the amount of testing compound required such that only microgram quantities are needed. This is advantageous and permits certain natural products that are difficult to isolate and purify and compounds that are difficult to synthesize to be assayed. It also reduces screening costs and enhances screening capability significantly—numerous compounds can be tested multiple times in different screening models, thereby increasing the probability of discovering new leads and drug candidates. HTS as a primary assay system (including in vitro biological, pharmacological, toxicological, and metabolic assessments) provides an effective tool in the early stage of drug discovery and development process. The results obtained from HTS must be further verified in organ and/or intact animal models. The therapeutic property of any lead compounds can only be established via well-controlled clinical studies that give rise to a new drug.[17]

Sample Preparation and Sample Library

HTS can test hundreds of thousands of compounds every day, so a large collection of samples becomes a basic necessity. Apart from a significant number of samples collected, functions such as storage conditions, ease of search, available analytical data, sample use, and information handling must be integrated into the library organization. This often requires a good computerized information management system. The application of HTS technologies and establishment of large-scale sample libraries have accelerated the development of sample preparation techniques, e.g., combinatorial chemistry, rapid extraction, and isolation methods from natural products, combinatorial biosynthesis. These advances have widened the scope of drug screening and range of materials to be assayed.[18]

Automation Technology

Conventional drug screening methods are slow, costly, and require large amount of samples. In addition, they cannot assay one sample in different models or many samples in one model simultaneously. However, one of the key features of HTS is the use of the automation technology. Based upon a computerized workstation, most HTS assay procedures are automated, including sample loading, mixing, eluting, incubating, and separating; endpoints measurement, data acquisition, reduction, and analysis are managed by a computer, thereby significantly enhancing screening efficiency and reducing assays errors and labor intensity.

Experimental Methods and Technology for Detection

HTS applies trace amounts of materials for experimentation that is carried out in microtiter plates of 96, 384, 864, or 1536 wells with a solution volume between 20 and 200 μL. Since combinatorial chemistry synthesis can also be conducted in 96-well microtiter plates, this makes the 96-well plate a standard platform in most HTS operations. HTS using higher density plates (with the same dimension) is largely employed in industrial settings. In addition, the assay procedure should be as simple as possible with reaction end point distinct enough to measure. Both isotopic and nonisotopic methods have been used in HTS, such as colorimetry, electrophysiology, liquid scintillation, and radioimmunoassay. A number of new techniques have been developed in recent years to better suit the needs of HTS, including sensitivity, convenience, rapidity, and sample-friendliness. Luminescence detection has significant advantages in terms of low cost, safety, and lack of disposal concerns commonly associated with radiochemistry approaches. Resonance energy transfer was one of the earliest methods developed for HTS. Later advances in this area included applications of homogeneous time-resolved fluorescence (HTRF) technology, fluorescence resonance energy transfer (FRET) procedure (for sensing voltage across cell membranes), fluorescence polarization assay (FPA), and fluorescence correlation spectroscopy (FCS).[19]

Computer Technology

In HST, computers are mainly involved in information management, automation control, data processing, and computer-assisted drug screening.

Samples Library Management

A samples library manager can record, store, and manage sample-related information including the name, origin, physiochemical property, molecular formula, molecular weight, structure, as well as preparation method, purity, quality, quantity, storage condition, and relevant literatures. Among them, accurate recording of sample structures to ensure no repetitive collection is of particular importance. Through comprehensive analysis of the information contained in the library, sample characteristics and their biological activities can be studied, thereby improving the screening efficiency.

Data Processing

Acquisition, calculation, reduction, and analysis of a large volume of data points generated by HTS can be managed only by a computer system with relevant software. According to different screening models, appropriate mathematical methods should be adopted in computer-assisted data processing to reflect assay veracity and result validity.

Assay Data Management

Positive results obtained from HTS are accumulating with time. Not only are the contents very complicated but the wealth of information on bioactive samples can also be overwhelming. Thus, proper information management (e.g., data storage, retrieval, and analysis) via a computer database is a critical component in HTS relative to effective use of bioactive results to guide further SAR studies.

Computer-Assisted Drug Screening

A screening information database serves as a tool for scientists to use to explore, interpret, evaluate, and summarize various assay results. Based on the structure of a target protein and compounds to be tested, simulated binding studies can be performed on the computer, thereby predicting potential active hits. This approach could improve screening efficiency and is complementary to HTS. For large-scale compound libraries, computer-assisted virtual screening can reduce assay costs, improve quality, and increase rate of hits. Meanwhile, it can be used to design compounds with novel structures or assist in optimization of lead compounds.

FUTURE PROSPECTS

In the past decade, tremendous progress has been made in HTS technology. For example, the number of assays has increased from 100,000 per year to 100,000 per day, or to even higher numbers in industrial organizations. This implies an enormous demand for structurally diversified chemical compounds. Combinatorial chemistry is an effective method for solving this problem. It is a general term for the approach to synthesize compounds in parallel rather than sequentially. Various techniques have been developed, and some of them are capable of generating vast numbers of different compounds very rapidly. These methods tend to be based on peptides or oligonucleotides. Therefore, although biological activity could be found on HTS, the active compound is unlikely to have the physiochemical properties of a drug. In contrast, natural products are expected to provide the necessary chemical diversity and druglike properties (i.e., they can be absorbed and metabolized in vivo). Bioactive natural products often appear as part of a family of related molecules so that it is possible to isolate a number of homologs and obtain SAR information. Of course, lead compounds discovered from the screening of natural products can be optimized by conventional medicinal chemistry or by application of combinatorial approaches. Since only a small fraction of plant diversity in the world has been tested for biological activities, it can be assumed that natural products will continue to offer novel leads for drug discovery if they are available for screening. However, natural products are unattractive to many pharmaceutical companies because of perceived difficulties related to the complexities of phytochemistry and to their continued access and supply. The technical difficulties concerning isolation and structural elucidation of bioactive natural products are being solved with contributions by chemists worldwide. For instance, extracts can be processed before use via bioassays in order to remove many of the reactive compounds that are likely to cause false-positive results. Fractionation of extracts that are active in screening can be performed rapidly by using high performance liquid chromatography (HPLC), and fractions can be passed directly for analysis by liquid chromatography/mass spectrometry (LC/MS) or even nuclear magnetic resonance (NMR) spectroscopy. By comparing MS data with those from libraries of known compounds, novel molecules in the extract can be distinguished from previously identified compounds. With automated sample injection and fraction collection, the HPLC system can readily and rapidly be used to isolate tens of milligrams pure compounds whose structure is usually resolved by NMR spectroscopy. The entire procedure from an crude extract to a defined molecule can be a matter of days rather than the several months that was routine a few years ago.[20] Such an approach is presently employed in many pharmaceutical research institutions in conjunction with HTS technology aiming at collection of active effluents, isolation of active constituents, and identification of lead compounds from natural products. Structural modification will follow to develop drug candidates. The principal procedures of this method include

1. collection of plant materials;
2. preparation of active fractions guided by HTS;
3. isolation and identification of active components;
4. verification of in vitro bioactivities;
5. evaluation of in vivo bioactivities; and
6. chemical synthesis, semi-synthesis or/and structural modification (see Figure 11.9).

Ultra high-throughput screening (uHTS) is a new concept developed following HTS practices. The basic format of HTS is to screen a large number of compounds on a single drug target, not multiple targets, at a given time, while uHTS is based on the application of biochip technology innovated in recent years that can screen a compound library against hundreds of thousands of targets simultaneously.[21] In certain instances, it also refers to HTS systems that assay extra-large volumes of samples using 3,800-well microtiter plates and robotic workstations.

The biochip was developed during the process of implementation of the Human Genome Project. Its key technology is to place a large number of different genes, protein, or cells on a microcarrier at extremely high spatial densities. The engineering methods are gradually maturing and some research and diagnostic biochips are commercially available on the market. Because the size of the biochip is small but contains numerous potential

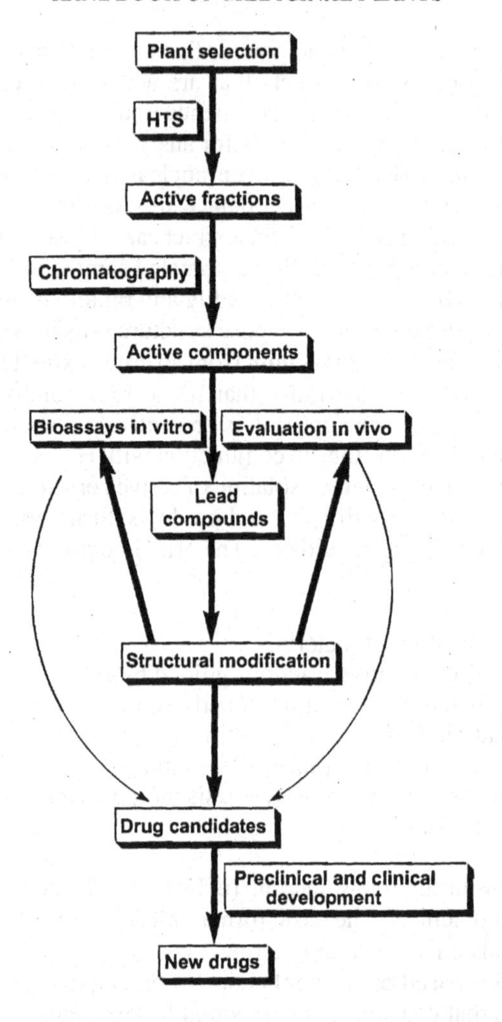

FIGURE 11.9. Drug discovery and development from medicinal plants.

drug targets, the combination of biochip and HTS technologies should, in theory, not only be able to screen a large number of compounds simultaneously, but also to identify and validate new drug targets with known therapeutic agents. This will certainly lead to a revolution in both the methodology and pace of drug screening. Indeed, the latest progress in assay miniaturization marks a transition from HTS to uHTS, ensuring that identification of lead compounds will not be the bottleneck in discovering new drugs. Thus, we are likely to witness a movement toward even greater screening throughput by miniaturization and increased reliability on robotics in production-scale or "industrialized" drug discovery efforts. One of the most challenging aspects for production-scale miniaturization is the microfluidics required for compound library reformatting and reagent addition onto high-density assay formats (e.g., microtiter plates or biochips) in uHTS. Although these screening technologies still require improvement and optimization, the potential advantages cannot be overlooked. All of these options, taken together, will help us discover and develop new and safe drugs quickly and more efficiently.[22]

NOTES

1. Bristol-Myers Squibb Company. <http://www.taxol.com>.

2. Guan, W.B., Huang, W.J., Zhou, Y.C., and Gong, J. Z., 1982, Effect of artemisinin and its derivatives on *Plasmodium falciparum* in vitro, *Acta Pharmacologica Sinica,* 3(2): 139-141.

3. Klayman, D.L., 1985, Qinghaosu (artemisinin): An antimalarial drug from China, *Science,* 228(4703): 1049-1055.

4. Li, Y., Yu, P.L., Chen, Y.X., Li, L.Q., Gai, Y.Z., Wang, D.S., and Zheng, Y.P., 1981, Studies on analogues of artemisinin, *Acta Pharmaceutica Sinica,* 16(6): 429-439.

5. Brossi, A., Venugopalan, B., Gerpe, L.D., Yeh, H.J.C., Flippen-Anderson, J.L., Buchs, P., Luo, X.D., Milhous, W., and Peters, W., 1988, Arteether, a new antimalarial drug: Synthesis and antimalarial properties, *Journal of Medicinal Chemistry,* 31(3): 645-650.

6. Dhingra, V., Vishweshwar, R.K., and Lakshmi, N.M., 2000, Current status of artemisinin and its derivatives as antimalarial drugs, *Life Sciences,* 66(4): 279-300.

7. Kozikauski, A.P., Xia, Y., and Rajirathnam, R.E., Tuckmantel, W., Hanin, I., and Tang, X.C., 1991, Synthesis of huperzine A and its analogues and their antiacetylcholinesterase activity, *The Journal of Organic Chemistry,* 56(15): 4636-4645.

8. Campiani, G., Sun, L.Q., Kozikowski, A.P., Aagaard, P., and McKinney, M., 1993, A palladium-catalyzed route to huperzine A and its analogues and their anticholinesterase activity, *Journal of Organic Chemistry,* 58(27): 7660-7669.

9. Zhu, D.Y., Tang, X.C., Lin, J.L., Zhu, C., and Shen, J.K., 1994, Derivatives of Huperzine and their applications, Chinese patent number 1125725.

10. Wang, G.S., 1980, The clinical studies of cantharidin, *Yao Xue Tong Bao,* 15(2): 119-123.

11. Wang, C.C., Wu, C.H., Hsieh K.J., Yen, K.Y., and Yang, L.L., 2000, Cytotoxic effects of cantharidin on the growth of normal and carcinoma cells, *Toxicology,* 147(2): 77-87.

12. McCluskey, A., Bowyer, M.C., Collins, E., Sim, A.T.R., Sakoff, J.A., and Baldwin, M.L., 2000, Anhydride modified cantharidin analogues: Synthesis, inhibi-

tion of protein phosphatases 1 and 2A and anticancer activity, *Bioorganic and Medicinal Chemistry Letters,* 10(15): 1687-1690.

13. Maruyama, M., Terahara, A., Itagaki, Y., and Nakanishi, K., 1967, The Ginkgolides, I. Isolation and characterization of the various groups, *Tetrahedron Letters,* 8(4): 299-302; Maruyama, M., Terahara, A., Itagaki, Y., and Nakanishi, K., 1967, The ginkgolides, II. Derivation of partial structures, *Tetrahedron Letters,* 8(4): 303-308; Maruyama, M., Terahara, A., Itagaki, Y., and Nakanishi, K., 1967, The ginkgolides, III. The structure of Ginkgolides, *Tetrahedron Letters,* 8(4): 309-313; Weinges, K., Hepp, M., and Jaggy, H., 1987, Chemistry of ginkgolides, II. Isolation and structural elucidation of a new ginkgolide, *Liebigs Annalen der Chemie,* 6: 521-526.

14. Braquet, P.G., 1985, Bn-52021 and related compounds: A new series of highly specific PAF-acether receptor antagonists, *International Journal of Immunopharmacology,* 7(3): 384.

15. Chen, J.L., Hu, L.H., Jiang, H.L., Gu, J.D., Zhu, W.L., Chen, Z.L., Chen, K.X., and Ji, R.Y., 1998, A 3D-QSAR study on ginkgolides and their analogues with comparative molecular field analysis, *Bioorganic and Medicinal Chemistry Letter,* 8(11): 1291-1296.

16. Kell, D., 1999, Screensavers: Trends in high-throughput analysis, *Trends in Biotechnology,* 17(3): 89-91; Sittampalam, G.S., Kahl, S.D., and Janzen, W.P., 1997, High-throughput screening: Advances in assay technologies, *Current Opinion in Chemical Biology,* 1(3): 384-391; Sundberg, S.A., 2000, High-throughput and ultra-high-throughput screening: Solution- and cell-based approaches, *Current Opinion in Biotechnology,* 11(1): 47-53; Fernandes, P.B., 1998, Technological advances in high-throughput screening, *Current Opinion in Chemical Biology,* 2(5): 597-603.

17. Jayawickreme, C.K. and Kost, T.A., 1997, Gene expression systems in the development of high-throughput screens, *Current Opinion in Biotechnology,* 8(5): 629-634.

18. Houston, J.G. and Banks, M., (1997). The chemical-biological interface: Developments in automated and miniaturized screening technology, *Current Opinion in Biotechnology,* 8(6): 734-740.

19. Silverman, L., Campbell, R., and Broach, J.R., 1998, New assay technologies for high-throughput screening, *Current Opinion in Chemical Biology,* 2(3): 397-403; Sittampalam, G.S., Kahl, S.D., and Janzen, W.P., 1997, High-throughput screening: Advances in assay technologies, *Current Opinion in Chemical Biology,* 1(3): 384-391; Weinstein, J.N., Myers, T.G., O'Connor, P.M., Friend, S.H., Fornace, A.J. Jr., Kohn, K.W., Fojo, T., Bates, S.E., Rubinstein, L.V., Anderson, N.L., et al., 1997, An information intensive approach to the molecular pharmacology of cancer, *Science,* 275(5298): 343-349.

20. Harvey, A.L., 1999, Medicines from nature: Are natural products still relevant to drug discovery? *Trends in Pharmacological Sciences,* 20(5): 196-198; Lawrence, R.N., 1999, Rediscovering natural product biodiversity, *Drug Discovery Today,* 4(10): 449-451; Hughes, D., 1998, New HTS imaging technology deal, *Drug Discovery Today,* 3(10): 438-439.

21. Sittampalam, Kahl, and Janzen, 1997, New assay technologies for high-throughput screening.

22. Hughes, 1998, New HTS imaging technology deal; Dunn, D.A. and Feygin, I., 2000, Challenges and solutions to ultra-high-throughput screening assay miniaturization: Submicroliter fluid handling, *Drug Discovery Today,* 5(12): 84-91; Debouck, C. and Goodfellow, P.N., 1999, DNA microassays in drug discovery and development, *Nature Genetics,* 21(1): 48-50.

Chapter 12

Biotechnology in Medicinal Crop Improvement

Efraim Lewinsohn
Ya'akov Tadmor

INTRODUCTION

Recent advances in molecular biology have offered promising tools for the creation of many novel crop varieties with improved nutritional and health values, resistance to herbicides, pests, diseases, pollutants, and adverse climatic conditions. These improvements have been accomplished by the use of molecular markers to monitor breeding programs and by using genetic engineering to introduce new traits to plants.

It is possible to manipulate the levels and composition of active pharmaceuticals in medicinal plants using genetic engineering. To efficiently accomplish this task, it is important to elucidate the biochemical pathways and the pattern of expression of the genes responsible for the synthesis of the specific natural products to be modified. To confer plants with desired traits, novel pathways have been introduced into target plants and, in parallel, endogenous genes have been "turned off" using proper constructs. Thus, it has been possible to manipulate biosynthetic pathways to cause either the accumulation of valuable metabolites either by enhancing their synthesis or by preventing their degradation. In addition, the organ specificity or temporal restrictions for the production of natural products can be overcome using genetic engineering. The emerging opportunities for the incorporation of marker-assisted breeding and genetic engineering into existing breeding programs for the full exploitation of the biosynthetic potential of medicinal plants are described in this chapter. In addition, a few examples demonstrating useful applications of these new technologies are presented.

This paper is dedicated to our beloved spouses, Dalia and Ofra, for their long-term friendship, support, and dedication.

Medicinal crops are valuable as sources of fresh and dry herbs, essential oils, botanicals, food supplements, and pharmaceuticals. The main approach to improve the agronomic and quality traits of plants has been through classical genetics and breeding. This approach has been tremendously successful and has yielded a huge number of crop varieties with unique agronomical attributes. However, the implementation of classic genetic methodologies is a relatively slow process and depends on the relatively narrow genetic variability within many crop plants. A transfer of a trait from one genetic background to another via sexual hybridizations often results in hybrids containing some undesired traits of the donor parent. Several backcrosses to the recurrent parent are necessary to eliminate these unfavorable traits. This is often a long, costly, and tedious process.[1]

Molecular markers are utilized to track the linked gene of interest and accelerate recovery of the recurrent parent genome in backcross programs.[2] More recently DNA molecular markers have also been utilized to create DNA fingerprints of plants and have become useful tools for genotype identification, pedigree analysis, and estimating genetic distance between organisms. DNA fingerprints are also utilized for plant variety protection and purity, ensuring breeders' rights. Classical breeding programs usually utilize phenotypic selection, while the utilization of molecular markers often assists in the development of the desired phenotype by the selection at the genotypic level. The utilization of molecular markers has significantly reduced the time and costs of the backcross processes, increased the efficiency of precise selection of desired phenotypes, and thus has increased the efficacy and profitability of breeding programs. However, the utilization of molecular markers does not eliminate the necessity to conduct crosses, and thus the source of contributing genes is limited by reproductive barriers. Transfer of genes is limited to the same biological species or to very close relatives.

A completely different approach to crop breeding has been the implementation of genetic engineering. Since the 1970s, technologies to incorporate one or several new genes into existing varieties are beginning to eliminate the need for backcrosses. Genetic engineering is the transfer and manipulation of genes among organisms in order to obtain useful products. The implementation of genetic engineering into crop breeding programs has become a reality, gaining much importance (and controversy) in the past twenty years. Through genetic engineering, the gene that is transferred may either originate from the same species or from a completely unrelated organism. Moreover, genetic engineering enables researchers to control and modify the expression pattern of the transferred genes. For example, it is possible to transfer a gene that is normally expressed in a flower of one plant species and force its expression in the fruit of the target plant.[3] Genetic engineering has therefore an unprecedented ability to complement existing breeding programs.

The commercialization of many genetically engineered plants and plant products is currently being actively pursued by biotechnology and seed companies, and many of the genetically engineered plants are presently field-tested to determine their potential for commercialization.[4] Examples of such comprise genetically engineered crops resistant to pests and pathogens, including nematodes, insects, viruses, and fungi; products with extended shelf life; seeds with improved amino acid or oil composition, increased starch content, and improved flavors; and crops tolerant to environmentally friendly herbicides.[5] In fact, genetically engineered soybean, corn, and cotton are increasingly produced in the United States. According to the U.S. Department of Agriculture (USDA), 63 percent of U.S. soybean acreage, 24 percent of corn acreage, and 64 percent of cotton acreage are planted with genetically modified organisms (GMOs).[6]

Medicinal plants constitute a group of crops presently gaining increasing popularity. To meet the demands of the growing markets for fresh and dried herbs and natural alternative medicine, better varieties of medicinal crops are constantly being produced. Until very recently the utilization of biotechnological tools has been very limited in medicinal plant as compared to other crops.

UTILIZATION OF DNA MARKERS IN MEDICINAL PLANTS

Many traits of interest to modern agriculture are very hard to score and select for because they are expressed at a certain developmental or physiological state and their analysis might be complicated and costly. For example, in the case of breeding for resistance to a disease or a pest, it is imperative to inoculate the population under study. This might be a long, costly, and complicated step, especially if containment of the pathogen or pest is needed to avoid the spread of disease. At times, a trait that is easy to determine is genetically linked to the trait of interest that is hard to determine. In such cases, and if the linkage is strong enough, it is advantageous to utilize the first trait as a genetic marker in lieu of selecting directly for the trait of interest. Morphological markers have been utilized in animal and plant breeding since ancient times.[7] Easily identifiable morphological markers have several major limitations. They are not only very limited in numbers but also often based on a phenotype that is not always desired. To alleviate this limitation, biochemical markers based on differences in protein migration patterns during electrophoresis (isozymes and other proteins) were developed.[8] Nevertheless, biochemical markers suffer from being specific to developmental and physiological states of specific tissues, and they reflect phenotypic differences. Thus, the homozygotes cannot always be easily distinguished from the dominant heterozygotes. Recombinant DNA technology has allowed the generation of a large number of genetic markers based directly on differences at the DNA sequence level. DNA markers are alleles of loci at which there is sequence variation—or polymorphism—in DNA that is usually neutral in terms of phenotype. Some advantages of DNA markers are that they are practically unlimited in their number, they are not influenced by the environment, and they can be scored at all stages of plant growth. Moreover, frequently only minimal quantity of DNA is required for analysis. DNA can be extracted from most tissues, fresh, dry, pulverized, and even fossilized. These advantages make DNA markers ideal for research in medicinal and aromatic plants since many of the quality characteristics of these plants are usually difficult to score. In addition, many medicinal and aromatic plants are either still collected from the wild or are only at first stages of cultivation, and as such, the utilization of molecular markers could enhance the efficiency of introduction, cultivation, and selection as well as in establishing reliable methods for product authentication. In general, utilization of molecular markers can assist in variety identification (variety purity and protection), phylogenetic studies, identifying, tagging, and mapping genes of interest, and marker-assisted breeding DNA markers including markers that require prior sequence knowledge and those that do not require such knowledge. The former are usually species specific and include simple sequence repeats (SSR) and restriction fragment length polymorphisms (RFLP). Markers that do not require prior sequence data include random amplified polymorphic DNA sequence (RAPD), inter-simple sequence repeats (ISSR) and amplified fragment length polymorphism (AFLP). Markers that do not require prior sequence data are therefore more universal *Anthus* and are more commonly utilized in studies involving medicinal plants.

DNA Fingerprinting

DNA fingerprinting is a powerful new procedure that is used to gain information on the genetics of living organisms. DNA fingerprints are essential tools for the identification and validation of plant materials as well as for the protection of breeders' rights. The data from DNA fingerprinting studies could also assist in the production of new medicinal plant varieties with enhanced levels of bioactive compounds. Fingerprinting data are utilized in biodiversity studies and for the assessment of genetic relatedness among accessions from various origins.[9] The use of DNA fingerprinting for germplasm characterization facilitates the conservation and utilization of plant genetic resources, permitting the identification of unique accessions or sources of genetically diverse germplasm. This is crucial for the production of novel improved crop varieties. Polymerase chain reaction (PCR)–based methods including ISSR and RAPD are highly informative techniques that have been used to fingerprint a wide range of crop DNA. Several examples for the utilization of DNA fingerprinting in the characterization, improvement and quality control, and assurance of medicinal plants and their products are described in the next paragraphs.

Genetic Diversity Studies

Medicinal plants still lack the biotechnological infrastructure necessary to utilize sequence-based markers. Thus, molecular markers that do not require prior sequence data are preferentially utilized. Fingerprinting data is collected and analyzed with a special algorithm that is based on the comparison of the allelic similarities. The genetic variation within and between the samples under study and the genetic distances are estimated. The genetic diversity data assist in the understanding of pedigree and genetic relatedness among species and accessions.

Echinecea is an important medicinal crop worldwide. Wolf et al. has applied the RAPD technique for analyzing the genetic diversity of three commercially important *Echinacea* species.[10] Two RAPD markers have been determined to be useful to aid in discrimination between the species. A more detailed study identified seventeen additional RAPD markers as discriminators among the three species.[11] The markers data were utilized to characterize the interspecific relationships and the intraspecific diversity, and to identify unique accessions at the genomic level. These markers can now be utilized for routine cultivar and species identification and for adulteration studies of dried plant material.

Similarly, Nebauer et al. assessed the genetic relationships within the genus *Digitalis* based on PCR-generated RAPD markers.[12] The leaves of *Digitalis purpurea* (purple foxglove) contain a large number of pharmacologically active cardiac glycosides, the main one being digitoxin. Other commercial species of *Digitalis,* such as the Grecian foxglove *(D. lanata),* also contain cardiac glycosides, but they differ in composition and contain mainly digoxin.[13] RAPD markers were efficient in detecting interspecific variation among six species of the genus *Digitalis: D. obscura, D. lanata, D. grandiflora, D. purpurea, D. thapsi,* and *D. dubia,* and the hybrid *D. excelsior* (*D. purpurea* × *D. grandiflora*). The species relationships revealed by the RAPD were fully consistent with those previously obtained using morphological affinities corroborating previous taxonomic data.

Similar genetic diversity studies differentiated cultivars, populations, and species of marjoram (*Origanum majorana* L.) and *Hypericum perforatum*, among others.[14] These studies demonstrate that a rapid, simple, but reliable molecular protocol can be efficiently utilized for the characterization of plant collections gaining important genetic background and evolutionary information.

Cultivar Identification and Characterization with DNA Molecular Markers

Morphological or physiological and chemical characters are not always sufficient for quick, easy, and precise cultivar identification. Many cultivars have similar flower and foliage color and shape, and can be difficult to distinguish solely based on morphology traits. Morphological and chemical characteristics may be expressed late in development, or they may be influenced by environmental factors. These factors make cultivar identification based on morphology and chemistry difficult and often unreliable, as plants must be maintained until all the characteristics in question develop. Molecular markers provide a reliable and cost-effective alternative for accurate cultivar identification. We describe here several examples of the utilization of DNA molecular markers for cultivar identification in medicinal plants.

Currently two major herbal products are referred to as ginseng. They are Korean or Chinese ginseng *(Panax ginseng)* and American ginseng *(P. quinquefolium)*. The market for ginseng in the United States is estimated to be $300 million annually. Current tests for ginseng species identification rely on expert botanical identification of fresh plant/root specimens or on biochemical characterization of active and marker compounds (e.g., ginsenosides). Mihalov et al. developed several successful DNA-based protocols for the identification of commercial samples of ginseng. The methods are based on sequence differences in the nuclear ribosomal internal transcribed spacer region or the chloroplast ribulose 1,5-bisphosphate carboxylase large subunit gene and DNA fingerprinting by RAPD.[15]

Skullcap is the common name for three different *Scutellaria* medicinal plants: *S. galericulata, S. lateriflora,* and *S. baicalensis*. Dried aerial parts of *S. galericulata* and *S. lateriflora* are difficult to distinguish morphologically. Hosokawa et al. utilized RAPD to distinguish between these species. They analyzed 10 primers that produced 92 bands. Cluster analysis produced a dendrogram of genetic relatedness that was in good agreement with the taxonomic designations of the three species and discovered primers that produced specific bands to each of the three species.[16] Thus, the RAPD markers should be useful for the future identification and differentiation of members of the three species.

Authentication and Adulteration of Plant Natural Products

Dried herbs obtained from herb wholesalers can be adulterated with other material; for example, other less expensive plant species can sometimes be added, reducing quality, safety, and efficacy. Methods for extracting DNA from fresh and dried herbs have been developed and thus DNA markers can be utilized for authentication and adulteration detection. The following section provides several examples for such usage of molecular markers.

Dendrobium (herb) (Shi Hu) is a commonly used Chinese medicine derived from the stem of several orchid species belonging to the genus *Dendrobium*. Shi Hu is an expensive commodity and adulteration is frequent. Proper authentication of the medicinal species is necessary to protect consumers and support conservation measures. Lau et al. utilized DNA variation to differentiate among *Dendrobium* species and common adulterants.[17] The intraspecific variation among the *Dendrobium* species studied was only about 1 percent. This study demonstrated the power of the utilization of molecular markers for the detection of adulteration in herbal preparations.

Da Huang (*Rheum* spp. [root and rhizome], medicinal rhubarb), a famous and important traditional Chinese medicine, has often been confused with an adulterant species in the same genus, *Rheum*. Through sequencing specific regions of the chloroplast DNA of 13 species of *Rheum* (3 medicinal rhubarb species and 10 adulterant ones), a specific sequence for the medicinal species was found. A molecular marker specific to medicinal rhubarb was thus designed and proved to be highly effective for differentiation between medicinal rhubarb and adulterants.[18] Molecular markers have also been utilized for the authentication of other medicinal plant preparations.

CONSTRUCTION OF GENETIC MAPS UTILIZING DNA MARKERS

Molecular markers are useful tools for developing detailed linkage maps of species that previously were very poorly mapped. These maps have obvious utility for the identification of markers linked to genes of agronomic, horticultural, or medicinal importance. A linkage map is a graphical presentation of the relative position of the genes on the chromosomes based on crossovers and recombination frequencies. Today's technologies enable the detection of enough polymorphic DNA markers in a segregating population to create a saturated linkage map. The development of genetic linkage maps provides a direct method for selecting desirable genes by their linkage to an easily detectable molecular marker. Linkage maps are essential for the identification of genes involved in multigenic traits and serve as a powerful tool for studying the inheritance of complex traits. Molecular maps can provide detailed information regarding the structure of the genome of a species, the evolutionary relatedness of species within a family, sequence rearrangements during evolution, and the relationship between physical and genetic distances. Marker-assisted selection (MAS) is the process by which selection for a desired genotype is assisted by markers[19] (see Figure 12.1). Linkage of DNA markers to important traits followed by MAS has been proved and utilized in several crops including maize, wheat, rice, and tomato. Linkage maps have not been fully developed in medicinal plants. Many of the medicinal plants are collected from the wild or were only recently introduced into cultivation and thus are still genetically very heterogeneous. Nevertheless, new analytical methodologies developed for the analysis of heterozygous material present in timber and fruit species, coupled with the decrease in the costs for DNA marker analysis have opened new opportunities for the construction of saturated linkage maps in medicinal plants. Yet there are only a very few examples of linkage maps developed in medicinal plants.

The opium poppy *(Papaver somniferum)* is the source of opium. Morphine is the main alkaloid of opium; while codeine is found in a lesser concentration.[20] There is a growing market for low-morphine poppy. For that, a practicable verification test for plants with low morphine content is advantageous. A partial linkage map in opium poppy has been constructed by the analysis of the segregation of AFLP, RAPD, and morphological markers in an F_2 population derived from a cross between a high- and a low-morphine poppy line.[21] Linkage analysis revealed 16 linkage groups, although in poppy only 11 pairs of chromosomes are present. Nevertheless, this preliminary map has the potential to provide DNA markers for breeding for morphine content once association with particular markers is found.

UTILIZATION OF GENETIC ENGINEERING IN MEDICINAL PLANTS

The production of genetically engineered plants depends on the stable introduction of foreign DNA into the plant genome, followed by regeneration to produce intact plants and expression of the introduced gene(s). Once

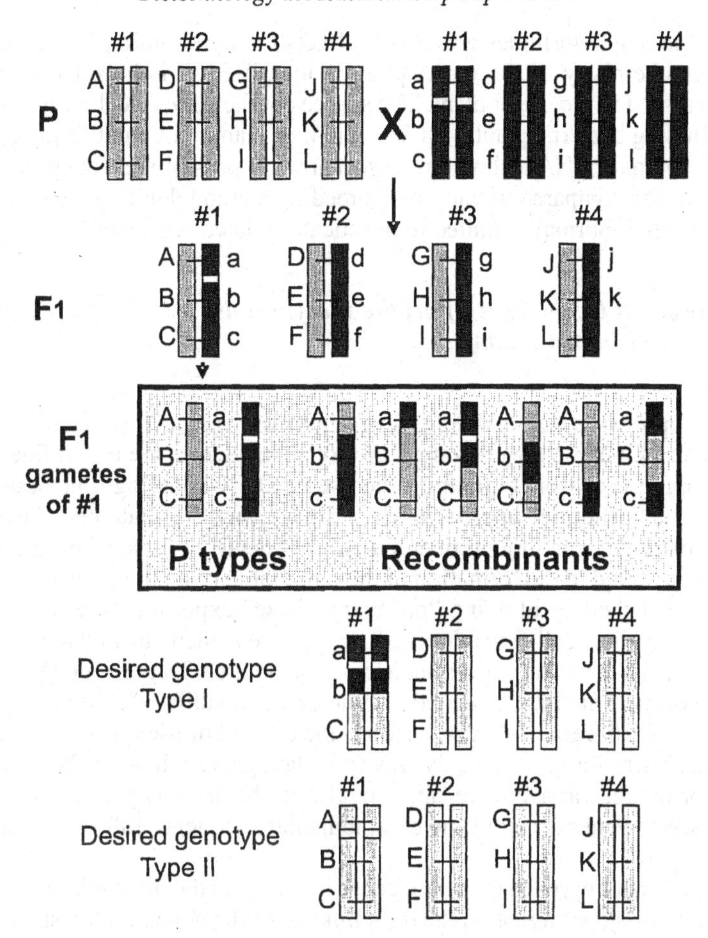

FIGURE 12.1. Schematic representation of marker-assisted selection (MAS). Assume we are breeding an organism in which $n = 4$. A receptor elite line (gray chromosomes) and a donor of the white trait located on chromosome #1 (black chromosomes) are crossed. Both parental and recombinant gametes are produced in the F_1. In the figure we observe only recombinant gametes of chromosome #1, but similarly recombination occurs in all other chromosomes. Assume we have 12 DNA markers, *A* to *L*, which are evenly distributed on the chromosomes. Gray chromosomes have the capital alleles, *A, B, C*, etc., while the black chromosomes carry the small-letter alleles, *a, b, c*, etc. We select genotypes that carry the *a* and *b* alleles (the flanking markers for the trait of interest) while having the *C*, and *D* to *L* capital alleles to get desired genotype Type I. Fine mapping of the *AB* region enables the precise introgression of the gene of interest with minimal linkage drag yielding the Type II desired genotype.

a gene that determines a trait that affects the agronomic parameters or the phytochemistry of the target plant is identified, it is introduced into and expressed in the target plant. The sources of the genes can be any organism, including bacteria, plants, insects, fungi, or mammals, and this in fact grants recombinant DNA technology a great array of possibilities for crop improvement—as compared to traditional breeding methodologies, in which transfer of genes is normally limited to genetically related organisms.

Sources of Genes That Affect the Biosynthesis of Plant Natural Products

Approaches for isolating genes involved in the formation of natural products include the utilization of protein sequence, the use of fine genetic mapping (positional cloning), the use of DNA-homology methodologies, and the "genomic" approach. The primary way to isolate key genes in the formation of a natural product initiates by identifying a key enzymatic reaction involved in the formation of the metabolite of interest. This is normally accomplished by utilizing "pulse and chase" experiments as well as catalytic assays in cell-free extracts. The enzyme is then purified to homogeneity and a partial amino acid sequence is generated. Degenerate DNA probes and primers are then designed, synthesized, and utilized to screen appropriate cDNA libraries in order to isolate the desired full-length clones. This approach was initially the only way to isolate genes when no DNA sequence information was available, and it might still be the only practical alternative to isolate genes for which no complementary sequence information is available.

Another approach to isolate genes involved in the biosynthesis of natural products is contingent on the tight linkage of the gene of interest with DNA markers. This approach is termed "map-based positional cloning." Map-based cloning is an iterative approach that identifies the underlying genetic cause of a mutant phenotype. The major strength of this approach is the ability to tap into a nearly unlimited resource of natural and induced genetic variation without prior assumptions or knowledge of specific genes. This approach is particularly promising to identify genes from species in which highly saturated genetic maps are available.[22] An excellent example to this approach is the use of introgression lines.[23] Introgression lines are generated by hybridization of individuals belonging to different species, followed by marker-assisted backcrosses. In introgression-line populations, defined chromosomal segments derived from one species (usually a wild relative) are integrated in known regions of the chromosomes of another species (usually an elite crop variety).[24] Tomato introgression lines displaying the typical discoloration of the previously characterized phenotypes 'Beta', 'Old Gold', and 'Tangerine' led to the identification and the characterization of genes coding for novel enzymes in the well-studied carotenogenesis pathway.[25] In a similar way, by systematically analyzing the volatile profiles of tomato introgression lines, a chromosomal area that codes for the production of 2-phenylethyl alcohol and phenylacetaldehyde has been mapped and characterized.[26] These compounds are normal constituents of the tomato aroma, but they are detrimental to tomato aroma when present in high levels. This trait *(malodorous)* is present in the wild tomato relative *(Lycopersicon pennellii)* used to construct the introgression lines population and was apparently lost during domestication of the modern tomato varieties.[27] Other chromosomal regions that code for the production of key aroma compounds in tomatoes have also been identified.[28] The use of positional cloning will ultimately lead to the isolation of these genes. Most of these genes could be utilized for the modification of essential oil composition and the aroma of herbs.

A third approach, possibly the most common today to isolate novel genes, takes advantage of discrete areas of similarity at the DNA level often found in genes that belong to particular multigene families, or in genes with related biochemical or structural function. By taking advantage of the existing conserved areas of similarity in such genes, it is possible to isolate novel genes with a related function, or that code for the production of related metabolites. The phenomenon of sequence conservation during gene evolution has been very useful in studies aimed at discovering new *O*-methyltransferase genes acting on novel substrates, novel sesquiterpene synthases and other genes involved in secondary metabolism.[29]

The Genomic Approach

During the past few years, "genomic" methodologies, aimed at analyzing the whole genome of organisms, have been developed. In other words, the genomic approach attempts to examine the totality of the genetic information including all the genes as well as the nontranscribed regions of the DNA, in contrast to the more classical approach that dissects individual traits. These novel methodologies include specially developed high-throughput technologies, through which a large number of sequences, genes, and gene products can be examined and characterized.[30]

Whole Genome Sequence Analysis

Sequence information of complete genomes is currently available. These include information on 90 fully sequenced prokaryotic genomes, and 8 fully sequenced eukaryotic genomes, including the human, mouse, *Arabidopsis thaliana,* and rice, while many genome projects of other important plants such as maize, tomato, and *Medicago truncatula* (an annual relative of alfalfa but with a genome size of only half that of alfalfa) are in progress. From these studies we have learned that in spite of the high phenotypic diversity among organisms, many similarities in genome organization may be found.[31] Therefore, the genomic information obtained for one organism is of relevance to another less-studied organism. No substantial effort has yet been made to obtain the whole genome sequence of a medicinal crop; however, tomato genomic data, for example, might be of high relevance to medicinal plants belonging to the Solanaceae family and to other plant families as well.

Expressed Sequence Tags (ESTs)

ESTs are generated by massive sequencing of cDNA generated from the mRNA of the tissue of interest. ESTs are typically short (normally only partially represent the full-length clones) and are of relatively low sequencing quality. Nevertheless, ESTs provide a profile of the mRNA population at a particular time and offer a quick method for cloning and examining a large number of genes known to be expressed in a particular cell population or tissue. Sequence information for more than 7 million ESTs is publicly available. The information includes more than 300,000 ESTs from barley (*Hordeum vulgare* subsp. *vulgare*), 290,000 ESTs from soybean *(Glycine max),* 260,000 from wheat *(Triticum aestivum),* about 150,000 from tomato *(Lycopersicon esculentum),* and many other plants. In the case of medicinal plants, EST information is still lacking, although more than 20,000 are known from *Capsicum annuum* and 5,500 from *Stevia rebaudiana.*

These promising technologies will undoubtedly increase our knowledge concerning processes taking place in medicinal plants. The information gained so far is readily searchable and has enlightened our understanding of the expression patterns and the control of genes involved in simple physiological processes, biosynthetic pathways, and other biological processes as well.

Transcriptome, Proteome, and Metabolome Analyses

Not only information on the DNA, genome organization, and gene expression patterns are being made available. Systematic modifications of other existing biochemical and molecular biology methodologies have been designed to allow for the processing of thousands of samples simultaneously. This has effectively shortened the time and cut analysis cost of thousands of biological samples. The new developments include the use of microarrays, or cDNA chips, that allow monitoring the expression levels of tens of thousands of genes simultaneously, generating what are called "transcriptome" profiles.[32] Developments in automation of two-dimensional polyacrylamide gel electrophoresis, coupled with mass spectrometry adapted to proteins, has resulted in a technique often described as "proteomics" by which it is possible to monitor abundance patterns and often identify thousands of proteins simultaneously.[33] In addition, developments in mass spectrometry have resulted in an approach often called "metabolomics" or "metabolic profiling" by which the levels of thousands of low molecular weight metabolites including sugars, acids, and other volatile and nonvolatile plant constituents can be monitored simultaneously.[34]

Bioinformatics

Bioinformatics provides the tools to process the unprecedented high volume of biological information accumulated through genomic studies. Many useful computer programs and algorithms to process the genomic data have been developed and are readily available. In utilizing bioinformatics, many putative biochemical functions are assigned to thousands of individual sequences, based on similarity to known genes and their expression patterns. For most cases this is sufficient to predict the role of a new available gene with a great degree of accuracy. Nevertheless, the final confirmation of the biological function of a sequence is normally made by functional expression utilizing homologous or heterologous systems. This process is not always simple or trivial, especially when annotating genes involved in secondary metabolism.

Combined genomic research has greatly advanced our understanding of gene regulation and the processes that control physiological phenomena. In a way, these developments have allowed researchers to examine biological processes utilizing a broader multidisciplinary perspective, which is much more difficult to achieve using conventional techniques. The simultaneous and concerted analyses of a large number of samples has corroborated much of the previously known information but also has yielded much new information. Assessment of this information has allowed us to conclude that many biosynthetic pathways to secondary metabolites are not only similar to one another but are probably evolutionarily derived from primary metabolism.[35]

Genomic Approaches Applied to Medicinal Plants Research

To take optimal advantage of the high-throughput technologies available to discover genes involved in secondary metabolism, tissue located where the genes of interest are expressed must be identified to serve as the source of genes. For example, mRNA obtained from isolated glandular trichomes from sweet basil has been used to generate sequence information on more than 1,200 ESTs, and this has been the key in identifying many genes involved in the biosynthesis of the essential oil components of basil.[36] Similar projects aimed at elucidating genes involved in secondary metabolism in many plants are currently been pursued. They include studies in peppermint, rose petals, opium poppy *(Papaver somniferum), Stevia rebaudiana,* and tea tree *(Melaleuca alternifolia),* among others.[37] These studies have corroborated previous evidence on the involvement of specific genes and biosynthetic pathways in the formation of mono-, di-, and tetraterpenes; volatile phenolics alkaloids; and other secondary metabolites; but they have also uncovered genes coding for novel methyltransferases, hydroxylases, acetyltransferases, sesquiterpene synthases, and many other genes involved in the formation of natural products.

These examples point to the necessity to functionally express the putative genes in order to verify and/or assess their biochemical function. Moreover, these functions represent those obtained in vitro under specific conditions and do not necessarily reflect the real biochemical role of the specific EST in vivo. The biological function of the sequence in question can also be determined in experiments aiming at overexpression of or silencing the gene in the original plant-utilizing sense, antisense, cosuppression, or RNA interference constructs.[38]

The utilization of high-throughput genomics analyses, including sequence data, transcriptomics, proteomics, and metabolomics, coupled to bioinformatics tools and database mining has allowed us to identify many novel genes that can be used to modify the phytochemistries of medicinal plants using metabolic engineering.

A Few Examples of Metabolically Engineered Organisms

Biotechnology advances have provided the tools to isolate novel genes that affect secondary metabolism and the means to introduce and express these genes in medicinal plants. These genes include members of the terpenoid, phenolic, and alkaloid pathways. Utilizing genetic engineering, it has been possible to modify these pathways and generate metabolically engineered plants with novel pharmacoactive components.

Terpenoid Metabolism

More than 25,000 different terpenoid structures have been described, and although particular terpenoids are restricted to single or a few organisms, terpenoids, as a group, have common biosynthetic origins.[39] Thus the implementation of recombinant DNA technology to modify terpenoid pathways is particularly promising. We describe next a few successful examples of metabolic engineering of the terpenoid pathway.

Monoterpene-improved quality and yield of essential oil in peppermint. Members of the genus *Mentha* (mints) are among the best known and economically important essential-oil-producing plants. Peppermint essential oil mainly consists of menthol and menthone, while high levels of menthofuran are considered detrimental to its quality. Introduction and expression of the gene that codes for the enzyme that catalyzes the conversion of geranyl diphosphate (GPP) to the cyclic monoterpene 4*S*-limonene (4*S*-limonene synthase; see Figure 12.2) have yielded transgenic mint plants with altered essential oil compositions. Limonene and total monoterpene levels varied only slightly as compared to the nontransformed controls. Nevertheless, altered levels of other monoterpenes (both derived directly from GPP or from limonene) were observed in transgenic lines, as compared to controls.[40] Mahmoud and Croteau have recently antisensed the gene coding for the cytochrome P450 enzyme menthofuran synthase in peppermint (see Figure 12.2), generating plants harboring an essential oil with a much lower menthofuran level and therefore a much higher quality than the essential oil from nontransgenic control plants.[41] In addition, ectopic expression of the gene coding for deoxyxylulose phosphate reductoisomerase, a key enzyme in the plastid terpenoid pathway, caused a 50 percent increase in the accumulation of the total essential oil levels in transgenic peppermint.[42] Another example for the metabolic engineering of the monoterpene pathway has been demonstrated in tomato fruit. Expression of the *Clarkia breweri* floral gene linalool synthase (LIS) has resulted in significant enhanced linalool levels in transgenic fruits.[43] These experiments

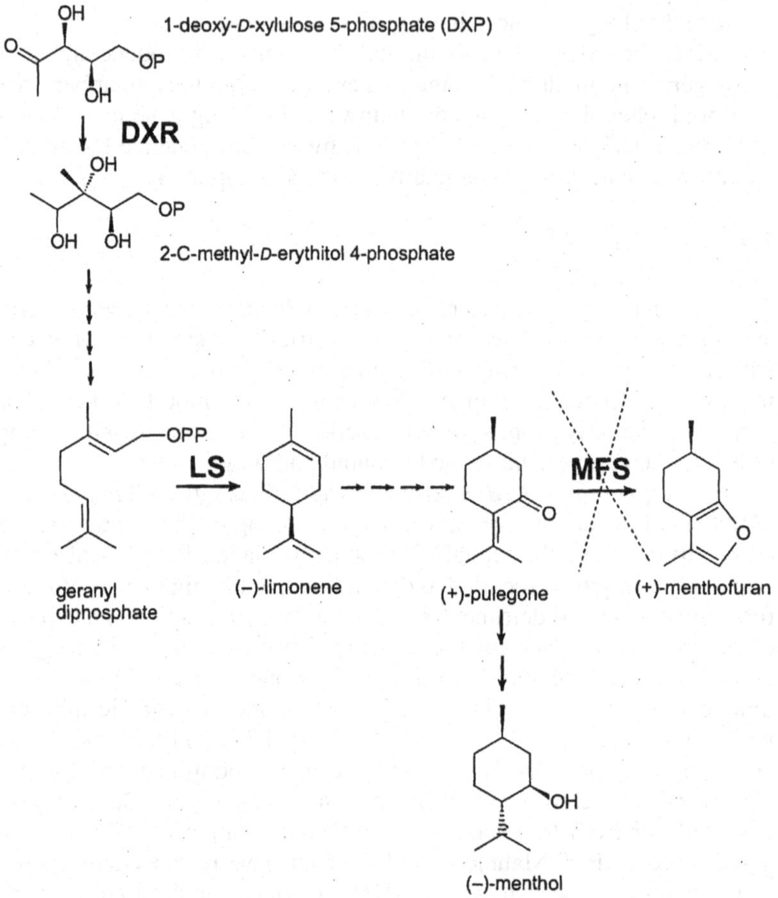

1-deoxy-D-xylulose 5-phosphate (DXP)

DXR

2-C-methyl-D-erythitol 4-phosphate

geranyl diphosphate

LS

(–)-limonene

(+)-pulegone

MFS

(+)-menthofuran

(–)-menthol

FIGURE 12.2. Metabolic engineering of the monoterpenoid pathway in peppermint. Menthol is biosynthesized in peppermint glandular trichomes from geranyl diphosphate (GPP) and limonene by a series of committed enzymes. Increases in limonene synthase (LS) expression have resulted in modified levels of other monoterpenes derived from GPP or from limonene. Silencing of menthofuran synthase (MFS) has resulted in decreased menthofuran levels and increased menthol proportions, and increases in deoxyxylulose phosphate reductoisomerase (DXR) has resulted in a 50 percent increase in the total essential oil levels in the plants as compared to controls.

could be achieved only with prior deep and detailed understanding of the biochemistry and the molecular regulation of monoterpene pathway.[44]

Sesquiterpenes. Some of the sesquiterpenes possess important antifungal and medicinal properties and therefore they have been the target of many biological studies. Novel sesquiterpenes have been produced in transgenic tobacco plants and cell cultures, as well as in bacteria by overexpressing heterologous sesquiterpene synthase genes.[45]

Wormwood (*Artemisia annua* L., family Asteraceae), known in China as qinghao, is an annual herb that contains artemisinin, a sesquiterpene lactone. Artemisinin is available commercially as an antimalarial drug efficacious against drug-resistant strains of *Plasmodium,* the malarial parasite. Malaria is one of the most devastating human diseases and affects more than 200 million people annually. Artemisinin is formed by a series of biosynthetic steps initiated by cyclization of farnesyl diphosphate, the precursor of all sesquiterpenes. Overexpression of a cotton gene encoding farnesyl diphosphate synthase (the enzyme catalyzing farnesyl diphosphate synthesis) in transgenic wormwood plants resulted in two- to threefold increased production of artemisinin as compared to nontransformed controls, probably due to an increase in the pool of available farnesyl diphosphate.[46]

At times, it is important to reduce the levels of a nondesired natural product to improve the value of a given crop. The cadinane sesquiterpenoids gossypol, hemigossypolone, and heliocides (sesterterpenoids, 25-carbon terpenoids) play an important role in pathogen and pest resistance in cotton. Yet gossypol is toxic to humans and is usually removed by an expensive method from cotton seed and cotton seed oil before consumption. Antisense (+)-δ-cadinene synthase constructs were used to transform cotton plants. The transformed cotton contained lower levels of the toxic compounds in their seeds and leaves.[47] These studies have helped elucidate key limiting steps in the production of important metabolites and have opened exciting avenues to efficiently implement the production of the desired compound via metabolic engineering.

Phenolic Compounds: Stilbenes and Phytoestrogenic Flavonoids

Phenolic compounds are widespread in the plant kingdom and have important ecological roles such as defense, pigmentation, and mediators in symbiotic interactions, as well as serve as aroma and taste compounds. Many phenolic compounds have been recognized as beneficial to human health due to their general antioxidant capacities. Some have been found to have specific health-promoting and medicinal properties.

Resveratrol. Several plants, including grapevine, produce the stilbene-type phytoalexin resveratrol when attacked by pathogens. This compound appears to be one of the health-promoting factors of grapevine, which are associated with reduced risk of heart diseases (popularly known as "the French paradox") and long recognized by folk medicine. Clinical studies have demonstrated the beneficial effects of resveratrol, isolated from red wine, on cardiovascular disease and confirmed the involvement of resveratrol in fighting arteriosclerosis and vascular tissue diseases. Resveratrol is synthesized from the ubiquitous precursors malonyl CoA and coumaryl CoA by the enzyme stilbene synthase. The gene encoding this enzyme was cloned from grapevine and introduced into tobacco.[48] Resveratrol was readily accumulated upon induction of the transgenic tissues, rendering the transgenic plants more resistant to fungal attack than the nontransgenic control.[49] Using a similar approach, the stilbene synthase gene could be also used to produce resveratrol in foods already associated with anticancer properties, or to create "functional foods" with novel health benefits. Moreover, since resveratrol can be generated in grape cell suspension cultures, it may also be possible to produce resveratrol to be marketed as a food supplement.

Flavonoids. Many members of the Fabaceae family accumulate a number of isoflavonoid compounds, such as the isoflavones genistein and daidzein as well as their glycosides that exist in soybeans.[50] Several health benefits have been assigned to these compounds, at times referred to as phytoestrogens. Phytoestrogens are associated with relief of menopausal symptoms, reduction of osteoporosis, improvement of blood cholesterol levels, and lowering the risk of certain hormone-related cancers and coronary heart disease.[51] The biochemical basis of these effects has not been fully established but the weak estrogenic activity of isoflavones may be a factor conferring these properties. Isoflavones are synthesized by a branch of the phenylpropanoid pathway and normally play a role in plant defense against fungal attacks. The branching of the flavonoid metabolic pathway to isoflavones occurs by the action of the enzyme isoflavone synthase. Isoflavone synthase, a member of the cytochrome P450 family, oxidizes the flavanone intermediates naringenin and liquiritigenin into genistein and daidzein, respectively.[52] Naringenin is synthesized by most plants as an intermediate to other flavonoids, such as the common anthocyanin pigments.[53] Overexpression of the soybean isoflavone synthase gene in transgenic *Arabidopsis*, tobacco, and maize plants, which naturally do not produce isoflavones, resulted in the production of genistein and its derivatives, possibly through the conversion of endogenous naringenin.[54] Although isoflavones are normally restricted to legumes, these results prove that by metabolic engineering, it is possible to produce health-associated isoflavones in nonlegume plants. This illustrates the feasibility to metabolically engineer plants to generate novel plants with new phytochemistries enhancing their health benefits.

Metabolic Engineering of Alkaloid Metabolism

Alkaloids are low-molecular-weight nitrogen-containing molecules that are an important group of natural products present in plants. Many alkaloids are pharmacologically active and are often the active principles of medicinal plants. Although the biosynthetic pathways to alkaloids in plants are complex and involve several biosynthetic steps, many of the key genes involved in the biosynthetic pathways for some alkaloids, such as scopolamine, nicotine, morphine, and berberine, have been isolated and identified.[55] The prospects and limitations of engineering plant alkaloid metabolism have been discussed.[56] A few examples describing metabolic engineering for the modification of the pharmacoactive principles of medicinal plants are described next.

Terpenoid indole alkaloid. Terpenoid indole alkaloids are biosynthesized by many medicinal plants, including *Rauwolfia serpentina, Catharanthus roseus,* and *Strychnos nux-vomica,* and have many pharmacological uses. They include reserpine which lowers blood pressure, ajmaline which is used to treat heart arrhythmias, and strichnine, a potent poison, formerly used as an invigorating tonic in lower doses due to its effect on the central nervous system.[57] The biosynthetic pathway to indole alkaloids is involved and includes more than 20 enzymatic steps. Nevertheless, the enzyme strictosidine synthase catalyzes a key step in the biosynthesis of terpenoid indole alkaloids: the condensation of tryptamine and secologanin to form strictosidine.[58] Strictosidine is a precursor to many important alkaloids such as strichnine, quinine, vincristine, vinblastine, ajmaline, and ajmalicine.[59] Cell cultures of *Catharanthus roseus* have been transformed with the gene strictosidine synthase *(Str)*. Ectopic expression of the *Str* transgene caused the accumulation of the glucoalkaloid strictosidine and some of its derivatives, including ajmalicine, catharanthine, serpentine, and tabersonine.[60]

Tropane alkaloids: Hyoscyamine and scopolamine. Belladonna extracts were used in Medieval Europe by ladies who wanted their pupils dilated, and hence the name belladonna ("beautiful woman" in Italian). Belladonna (or deadly nightshade, *Atropa belladonna*) accumulates the tropane alkaloid hyoscyamine, which together with its racemic form (atropine) are used as a remedy against spasms in skeletal muscles and as an antidote against cholinesterase inhibitors (organophosphates) poisoning.[61] The epoxidated derivative scopolamine is also an anticholinergic agent, but unlike hyoscyamine it is primarily a CNS depressant. Scopolamine is used prophylactically against motion sickness. Scopolamine is biosynthesized from hyoscyamine by the action of hyoscyamine 6-β-hydroxylase present in *Hyoscyamus niger,* a producer of both hyoscyamine and scopolamine. This gene was transferred to belladonna and constitutively expressed. The transgenic belladonna and the progeny that inherited the transgene accumulated almost exclusively scopolamine in their aerial parts.[62] This study has demonstrated the utilization of transgenic plants in producing an important alkaloid usually produced by a different plant species.

CONCLUSION

We are in the midst of a biotechnological revolution. By the accomplishment of the inherent potential of biotechnology, novel medicinal plants, as well as many plant-derived nutraceuticals, will soon be available to consumers worldwide. Biotechnology has had a profound influence on breeding methodologies, germplasm selection and variety identification in most crops. Nevertheless, the utilization of molecular markers in medicinal plants has been thus far limited for several reasons, including the relative lack of detailed genetic background for the important traits of medicinal plants, the minor market volume for each individual medicinal crop and the allocation of fewer resources to the study of medicinal plants as compared to other crops. Nevertheless, the increasing demand for high-quality medicinal and aromatic plant products as well as the development of less expensive PCR-based markers, has made molecular markers highly attractive tools for the improvement of medicinal plants.

Genetic engineering has allowed the generation of many modified crops with improved agricultural traits. Moreover, metabolic engineering has made it possible to change the photochemistry of many crops, enhancing their nutritional and pharmaceutical value. Nevertheless, many obstacles still need to be addressed before the full economical potential of these novel technologies can be implemented. Most of the obstacles will be overcome as a result of research aimed at solving both technical and fundamental problems, while others involve gaining public acceptance and tackling legal issues. Biotechnology will not replace conventional breeding programs, but it is complementing, augmenting, and making them more efficient.

NOTES

1. Agrawal, R.L., *Fundamentals of plant breeding and hybrid seed production* (Enfield, NH, USA: Science Publishers, Inc. 1998).

2. Frisch, M. and Melchinger, A.E. (2001). Marker-assisted backcrossing for introgression of a recessive gene, *Crop Science,* 41(5): 1485-1494.

3. Lewinsohn, E., et al., 2001, Enhanced levels of the aroma and flavor compound *S*-linalool by metabolic engineering of the terpenoid pathway in tomato fruits, *Plant Physiology,* 127: 1256-1265.

4. International Field Test Sources, 2004. Available online at: <http://www.nbiap.vt.edu/cfdocs/globalfieldtests.cfm>.

5. Galili, G., Galili, S., Lewinsohn, E., and Tadmor, Y., 2002, Utilization of genetic, molecular and genomic approaches to improve the value of plant food and feeds, *Critical Reviews in Plant Sciences,* 21(3): 167-204.

6. USDA Economics, Statistics and Market Information System, 2004, Available online at: <http://usda.mannlib.cornell.edu/>.

7. Agrawal, 1998, *Fundamentals of plant breeding.*

8. Tadmor, Y., Zamir, D., and Ladizinsky, G., 1987, Genetic mapping of an ancient translocation in the genus *Lens, Theoretical and Applied Genetics,* 73: 883-892.

9. Westman, A.L. and Kresovich, S., Use of molecular marker techniques for description of plant genetic variation, in J.A. Callow, B.V. Ford-Lloyd, and H.J. Newbury (eds.), *Biotechnology and plant genetic resources* (Wallingford, UK: CAB International, 1997), pp. 9-48.

10. Wolf, H.T., Zundorf, I., Winckler, T., Bauer, R., and Dingermann, T., 1999, Characterization of *Echinacea* species and detection of possible adulterations by RAPD analysis, *Planta Medica,* 65: 773- 774.

11. Kapteyn, J., Goldsbrough, P.B., and Simon, J.E., 2002, Genetic relationships and diversity of commercially relevant *Echinacea* species, *Theoretical and Applied Genetics,* 105: 369-376.

12. Nebauer, S.G., del Castillo-Agudo, L., and Segura, J., 2000, An assessment of genetic relationships within the genus *Digitalis* based on PCR-generated RAPD markers, *Theoretical and Applied Genetics,* 100: 1209-1216.

13. Samuelsson, G., *Drugs of natural origin: A textbook of pharmacognosy* (Stockholm: Swedish Pharmaceutical Press, 1991); Bruneton, J., *Pharmacognosy, phytochemistry, medicinal plants* (Andover, UK: Intercept Ltd, 1995).

14. Mihalov, J.J., Marderosian, A.D., and Pierce, J.C., 2000, DNA identification of commercial ginseng samples, *Journal of Agricultural and Food Chemistry,* 48: 3744-3752; Schluter, C. and Punja, Z.K., 2002, Genetic diversity among natural and cultivated field populations and seed lots of American ginseng (*Panax quinquefolius* L.) in Canada, *International Journal of Plant Sciences,* 163: 427-439; Negi, M.S., Singh, A., and Lakshmikumaran, M., 2000, Genetic variation and relationship among and within *Withania* species as revealed by AFLP markers, *Genome,* 43: 975-980; Klocke, E., Langbehn, J., Grewe, C., and Pank, F., 2002, DNA fingerprinting by RAPD of *Origanum majorana* L., *Journal of Herbs Spices and Medicinal Plants,* 9: 171-176; Arnholdt-Schmitt, B., 2002, Characterization of *Hypericum perforatum* L. plants from various accessions by RAPD fingerprinting, *Journal of Herbs, Spices, and Medicinal Plants,* 9: 163-170.

15. Mihalov, Marderosian, and Pierce, 2000, DNA identification of commercial ginseng.

16. Hosokawa, K., Minami, M., Kawahara, K., Nakamura, I., and Shibata, T., 2000, Discrimination among three species of medicinal *Scutellaria* plants using RAPD markers, *Planta Medica,* 66: 270-272.

17. Lau, T., Shaw, P., Wang, J., and But, P., 2001, Authentication of medicinal *Dendrobium* species by the internal transcribed spacer of ribosomal DNA, *Planta Medica,* 67: 456-460.

18. Yang, M., Zhang, D., Liu, J., and Zheng, J., 2001, A molecular marker that is specific to medicinal rhubarb based on chloroplast *trnL/trnF* sequences, *Planta Medica,* 67: 784-786.

19. Ribaut, J.M. and Hoisington, D., 1998, Marker-assisted selection: New tools and strategies, *Trends in Plant Science,* 3: 236-239.

20. Samuelsson, 1991, *Drugs of natural origin.*

21. Straka, P. and Nothangel, T., 2002, A genetic map of *Papaver somniferum* L. based on molecular and morphological markers, *Journal of Herbs Spices and Medicinal Plants,* 9: 235-241.

22. Yano, M., 2001, Genetic and molecular dissection of naturally occurring variation, *Current Opinion in Plant Biology,* 4: 130-135.

23. Zamir, D., 2001, Improving plant breeding with exotic genetic libraries, *Nature Reviews Genetics,* 2: 983-989.

24. Ibid.

25. Isaacson, T., Ronen, G., Zamir, D., and Hirschberg, J., 2002, Cloning of tangerine from tomato reveals a carotenoid isomerase essential for the production of beta-carotene and xanthophylls in plants, *Plant Cell,* 14: 333-342.

26. Tadmor, Y., Fridman, E., Gur, A., Larkov, O., Lastochkin, E., Ravid, U., Zamir, D., and Lewinsohn E., 2002, Identification of *malodorous,* a wild species allele affecting tomato aroma that was selected-against during domestication, *Journal of Agricultural and Food Chemistry,* 50: 2005-2009.

27. Ibid.

28. Causse, M., Saliba-Colombani, V., Lesschaeve, I., and Buret, M., 2001, Genetic analysis of organoleptic quality in fresh market tomato, 2. Mapping QTLs for sensory attributes, *Theoretical and Applied Genetics,* 102: 273-283.

29. Ibrahim, R.K. and Muzac, I., The methyltransferase gene superfamily: A tree with multiple branches, in J.T. Romeo, R. Ibrahim, L. Varin, and V.D. Luca (eds.), *Evolution of metabolic pathways: Recent advances in phytochemistry,* Volume 34 (Kidlington, Oxford: Elsevier Science Ltd, 2000), pp. 349-364; Bohlmann, J., Meyer-Gauen, G., and Croteau, R., 1998, Plant terpenoid synthases: Molecular biology and phylogenetic analysis, *Proceedings of the National Academy of Sciences U.S.A.,* 95: 4126-4133.

30. Richmond, T. and Somerville, S., 2000, Chasing the dream: plant EST microarrays, *Current Opinion in Plant Biology,* 3: 108-116.

31. McCouch, S.R., 2001, Genomics and synteny, *Plant Physiology,* 125: 152-155.

32. Oliver, D.J., Nikolau, B., and Wurtele, E.S., 2002, Functional genomics: High-throughput mRNA, protein and metabolite analyses, *Metabolic Engineering,* 4: 98-106.

33. Ibid.; Griffin, T.J., Goodlett, D.R., and Aebersold, R., 2001, Advances in proteome analysis by mass spectrometry, *Current Opinion in Biotechnology,* 12: 607-612.

34. Fiehn, O., Kopka, J., Dormann, P., Altmann, T., Trethewey, R.N., and Willmitzer, L. 2000, Metabolite profiling for plant functional genomics, *Nature Biotechnology,* 18: 1157-1161; Oliver, Nikolau, and Wurtele, 2002, Functional genomics.

35. Pichersky, E. and Gang, D.R., 2000, Genetics and biochemistry of specialized metabolites in plants: An evolutionary perspective, *Trends in Plant Science,* 5: 439-445.

36. Gang, D., Lavid, N., Zubieta, C., Cheng, F., Beuerle, T., Lewinsohn, E., Noel, J.P., and Pichersky, E., 2002, Characterization of phenylpropene *O*-methyltransferases from sweet basil: Facile change of substrate specificity and convergent evolution within a plant *O*-methyltransferase family. *Plant Cell* 14: 505-519; Gang, D.R., Beuerle, T., Ullmann, P., Werck-Reichhart, D., and Pichersky E., 2002, Differential production of *meta* hydroxylated phenylpropanoids in sweet basil peltate ular trichomes and leaves is controlled by the activities of specific acyltransferases and hydroxylases, *Plant Physiology,* 130: 1536-1544; Gang, D.R., Wang, J., Dudareva, N., Nam, K.H., Simon, J., Lewinsohn, E., and Pichersky, E., 2001, An investigation of the storage and biosynthesis of phenylpropenes in sweet basil (*Ocimum basilicum* L.), *Plant Physiology,* 125: 539-555.

37. Lange, B.M., Wildung, M.R., Stauber, E.J., Sanchez, C., Pouchnik, D., and Croteau, R., 2000, Probing essential oil biosynthesis and secretion by functional evaluation of expressed sequence tags from mint glandular trichomes, *Proceedings of the National Academy of Sciences U.S.A.,* 97: 2934-2939; Lavid, N., Guterman, I., Shalit, M., Menda, M., Piestun, D., Dafny-Yelin, M., Shalev, G., Bar, E., Davydov, O., Ovadis, M., et al., 2002, *O*-Methyltransferases involved in the biosynthesis of volatile phenolic derivatives in rose petals, *Plant Physiology,* 129: 1899-1907; Guterman, I., Shalit, M., Menda, M., Piestun, D., Dafny-Yelin, M., Shalev, G., Bar, E., Davydov, O., Ovadis, M., Emanuel, M., et al., 2002, Rose scent: Genomics approach to discovering novel floral fragrance-related genes, *Plant Cell* 14: 2325-2338; Kutchan, T.M., Sequence based approaches to alkaloid biosynthesis gene identification, in J.T. Romeo and R.A. Dixon (eds.), *Phytochemistry in the genomics and post genomics eras: Recent advances in phytochemistry,* Volume 36 (Amsterdam: Pergamon Press, 2002), pp. 163-178; Brandle, J.E., Richman, A., Swanson, A.K., and Chapman, B.P., 2002, Leaf ESTs from *Stevia rebaudiana*: A resource for gene discovery in diterpene synthesis, *Plant Molecular Biology,* 50: 613-622; Shelton, D., Leach, D., Baverstock, P., and Henry R., 2002, Isolation of genes involved in secondary metabolism from *Melaleuca alternifolia* (Cheel) using expressed sequence tags (ESTs), *Plant Science,* 162: 9-15.

38. Dudley, N.R., Labbe, J.C., and Goldstein, B., 2002, Using RNA interference to identify genes required for RNA interference, *Proceedings of the National Academy of Sciences U.S.A.,* 99: 4191-4196.

39. Croteau, R., Kutchan, T.M., and Lewis, N.G., Natural products (secondary metabolites), in B. Buchanan, W. Gruissem, R. Jones (eds.), *Biochemistry and molecular biology of plants* (Rockville, MD: American Society of Plant Physiologists, 2000), pp. 1250-1318.

40. Krasnyanski, S., May, R.A., Loskutov, A., Ball, T.M., and Sink, K.C., 1999, Transformation of the limonene synthase gene into peppermint (*Mentha piperita* L.)

and preliminary studies on the essential oil profiles of single transgenic plants, *Theoretical and Applied Genetics,* 99: 676-682; Diemer, F., Caissard, J.C., Moja, S., Chalchat, J.C., and Jullien, F., 2001, Altered monoterpene composition in transgenic mint following the introduction of 4*S*-limonene synthase, *Plant Physiology and Biochemistry,* 39: 603-614.

41. Mahmoud, S.S. and Croteau, R.B., 2001, Metabolic engineering of essential oil yield and composition in mint by altering expression of deoxyxylulose phosphate reductoisomerase and menthofuran synthase, *Proceedings of the National Academy of Sciences U.S.A.,* 98: 8915-8920.

42. Zook, M., Hohn, T., Bonnen, A., Tsuji, J., and Hammerschmidt, R., 1996, Characterization of novel sesquiterpenoid biosynthesis in tobacco expressing a fungal sesquiterpene synthase, *Plant Physiology,* 112: 311-318.

43. Lewinsohn et al., 2001, Enhanced levels of the aroma and flavor.

44. Croteau, Kutchan, and Lewis, 2000, Natural products (secondary metabolites).

45. Zook et al., 1996, Characterization of novel sesquiterpenoid biosynthesis.

46. Chen, D.H., Ye, H.C., and Li, G.F., 2000, Expression of a chimeric farnesyl diphosphate synthase gene in *Artemisia annua* L. transgenic plants via *Agrobacterium tumefaciens*-mediated transformation, *Plant Science,* 155: 179-185.

47. Martin, G.S., Liu, J., Benedict, C.R., Stipanovic, R.D., and Magill. C.W., 2003, Reduced levels of cadinane sesquiterpenoids in cotton plants expressing antisense (+)-α-cadinene synthase, *Phytochemistry,* 62: 31-38.

48. Hain, R., Reif, H.J., Krause, E., Langebartels, R., Kindl, H., Vornam, B., Wiese, W., Schmelzer, E., Schrier, P.H., Stocker, R.H., et al., 1993, Disease resistance results from foreign phytoalexin expression in a novel plant, *Nature,* 361: 153-156.

49. Ibid.

50. Jung, W., Yu, O., Lau, S.M.C., O'Keefe, D.P., Odell, J., Fader, G., and McGonigle, B., 2000, Identification and expression of isoflavone synthase, the key enzyme for biosynthesis of isoflavones in legumes, *Nature Biotechnology,* 18: 208-212.

51. Dixon, R.A. and Steele, C.L., 1999, Flavonoids and isoflavonoids—A gold mine for genetic engineering, *Trends in Plant Science,* 4: 394-400.

52. Jung et al., 2000, Identification and expression of isoflavone synthase.

53. Croteau, Kutchan, and Lewis, 2000, Natural products (secondary metabolites).

54. Yu, O., Jung, W., Shi, J., Croes, R.A., Fader, G.M., McGonigle, B., and Odell, J.T., 2000, Production of the isoflavones genistein and daidzein in non-legume dicot and monocot tissues, *Plant Physiology,* 124: 781-793.

55. Verpoorte R., Contin, A., and Memelink, J., 2002, Biotechnology for the production of plant secondary metabolites, *Phytochemistry Reviews,* 1: 13-25; Croteau, Kutchan, and Lewis, 2000, Natural products (secondary metabolites).

56. Verpoorte, Contin, and Memelink, 2002, Biotechnology for the production of plant secondary metabolites; Kutchan, T.M., 1995, Alkaloid biosynthesis—The basis for metabolic engineering of medicinal plants, *Plant Cell,* 7:1059-1070; Sato, F., Hashimoto, T., Hachiya, A., Tamura, K., Choi, K.B., Morishige, T., Fujimoto, H., and Yamada, Y., 2001, Metabolic engineering of plant alkaloid an der Heijden, R., Hoge, J.H.C. and Verpoorte, R. (eds.), Effects of over-expression of strictosidine

synthase and tryptophan decarboxylase on alkaloid biosynthesis, *Proceedings of the National Academy of Sciences U.S.A.*, 98(1): 367-372; Verpoorte, R. and Memelink, J., 2002, Engineering secondary metabolite production in plants, *Current Opinion in Biotechnology,* 13: 181-187.

57. Samuelsson, 1991, *Drugs of natural origin.*

58. Croteau, Kutchan, and Lewis, 2000, Natural products (secondary metabolites).

59. Ibid.

60. Canel, C., Lopes-Cardoso, M.I., Whitmer, S., van der Fits, L., Pasquali, G., van der Heijden, R., Hoge, J.H.C., and Verpoorte, R., 1998, Effects of over-expression of strictosidine synthase and tryptophan decarboxylase on alkaloid production by cell cultures of *Catharanthus roseus, Planta,* 205: 414-419.

61. Samuelsson, 1991, *Drugs of natural origin.*

62. Yun, D.J., Hashimoto, T., and Yamada, Y., 1992, Metabolic engineering of medicinal plants: Transgenic *Atropa belladonna* with an improved alkaloid composition, *Proceedings of the National Academy of Sciences U.S.A.*, 89: 11799-11803.

Chapter 13

In Vitro Cultivation of Medicinal Plants

Christoph Wawrosch

INTRODUCTION

In the past decades the demand for plant-based medicines has increased significantly in Western countries. In addition, large amounts of medicinal plants are used by companies for the preparation of various pure compounds either for direct use or for the synthesis of other products.[1] This trend involves, on one hand, plant taxa native to these countries, and as a result of enhanced harvesting from wild populations several species are now endangered.[2] On the other hand, due to increasing interest in complementary medicinal systems, such as Ayurveda, developing countries are affected, too. The expanding international market along with the rapidly rising populations lead to ecological problems—not only because of overharvesting of medicinal plants for home use and export purposes but also because of deforestation by logging and conversion of backwoods to pasture and agriculture.[3] This global development, together with increasingly restrictive conservation laws, will finally lead to considerable reduction of the still-dominant exploitation of natural resources in favor of cultivation.[4] In Germany, for example, producers of phytopharmaceuticals are undertaking significant initiatives in developing field culture procedures for plants that are collected from the wild—not only for reasons of species protection but also because companies are interested in a stable, long-term supply of raw material of high and standardized quality.[5]

Since the beginning of the twentieth century the first attempts to cultivate plant cells in vitro were reported by the botanist Haberlandt.[6] Since then enormous progress has been made in the field of plant biotechnology. A whole range of different in vitro techniques have been established that allow manifold manipulations of plant organs, tissues, and cells. Today plant tissue culture laboratories and production facilities are common in industrialized countries all around the world. Micropropagation is now possible for small companies in developing countries, too.[7]

The present chapter deals with the application of plant tissue culture techniques to medicinal plants. In this it distinguishes between methodologies for the production of plants and practicabilities of the in vitro production of active compounds. The main emphasis is put on the classical techniques; methods of plant genetic engineering are dealt with in Chapter 12 of this book.

PLANT TISSUE CULTURE TECHNIQUES

Within the scope of this handbook the following section provides basic information about general procedures and more specific plant tissue culture techniques with potential impact on production and improvement of medicinal plants, as well as in vitro manufacturing of crude drugs and active compounds.

Establishment of Sterile Cultures

Basically, cultures can be initiated from a variety of different parts of a plant. Ideally, young tissue of healthy plants should be used whenever possible. Tissues such as bark or highly lignified, physiologically old structures generally are not suitable. Explants can be collected from field-grown or greenhouse-grown plants, the latter being of advantage because of the controlled environment with the possibility of specific application of fertilizer and/or fungicides, for example. Depending on the specific project, seeds may be used, too. All of these materials are usually contaminated by various microorganisms, thus making a surface sterilization necessary. The type and concentration of the sterilizing agent (commonly ethanol, sodium hypochlorite, mercuric chloride) and the time of exposure will depend on the material (provenance, part of the plant, nature). However, a proper sterilization ensures that most of the contaminants are eradicated with the plant tissue being damaged as little as possible. This also requires a repeated rinsing of the material with sterilized water to completely remove the sterilizing agent.

Subsequently, the explants are inoculated on a semisolid nutrient medium, mostly based on the formulation of Murashige and Skoog (MS medium).[8] The medium basically contains all of the essential major and minor plant nutrient elements, a carbon source (typically 2 to 3 percent sucrose), vitamins, and plant growth regulators (PGRs). The concentrations of these compounds and optional additional ingredients can vary and a large number of different formulations have been described.[9] Initial growth as well as later morphogenetic processes can be largely controlled through the types and concentrations of the PGRs, but other such parameters as ionic strength and culture conditions (temperature, light, culture vessel size, etc.) can be critical to success, too.[10] After a culture period which, depending on the species, may last from four to eight or even more weeks, the tissue must be transferred to fresh medium. Usually, for this subculture, the material is dissected to new explants so that several new cultures originate from the old one.

Organized Growth

A rough classification in organized and unorganized growth (or cultures) has been made by George.[11] Organized cultures (or organ cultures) comprise any culture of organs or, more generally, of organized structures whether these structures are maintained or are created in vitro. In the latter case the organs may form on organized explants taken in culture (e.g., formation of shoots on a leaf section) or they may regenerate from previously unorganized tissues or cells. The use of organ cultures is the most common method of mass production (or micropropagation) of plants. In respect to the production of secondary metabolites they can offer advantages over unorganized cultures, too.

In practice, it can be distinguished between cultures of (axillary) buds, in other words, existing meristems, and adventitious shoots or structures which are formed on any type of explant in culture. In the first case shoot tips or axillary buds are inoculated on a suitable medium and grow to form shoots. Subsequently, these are divided into new explants (e.g., sections of the stem-bearing nodes) and subcultured, thus leading to an exponentially increasing number of cultures or, at long last, plants. This procedure can also be carried out with adventitious shoots that, in a first step, have been regenerated from a suitable explant.

Another pathway is the induction of somatic embryos which, compared to zygotic embryos, are identical in appearance and morphogenetic potential. Somatic embryos can be germinated and grown to plantlets in vitro, or they are encapsulated in a suitable gel to form so-called synthetic seeds, the practical uses of which are being intensively studied.[12] Finally, in appropriate plants the formation of such storage organs as tubers or bulbs can be induced, too. These microtubers or bulblets can be directly sown to soil in a greenhouse and this is currently an important propagation technique, for example, for potatoes.

For the purpose of micropropagation, shoots obtained through one of the previously mentioned cultures must be rooted—only in some species rooted plantlets are formed in the multiplication stage. The formation of adventitious roots can be induced either in vitro or ex vitro during the acclimatization stage. This last step is necessary to have the in vitro–grown plantlets hardened and adapted to the environment outside of the culture vessel. In some species this procedure may be difficult and the feasibility without excessive losses of plants often dictates whether the whole procedure is cost-effective.

The in vitro cultivation of organized structures may be used not only for the propagation of plants; organ cultures also offer interesting opportunities for the production of active compounds.

Unorganized Growth

Cultures of unorganized tissues first of all comprise callus (or generally tissue) cultures, i.e., masses of unorganized or largely undifferentiated cells. These cultures can be established from previously cultured cells or from callus, which is formed on explants inoculated in a suitable callus medium. From such cultures suspension (cell) cultures can be derived by keeping the cells and cell aggregates dispersed in liquid medium. By definition, cultures of unorganized tissues also include protoplast cultures in which the cell wall usually has been removed enzymatically) and anther or pollen cultures.[13]

Callus and suspension cultures have been intensely studied since the 1960s. In the beginning it was assumed that undifferentiated cells lack the ability to produce secondary compounds. However, it can now be stated that after selection of high-yielding cell lines and manipulation and optimization of the cultural environment, the production of metabolites at similar or even higher levels than in the intact plant is feasible.[14] Certainly plant cell cultures have also become a very important tool for basic and applied research in plant physiology and biochemistry.[15] Some examples of highly productive cultures as well as the limitations that still exist for a large-scale industrial application are discussed next.

In most species investigated so far, it is also possible to regenerate plants from callus tissue, either through shoot organogenesis or somatic embryo formation. Although this is an interesting alternative to micropropagation with organized cultures, problems concerning genetic uniformity may arise due to a phenomenon termed *somaclonal variation.* This term describes genetic variations occurring between plants derived from any type of tissue culture and while, as mentioned, it might be undesired when genetically identical plants are to be produced, it is considered to be exploitable in finding new useful varieties.[16] This strategy of inducing variations will be briefly presented later. Regardless, when callus stages are a necessary part of a propagation protocol they should be kept as short as possible, because it is known that callus is particularly susceptible to genetic variation.

The isolation of protoplasts has been described for numerous plants, including taxa of medicinal importance, and regeneration of plants is feasible either through an intermittent callus stage or in some cases through direct somatic embryogenesis.[17] One the one hand, protoplasts are a convenient tool for genetic manipulation. On the other hand, under appropriate conditions protoplasts of different varieties or even species can be fused, leading, after regeneration, to somatic hybrids. Although this breeding technique has been applied mainly to crops and ornamentals, it has also been reported for some medicinal species and offers interesting opportunities for the improvement of plants when sexual incompatibilities exist.[18]

The in vitro cultivation of anthers and pollen are not further covered in the present chapter. It should, however, be briefly mentioned that this technique, which allows the establishment of homozygous lines for breeding purposes, offers some advantages over the conventional process of repeated self-pollination and has occasionally been applied to medicinal plants.[19]

Cultivation Systems

The use of surface cultures on media solidified with a gelling agent, such as agar or a gellan gum, is still the most adopted method, be it in research laboratories or in facilities for the mass production of plants. However, a number of factors are limiting to the cheap mass production of plants of various species. Basically this concerns the cultural environment itself and matters affecting production costs.[20] To sum up, the current techniques necessitate a large number of comparatively small containers that have to be filled by hand. The plant material has to be periodically transferred to fresh medium in new containers, which again has to be done manually. Although millions of plants are produced annually through micropropagation, the high labor input and the resulting price per plant prevents the commercial application of micropropagation for many plant species. The use of liquid media offers several advantages. The culture containers can be scaled up to the size of bioreactors, exhausted media can be replaced by fresh ones while keeping the plant material in the same container, and various possibilities for automation are available. In many species the continuous submerged culture of shoots in liquid medium results in physiological and morphological disorders such as hyperhydricity (formerly termed vitrification). To overcome this problem a variety of special culture systems has been developed. In recent years temporary immersion systems with only periodical contact of the plant material with liquid medium have gained increased interest, and the current commercialization of these techniques offers new opportunities for the mass production of healthy plants at reduced costs.[21]

IN VITRO PLANT PROPAGATION

Since the introduction of plant tissue culture to commercial plant production in the 1960s micropropagation has become a multibillion-dollar industry. The current market for tissue-culture-derived plants is estimated at 500 million to 1 billion plants annually, which is quite small when compared to other such products as seeds or cuttings.[22] As mentioned, production costs are one of the limitations to larger marketing of in vitro–produced plants. Although actual data are difficult to obtain, a report about the status in the United States showed that in 1996 the approximately 121 million micropropagated plants were grouped in foliage plants, greenhouse flowers and orchids, herbaceous annuals and perennials, trees and shrubs, vegetables, fruits, and miscellaneous.[23] Medicinal plants (peppermint, spearmint, and *Citrus* spp.) are only mentioned as a small portion of the miscellaneous group that comprises as few as 4.5 million plants. What is more, these three genera are most likely produced for the aroma and fruit industry, respectively.

A similiar situation has been described for India by Govil and Gupta: In 1996, commercial units produced fruit crops, forest trees, ornamentals, vegetables and plantation crops at a total amount of 190 million plants.[24] Again, although cardamom, tea, black pepper, and ginger (for example) are micropropagated in small amounts, presumably these are mostly used as spices and aroma compounds rather than for medicinal purposes. The report clearly shows that the amount of ornamentals by far exceeds all other categories. Although dating back more than one decade, an analysis of commercial micropropagation in Western Europe and Israel showed the same figure as in the United States and India.[25]

It seems clear that the very limited commercial in vitro propagation of medicinal plants is due mainly to economic and technical problems but surely not due to lack of scientific knowledge. A look at relevant databases, books, and journals reveals that a wealth of contributions to tissue culture of medicinal plants has been published since the 1960s. It is not within the scope of this chapter to list each species that has been successfully propagated in vitro, but the very small selection listed in Table 13.1 should clarify that widely used as well as lesser-known medicinal species of all categories (herbs, trees, bulbous plants) have been successfully propagated in vitro.

TABLE 13.1. Selected examples of medicinal plant species that have been multiplied in vitro.

Plant name	Type of culture	Reference
Achillea asplenifolia	NC	Wawrosch et al., 1994[a]
Aconitum carmichaelii	SC	Hatano et al., 1988[b]
Aloe barbadensis	SC	Castorena Sanchez et al., 1988[c]
Atropa belladonna	SC	Benjamin et al., 1987[d]
Azadirachta indica	SE	Su et al., 1996[e]
Bacopa monniera	AS	Tiwari et al., 2001[f]
Catharanthus roseus	NC	Alen and Jain, 1997[g]
Cephaelis ipecacuanha	SE	Rout et al., 2001[h]
Charybdis (Urginea) sp.	AS	Kongbangkerd et al., 2002[i]
Dioscorea composita	NC	Ammirato, 1982[j]
Eleutherococcus sessiliflorus	SE	Choi et al., 2002[k]
Ephedra sp.	NC	O'Dowd and Richardson, 1993[l]
Ginkgo biloba	SE	Laurain et al., 1996[m]
Panax ginseng	SE	Arya et al., 1993[n]
Trichopus zeylanicus	SC (shoot tips)	Krishnan et al., 1995[o]
Vinca minor	NC	Stapfer and Heuser, 1985[p]

AS = adventitious shoots; NC = nodal culture; SC = shoot culture; SE = somatic embryogenesis.

[a]Wawrosch, C., Kopp, B., and Kubelka, W., 1994. In vitro propagation of *Achillea asplenifolia* Vent. through multiple shoot regeneration, *Plant Cell Reports,* 14(2/3): 161-164.

[b]Hatano, K., Kamura, K., Shoyama, Y., and Nishioka, I., 1988. Clonal multiplication of *Aconitum carmichaeli* by tip tissue culture and alkaloid contents of clonally propagated plant, *Planta Medica,* 54: 152-155.

[c]Castorena Sanchez, I., Natali, L., and Cavallini, A., 1988. In vitro culture of *Aloe barbadensis* Mill.: Morphogenetic ability and nuclear DNA content, *Plant Science,* 55: 53-59.

[d]Benjamin, B.D., Roja, P.C., Heble, M.R., and Chadha, M.S., 1987. Multiple shoot cultures of *Atropa belladonna*: Effect of physico-chemical factors on growth and alkaloid formation, *Journal of Plant Physiology,* 129(1/2): 129-135.

[e]Su, W.W., Hwang, W.-I., Kim, S.Y., and Sagawa, Y., 1996. Induction of somatic embryogenesis in *Azadirachta indica, Plant Cell, Tissue and Organ Culture,* 50(2): 91-95.

[f]Tiwari, V., Tiwari, K.N., and Singh, B.D., 2001. Comparative studies of cytokinins on in vitro propagation of *Bacopa monniera, Plant Cell, Tissue and Organ Culture,* 66(1): 9-16.

[g]Alen, K.H. and Jain, S.M., 1997. In vitro multiplication of *Catharanthus roseus, Acta Horticulturae,* 447: 167-169.

[h]Rout, G.R., Saxena, C., and Das, P., 2001. Somatic embryogenesis in *Cephaelis ipecacuanha* A. Richard: Effect of growth regulators and culture con-

TABLE 13.1 *(continued)*

ditions, *Journal of Herbs, Spices & Medicinal Plants,* 8(1): 59-68.
[i]Kongbangkerd, A., Wawrosch, C., and Kopp, B., 2002. In vitro clonal propagation of *Charybdis* sp. through nodule culture in liquid medium, *Revista de Fitoterapia,* 2(Suppl. 1): 330.
[j]Ammirato, P.V. Growth and morphogenesis in cultures of the monocot Yam, in A. Fujiwara (ed.) *Plant tissue culture 1982, Dioscorea* (Tokyo: Marizen Co., Ltd., 1982), pp. 169-170.
[k]Choi, Y.E., Ko, S.K., Lee, K.S., and Yoon, E.S., 2002. Production of plantlets of *Eleutherococcus sessiliflorus* via somatic embryogenesis and successful transfer to soil, *Plant Cell, Tissue and Organ Culture,* 69(2): 201-204.
[l]O'Dowd, N.A. and Richardson, D.H.S., 1993. In vitro micropropagation of *Ephedra, Journal of Horticultural Science,* 68: 1013-1020.
[m]Laurain, D., Chenieux, J.-C., and Tremouillaux-Guiller, J., 1996. Somatic embryogenesis from immature zygotic embryos of *Ginkgo biloba, Plant Cell, Tissue and Organ Culture,* 44(1): 19-24.
[n]Arya, S., Arya, I.D., and Eriksson, T., 1993. Rapid multiplication of adventitious somatic embryos of *Panax ginseng, Plant Cell, Tissue and Organ Culture,* 34: 157-162.
[o]Krishnan, P.N., Sudha, C.G., and Seeni, S., 1995. Rapid propagation through shoot tip culture of *Trichopus zeylanicus* Gaertn., a rare ethnomedicinal plant, *Plant Cell Reports,* 14: 708-711.
[p]Stapfer, R.E. and Heuser, C.W., 1985. In vitro propagation of periwinkle, *HortScience,* 20(1): 141-142.

Hereafter I will relate some recent applications of plant tissue culture to medicinal plants with the aims of quality improvement and conservation of species which, mostly due to overharvesting, have become threatened.

Quality Improvement Through Selection and Induced Variation

Valerian (*Valeriana officinalis* L., Valerianaceae) is a well-known and important medicinal plant whose roots are used. Despite extensive research it is still unknown which of the various secondary compounds of valerian are responsible for its sedative activity. However, the pharmaceutical industry tends to use varieties with high contents of essential oil and valerenic acid derivatives, and low amounts of valepotriates.[26] For the latter compounds it has furthermore been recommended not to apply preparations with high concentrations of valepotriates until clear data about possible side effects are available.[27] Due to the high degree of outcrossing and heterozygoty, traditional breeding methods are time-consuming and labor intensive. In vitro techniques have been suggested as interesting alternatives. Exploiting the often increased occurrence of variation in undifferentiated tissue (somaclonal variation as mentioned earlier), a number of plants regenerated from callus culture showed substantial chemical and also morphological differences.[28] When compared to the original plants from which the callus cultures were initiated, some of the regenerants of the respective callus line had higher concentrations of valerenic acid, lower concentrations of valepotriates, and higher root weight. These findings indicate that plant tissue culture can be helpful in the production of new breeds with improved traits. However, it is known that somaclonal variation may be uninheritable, so further studies are needed to investigate the stability of the new attributes.[29] Besides the introduction of new traits it has also been shown for valerian that micropropagation can be a useful biotechnological tool in the frame of selection breeding.[30]

The Asteraceae *Artemisia annua* L., originally native to Asia, contains the endoperoxide sesquiterpene lactone artemisinin, which is well known for its antimalarial activity. The total synthesis of the molecule is not economically feasible and thus the plant remains the only source. However, the contents in artemisinin have been found to vary substantially in different accessions of *A. annua.*[31] Micropropagation can be used for storage of valuable genotypes and production of plants for breeding purposes.[32] In vitro derived clones established in Switzerland have recently been included in a breeding program. New hybrid lines that were adapted to Brazilian climatic conditions were shown to yield approximately 25 $kg{\cdot}ha^{-1}$ artemisinin instead of 5 $kg{\cdot}ha^{-1}$ obtained with the base population.[33] This is an impressive example of results obtained through the integration of biotechnological steps in domestication processes.

Coltsfoot (*Tussilago farfara* L., Asteraceae) is an old medicinal plant that is used against cough due to its polysaccharide content. The plant also contains toxic pyrrolizidine alkaloids (PA), namely senkirkine and senecionine. Although the concentration of these compounds is usually low, in Germany limit values for these alkaloids were defined by the law in 1992, and in Austria the sale of PA-containing plants and preparations has been forbidden since 1994. Subsequently, few drugs meeting these requirements were available. To reintroduce coltsfoot to the market, a screening program was started that was based on micropropagation for germplasm storage and quick true-to-type multiplication.[34] The screening of more than 120 wild types resulted in the identification of one genotype with no detectable amounts of PA, which subsequently was multiplied in vitro and tested for several years in the field. The registered variety *Tussilago farfara* Wien is now cultivated on a regular basis in Germany and Chile, and preparations as well as the drug itself are again available to the consumer.[35]

An interesting as well as efficient combination of in vitro techniques and hydroponic cultivation has recently been described for St. John's wort (*Hypericum perforatum* L., Hypericaceae). Preparations of this well-known medicinal plant for the treatment of depression play an important commercial role. Murch and co-workers established germplasm lines from shoots that were regenerated from hypocotyl sections.[36] After acclimatization the plants were grown in a nutrient film technique hydroponic system in the greenhouse. Flowering plants were harvested and the analysis of three main compounds—hypericin, pseudohypericin, and hyperforin—revealed that the in vitro–generated and greenhouse-grown plants were of equal to and often better quality than field-grown ones.

A last example to clarify the various advantages of integrating modern techniques in commercial plant production is *Baptisia tinctoria* (L.) R.Br. (Fabaceae). This plant was traditionally used by Native Americans, and today root extracts are employed in immune-modulating herbal medicines. Plant propagation through tissue culture has been recommended, among other things, because of the low seed germination rate, and some in vitro–derived clones with, for example, resistance to powdery mildew have been described.[37] A German company recently has adapted micropropagation, acclimatization, field cultivation, harvest, and postharvest techniques for *B. tinctoria* on a commercial scale.[38] The economical feasibility of the procedure is possible through the application of arbuscular mycorrhizal fungi (AMF) to the micropropagated plants when they are transferred ex vitro. The integration of AMF in both conventional and in vitro plant propagation is considered to be an important step toward an ecologically and economically more efficient production of healthy plants.[39]

Micropropagation of Endangered Medicinal Plant Species

As stated earlier, the use of medicinal plants has substantially increased since the 1970s. The large-scale collection of plants from natural stands, but also deforestation and erosion, resulted in the rapid depletion of wild populations, which is still going on. Most obviously general strategies such as appropriate information policies, environmental and resource management, and in situ/ex situ conservation programs are necessary. Domestication of intensely used wild-growing species is needed for a sustainable use as well as for the maintenance of biological diversity. The application of plant tissue culture techniques makes sense with species which are difficult to propagate by conventional ways. This may include seldomly flowering species, those with bad or inhomogenous seed quality, or slowly-growing species such as woody plants.

Domestication of a wild-growing species requires several steps and may take several years. In vitro culture techniques may be included in the process, in other words, in the storage of plant material, plant propagation, and genetic improvement through (selection) breeding.[40] In addition, in vitro collection has been described as an adaptation of plant tissue culture to field collection and might be useful during long collection trips when seeds are not available or when plants should remain undisturbed.[41]

Numerous reports have been published since the 1960s describing the successful application of plant tissue culture to endangered medicinal plant species. As mentioned, such taxa are especially threatened in developing countries. Recent reports include plants such as *Allium wallichii* Kunth (Alliaceae), *Holostemma ada-kodien* Schult. (Asclepiadaceae), *Lilium nepalense* D. Don (Liliaceae), *Swertia chirata* Buch.-Ham. ex Wall. (Gentianaceae), or *Picrorhiza kurroa* Royle ex Benth. (Scrophulariaceae), species native to Nepal and/or India.[42] However, an increased integration of tissue culture in the preservation and domestication of threatened plants will, as noted earlier, again depend on the availability of cost-efficient systems for the mass production of plantlets for further field culture.

IN VITRO PRODUCTION OF SECONDARY METABOLITES

Since the early days of plant tissue culture in the past century, expectations for the commercial in vitro production of plant secondary metabolites were high. Independence of season, climate, soil conditions, need for irrigation and fertilizing, and more were attractive perspectives which led to decades of intense research in the field of secondary plant product formation in unorganized and organized cultures. A wealth of reports about hundreds of secondary products in tissue culture have been published, and research is going on. Just to give an example, the regulation and the biosynthesis of secondary metabolites in cell and tissue cultures of Madagascar periwinkle, *Catharanthus roseus* (L.) G. Don, has been intensively studied; two reviews covering the literature up to 1993 included a total of 417 references.[43] Yet the alkaloid levels obtained in vitro so far are still too low for an industrial large-scale production, and the authors conclude that a better understanding of the regulation of secondary metabolism is needed.[44] This is generally true for most plant species and therefore only very few commercial processes are established nowadays.[45]

Most commonly, callus cultures and cell suspension cultures derived from a given plant show only low concentrations of the secondary products typical for the donor plant. What is more, in some cases the profiles of secondary metabolites in tissue cultures are unlike those in the intact plant.

Strategies necessary to increase yields of natural products in cell cultures include selection of high producing cell lines, modification of the nutrient media, elicitation, immobilization of cells, and the adaptation of bioreactors.[46] Selection and media modification were two important steps in the development of the first and most successful commercial process for the production of the naphthoquinone shikonin with cell cultures of *Lithospermum erythrorhizon* Sieb. & Zucc. (Boraginaceae). Callus cultures and suspension cultures derived thereoff usually are inhomogenous, and the selection of true single-cell clones is not easy because as a general rule suspensions are mostly composed of cell aggregates. In the case of *Lithospermum* protoplasts were isolated and subsequently plated on semisolid medium. After cell wall regeneration and cell division the resulting colonies were screened for high contents of shikonin derivatives and it was possible to select cell lines with higher and more stable shikonin production.[47] In the production process two different nutrient media are used for cell growth and shikonin production, respectively—this two-stage culture is known to increase the yield of secondary metabolites in a number of other species, too.[48]

The phenolic compound rosmarinic acid is found, among others, in various members of the Lamiaceae family. Rosmarinic acid production in cell cultures of *Coleus blumei* Benth. (Lamiaceae) was described in 1977.[49] It was shown that, in an optimized two-stage culture system, high-yielding cell lines form up to 21.4 percent of dry weight of the secondary compound as compared to 1.1 percent of the mother plant.[50] These impressive results clearly demonstrate the great potential of plant cell cultures. Some other examples for which noteworthy production rates have been achieved in cell cultures are listed in Table 13.2. One interesting application involving cell cultures is the production of ginseng (*Panax ginseng* CA Meyer) biomass in Japan. Since 1988, fermentation of ginseng cells is performed in 20,000, to 25,000-liter bioreactors and the products are used commercially.[51]

Organ cultures are an alternative to cell suspension cultures for the production of secondary metabolites or biomass. Recently, the production of ginseng roots in 500-liter bioreactors has been described, and a method for the mass propagation of adventitious roots with a saponin profile similar to that of field-grown roots has been patented.[52] The antimalarial sesquiterpene artemisinin is formed both in shoot and transformed root cultures of *Artemisia annua* L.[53] The latter organs result from the infection of plant tissue with *Agrobacterium rhizogenes* and exhibit quick and vigorous growth in simple media while usually showing a secondary compounds pattern similar to that of the intact plant. Due to these characteristics, hairy roots are regarded as a promising tool for, besides other applications, the production of plant chemicals.[54] The number of reports on the production of natural products by hairy

TABLE 13.2. Production of secondary metabolites in plant cell cultures.

Plant name	Compound	Reference
Catharanthus roseus	ajmalicine	Schlatmann et al., 1995[a]
Chrysanthemum cinerariafolium	pyrethrins	Dhar and Pal, 1993[b]
Coptis japonica	berberine	Hara et al., 1988[c]
Digitalis lanata	ß-methyldigoxin*	Alfermann et al., 1985[d]
Papaver somniferum	sanguinarine	Park et al., 1992[e]
Taxus chinensis	taxuyunnanine C	Wang and Zhong, 2002[f]

* biotransformation of ß-methyldigitoxin.

[a]Schlatmann, J.E., Moreno, P.R.H., Selles, M., Vinke, J.L., ten Hoopen, H.J.G., Verpoorte, R., and Heijnen, J.J., 1995. Two-stage batch process for the production of ajmalicine by *Catharanthus roseus:* The link between growth and production stage, *Biotechnology and Bioengineering,* 47(1): 53-59.

[b]Dhar, K. and Pal, A., 1993. Factors influencing efficient pyrethrin production in undifferentiated cultures of *Chrysanthemum cinerariaefolium, Fitoterapia,* 64(4): 336-340.

[c]Hara, Y., Yoshioka, T., Morimoto, T., Fujita, Y., and Yamada, Y., 1988. Enhancement of berberine production in suspension cultures of *Coptis japonica* by gibberellic acid treatment, *Journal of Plant Pysiology,* 133(1): 12-15.

[d]Alfermann, A.W., Spieler, H., and Reinhard, E., Biotransformation of cardiac glycosides by *Digitalis* cell cultures in airlift reactors, in K.-H. Neumann, W. Barz, and E. Reinhard (eds.), *Primary and secondary metabolism of plant cell cultures* (Berlin, Heidelberg: Springer, 1985), pp. 316-322.

[e]Park, J.M., Yoon, S.Y., Giles, K.L., Songstad, D.D., Eppstein, D., Novakovski, D., Freisen, L., and Roewer, I., 1992. Production of sanguinarine by suspension culture of *Papaver somniferum* in bioreactors, *Journal of Fermentation and Bioengineering,* 74(5): 292-296.

[f]Wang, Z.-H. and Zhong, J.-J., 2002. Repeated elicitation enhances taxane production in suspension cultures of *Taxus chinensis* in bioreactors, *Biotechnology Letters,* 24: 445-448.

root cultures is ever increasing and includes, for example, valepotriates (*Valeriana wallichii* DC), cardenolides (*Digitalis purpurea* L.), anthraquinones (*Rubia tinctorum* Lapeyr.), or tropane alkaloids (*Atropa belladonna* L.).[55]

CONCLUSION

Plant-based medicines have been used throughout the ages and are still of great importance. A number of commercially important chemicals are still extracted from plant material. However, increasing demand as well as strengthened quality requirements create the need for new strategies concerning the provision of raw material. In the process of plant domestication as an alternative to collection from the wild-plant tissue culture can be helpful. Over the past decades, remarkable advances have been made in the field of plant biotechnology. Protocols for the propagation of a wealth of plant species have been worked out and include numerous medicinal plants. But mostly due to economic reasons only a limited number of these species are presently propagated in vitro on a commercial scale. However, the methodology is nowadays much more sophisticated than in the early days of plant tissue culture. Techniques for low-cost mass propagation are currently being commercialized, and it can be expected that they will be applied to an increasing number of species in the near future.

Similarly, the formation of secondary compounds in vitro has also been an important research topic. While the commercial manufacture of chemicals through cell cultures so far has been an exception, considerable advantages have been achieved in the understanding of biosynthetic pathways and mechanisms of regulation. Research in this area is ongoing and, together with progress in cultivation technology, industrial applications should be possible in the near future. Furthermore, the recent emergence of recombinant DNA technology makes it possible to directly modify the expression of key genes in the biosynthetic pathway, as well as to insert foreign genes.[56] It is expected that molecular farming will play an important role in the production of pharmaceuticals.[57]

NOTES

1. Bonati, A., 1988, Industry and the conservation of medicinal plants, *Fitoterapia*, 59(5): 367-370.

2. Lange, D., *Europe's medicinal and aromatic plants: Their use, trade and conservation* (Cambridge, United Kingdom: TRAFFIC International, 1998).

3. Srivastava, J., Lambert, J., and Vietmeyer, N., *Medicinal plants: An expanding role in development,* World Bank Technical Paper 320 (Washington, DC: The World Bank, 1996); Leaman, D.J., Conservation, trade, sustainability and exploitation of medicinal plant species, in P.K. Saxena (ed.), *Development of plant-based medicines: Conservation, efficacy, and safety* (Dordrecht, the Netherlands: Kluwer Academic Publishers, 2001), pp. 1-15.

4. Pank, F., 1997, Zweiter Weltkongress für Arznei- und Aromapflanzen, *Zeitschrift für Arznei- und Gewürzpflanzen,* 3(1): 36-39.

5. Franke, R., 1999, Neuentstehung von Kulturpflanzen heute—Beispiele aus dem Bereich Arznei- und Gewürzpflanzen (Evolution of cultivated plants today—Examples from medicinal and spice plants), *Zeitschrift für Arznei- und Gewürzpflanzen,* 4(1): 24-38.

6. Haberlandt, G., 1902, Culturversuche mit isolierten Pflanzenzellen, *Sitzungsberichte der kaiserlichen Akademie der Wissenschaften Wien, Mathematisch-naturwissenschaftliche Classe,* 111: 69-92.

7. Leaman, 2001, Conservation, trade, sustainability and exploitation of medicinal plant species.

8. Murashige, T. and Skoog, F., 1962, A revised medium for rapid growth and bioassays with tobacco tissue cultures, *Physiologia Plantarum,* 15: 473-497.

9. George, E.F., Puttock, D.J.M., and George, H.J., *Plant culture media,* Volume 1: *Formulations and uses* (Edington, England: Exegetics Limited, 1987).

10. Rout, G.R., Samantaray, S., and Das, P., 2000, In vitro manipulation and propagation of medicinal plants, *Biotechnology Advances,* 18(2): 91-120.

11. George, E.F., *Plant propagation by tissue culture,* Part 1: *The technology* (Edington, England: Exegetics Limited, 1993).

12. Dupuis, J.-M., Millot, C., Teufel, E., Arnault, J.-L., and Freyssinet, G., Germination of synthetic seeds, in W.Y. Soh and S.S. Bhojwani (eds.), *Morphogenesis in plant tissue cultures* (Dordrecht, the Netherlands: Kluwer Academic Publishers, 1999), pp. 419-441.

13. George, 1993, *Plant propagation by tissue culture.*

14. Yeoman, M.M., Holden, M.A., Corchet, P., Holden, P.R., Goy, J.G., and Hobbs, M.C., Exploitation of disorganized plant cultures for the production of secondary metabolites, in B.V. Charlwood and M.J.C. Rhodes (eds.), *Secondary products from plant tissue culture* (Oxford, United Kingdom: Oxford University Press, 1990), pp. 139-166; Zenk, M.H., 1991, Chasing the enzymes of secondary metabolism: Plant cell cultures as a pot of gold, *Phytochemistry,* 30(12): 3861-3863.

15. Zenk, 1991, Chasing the enzyme of secondary metabolism.

16. Larkin, P.J. and Scowcroft, W.R., 1981, Somaclonal variation—A novel source of variability from cell cultures for plant improvement, *Theoretical and Applied Genetics,* 60: 197-214.

17. Roest, S. and Gilissen, L.J.W., 1983, Plant regeneration from protoplasts: A literature review, *Acta Botanica Neerlandica,* 38(1): 1-23; Roest, S. and Gilissen, L.J.W., 1993, Regeneration from protoplasts—A supplementary literature review, *Acta Botanica Neerlandica,* 42(1): 1-23.

18. Negrutiu, I., Hinnisdaels, S., Mouras, A., Gill, B.S., Gharti-Chhetri, G.B., Davey, M.R., Gleba, Y.Y., Sidorov, V., and Jacobs, M., 1989, Somatic versus sexual hybridization: Features, facts and future, *Acta Botanica Neerlandica,* 38(3): 253-272; Li, Y.-G., Stoutjestijk, P.A., and Larkin, P.J., Somatic hybridization for plant improvement, in W.Y. Soh and S.S. Bhojwani (eds.), *Morphogenesis in plant tissue cultures* (Dordrecht, the Netherlands: Kluwer Academic Publishers, 1999), pp. 363-418.

19. Depaepe, R., Nitsch, C., Godard, M., and Pernes, J., Potential from haploid and possible use in agriculture, in W. Barz, E. Reinhard, and M.H. Zenk (eds.), *Plant tissue culture and its bio-technological application* (Berlin: Springer, 1977), pp. 341-352; Diekmann, W., 1997, Biotechnologische Methoden in der Arzneipflanzenzüchtung [Biotechnological methods in medicinal plant breeding], *Drogenreport,* 10(18): 80-85.

20. Aitken-Christie, J., Automation, in P.C. Debergh and R.H. Zimmerman (eds.), *Micropropagation: Technology and application* (Dordrecht, the Netherlands:

Kluwer Academic Publishers, 1991), pp. 363-388; Ziv M., 2000, Bioreactor technology for plant micropropagation, *Horticultural Reviews,* 24: 1-30.

21. Etienne, H. and Berthouly, M., 2002, Temporary immersion systems in plant micropropagation, *Plant Cell, Tissue and Organ Culture,* 69(3): 215-231.

22. Ilan, A. and Khayat, E., 1997, An overview of commercial and technological limitations to marketing of micropropagated plants, *Acta Horticulturae,* 447: 643-648.

23. Hartman, R.D. and Zimmerman, R.H., Commercial micropropagation in the United States, 1965-1998, in A. Altman, M. Ziv, and S. Izhar (eds.), *Plant biotechnology and in vitro biology in the 21st century* (Dordrecht, the Netherlands: Kluwer Academic Publishers, 1999), pp. 699-707.

24. Govil, S. and Gupta, S.C., 1997, Commercialization of plant tissue culture in India, *Plant Cell, Tissue and Organ Culture,* 51(1): 65-73.

25. Pierik, R.L.M., Commercial micropropagation in Western Europe and Israel, in P.C. Debergh and R.H. Zimmerman (eds.), *Micropropagation: Technology and application* (Dordrecht, the Netherlands: Kluwer Academic Publishers, 1991), pp. 155-165.

26. Diekmann, W., 1997, Auslese von valepotriatarmen Einzelpflanzen des Baldrians mittels biotechnologischer und mikrochemischer Methoden [Selection of individuals of valerian with a reduced valepotriate content by means of biotechnological and microchemical methods], *Drogenreport,* 10(17): 40-47.

27. Hänsel, R., Sticher, O., and Steinegger, E., *Pharmakognosie—Phytopharmazie* (Berlin: Springer, 1999), pp. 463-464.

28. Gao, X. and Bjork, L., 2000, Chemical characters, height and root weight in callus regenerated plants of *Valeriana officinalis* L., *Journal of Herbs, Spices and Medicinal Plants,* 7(4): 31-39.

29. Ibid.

30. Diekmann, 1997, Auslese von valepotriatarmen.

31. Delabays, N., Simonnet, X., and Gaudin, M., 2001, The genetics of artemisinin content in *Artemisia annua* L. and the breeding of high yielding cultivars, *Current Medicinal Chemistry,* 8(15): 1795-1801.

32. Lê, C. and Collet, G., 1991, The in vitro culture of *Artemisia annua* L., *Recherche agronomique en Suisse,* 31(3): 111-116.

33. De Magalhães, P.M., Pereira, B., Sartoratto, A, de Oliveira, J., and Debrunner, N., 1999, New hybrid lines of the antimalarial species *Artemisia annua* L., *Acta Horticulturae,* 502: 377-381.

34. Kopp, B., Wawrosch, C., Lebada, R., and Wiedenfeld, H., 1997, PA-freie Huflattichblätter. Teil 1. In-vitro-Kultivierung und Selektionszüchtung [Pyrrolizidine alkaloid (PA) free coltsfoot leaves. Part 1. In vitro cultivation and selective breeding], *Deutsche Apotheker Zeitung,* 137(45): 4066-4069.

35. European Union, Community Plant Variety Office, 1999, Grant number EU5559 of 06/12/1999.

36. Murch, S.J., Rupasinghe, H.P.V. and Saxena, P.K., 2002, An in vitro and hydroponic growing system for hypericin, pseudohypericin, and hyperforin production of St. John's wort (*Hypericum perforatum* CV New Stem), *Planta Medica,* 68(12): 1108-1112.

37. Mevenkamp, G., Lieberei, R., and Harnischfeger, G., *Baptisia tinctoria* (L.) R. Brown: Micropropagation, in vitro culture and production in direction of phar-

maceutically used root biomass, in Y.P.S. Bajaj (ed.), *Biotechnology in agriculture and forestry,* Volume 28: *Medicinal and aromatic plants VII* (Berlin: Springer, 1994), pp. 43-55.

38. Grotkass, C., Hutter, I., and Feldmann, F., 2000, Use of arbuscular mycorrhizal fungi to reduce weaning stress of micropropagated *Baptisia tinctoria* (L.) R. Br., *Acta Horticulturae,* 530: 305-311.

39. Lovato, P.E., Schüepp, H., Trouvelot, A., and Gianinazzi, S., Application of arbuscular mycorrhizal fungi (AMF) in orchard and ornamental plants, in A. Varma and B. Hock (eds.), *Mycorrhiza* (Second edition) (Berlin: Springer, 1999), pp. 443-467; Feldmann, F., Hutter, I., Niemann, P., Weritz, J., Grotkass, C., and Boyle, C., 1999, Einbindung der Mykorrhizatechnologie in die Heil- und Zierpflanzenproduktion sowie den Endverkauf [Integration of the mycorrhizal technology into plant production process of medicinal and ornamental plants as well as commercialisation], *Mitteilungen der Biologischen Bundesanstalt,* 363: 6-38.

40. Franz, C., 1991, Kultivierung bedeutender wildwachsender Arznei- und Gewürzpflanzen [Cultivation of important wild growing medicinal and aromatic plants], *entwicklung + ländlicher raum,* 4: 3-7; Bohr, C., 1997, Inkulturnahme von bisher aus Wildsammlungen stammenden Wirkstoffpflanzen [Domestication of plant species with valuable active substances being hitherto collected from their natural habitat], *Drogenreport,* 10(17): 37-39.

41. Pence, V.C., In vitro collection (IVCN), in B.G. Bowes (ed.), *A colour atlas of plant propagation and conservation* (London: Manson Publishing, 1999), pp. 87-96.

42. Wawrosch, C., Malla, P.R., and Kopp, B., 2001, Micropropagation of *Allium wallichii* Kunth, a threatened medicinal plant of Nepal, *In Vitro Cellular and Developmental Biology—Plant,* 37: 555-557; Martin, K.P., 2002, Rapid propagation of *Holostemma ada-kodien* Schult., a rare medicinal plant, through axillary bud multiplication and indirect organogenesis, *Plant Cell Reports,* 21(2): 112-117; Wawrosch, C., Malla, P.R., and Kopp, B., 2001, Clonal propagation of *Lilium nepalense* D.Don, a threatened medicinal plant of Nepal, *Plant Cell Reports,* 20(4): 285-288; Wawrosch, C., Maskay, N., and Kopp, B., 1999, Micropropagation of the threatened Nepalese medicinal plant *Swertia chirata* Buch.-Ham. ex Wall., *Plant Cell Reports,* 18(12): 997-1001; Wawrosch, C., Rührlinger, S., Grauwald, B., Sturm, S., Stuppner, H., and Kopp, B., 2002, Clonal propagation of the Himalayan medicinal plant *Picrorhiza kurroa* (Scrophulariaceae), *Revista de Fitoterapia,* 2(Suppl. 1): 330.

43. Van der Heijden, R., Verpoorte, R., and Ten Hoopen, H.J.G., 1989, Cell and tissue cultures of *Catharanthus roseus* (L.) G. Don: A literature survey, *Plant Cell, Tissue and Organ Culture,* 18: 231-280; Moreno, P.R.H., Van der Heijden, R., and Verpoorte, R., 1995, Cell and tissue cultures of *Catharanthus roseus:* A literature survey II—Updating from 1988 to 1993, *Plant Cell, Tissue and Organ Culture,* 42: 1-25.

44. Moreno, Van der Heijden, and Verpoorte, 1995, Cell and tissue cultures of *Catharanthus roseus.*

45. Bourgaud, F., Gravot, A., Milesi, S., and Gontier, E., 2001, Production of plant secondary metabolites: A historical perspective, *Plant Science,* 161(5): 839-851.

46. Ibid; Collin, H.A. and Edwards, S., *Plant cell culture* (Oxford, United Kingdom: BIOS Scientific Publishers Limited, 1998), pp. 103-120.

47. Fujita, Y., Takahashi, S., and Yamada, Y., 1985, Selection of cell lines with high productivity of shikonin derivates by protoplast culture of *Lithospermum erythrorhizon, Agricultural and Biological Chemistry,* 49(6): 1755-1759.

48. Fuyita, Y. and Tabata, M., Secondary metabolites from plant cells—Pharmaceutical applications and progress in commercial production, in C.E. Green, D.A. Somers, W.P. Hackett, and D.D. Biesboes (eds.), *Plant tissue and cell culture* (New York: Alan R. Liss, Inc., 1987), pp. 169-185.

49. Razzaque, A. and Ellis, B.E., 1977, Rosmarinic acid production in *Coleus* cell cultures, *Planta,* 137(3): 287-291.

50. Ulbrich, B., Wiesner, W., and Arens, H., Large-scale production of rosmarinic acid from plant cell cultures of *Coleus blumei* Benth., in K.-H. Neumann, W. Barz, and E. Reinhard (eds.), *Primary and secondary metabolism of plant cell cultures* (Berlin: Springer, 1985), pp. 293-303.

51. Hibino, K. and Ushiyama, K., Commercial production of ginseng by plant tissue culture technology, in T.-J. Fu, G. Singh, and W.R. Curtis (eds.), *Plant cell and tissue culture for the production of food ingredients* (Dordrecht, the Netherlands: Kluwer Academic Publishers, 1999), pp. 215-224.

52. Choi, S.M., Son, S.H., Yun, S.R., Kwon, O.W., Seon, J.H., and Paek, K.Y., 2000, Pilot-scale culture of adventitious roots of ginseng in a bioreactor system, *Plant Cell, Tissue and Organ Culture,* 62(3): 187-193; Paek, K.Y., 2002, Method for the mass propagation of adventitious roots of ginseng, camphor ginseng and wild ginseng by tissue culture and the improvement of their saponin content, Patent No. US 2002142463.

53. Liu, C.Z., Wang, Y.-C., Guo, C., Ouyang, F., Ye, H.-C., and Li, G.-F., 1998, Production of artemisinin by shoot cultures of *Artemisia annua* L. in a modified inner-loop mist bioreactor, *Plant Science,* 135(2): 211-217; Weathers, P.J., Hemmavanh, D.D., Walcerz, D.B., Cheetham, R.D., and Smith, T.C., 1997, Interactive effects of nitrate and phosphate salts, sucrose, and inoculum culture age on growth and sesquiterpene production in *Artemisia annua* hairy root cultures, *In Vitro Cellular and Developmental Biology—Plant,* 33: 306-312.

54. Doran, P.M., *Hairy roots: Culture and applications* (Amsterdam, the Netherlands: Harwood Academic Publishers, 1997).

55. Banerjee, S., Rahman, L., Uniyal, G.C., and Ahuja, P.S., 1998, Enhanced production of valepotriates by *Agrobacterium rhizogenes* induced hairy root cultures of *Valeriana wallichii* DC, *Plant Science,* 131(2): 203-208; Saito, K., Yamazaki, M., and Murakoshi, I., 1992, Transgenic medicinal plants: *Agrobacterium*-mediated foreign gene transfer and production of secondary metabolites, *Journal of Natural Products,* 55(2): 149-162; Saito, K., Yamazaki, T., Okuyama, E., Yoshihira, K., and Shimomura, K., 1991, Anthraquinone production by transformed root cultures of *Rubia tinctorum*: Influence of phytohormones and sucrose concentration, *Phytochemistry,* 30(5): 1507-1509; Kamada, H., Okamura, N., Satake, M., Harada, H., and Shimomura, K., 1986, Alkaloid production by hairy root cultures in *Atropa belladonna, Plant Cell Reports,* 5(4): 239-242.

56. Verpoorte, R. and Alfermann, A.W. (eds.), *Metabolic engineering of plant secondary metabolism* (Dordrecht, the Netherlands: Kluwer Academic Publishers, 2000).

57. Fischer, R., Emans, N., Schuster, F., Hellwig, S., and Drossard, J., 1999, Toward molecular farming in the future: Using plant-cell-suspension cultures as bioreactors, *Biotechnology and Applied Biochemistry,* 30(2): 109-112.

PART IV: LATEST DEVELOPMENTS IN MEDICINAL APPLICATIONS

Chapter 14

Medicinal Plants in the Prevention and Therapy of Cancer

David Mantle
Richard M. Wilkins

INTRODUCTION

Medicinal plants represent a vast potential resource for anticancer compounds. As with all areas of phytomedicine, the value of medicinal plants lies in the potential access to extremely complex molecular structures that would be difficult to synthesize (or conceptualize) in the laboratory. The disadvantage of this phytomedical approach lies in the extensive screening procedures required to identify the biological activities of interest associated with such compounds. The antitumor activity of medicinal-plant-derived compounds may result from a number of mechanisms, including effects on cytoskeletal proteins that play a key role in cell division, inhibition of DNA topoisomerase enzymes, antiprotease or antioxidant activity, stimulation of the immune system, etc. Examples of plant-derived compounds with anticancer activity used clinically include the following: the *Vinca* alkaloids (which induce metaphase arrest by interfering with microtubule assembly) vinblastine, vincristine, and vincamine derived from *Vinca rosea,* and their synthetic derivates vindesine and vinorelbine; the topoisomerase inhibitors campothecin, an alkaloid isolated from *Camptotheca acuminata*, and etoposide, a semisynthetic derivative of podophyllotoxin, an antineoplastic glucoside obtained from *Podophyllum peltatum*; the taxoids paclitaxel (Taxol) and the synthetic analog docetaxel (taxotere), which stabilize microtubules and disrupt mitosis. The various compounds just described are associated primarily with the treatment of breast cancer but are also active against other forms of malignancy. For example, recent Phase II/III trials have established docetaxel as the most active single agent in the treatment (first or second line) of advanced metastatic breast cancer, while paclitaxel and platinum compound combination therapy is used as the standard regimen in the treatment of ovarian cancer, and both agents are active in the treatment of lung cancer.[1]

Medicinal plants continue to be subject to extensive screening worldwide, in an attempt to develop still more effective anticancer treatments. Examples of 2002 reports of plant-derived extracts with anticancer activity are listed in Table 14.1; in many cases the structure of these active compounds has been determined, and in some cases the potential mechanisms of action elucidated.[2] However, the anticancer activity of these compounds has (in general) been established in cultured cell systems only, and their in vivo anticancer potential in animal model systems or human studies has yet to be established. In this chapter, we therefore focus on the novel anticancer activity of compounds derived from three medicinal plants, mistletoe *(Viscum album)*, ginseng *(Panax ginseng)*, and garlic *(Allium sativum)*, the anticancer action of which have been established in cell culture, animal models, and human studies.

MISTLETOE

The anticancer properties of mistletoe have been known since antiquity. More recently, extracts of mistletoe have been widely used (since the 1960s) in the treatment of cancers, although the mechanisms underlying the antitumor activity have still to be fully elucidated. Mistletoe extracts stimulate nonspecific immune function, via increasing the activity of natural killer cells and large granular leukocytes, as well as inducing cytokines such as TNFα, IFN-8, and ILS1, 6, and 10. Mistletoe extracts inhibit the growth of various tumor cell lines and are cytotoxic against animal tumors.

Much of this work was carried out using European mistletoe *(Viscum album)*, with somewhat less being known about Korean mistletoe *(Viscum album coloratum)*, another subspecies. More than 1,000 proteins are present in mistletoe extract, the best characterized components being the mistletoe lectins. These are ribosome-inactivating, protein-synthesis-inhibiting proteins with some similarities to plant-derived toxins such as ricin. Three types of lectin with different sugar specificities have been identified: lectin I, which binds D-galactose; lectin II, which binds D-galactose and NAC-galactosamine; and lectin III, which binds NAC-galactosamine. These lectins are highly toxic to tumor cells, in which they may induce apoptosis. Polysaccharide and viscotoxin fractions from mistletoe extracts have also been shown to be cytotoxic. In the following sections, we review recent work on the anticancer effects of mistletoe extracts in cultured cells, animal model systems, and human studies.

TABLE 14.1. Novel anticancer compounds isolated from medicinal plants in 2002.

Plant species	Class of compound	Specific compound	Cultured cell type	IC_{50}
Amoora rohitaka[a]	Triterpene	Amaronin	MCF-7, HeLa	1.8-3.4 μg/mL
Kageneckia oblonga[b]	Whole extract	–	P388, A549	
Neolitsea acuminatissima[c]	Eudesmanolide sesquiterpenes	Neolitacumones A-C	Hep 2, 2, 15	0.04-0.24 μg/mL
Isodon pseudo-irrorata[d]	Ent-kaurene diterpene	Psuedoirroratin A	Lul, SW626, LN Cap	0.20-0.90 μg/mL
Hyoscyamus niger[e]	Phenol amide liganamide	Grossamide cannabisins D & G	LN Cap	33-81 μM
Cratoxylum sumatranum[f]	Xanthenes	Cratoxyarborenones A-F	KB	1.0-4.3 μg/mL
Pittosporum Hosporum viridiflarum[g]	Triterpenoid saponin	Pittoviridoside	A2780	10 μg/mL
Tacca chantrieri[h]	Diarylheptanoids	–	HL60, HSC-2	1.8-6.6 μg/mL
Casearia sylvestris[i]	Clerodane diterpenoids	Casearvestrihs A-C	KB, LX1, ACT 116, A2780	0.3-14 μg/ml 0.2-0.8μM
Caseaia lucida[j]	Clerodane diterpenoids	Casearlucins A-K	A2780	10.8 μg/mL
Acacia tenuifolia[g]	Saponin	Albiziatroside A	M109	1μg/mL
Annona muricata[k]	Monotetra-hydrofuran acetogenins	Murcins I & H annomontacin *Cis*-corrosolone annocatolin	hep G2 hep 2.2.15	0.095-0.16 μg/mL 0.003-0.22 μg/mL

TABLE 14.1 *(continued)*

Key to cell types: hepG2 = human hepatoma; hep 2.2.15 = hepG2 transfected with HBV; M109 = murine lung carcinoma, A2780 = human ovarian cancer; KB = human oral epidermoid cancer, LX1 = human lung cancer; HCT 116 = human colon cancer, HL-60 = human promyelocytic leukemia, HSC2 = human oral squamous carcinoma; LN Cap = human prostate cancer; SW 626 = human ovarian cancer; P388 = murine leukemia; A549 = human lung cancer; MCF7 = human breast cancer; HeLa-human cervical cancer.

[a] Rabi, T., Karunagaran, D., Krishnan Nair, M., and Bhattathiri, V.N., 2002, Cytotoxic activity of amooranin and its derivatives, *Phytotherapy Research,* 16(Suppl. 1): S84-S86.

[b] Delporte, C., Munoz, O., Rojas, J., Ferrandiz, M., Paya, M., Erazo, S., Negrete, R., Maldonado, S., San, F.A., and Backhouse, N., 2002, Pharmaco-toxicological study of *Kageneckia oblonga*, Rosaceae, *Zeitschrift Naturforschung [C],* 57(1-2): 100-108.

[c] Chang, F.R., Hsieh, T.J., Huang, T.L., Chen, C.Y., Kuo, R.Y., Chang, Y.C., Chiu, H.F., and Wu, Y.C., 2002, Cytotoxic constituents of the stem bark of *Neolitsea acuminatissima, Journal of Natural Products,* 65(3): 255-258.

[d] Zhang, H., Fan, Z., Tan, G.T., Chai, H.B., Pezzuto, J.M., Sun, H., and Fong, H.H., 2002, Pseudoirroratin A, a new cytotoxic ent-kaurene diterpene from Isodon pseudo-irrorata, *Journal of Natural Products,* 65(2): 215-217.

[e] Ma, C.Y., Liu, W.K., and Che, C.T., 2002, Lignanamides and nonalkaloidal components of *Hyoscyamus niger* seeds, *Journal of Natural Products,* 65(2): 206-209.

[f] Seo, E.K., Kim, N.C., Wani, M.C., Wall, M.E., Navarro, H.A., Burgess, J.P., Kawanishi, K., Kardono, L.B., Riswan, S., Rose, W.C., et al., 2002, Cytotoxic prenylated xanthones and the unusual compounds anthraquinobenzophenones from *Cratoxylum sumatranum, Journal of Natural Products,* 65(3): 299-305.

[g] Seo, Y., Berger, J.M., Hoch, J., Neddermann, K.M., Bursuker, I. Mamber, S.W., and Kingston, D.G., 2002, A new triterpene saponin from *Pittosporum viridiflorum* from the Madagascar rainforest, *Journal of Natural Products,* 65(1): 65-68.

[h] Yokosuka, A., Mimaki, Y., Sakagami, H., and Sashida, Y., 2002, New diarylheptanoids and diarylheptanoid glucosides from the rhizomes of *Tacca chantrieri* and their cytotoxic activity, *Journal of Natural Products,* 65(3): 283-289.

[i] Oberlies, N.H., Burgess, J.P., Navarro, H.A., Pinos, R.E., Fairchild, C.R., Peterson, R.W., Soejarto, D.D., Farnsworth, N.R., Kinghorn, A.D., Wani, M.C., and Wall, M.E., 2002, Novel bioactive clerodane diterpenoids from the leaves and twigs of *Casearia sylvestris, Journal of Natural Products,* 65(2): 95-99.

[j] Sai Prakash, C.V., Hoch, J.M., and Kingston, D.G., 2002, Structure and stereochemistry of new cytotoxic clerodane diterpenoids from the bark of *Casearia lucida* from the Madagascar rainforest, *Journal of Natural Products,* 65(2): 100-107.

[k] Liaw, C.C., Chang, F.R., Lin, C.Y., Chou, C.J., Chiu, H.F., Wu, M.J., and Wu, Y.C., 2002, New cytotoxic mono-tetrahydrofuran annonaceous acetogenins from *Annona muricata, Journal of Natural Products,* 65(4): 470-475.

Cell Culture Studies Using Mistletoe Extracts

Lectins from Korean mistletoe have been shown to be strongly cytotoxic against various types of tumor cells, inducing morphological changes and DNA fragmentation characteristic of Ca^{2+}/Mg^{2+} endonuclease-mediated apoptosis.[3] Three distinct components of this mistletoe extract, galactose/NAC-galactosamine specific lectin II, polysaccharide, and viscotoxin, induce apoptosis in human myeloleukemic U937 cells. Lectin II is most potent in inducing apoptosis in various cancer cell lines but not in normal lymphocytes.[4] Apoptosis of U937 cells induced by mistletoe lectin II has subsequently been shown to involve activation of C-Jun-N-terminal kinase 1, and activation of caspase cascades, via specific activation of caspases 3, 8, and 9, but not caspase 1.[5] Peptidic inhibition of caspase 3 significantly inhibited lectin II-induced apoptosis. A synergistic action between lectin II and IFNγ in the apoptosis of U937 cells has also been demonstrated.[6] Prior differentiation of U937 cells induced by IFNγ resulted in increased expression of Fas (CD95/APO) and Fas L, increased activity of JNK1, and increased activity of caspases 3 and 9, leading to greatly increased susceptibility to lectin II induced apoptosis. Korean mistletoe agglutinin lectin has been shown to induce apoptosis in human hepatoma and cell lines SK Hep 1 (p53 positive) and Hep 3B (p53 negative) via p53- and p21-independent pathways.[7] The lectin-induced apoptosis by down-regulation of Bel-2 and up-regulation of Bax functioning upstream of caspase 3 in both cell lines, as well as a down-regulation of telomerase activity via a mitochondrial controlled pathway independent of p53. Mistletoe lectin I has also been shown to induce apoptosis in leukemic T or B cell lines, via activation of caspase 3, 8, and 9, with lectin-induced apoptosis again prevented by caspase peptide inhibitors.[8] There was no evidence for receptor involvement in this process, since cell death induced by lectin I was neither attenuated in cell clones resistant to CD 95 nor in cells that were rendered refractory to other cell death receptors by overexpressing a dominant negative FADD mutant. In contrast, lectin I triggered a receptor independent mitochondrial controlled apoptotic pathway, since cytochrome C was rapidly released into the cytosol.

The cytotoxic activity of a recombinant fusion protein (bFGF-rMLA) containing the mitogen basic fibroblast growth factor (bFGF) and the cytotoxic A-chain component of recombinant mistletoe lectin (rML) was evaluated in mouse B16 melanoma cells expressing the FGF receptor. While rMLA showed no significant effect on B16 cell viability at concentrations up to 140 nM, an IC_{50} value of about 1nM was obtained for bFGF-rMLA. Further association of bFGF-rMLA with recombinant lectin B chain resulted in an IC_{50} value of 134 pM.[9] Thus it is possible to increase the cytoxicity of mistletoe lectin via receptor-specific inactivation of the target cells, using the various recombinant protein combinations just discussed.

Synergistic cytotoxic interaction between mistletoe lectin I and conventional anticancer agents has been found in cultured tumor cells. Thus synergistic cytoxicity was observed with simultaneous treatment of human hepatocarcinoma SK-Hep1 cells with the lectin and paclitaxel.[10] Similarly, mistletoe agglutinin I was found to have synergistic action in combination with the conventional chemotherapeutic drugs cyclohexamide, doxorubicin, cisplatin, and taxol in human lung cancer cell line A549.[11]

Animal Model Studies Using Mistletoe Extracts

The anticancer activity of mistletoe extracts has been investigated against a variety of tumor types in murine or rodent animal model systems. Isorel, a commercially produced standardized aqueous extract of the whole mistletoe plant, has been shown to have both local and systemic beneficial effects, in terms of tumor growth inhibition, reduction of metastases, and immunomodulatory effects in tumor-bearing mice. Thus mice carrying mammary carcinomas in both hind limbs were also injected (i.v.) with tumor cells, resulting in the development of lung metastases. Isorel applied to one limb resulted in the effective inhibition of tumor growth in approximately 30 percent of treated animals, with a less pronounced effect on the contralateral limb tumors. Histological analysis showed Isorel treatment applied locally or systemically increased the incidence of tumor apoptosis and necrosis, while a reduction of mitosis was observed only with locally applied Isorel. In addition, animals treated with Isorel had approximately threefold fewer lung metastases than untreated controls. Isorel treatment also resulted in immunostimulatory action, increasing the reactivity of tumor-bearing mouse lymphocytes to mitogens (conA and LPS) in vitro, and inhibited protein synthesis in cultured cells from a variety of malignant cell lines.[12]

The antimetastatic activity of lektinol, a mistletoe-derived (agglutinin-standardized) immune-response modifying drug, against B16 melanoma lung colonization has been shown in vivo. Following the injection (i.v.) of B16 melanoma cells into mice, histological analysis of lung tissue or bronchoalveolar lavage fluid cells showed 60 to 90 percent inhibition of pulmonary metastatic colonization, after daily injection (i.v.) of Lektinol (up to $150 ng \cdot kg^{-1}$) over a three-week period. Analysis of lavage fluid from tumor-bearing mice showed a five- to sixfold increase in the proportion of MAC^{-1}+(CD116/CD18) immunocompetent macrophages, and an increase in double-positive immature CD4+8+ thymocytes (compared to untreated controls), with no evidence of treatment related toxicity.[13]

Aqueous mistletoe extract showed anticancer activity against a variety of transplantable, subcutaneously growing syngeneic murine tumors, transplanted into immunocompetent mouse strains (C56BL16 or BALB/C). Repeated parenteral treatment (i.p. or s.c.) over a four-week period with the mistletoe extract (30 to 300 ng lectin equivalent per Kg per day) resulted in significant tumor growth inhibition (assessed via caliper measurements) with Renca renal, C8 colon 38, and F9 testicular carcinomas, but had no inhibitory effect on Lewis lung carcinoma or B16 melanoma.[14] The authors concluded that these effects were most likely due to immunostimulatory rather than cytotoxic effects in view of the low-dose levels at which the extracts were effective.

Administration (i.v.) of Korean mistletoe extract (100 μg) two days before tumor inoculation significantly inhibited lung metastasis of B16-BL6 melanoma and 26-M3.1 colon cells, and liver and spleen metastasis of L5178Y-ML25 lymphoma cells. The prophylactic effect of the mistletoe extracts was obtained using various routes (i.e., i.v., s.c., oral, intranasal) of administration. Natural killer cell activity was significantly increased one to three days after extract administration, and natural killer cell depletion (via injection of rabbit antiasialo GM1 serum) completely abolished the inhibitory effect on lung metastasis of colon 26-M3.1 cells.[15] These data therefore show that the prophylactic effect of the extract on tumor metastasis results from immunopotentiating activity, via the activation of natural killer cells.

Evidence that mistletoe lectin inhibits tumor growth and metastasis by increasing apoptosis and inhibiting angiogenesis has been obtained in C57B6 mice inoculated with B16-BL6 melanoma cells. Treatment with purified Korean mistletoe lectin resulted in the inhibition of lung metastases and increased survival times, compared to untreated controls. Treatment of cells with the lectin resulted in growth suppression, nuclear morphological changes, DNA fragmentation, and an increased fraction of cells in the sub-G1 phase, consistent with the induction of apoptosis; in addition, antiangiogenic activity (assessed via the choriallantoic membrane assay) was increased after lectin treatment.[16] Following the intracerebral implantation of F98 gliomal cells in Fischer 344 rats, local or systemic treatment with galactosidase specific mistletoe lectin resulted in a reduction of tumor volume assessed histologically; systemic low dosage (1ng/Kg s.c.) was more effective than higher dose (100 ng) local application, suggesting an immunomodulatory beneficial effect.[17]

The antitumoral effects of aqueous mistletoe extract on the growth of urinary bladder cancer have been investigated in an orthoptic murine model. Ten days after the implantation of MB 49 carcinoma cells into the bladders of C57BL/6J mice, animals were treated (i.v.) for three days per week over a four-week period with aqueous mistletoe extract (30-300 ng lectin equivalent/ml, 0.1 mL total volume). At the higher dose rate, treated mice showed an increased survival rate compared to untreated controls (85 percent versus 39 percent), with 18 percent of treated mice showing evidence of solid bladder tumors, compared to an 80 percent incidence in untreated animals.[18] The beneficial action of the mistletoe extract under these experimental conditions was considered to result from cytotoxic properties of the mistletoe lectin.

The anticancer effect of extracts of mistletoe grown on fir or pine trees, or of mixed fir tree/mistletoe extracts has been described. Treatments with fir or pine tree derived aqueous mistletoe extracts (5-50 μg/injection, three times weekly for 14 days), resulted in a significant reduction in liver or lung metastases, following innoculation (i.v.) of RAW 117H 10 lymphosarcoma or L-1 sacroma cells into BALB/C mice, to establish liver or lung colonization respectively. In addition to this antimetastatic activity, immunomodulatory action of the extracts was noted, via up-regulation of thymus weight and peripheral blood leukocyte counts in tumor-bearing mice.[19] A traditional remedy, based on an aqueous mixed extract of pine tree *(Abies alba)* and mistletoe, has been found to be effective against chemically induced tumors in rats. Tumors were induced in Wistar rats following injection (s.c.) of 10 μg of benzoalphapyrene (BAP), a dose inducing 100 percent carcinogenesis. Animals were administered the mixed extract ($50 mL \cdot kg^{-1}$) orally, either immediately following BAP injection, or 120 days thereafter. Animals treated immediately following the BAP injection showed significant inhibition of tumor induction, reduction in tumor growth rate, and prolongation of life, compared to untreated controls. For animals in which tumors had been established prior to treatment, with the mixed extract, there was evidence of tumor necrotic effects and life prolongation.[20] The antiproliferative and anticarcinogenic effects of the mixed extract may result from active compounds derived both from V. *album* (lectins, thionins) and *Abies alba* (monoterpenes). The growth of murine non-Hodgkin lymphoma tumors has been shown to be reduced by incorporating mistletoe lectins in the diet. Tumors from such mice showed altered morphological characteristics compared to control-fed animals, with a reduced nuclear area and reduced mitotic activity, as well as increased evidence of apoptic bodies and increased lymphocyte infiltration of the tumor. Mice fed the lectin supplemented diet showed a poorly developed tumor blood supply compared to control-fed mice, indicating an antiangiogenic response.[21]

Animal-model-based investigations into the anticancer potential of recombinant mistletoe lectins (rml, r viscumin) have recently been reported. After innoculation (i.v.) of Balb/c mice with RAW-117P or L1 sarcoma cells, r viscumin was administered at nontoxic doses ranging from 0.3-150 ng·Kg^{-1} (s.c.). Treatment with r viscumin resulted in a prolonged survival time, a reduction in the number of tumor colonies infiltrating lungs (RAW-117P) or liver (L1), and the activation of immune cell subsets, in other words, increased numbers of T-lymphocytes, pan natural killer cells, and activated monocytes.[22] Recombinant lectin has been administered (30-50 mg·Kg^{-1} i.p., daily for 80 days) to SCID mice, following injection (i.p.) of human ovarian cancer cells. In animals treated with the higher does of rml, 13 of 25 animals survived the experiment, with no evidence of tumor cells in the peritoneum, compared to 2 of 25 animals in the untreated control group.[23] Antitumor activity of recombinant galactoside specific mistletoe lectin on chemically induced bladder carcinogenesis has also been reported in a rat model. Following tumor induction over a four-week period via *N*-methyl-*N*-nitrosourea administration (i.v.), rats were treated with rml (30-150 ng, twice weekly for 20 weeks). At the end of the experimental period, histological analysis showed treated animals had significantly lower rates of hyperplasia and neoplastic transformation (40 to 50 percent), compared to untreated controls (82 percent). The anticarcinogenic action of rml in this model system was not due to immunomodulatory effects involving interferon γ or IL-10 dependent mechanisms, since mRNA expression levels for these parameters were comparable in treated and control animals groups.[24]

In contrast to the various investigations just outlined, there have been several reports relating to the ineffectiveness of mistletoe lectins as antitumor agents in animal models. Thus immunomodulatory doses (1ng·Kg^{-1}) of galactoside specific lectin, given twice weekly over a 15-month period, did not significantly reduce the incidence of epithedial bladder tumors (compared to control animals), following chemical induction of the latter in rats using a single dose (i.v.) of *N*-nitrosourea. In addition, quantitative immunohistochemical analysis of the bladder wall tissue for a range of immune cell types, including T-lymphocytes, T-helper/inducer cells, T-suppressor/cytotoxic cells, lymphocytes, macrophages, and natural killer cells, showed no significant differences in levels between treated and untreated animal groups.[25] The authors concluded there was no evidence that mistletoe lectin was effective in inhibiting chemically induced bladder carcinogenesis or trigging a local cellular immune response after prolonged application. The same authors have subsequently demonstrated a lack of protection against *N*-butyl-*N*-hydroxybutyl nitrosamine induced bladder carcinogenesis, and no induction of local cellular immune response, following long-term administration of galactoside-specific lectin in rats.[26] They concluded that galactoside-specific mistletoe lectin fails to inhibit or reduce the development of chemically induced bladder cancer, even after long-term administration in a clinically recommended schedule, and questioned the value of mistletoe extract adjuvant therapy in the corresponding clinical setting.

Finally, galactoside-specific lectin has been investigated as part of a combination therapy with interleukin-2 in CH3/HeJ mice bearing transplants of the highly metastatic C3L mammary adenocarocarcinoma.[27] The object of this investigation was to establish if coadministration of the lectin might reduce capillary leakage, a major adverse effect limiting IL-2-based cancer immunotherapy. Although IL-2 therapy alone reduced tumor growth and metastases, addition of the lectin had no beneficial effect on capillary leakage induced by IL-2. However, an unexpected finding from this investigation was that lectin therapy alone stimulated tumor growth and lung metastases, indicating a need for caution in the use of this lectin as an immunoadjuvant in human cancer therapy.

Clinical Studies Using Mistletoe Extracts

In a prospective randomized study of breast cancer patients, subcutaneous administration of mistletoe lectin (0.5-1mg·Kg^{-1}, twice weekly) induced plasma endorphin levels and enhanced activity of peripheral blood natural killer cells and T-lymphocytes (expression of CD-25/IL-2 receptors and HLA/DR antigens).[28] Mistletoe extract therapy (16 weeks s.c.) in eight breast cancer patients resulted in a strong initial proliferation of peripheral blood mononuclear cells in all cases, which, however, subsequently decreased, indicating that both activating and inhibitory mechanisms had been induced. Extract stimulated cultured cells produced TNF-α and IL-6, indicating activation of monocyte/macrophage cells. At the end of the therapy, a shift to Th1-related cytokines was observed with in vitro cell culture.[29] Thus, mistletoe extract can influence the Th1/Th2 cellular/humoral immunity balance, and in the case of a Th1 shift, may favorably influence tumor growth. In addition, mistletoe extract has been shown to promote DNA repair in peripheral blood mononuclear cells from ten breast cancer patients, after exposure to damaging agents in vitro; exposure of cells to γ-radiation or 4-hydroxycyclophosphamide resulted in improved DNA repair and increased IFN-γ production in the presence of mistletoe extract.[30]

Evidence for an immunomodulatory effect of mistletoe extract in healthy volunteers was obtained.[31] Injection of Iscador (0.1-1 mg at defined intervals) resulted in temporary improvements in microcirculation and increased adhesion and transmigration of white blood cells (assessed microscopically) in dermal and intestinal target tissues.

The effect of mistletoe extract on the serum levels of IL-12 and production of IFN-γ, IL-2, and IL-4 in peripheral blood mononuclear cells in cancer patients and healthy controls has been investigated.[32] Cytokine levels were determined by ELISA at various time points over a 30-day treatment period. In cancer patients, the production of IFN-γ and IL-2 were threefold and ninefold lower than in controls respectively, and increased significantly during treatment. IL-4 production in cancer and control patients was similar pretreatment and did not alter during treatment; increased serum IL-12 levels in cancer patients (initially threefold higher than in controls) following treatment was of borderline significance. These data again indicate that mistletoe extract positively affects cell-mediated immunity via increased Th1 cytokine levels.

Prospective nonrandomized and randomized matched pair studies have been undertaken on more than 10,000 cancer patients, approximately 1,600 of whom had been treated with Iscador. In the nonrandomized matched pair study, the survival time of patients treated with Iscador was longer for all the types of cancer studied (colon, rectum, stomach, breast, lung). In the pool of 396 matched pairs, mean survival time in the Iscador groups was approximately 40 percent longer than in the control groups (4.28 years versus 3.05 years). Results from the two randomized matched pairs studies largely confirmed the results of the nonrandomized studies, showing a clinically relevant prolongation of survival time of cancer patients treated with Iscador.[33]

A prospective clinical trial of patients with malignant glioma has been carried out, in which patients undergoing standard oncological treatment (neurosurgery, radiation therapy) were randomly assigned into a treatment group, receiving complementary immunotherapy with galactoside-specific mistletoe lectin, and a control group without complementary treatment.[34] Nonstratified analysis of all patients did not show a significant prolongation of relapse-free interval or survival time for the treated (versus control) group. However, stratified analysis of stage III/IV glioma patients showed a significant prolongation of relapse-free survival for treated versus control patients (17.4 versus 10.5 months), and a prolongation of overall survival for the treatment group compared to control patients (20 versus 9.9 months).

A case report of the favorable long-term outcome using mistletoe therapy in a patient with centroblastic/centrocytic non-Hodgkin's lymphoma has been described. Clinical examination (including CT analysis) of a 44-year-old patient confirmed multiple thoracic, abdominal, and inguinal lymphoma, designated as follicular non-Hodgkin's stage IV, with bone marrow infiltration. The efficacy of treating lymphoma with Iscador over a 12-year period was demonstrated, with phases of uninterrupted treatment resulting in lymphoma regression, while cessation of treatment resulted in progression.[35]

In contrast to the various reports summarized above describing the beneficial action of mistletoe extracts in cancer patients, several reports have found a lack of benefit of mistletoe extracts in the treatment of human cancers. Thus the effect of an adjuvant mistletoe extract treatment was investigated in a prospective, randomized controlled clinical trial involving almost 500 patients with head and neck squamous cell carcinoma. The patients were stratified into two treatment groups undergoing surgery or surgery plus radiotherapy, and both groups were randomized for additional treatment with standardized mistletoe lectin. Lectin-treated patients had no reduced risk of local recurrences, distant metastases, or second primaries, and five-year survival rates were no better than for the control patient group; in addition, no significant changes in cellular immune reaction or quality of life were noted following mistletoe treatment.[36]

Goebell and colleagues carried out a randomized Phase II trial to investigate the efficacy of mistletoe lectin in preventing recurrence of superficial bladder tumors after transurethral resection; 45 patients were randomly assigned to mistletoe lectin treated (s.c.) and untreated control groups, with assessment (including urethrocystoscopy) after 18 months. Adjuvant therapy with mistletoe lectin did not affect the time to first recurrence, total number of recurrences, or recurrence-free outcome.[37]

The effect of treating patients with malignant haematological and lymphatic diseases with mistletoe extract has been investigated in a retrospective analysis of several hundred patients.[38] There was no significance difference on median survival time between mistletoe-treated (11.4 years) and untreated (8.6 years) patients, although the authors noted that neither was there any adverse potential risk of using mistletoe therapy in these patients. During treatment with whole plant mistletoe extracts (e.g., Iscador), an inflammatory reaction usually occurred at the site of injection.[39] While severe adverse effects resulting from mistletoe therapy appear to be relatively common, there have been several reports of nausea, gastrointestinal problems, etc., as well as cases of anaphylactic shock.[40]

GINSENG

Ginseng root has been used empirically in the traditional medicine of the Far East for maintaining physical vitality and promoting long life, and it has been proposed that one of the ways in which ginseng may promote longevity may be its protective activity against the development of cancers. There are 13 species of ginseng varying in geographical distribution, the most commonly studied being *Panax ginseng* C.A. Meyer (Korea, China). *Panax japanicus* (Japan), and *Panax quinquefolium* (United States). The constituents and pharmacological action differ in the various species and also depend on the age at harvest, the extraction method, and physical form, in other words, fresh ginseng (< 4 years), white ginseng (peeled and dried), red ginseng (steamed and dried), which may be sliced, juiced, powdered, served as a tea, etc. Most of the pharmological actions of ginseng have been attributed to ginsenosides, a structurally diverse group of steroidal saponins, more than 30 of which have been identified and their structures determined. The pharmacology of ginsenosides is complex, since each compound may initiate multiple actions within the same tissue, and these responses differ for each ginsenoside. Ginsenosides are amphiphilic molecules, able to intercalate the plasma membrane, thereby affecting fluidity and function; they are able to interact directly with plasma membrane receptors and activate intracellular steroid receptors. Ginseng has been shown to have beneficial effects on the central nervous system (CNS) (e.g., protection of neurons against ischemic damage), the cardiovascular system (e.g., antiplatelet aggregation), and the gastrointestinal tract (e.g., protection against ulceration).[41] However, it is the anticancer effects of ginseng which are of relevance to this review. Whole extracts or individual compounds prepared from ginseng species have antimutagenic and anticarcinogenic properties, acting at various stages of the multistep cancer process via growth-inhibitory, cytoxic, and immunomodulatory mechanisms. These include protective effects at the initiation and promotion stages of carcinogenesis, inhibition of cell growth, induction of differentiation/reverse transformation of malignant cells, inhibition of metasases, and antitumor angiogenesis. These effects may be manifest via antioxidant, anti-inflammatory, or apoptotic action of ginseng-derived compounds, mediated via effects on specific enzymes or proteins such as ornithine decarboxylase, cycloxygenase-2, mitogen-activated protein kinases, transcription factor NFk, etc., which play a key role in cell proliferation and malignant transformation. Evidence for these effects obtained in cell culture, animal models, and clinical studies are reviewed in detail in the following sections.

Cell Culture Studies Using Ginseng

Ginseng extracts exert anticarcinogenic effects in vitro via a variety of mechanisms. Several ginsenosides show direct growth inhibitory effects against tumor cells or induce differentiation and inhibit metastasis. Thus in B16-BL6 melanoma cells, the ginsenoside Rh_2 inhibited growth and stimulated melanogenesis, arresting cell cycle progression at the G1 stage.[42] In association with G1 arrest, there was suppression of cyclin dependent kinase 2 activity. The proliferation of three human renal carcinoma cell lines (A491, Caki-1, CURC II) was inhibited by lipid-soluble components of *P. ginseng,* by blocking cell cycle progression at the G1 to S phase transition.[43] In human breast cancer MCF-7 cells, ginsenoside Rh2 also significantly inhibited cell growth in a concentration-dependent manner, inducing G1 arrest in the cell cycle progression.[44] The data obtained suggested ginseng inhibited MCF-7 cell growth by inducing p21 protein expression and reducing levels of cyclin D, resulting in down-regulation of cyclin D/cyclin-dependent kinase complex activity, and inhibition of E2F release. Extracts of American ginseng *(P. quinquefolium)* showed a close dependent inhibition of MCF-7 cell proliferation, with a further synergistic action with conventional therapeutic agents including doxorubicin, tamoxifen, Taxol, and fluorouracil against the latter cell type.[45] The ginseng extract was subsequently shown to inhibit the proliferation of both hormone-sensitive (MCF-7) and -insensitive (MDA.MB.23) breast cancer cells, via transcriptional activation of the cell cycle gene/protein p21, independently of p53.[46]

In an attempt to establish the cytotoxic/structural relationship of ginseng-derived panaxydol and related compounds, synthetic analogs of ginseng diyne have been shown to have antiproliferative activity against murine leukemia L1210 cells.[47] Steaming ginseng at high temperatures increases the cytotoxicity toward SK-Hep-1 hepatoma cells, due to the high action of ginsenosides Rg3, Rg5, RK1, Rs4, and RS5 (50 percent growth inhibition at 10-40 μm).[48] A novel ribonuclease has been isolated from *P. notoginseng* with potent RNA-ase and translational inhibitory activities, which showed antiproliferative activity against leukemia L1210 cells.[49]

A number of in vitro studies have provided evidence that the anticancer action of ginseng extracts is mediated via the induction of apoptosis in malignant cells. For example, ginsenoside Rh2 induces apoptosis in SK-HEP-1 cells, via increased caspase 3 activity which in turn cleaved the cyclin kinase inhibitor p21WAF1/C1P1.[50] A modified ginsenoside resulting from the metabolism of ginsenosides by intestinal bacteria inhibited the proliferation of B16-BL6 mouse melanoma cells (to a greater extent that the original ginsenoside), inducing apoptotic cell death at higher (40 μm) concentration.[51] The ginsenoside Rs4 induces apoptosis in human hepatoma SK-HEP-1 cells, via increasing levels of P53 and p21WAF1, and down-regulating cycline E and A kinase activity.[52] The ginsenoside Rh2 induced apoptosis in a rat C6 gliomal cell line, via a mechanism involving activation of caspases and production of reactive oxygen species, but independently of Bcl-x 68.[53] Ginsenoside Rg3 inhibited the growth of human prostate carcinoma LNCaP cells by activating expression of cyclin kinase inhibitors p21 and p27, arresting the cell cycle at the G1 phase, and inducing apoptosis via a Bcl-2/caspase 3-mediated mechanism.[54] The novel compound IH-901, an intestinal bacterial metabolite of ginsenosides Rb1, Rh2, and Rc, substantially suppressed the growth of human leukemia HL-60 cells via an apoptotic mechanism involving activation of caspase 3, by a process involving release of mitochondrial cytochrome C into the cytosol, without affecting the expression of Bcl-2 protein.[55]

Evidence has also been obtained that ginseng components can inhibit the invasiveness of malignant cells in vitro, via down-regulation of proteolytic (particularly matrix metalloproteinase) enzymes. Park and colleagues have shown that the purified ginseng components panaxadiol and panaxatriol were able to interact with nuclear glucocorticoid receptors, thereby down-regulating the activity of matrix metalloproteinase 9 (but not matrix metalloproteinase 2) and reducing the invasive capacity in vitro of HT1080 human fibrosarcoma cells.[56] In addition, ginsenoside Rb2 inhibited the in vitro invasiveness of HHUA or HEC-1-A endometrial cancer cell lines, via reducing the activity of matrix metalloproteinase-2 (but not matrix metalloproteinase1).[57] Other mechanisms through which ginseng extracts may exert anticancer activity include the transcriptional activation of superoxide dismutase and catalase gene expression (two- to threefold) in HepG2 human hepatoma cells by panaxadiol ginsenoside Rb2 (but not panaxatriol or total saponins), thereby mediating cell viability by increasing intracellular defenses against damaging reactive oxygen species.[58] Evidence also suggests that ginsenoside Rb2 may prevent cancer by preventing down-regulation of gap junction intracellular communication by TPA or hydrogen peroxide during the promotion stage, in rat liver epithelial cells.[59] Ginseng extracts may be useful in the treatment of cancers resistant to conventional drugs, since the ginseng metabolite IH-901 was able to inhibit the growth of a pulmonary cancer cell line resistant to CDDP, while ginseng saponins can increase the sensitivity of leukemic progenitor cells to cytotoxic drugs such as Adriamycin.[60]

Finally, in contrast to the various data just described, there has been a recent report that herbal therapies used by women to manage menopausal symptoms, including ginseng, may actually stimulate the growth of MCF-7 breast cancer cells.[61]

Animal Model Studies Using Ginseng Extracts

Extracts of red ginseng have been shown to have a significant inhibitory effect on skin cancer formation in a two-stage mouse model. The incidence of skin papilloma, latent period of tumor occurrence, and number of tumors per mouse induced by DMBA croton oil were significantly improved after treatment with ginseng extracts, in a dose-dependent manner (50-400 $mg \cdot Kg^{-1}$).[62] Extracts of the roots of *P. notoginseng* showed significant antitumor-promoting activity in two-stage carcinogenesis of mouse skin tumors induced by DMBA initiator and fumonisin B1 mycotoxin as promoter. The extract also showed antitumor activity in the two-stage skin tumor mouse model using the nitric oxide donor NOR-1 as initiator and TPA as promoter.[63] Anticarcinogenic effects of the majonoside-R2 isolated from Vietnamese ginseng has also been demonstrated in two-stage skin cancer mouse models.[64] Methanolic extracts of heat-treated ginseng significantly ameliorated subsequent skin papillomagenesis initiated by DMBA, when topically applied to the backs of ICR mice prior to exposure to TPA promoter.[65] Levels of ornithine decarboxylase were substantially reduced following ginseng treatment. Elevated levels of the latter enzyme are associated with tumor promotion because of their key role in the biosynthesis of polyamines (which are of central importance in cell proliferation). Similarly, expression of the pro-inflammatory enzyme cyclooxygenase-2, mitogen-activated protein kinase, and the transcription factor NFkβ induced in TPA stimulated mouse skin was suppressed by pre-treatment with the ginsenoside Rg3.[66]

Dietary administration of red ginseng in the initiation stage of carcinogenesis suppressed preneoplastic lesions induced by 1,-2-dimethylhydrazine (DMH) in the colons of rats.[67] Rats were fed diets containing 1 percent ginseng for five weeks, the latter four weeks of which the animals were injected with DMH at weekly intervals. The numbers of foci with at least four crypts (aberrant crypt foci [ACF], the precursor lesions of colon cancer), were significantly reduced with ginseng treatment, compared to untreated controls, suggesting that suppression of cell proliferation in the colonic mucosa may be responsible for inhibition of DMH-induced colon carginogenesis. Rats fed red ginseng powder (2 $mg \cdot Kg^{-1}$ or 0.5 $g \cdot Kg^{-1}$) for five weeks were injected during weeks two and three with azoxymethane (15 $mg \cdot Kg^{-1}$) to induce aberrant crypt foci in the colon (i.e., to investigate the effect of ginseng during the initiation phase). Rats were also fed ginseng in this way after ACFs had been allowed to develop in the colon for four weeks (to evaluate the effects of ginseng during outgrowth and progression). When ginseng was administered during the initiation phase of carcinogenesis, only a modest inhibition of ACF was noted, and only at the 0.5 $mg{\cdot}Kg^{-1}$ dose, indicating that the effect of ginseng during initiation of ACFs from normal mucosa was relatively weak. However, after ACFs had been established in the colon during the four-week outgrowth period, an additional four-week exposure to ginseng (2 $mg{\cdot}Kg^{-1}$) significantly reduced the incidence of ACFs, indicating an inhibitory effect on the growth of preneoplastic lesions.[68] A traditional Chinese medicine, Shikunshito-Kamiho, comprising extracts from eight plants including ginseng, has been shown to have anticarcinogenic activity on experimental murine colorectal cancer. Oral administration of the above (0.5 to 1.5 percent of diet) for five weeks significantly reduced the number of ACFs induced in the colons of mice following exposure to DMH, compared to animals treated with DMH alone.[69]

Red ginseng extract inhibited the induction and proliferation of lung tumors by DMBA or aflatoxin B1, using a medium-term mouse anticarcinogenicity model (termed Yun's model).[70] Using the latter model, the cancer preventative effects of ginseng were shown to vary according to the type and age of ginseng used. In addition, *P. notoginseng* extract was found to have antitumor-promoting activity in a mouse model of lung carcinogenesis induced by 4-nitroquinoline-*N*-oxide and glycerol.[71] An acidic polysaccharide purified from *P. ginseng,* ginsan, exhibited significant in vivo antitumor activity (at 1 $g{\cdot}Kg^{-1}$ i.p.) in a benzopyrene induced in vivo autochthonous lung tumor model, via activation of the immune system.[72] The incidence of lung tumors was also significantly reduced by administration of ginsan (1 $mg{\cdot}mL^{-1}$ ad lib) in drinking water after the third week of benzopyrene treatment. Two ginsenosides purified from red ginseng, Rg3 and Rg5, have subsequently been shown to significantly reduce the incidence of lung tumors (Yun's model), when administered at 80 $\mu g{\cdot}mL^{-1}$ in drinking water for six weeks after benzopyrene treatment.[73]

The anticarcinogenic action of ginseng against liver cancer was first demonstrated by Yun, Yun, and Han, where the incidence of hepatoma induced by aflatoxin B1 in mice was decreased by 75 percent following oral administration of red ginseng extract.[74] Ginseng extract was also shown to be effective in preventing the development of liver cancer induced by diethylnitrosamine (DEN) in rats.[75] In rats given ginseng preventatively (i.e., 3.8 $g{\cdot}Kg^{-1}$ per week for 15 weeks, coadministered with DEN), the development rate for liver cancer was approximately 14 percent, compared to 100 percent in animals given DEN only. In rats given ginseng curatively, i.e., after development of liver cancer nodules (15 $g{\cdot}Kg^{-1}$ per week for eight weeks, 20 weeks after DEN treatment), hepatoma nodules were smaller in ginseng-treated animals, with better structural preservation of hepatic tissue, than in animals given DEN alone. The average life span of the ginseng/DEN treated animal group was 73 days, compared to 42 days for DEN treated animals.[76] The majonoside R2 isolated from *P. vietnamensis* is a potent antitumor-promoting agent in the two-stage hepatic tumor mouse model using *N*-nitrosodiethylamine as initiator and phenobarbital as promoter (in addition to similar activity against mouse skin carcinogenesis previously noted).[77] Oral administration of red ginseng extract (1 percent of diet) for 40 weeks resulted in the significant suppression of liver tumor formation in C3H/He mice, which have a high incidence of spontaneous liver tumor development. The average number of tumors per mouse in the ginseng treated animals was 0.33, compared to 1.0 in untreated controls.[78] Long-term (70 weeks) dietary administration of lycopene (0.005 percent) and Sho-Saiko-To (1 percent), a traditional preparation containing seven plant extracts including ginseng, did not reduce the risk of hepatocarcinogenesis in Long-Evans Cinnamon rats, an established animal model of spontaneous liver tumor development.[79]

Adenocarcinomas of the mammary gland induced by intramammary injection of *N*-methyl-*N*-nitrosourea in rats, and the development of uterine, cervical, and vaginal tumors induced by intravaginal applications of DMBA in mice, were inhibited by extracts of ginseng obtained from tissue culture; the compounds were administered orally or intravaginally for up to 40 weeks.[80]

The development of tumors in nude mice, following inoculation of human ovarian cancer cells, was significantly inhibited by daily peroral treatment with ginsenoside Rh2, compared to treatment with *cis*-diaminedichloroplatinum II (CDDP). Combination treatment of ginsenoside Rh2 and CDDP did not potentiate tumor growth retardation. The survival time of ginsenoside-treated tumor-bearing animals was significantly longer than that for CDDP or untreated mice. However, administration of ginsenoside (i.p.) had little inhibitory effect on tumor growth, although when combined with CDDP, tumor growth was significantly inhibited compared to treatment with CDDP alone. Doses of ginsenoside used in the investigation (0.4-1.6 $mg \cdot Kg^{-1}$) did not result in any adverse side effects, in contrast to CDDP (4 $mg \cdot Kg^{-1}$), which had severe side effects.[81]

Clinical Studies Using Ginseng

Evidence for a protective effect of ginseng against cancer was initially reported by Yun and Choi.[82] A case control study of approx 2,000 pairs showed a decreased risk (odds ratio of 0.50) for ginseng intakers compared with nonintakers. For the site of cancer, the odds ratios were 0.15 for ovarian cancer, 0.18 for laryngeal cancer, 0.20 for oesophageal cancer, 0.22 for pancreatic cancer, 0.36 for stomach cancer, 0.42 for colorectal cancer, 0.47 for oral cancers, 0.48 for liver cancer, and 0.55 for lung cancer. No protective effect of ginseng against breast, cervical, bladder, or thyroid cancers was found. There was a dose-response relationship of decreased cancer risk with increasing frequency and duration of ginseng intake. Red ginseng was most effective (odds ratio 0.20), followed by white ginseng (0.30-0.57, powder or extract) and fresh ginseng (0.37); fresh ginseng slice or juice or white ginseng tea were ineffective in protecting against cancer.

These data were confirmed by the results of a prospective cohort population study, which strongly demonstrated that *P. ginseng* C.A. Meyer has a non-organ-specific preventative effect against cancer.[83] Based on a questionnaire detailing ginseng intake in a cohort of more than 4,600 Koreans over 40 years of age, ginseng consumers were found to have a decreased risk (relative risk [RR] of 0.40) compared with nonconsumers, with a reduced cancer risk with increasing frequency of ginseng intake. The RR for lung cancer and gastric cancer was 0.30 and 0.33 respectively. No cancer deaths occurred among red ginseng consumers, with RR values of 0.31 and 0.34 for fresh ginseng and combination intakers respectively.

Several case reports have reported cancer patients being treated with traditional herbal preparations containing ginseng. An elderly patient with lung carcinoma was treated with the traditional Japanese herbal medicine Ninjin Yoei To, based on extracts of ginseng, cinnamon bark, angelica root, astragalus root, paeoni root, citrus peel, rehmannia root, polygala root, atractylodes rhizome, schisandra fruit, poria sclerotium, and glycyrrhiza. Treatment with the above at 15 g per day for seven weeks decreased the levels of tumor markers CEA and CA19-9, with an improved clinical status of the patent.[84] Two cases with hormone-refractory metastatic prostate cancer showed a favorable response to therapy with a traditional herbal compound (PL-SPES) comprising ginseng, licorice, isatis, chrysanthemum, saw palmetto, and scuttellaria. Prostate-specific antigen levels were reduced to substantially lower levels, which were stable over a period of 4 to 12 months, and the treatment was well tolerated.[85]

The anticarcinogenic effect of three drugs derived from ginseng via tissue culture techniques—bioginseng, panaxel, and panaxel-5—has been evaluated in preliminary clinical trials in patients with precancerous lesions of the esophagus and endometrium.[86] Panaxel had a strong therapeutic effect in 64 patients with chronic erosive esophogitis. Treatment with panaxel (4 mL per day perorally for one month, up to three monthly cycles) for one to three months induced complete regression of erosions and inflammatory lesions in the esophageal mucosa in 73 percent of patients, whereas in an untreated control group of 19 cases, the spontaneous remission of erosions/lesions over three months was only 15 percent. Treatment with bioginseng (2.5 mL perorally per day over six months) induced complete regression of ademonatous cystic hyperplasia of the endometrium in a small number (3 of 11) of patients; untreated control cases (9) showed no instances of hyperphasic regression.

Extracts of *P. ginseng* enhanced natural killer and antibody-dependent cell cytotoxicity of peripheral blood nononuclear cells isolated from normal volunteers and patients with depressed cellular immunity and activated phagocytosis of peripheral blood polymorphonuclear leukocytes.[87]

With regard to safety, data from clinical trials have shown that treatments with *P. ginseng* is rarely associated with adverse effects or drug interactions. Documented effects are usually comparable with placebo, being mild and transient, the most commonly reported being headaches, gastrointestinal tract disorders, and sleep disturbance. Combination products containing ginseng have been associated with more serious adverse effects, presumably due to other components present. Four reports of possible drug interactions between ginseng and phenelzine (two), warfarin (one), and alcohol (one) have been received.[88]

GARLIC

The reputation of garlic *(Allium sativum)* as an effective treatment for malignant tumors extends back to the time of the ancient Egyptian civilizations, but scientific support for its therapeutic potential has been obtained only relatively recently. Cell culture, animal model, and epidemiological studies have shown evidence for the anticancer activity of garlic extracts or specific chemical compounds derived from garlic. These effects include the regulation of cell cycle progression, the modification of signal transduction pathways, the induction of apoptosis, antioxidant activity, stimulation of immune function, and the regulation of nuclear factors involved in inflammation. The anticancer potential of garlic may also be influenced by other dietary components such as selenium and vitamin A.

Evidence also suggests that garlic can suppress carcinogen (e.g., nitrosamine) formation and carcinogen bioactivation, via its effects on Phase I and II enzymes, etc. These aspects are reviewed in detail in the following sections.

Cell Culture/In Vitro Studies Using Garlic

Garlic oil induces the differentiation and apoptosis of human gastric cancer BGC-823 cells via increased expression of p21 and p53 genes.[89] Aqueous garlic extract caused cell cycle arrest at the G2/M phase boundary in human HT-29 adenocarcinoma cells;[90] evidence suggested that garlic extract induced differential gene expression, with increased expression of epidermal growth factor receptor and integrin-α6, and decreased expression of menin. Extracts of garlic or selenium-enriched garlic inhibited the growth of human gastric carcinoma MGC803 cells, with growth again arrested at the G2/M phase boundary, and increased levels of Cdk_2-cyclin E and Cdk_4-cyclin D1 complexes.[91] Aqueous/ethanolic garlic extracts inhibit mutagenesis in *Salmonella* strains induced by afflatoxin B1, and inhibit the growth of cultured *Helicobacter pylori* bacteria (as a suspected causitive agent of stomach cancer) at low concentration, compared to the effect on *Salmonella aureus*.[92]

Aged garlic extract inhibited the growth and migration of rat sarcoma cells in a dose-dependent (5-20 $mg \cdot mL^{-1}$) manner.[93] Pretreatment with aged garlic extract (1-4 $\mu g \cdot mL^{-1}$) also protected cultured pulmonary vascular endothelial cells from subsequent oxidative injury induced by hydrogen peroxide, and inhibits Cu^{2+} induced oxidative modification of low-density lipoproteins in vitro.[94] Using molecular biology techniques, the expression of several genes, including *FR alpha, calcyclin,* and *nexon,* has been shown to be up-regulated in human gastric cancer cells, following treatment with allitride garlic extract.[95]

The anticancer action of a number of specific chemical compound types obtained from garlic extracts has been extensively investigated in cell cultures. Both aqueous and lipid-soluble allyl sulphur compounds have anticancer properties, with the latter compounds in general being more effective than their water-soluble counterparts. The ability of these compounds to inhibit cell proliferation is likened to their ability to block cell cycle progression at the G2/M phase and induce apoptosis (characterized by increased intracellular Ca^{2+} reactive oxygen species levels and decreased p34 Cdk2 kinase activity).[96] Thus three of the major components of garlic—diallyl sulphide (DAS), diallyl disulphide (DADS), and diallyl trisulphide (DATS)—have been shown to inhibit the proliferation of human neoplastic A549 lung cells (by 20 to 50 percent using 1-10 μm DADS or DATS), inducing apoptosis, inhibit the activity of aryl amine-*N*-acetyltransferase (DAS or DADS) in human colon and bladder tumor cell lines, and inhibit the proliferation of HepG2 hepatoma cells (DAS).[97]

DAS and DADS induce apoptosis in non-small-cell lung carcinoma cells, with DAS inducing increased Bax and decreased Bcl-2 levels, and DADS increasing p53 protein levels.[98] Treatment of colon HT-29 carcinoma cells with DADS affected cell adhesion, cell cycle, and metabolic parameters; loss of cell adhesiveness to the substratum and cell cycle arrest at the G2/M phase was characterized by reduced protein and polyamine biosynthesis.[99] DADS (1.8-18 μM) inhibited the growth of hormone receptor positive (MCF-7, KPL-1) and negative (MDA-MB-231, MKL-F) breast cancer cell lines; growth inhibition due to apoptosis resulted from increased Bax and down-regulation of Bcl protein, and activation of caspase 3. In cultured MDA-MB-231 cells, DADS also antagonized the effect of linoleic acid (a potent stimulator of breast cancer cells), and synergised with the effect of eicosapentaenoic acid (a potent breast cancer cell suppressor).[100] In human colorectal HCT 116 cells, DADS (5-25 μm) inhibited cell proliferation, inducing apoptosis via a mechanism involving increased antitumorogenic nonsteroidal anti-inflammatory activated (NAG-1) and p53 expression.[101] Induction of apoptosis in human leukemia HL-60 cells by DADS (IC_{50} < 25 μm) has been shown to involve increased production of hydrogen peroxide, activation of caspase 3, and degradation of PARP prior to DNA fragmentation.[102] It is of note that the anticancer properties of garlic allyl sulphur compounds may be destroyed by microwave (1 min) or oven (45 min) heating.[103]

S-allylmercaptocysteine (SAMC), a stable-water-soluble organosulphur compound present in aged garlic extract, has antiproliferative activity against hormone-dependent breast (MCF-7) and prostate (CRC-1740) cancer cell lines and two erythroleukemia (HEL and OC1M-1) cell lines, but not against normal cultured human umbilical vein endothelial cells.[104] The dose-dependent inhibition of the growth of HEL and OC1M-1 cells (IC_{50} 46-93 μm) was characterized by blockage of cell cycle progression at the G2/M phase and induction of reduced protein synthesis/apoptosis.[105] SAMC (50 $mg \cdot L^{-1}$) inhibited human prostate LNCap cell growth, characterized by increased reduced glutathione levels and decreased polyamine synthesis, suggesting interaction between SAMC and the polyamine-synthesizing enzyme ornithine decarboxylase.[106] Treatment of LNCap cells with SAMC also showed a differential effect on prostate-specific antigen (decreased) and prostate-specific membrane antigen (increased) biomarkers similar to that produced by androgen deprivation, suggesting that the effect may be mediated by conversion of testosterone to metabolites less reactive toward androgen receptors.[107] SAMC (but not S-allylcysteine) inhibited the growth (at G2/M phase) of colon cancer SW-480 and HT-29 cell lines, inducing apoptosis characterized by increased caspase 3 activity, increased reduced glutathione levels and induction of jun kinase activity. Coadministration of SAMC and the established colon cancer chemopreventative agent sulindac sulphide resulted in a synergistic interaction in these cell lines.[108]

Another major compound of garlic, ajoene, induced apoptosis in human leukemia cells but not in normal peripheral mononuclear blood cells. The effect was dose and time dependent, and was characterized by increased levels of intracellular reactive oxygen species and activation of NFkβ transcription factor.[109] Ajoene has also been shown to significantly enhance the effects of the chemotherapeutic drugs cytarabine and fludarabine on human myeloid leukemia CD34+ resistant cells, by inhibiting bcl-2 expression and activating caspase 3 activity.[110] In addition, ajoene inhibits the release of lipopolysaccharide-induced prostaglandin E2 in RAW 264.7 macrophages in vitro (IC_{50} 2.4 μm), via induction of cyclooxygenase 2. In the absence of lipopolysaccharide, ajoene was unable to induce COX-2.[111] Since the nonsteroidal anti-inflammatory drug indomethacin acted similarly in this model system, the data suggested that ajoene may exert its effect via a similar pathway to that of the nonsteroidal anti-inflammatory drugs.

Another major (although less stable) garlic-derived compound (water soluble) is alliin and its derivative allicin. Allicin has been shown to inhibit the proliferation of a variety of tumor cells, via blockage of cell cycle progression at the S/G2M phase boundary and induction of apoptosis.[112] Allicin, but not alliin, is reported to inhibit the proliferation of mammary (MCF-7), endometrial (Ishikura), and colon (HT-29) cell lines (IC_{50} 10-25 μm); growth inhibition of MCF-7 cells was characterized by accumulation of cells in the Go/G1 and G2/M cell cycle phases but not by increased cell death.[113] Allicin induced a fall in intracellular glutathione levels, the extent of which correlated with the inhibition of cell growth. Garlic extract, in which the content of alliin was highly enriched, was unable to inhibit the growth of human HepG2 or colorectal Cac_{o2} cells in vitro. However, when garlic extract was supplemented with garlic powder (as a source of allinase), a concentration-dependent inhibition of tumor cell growth occurred (IC_{50} 330-480 $\mu g \cdot m^{-1}$). These data suggest that the antiproliferative effect of aqueous garlic extracts may be due to the conversion of alliin to allicin by allinase.

Animal Model Studies Using Garlic

C3H/He mice were injected with murine bladder MBT-2 carcinoma cells (s.c.) into the right thigh. Garlic extract was injected at the tumor site at intervals over the following 28 days. Animals were also treated orally with garlic extract for 30 days prior to tumor inoculation and for 30 days afterward. Animals treated with garlic s.c. (13 mg cumulative dose) had significantly reduced tumor incidence and tumor growth, and increased survival compared to untreated controls. Mice receiving garlic extract orally (50-

500 mg·100 mL^{-1} drinking water) also showed significant reductions in tumor incidence and mortality compared to untreated controls.[114]

In a two-step, diethylnitrosamine-induced model of hepatocarcinogenesis in rats, treatment with garlic (20 mg·Kg^{-1} per day) significantly reduced the number and extent of induced glutathione-*S*-transferase (GST) positive foci in liver (as a marker of carcinogenic potential).[115] In rat feeding studies, supplementation of the diet with selenium-enriched garlic (principal active compound γ-glutamyl-Se-methyl-selenocysteine) was effective in suppressing the development of premalignant lesions and the formation of adenocarcinomas in the mammary glands of carcinogen treated rats.[116] Garlic extract (25 mg cumulative dose over three to five weeks) injected into subcutaneous tumors in C3H/He mice, showed significant activity in inhibiting tumorigenesis and growth of MBT-2 transitional cell carcinoma, and demonstrated an additive antitumor effect when combined with suicide gene therapy.[117]

In terms of the anticancer activity of specific garlic-derived compounds in animal models, DAS showed anticarcinogenic activity against DMBA or benzo(a)pyrene-induced (BP) skin cancer in mice. For animals on which DAS was applied topically one hour prior to application of DMBA, a lower incidence of neoplasia was noted over the study period (28 weeks), compared to animals treated with DAS one hour after DMBA administration; in BP-exposed mice, the antitumorigenic potential of DAS was more evident in animals treated one hour after BP exposure, compared to those treated with DAS 1 hour prior to BP.[118] In this model system, the activity of γ-glutamyl transpeptidase (a marker of cell proliferation and neoplasia) was significantly inhibited by DAS.[119] Using a two-stage mouse skin carcinogenesis model (DMBA initiator, TPA promoter), topical application of DAS delayed the onset of tumorigenesis and the number of tumors per mouse. When DAS was applied prior to initiator or promoter, a significant proportion of animals remained tumor free during the course of the experiment.[120] Treatment with DADS (200 mg·Kg^{-1}) or allyl mercaptan (100 mg·Kg^{-1}) increased the acetylation of core nucleosomal histones in tumor-bearing rats, suggesting that garlic-derived compounds may induce the differentiation of malignant cells.[121] Another garlic-derived compound, scordinin, has chemopreventive action against liver neoplasia induced in rats by diethylnitrosamine or phenobarbital; the incidence of tumors and GST positive foci were significantly decreased by dietary scordinin treatment in the initiation and promotion phases.[122]

Several animal-model-based studies have suggested that the anticancer action of garlic is due to antioxidant activity. Thus administration of aqueous garlic extract (250 mg·Kg^{-1}, twice weekly for 14 weeks) effectively suppressed DMBA induced oral neoplasms in Syrian hamsters. The chemopreventive effect correlated with enhanced levels of intracellular antioxidants (glutathione, glutathione peroxidase, GST) and reduced lipid peroxidation in oral tumor tissue.[123] The same animal model was similarly used to demonstrate the antitumor/antioxidant activity of S-allylcysteine.[124] The inhibitory effect of garlic extract in preventation of 4-nitroquinoline-1-oxide induced rat tongue carcinogenesis also correlates with reduced lipid peroxidation and increased antioxidant levels (glutathione, glutathione peroxidase, GST) in tumor tissue.[125] Treatment of the rats with the carcinogenic nitrosamine *N*-methyl-*N'*-nitro-*N*-nitrosogu.anidine induced oxidative stress in stomach and liver tissues, characterized by increased levels of lipid peroxidation and reduced levels of glutathione, glutathione peroxidase, and GST. Oral administration of garlic extract for five days prior to i.p. injection of the carcinogen significantly reduced the formation of lipid peroxides and enhanced antioxidant levels.[126] Patient treatment with the powerful anticancer drug doxorubicin is limited by severe cardiotoxic effects resulting from the generation of reactive oxygen species and lipid peroxidation. Doxorubicin toxicity in mice was substantially reduced following treatment with the garlic derived antioxidant compound S-allylcysteine.[127]

Evidence suggests that the anticancer action of garlic may result from a reduced activity of Phase I (activation) enzymes or enhanced activity of Phase II (detoxifying) enzymes involved in carcinogen metabolism. For example, DADS and DATS were found to be potent inducers of the Phase II detoxifying enzymes quinone reductase and GST in a variety of rat tissues. Stomach, duodenum, and jejunum were most sensitive to enzyme induction, using a dose of only 0.3 $mg{\cdot}Kg^{-1}$ per day, comparable to that achievable through consumption of garlic in man.[128] Treatment of mice with DAS, DADS, or DATS resulted in significantly increased GST activity in liver (three- to fourfold) or stomach (1.5- to twofold) tissues, with GST activity in lung tissue unaltered; treatment with these compounds had relatively little effect on epoxide hydrolase activity in these tissue.[129] Using the same animal model, DADS or DATS treatment resulted in substantial activation of NADPH quinone reductase activity (an enzyme involved in detoxification of activated quinone metabolites of BP) in stomach (1.5- to twofold) or lung (threefold) tissues.[130] There were reductions in carbon tetrachloride, N-nitrosodimethyl amine, and acetaminophen-induced toxicity in rodents, and protection against 4-(methylnitrosamino)-1-(3-pyridyl)-1-butanone-induced lung tumorigenesis in mice, following treatment with DAS (or its metabolites diallyl sulphoxide or diallyl sulphone).[131] The protective action of these compounds appears to result from their inhibitory effect on the cytochrome P450 enzyme CYP2E1 during Phase 1 carcinogen activation.

Other mechanisms implicated in the anticancer action of garlic include modulation of cytokeratin expression in DMBA-induced hamster buccal pouch carcinogenesis; anticlastogenic effects against methyl-*N*-nitro-*N*-nitrosoguanidine in rats; and immunomodulatory activation of natural killer cells/T-lymphocytes in mice with oral precancer or sarcoma-bearing mice.[132] Garlic extract has also been demonstrated to protect against damage to the small intestine induced by the antitumor drugs methotrexate and 5-fluorouacil in rats.[133]

Clinical Studies Using Garlic Extracts

A number of case control studies have been reported which demonstrate a protective effect of garlic consumption against human cancers. In a case control study of diet and prostate cancer (328 cases), a significantly decreased risk of prostatic cancer was associated with dietary garlic intake.[134] In a French case control study of diet and breast cancer (345 cases), cancer risk was significantly decreased with increased dietary consumption of garlic and onions.[135] The protective effect of garlic consumption against the development of colorectal cancer was described in a case control study of 1,192 cases from ethnic groups in Hawaii.[136] A Swiss case control study of 223 patients with colorectal cancer also showed the protective effect of garlic consumption.[137] A case control study of 136 Korean patients showed a decreased risk of gastric cancer associated with garlic intake.[138] Frequent consumption of *Allium* vegetables, particularly garlic, has been shown from case control studies in low risk (386 cases) and high risk (234 cases) geographical areas of China, to protect against the development of gastric and oesophageal cancers.[139] Garlic consumption protected against the incidence of *H. pylori* infection and development of gastric cancer in a screening study of 214 Chinese patients, and a decreased risk of cancer associated with garlic intake was shown in an Italian case control study of 382 patients with gastric cancer.[140] A meta-analysis of the epidemiological literature on garlic consumption and the risk of various cancer types found a high intake of garlic to be associated with a protective effect against stomach and colorectal cancers.[141]

A blinded, randomized multifactorial intervention trial (incorporating garlic as one of the treatments) is in progress to assess the inhibition of the progression of precancerous gastric lesions in a region of China with a high risk of gastric cancer.[142] A randomized, controlled intervention trial using garlic for the prevention of familial colorectal cancer is also under way, to be completed in 2005.[143]

NOTES

1. Eisenhauer, E.A. and Vermorken, J.B., 1998, The taxoids: Comparative clinical pharmacology and therapeutic potential, *Drugs,* 55(1): 5-30; Miller, K.D. and Sledge, G.W. Jr., 1999, Taxanes in the treatment of breast cancer: A prodigy comes of age, *Cancer Investments,* 17(2): 121-136; Mantle, D., Lennard, T.W., and Pickering, A.T., 2000, Therapeutic applications of medicinal plants in the treatment of breast cancer: A review of their pharmacology, efficacy and tolerability, *Adverse Drug Reaction and Toxicological Review,* 19(3): 223-240.

2. Lee, J.H., Koo, T.H., Hwang, B.Y., and Lee, J.J., 2002, Kaurane diterpene, kamebakaurin, inhibits NF-kappa B by directly targeting the DNA-binding activity of p50 and blocks the expression of antiapoptotic NF-kappa B target genes, *Journal of Biological Chemistry,* 277(21): 18411-18420.

3. Yoon, T.J., Yoo, Y.C., Kang, T.B., Shimazaki, K., Song, S.K., Lee, K.H., Kim, S.H., Park, C.H., Azuma, I., and Kim, J.B., 1999, Lectins isolated from Korean mistletoe (*Viscum album coloratum*) induce apoptosis in tumor cells, *Cancer Letters,* 136: 33-40.

4. Park, R., Kim, M.S., So, H.S., Jung, B.H., Moon, S.R., Chung, S.Y., Ko, C.B., Kim, B., and Chung, H.T., 2000, Activation of c-Jun N-terminal kinase 1 (JNK1) in mistletoe lectin II-induced apoptosis of human myeloleukemic U937 cells, *Biochemical Pharmacology,* 60(11): 1685-1691.

5. Ibid; Kim, M.S., So, H.S., Lee, K.M., Park, J.S., Lee, J.H., Moon, S.K., Ryu, D.G., Chung, S.Y., Jung, B.H., Kim, Y.K., and G.R., 2000, Activation of caspase cascades in Korean mistletoe (*Viscum album* var. *coloratum*) lectin-II-induced apoptosis of human myeloleukemic U937 cells, *General Pharmacolpgy,* 34(5): 349-355.

6. Kim, M.S., Lee, J., So, H.S., Lee, K.M., Jung, B.H., Chung, S.Y., Moon, S.R., Kim, N.S., Ko, C.B., Kim, H.J., et al., 2001, Gamma-interferon (IFN-gamma) augments apoptotic response to mistletoe lectin-II via upregulation of Fas/Fas L expression and caspase activation in human myeloid U937 cells, *Immunopharmacology and Immunotoxicology,* 23(1): 55-66.

7. Lyu, S.Y., Choi, S.H., and Park, W.B., 2002, Korean mistletoe lectin-induced apoptosis in hepatocarcinoma cells is associated with inhibition of telomerase via mitochondrial controlled pathway independent of p53, *Archives of Pharmaceutical Research,* 25(1): 93-101.

8. Bantel, H., Engels, I.H., Voelter, W., Schulze-Osthoff, K., and Wesselborg, S., 1999, Mistletoe lectin activates caspase-8/FLICE independently of death receptor signaling and enhances anticancer drug-induced apoptosis, *Cancer Research,* 59(9): 2083-2090.

9. Schmidt, A., Mockel, B., Eck, J., Langer, M., Gauert, M., and Zinke, H., 2000, Cytotoxic activity of recombinant bFGF-rViscumin fusion proteins, *Biochemical and Biophysical Research Communications,* 277(2): 499-506.

10. Pae, H.O., Oh, G.S., Kim, N.Y., Shin, M.K., Lee, H.S., Yun, Y.G., Oh, H., Kim, Y.M., and Chung, H.T., 2001, Roles of extracellular signal-regulated kinase and p38 mitogen-activated protein kinase in apoptosis of human monoblastic leukemia U937 cells by lectin-II isolated from Korean mistletoe, *In Vitro Molecular Toxicology,* 14(2): 99-106.

11. Siegle, I., Fritz, P., McClellan, M., Gutzeit, S., and Murdter, T.E., 2001, Combined cytotoxic action of *Viscum album* agglutinin-1 and anticancer agents against human A549 lung cancer cells, *Anticancer Research,* 21(4A): 2687-2691.

12. Zarkovic, N., Vukovic, T., Loncaric, I., Miletic, M., Zarkovic, K., Borovic, S., Cipak, A., Sabolovic, S., Konitzer, M., and Mang, S., 2001, An overview on anticancer activities of the *Viscum album* extract Isorel, *Cancer Biotherapy and Radiopharmacy,* 16(1): 55-62.

13. Weber, K., Mengs, U., Schwarz, T., Hajto, T., Hostanska, K., Allen, T.R., Weyhenmeyer, R., and Lentzen, H., 1998, Effects of a standardized mistletoe preparation on metastatic B16 melanoma colonization in murine lungs, *Arzneimittelforschung,* 48: 497-502.

14. Burger, A.M., Mengs, U., Schuler, J.B., and Fiebig, H.H., 2001, Antiproliferative activity of an aqueous mistletoe extract in human tumor cell lines and xenografts in vitro, *Anticancer Research,* 21(3B): 1965-1968.

15. Yoon, T.J., Yoo, Y.C., Kang, T.B., Baek, Y.J., Huh, C.S., Song, S.K., Lee, K.H., Azuma, I., and Kim, J.B., 1998, Prophylactic effect of Korean mistletoe (*Viscum album coloratum*) extract on tumor metastasis is mediated by enhancement of NK cell activity, *International Journal of Immunopharmacology,* 20(4-5): 163-172.

16. Park, W.B., Lyu, S.Y., Kim, J.H., Choi, S.H., Chung, H.K., Ahn, S.H., Hong, S.Y., Yoon, T.J., and Choi, M.J., 2001, Inhibition of tumor growth and metastasis by Korean mistletoe lectin is associated with apoptosis and antiangiogenesis, *Cancer Biotherapy and Radiopharmacy,* 16(5): 439-447.

17. Lenartz, D., Andermahr, J., Plum, G., Menzel, J., and Beuth, J., 1998, Efficiency of treatment with galactoside-specific lectin from mistletoe against rat glioma, *Anticancer Research,* 18(2A): 1011-1014.

18. Mengs, U., Schwarz, T., Bulitta, M., and Weber, K., 2000, Antitumoral effects of an intravesically applied aqueous mistletoe extract on urinary bladder carcinoma MB49 in mice, *Anticancer Research,* 20(5B): 3565-3568.

19. Braun, J.M., Ko, H.L., Schierholz, J.M., Weir, D., Blackwell, C.C., and Beuth, J., 2001, Application of standardized mistletoe extracts augment immune response and down regulates metastatic organ colonization in murine models, *Cancer Letters,* 170(1): 25-31.

20. Karkabounas, S., Assimakopoulos, D., Malamas, M., Skaltsounis, A.L., Leonce, S., Zelovitis, J., Stefanou, D., and Evangelou, A., 2000, Antiproliferative and anticarcinogenic effects of an aqueous preparation of *Abies alba* and *Viscum album* se abies, on a L-1210 malignant cell line and tumor-bearing Wistar rats, *Anticancer Research,* 20(6B): 4391-4395.

21. Pryme, I.F., Bardocz, S., Pusztai, A., and Ewen, S.W., 2002, Dietary mistletoe lectin supplementation and reduced growth of a murine non-Hodgkin lymphoma, *Histology and Histopathology,* 17(1): 261-271.

22. Schaffrath, B., Mengs, U., Schwarz, T., Hilgers, R.D., Beuth, J., Mockel, B., Lentzen, H., and Gerstmayer, B., 2001, Anticancer activity of rViscumin (recombinant mistletoe lectin) in tumor colonization models with immunocompetent mice, *Anticancer Research,* 21(6A): 3981-3987.

23. Schumacher, U., Feldhaus, S., and Mengs, U., 2000, Recombinant mistletoe lectin (rML) is successful in treating human ovarian cancer cells transplanted into severe combined immunodeficient (SCID) mice, *Cancer Letters,* 150(2): 171-175.

24. Elsasser-Beile, U., Ruhnau, T., Freudenberg, N., Wetterauer, U., and Mengs, U., 2001, Antitumoral effect of recombinant mistletoe lectin on chemically induced urinary bladder carcinogenesis in a rat model, *Cancer,* 91(5): 998-1004.

25. Kunze, E., Schulz, H., and Gabius, H.J., 1998, Inability of galactoside-specific mistletoe lectin to inhibit N-methyl-N-nitrosourea-induced tumor development in the urinary bladder of rats and to mediate a local cellular immune response after long-term administration, *Journal of Cancer Research and Clinical Oncology,* 124(2): 73-87.

26. Kunze, E., Schulz, H., Adamek, M., and Gabius, H.J., 2000, Long-term administration of galactoside-specific mistletoe lectin in an animal model: no protection against N-butyl-N-(4-hydroxybutyl)-nitrosamine-induced urinary bladder carcinogenesis in rats and no induction of a relevant local cellular immune response, *Journal of Cancer Research and Clinical Oncology,* 126(3): 125-138.

27. Timoshenko, A.V., Lan, Y., Gabius, H.J., and Lala, P.K., 2001, Immunotherapy of C3H/HeJ mammary adenocarcinoma with interleukin-2, mistletoe lectin or their combination: effects on tumour growth, capillary leakage and nitric oxide (NO) production, *European Journal of Cancer,* 37(15): 1910-1920.

28. Heiny, B.M., Albrecht, V., and Beuth, J., 1998, Correlation of immune cell activities and beta-endorphin release in breast carcinoma patients treated with galactose-specific lectin standardized mistletoe extract, *Anticancer Research,* 18(1B): 583-586.

29. Stein, G.M., Schietzel, M., and Bussing, A., 1998, Mistletoe in immunology and the clinic (short review), *Anticancer Research,* 18(5A): 3247-3249.

30. Kovacs, E., 2002, The in vitro effect of *Viscum album* (VA) extract on DNA repair of peripheral blood mononuclear cells (PBMC) in cancer patients, *Phytotherapy Research,* 16(2): 143-147.

31. Klopp, R., Schmidt, W., Niemer, W., Werner, M., and Beuth, J., 2001, Changes in immunological characteristics of white blood cells after administration of standardized mistletoe extract, *In Vivo,* 15(6): 447-457.

32. Kovacs, E., 2000, Serum levels of IL-12 and the production of IFN-gamma, IL-2 and IL-4 by peripheral blood mononuclear cells (PBMC) in cancer patients treated with *Viscum album* extract, *Biomedical Pharmacotherapy,* 54(6): 305-310.

33. Grossarth-Maticek, R., Kiene, H., Baumgartner, S.M., and Ziegler, R., 2001, Use of Iscador, an extract of European mistletoe *(Viscum album),* in cancer treatment: Prospective nonrandomized and randomized matched-pair studies nested within a cohort study, *Alternative Therapies and Health Medicine,* 7(3): 57-66, 68-72, 74-76 passim.

34. Lenartz, D., Dott, U., Menzel, J., Schierholz, J.M., and Beuth, J., 2000, Survival of glioma patients after complementary treatment with galactoside-specific lectin from mistletoe, *Anticancer Research,* 20(3B): 2073-2076.

35. Kuehn, J.J., 1999, Favorable long-term outcome with mistletoe therapy in a patient with centroblastic-centrocytic non-Hodgkins lymphoma, *Deutsche Medizinische Wochenschrift,* 124(47): 1414-1418 (in German).

36. Steuer-Vogt, M.K., Bonkowsky, V., Ambrosch, P., Scholz, M., Neiss, A., Strutz, J., Hennig, M., Lenarz, T., and Arnold, W., 2001, The effect of an adjuvant mistletoe treatment programme in resected head and neck cancer patients: A randomised controlled clinical trial, *European Journal of Cancer,* 37(1): 23-31.

37. Goebell, P.J., Otto, T., Suhr, J., and Rubben, H., 2002, Evaluation of an unconventional treatment modality with mistletoe lectin to prevent recurrence of superficial bladder cancer: A randomized phase II trial, *Journal of Urology,* 168(1): 72-75.

38. Stumpf, C., Rosenberger, A., Rieger, S., Troger, W., and Schietzel, M., 2000, Mistletoe extracts in the therapy of malignant, hematological and lymphatic diseases—A monocentric, retrospective analysis over 16 years, *Forsch Komplementarmed Klass Naturheilkd.* 7(3): 139-146 (in German).

39. Gorter, R.W., van Wely, M., Stoss, M., and Wollina, U., 1998, Subcutaneous infiltrates induced by injection of mistletoe extracts (Iscador), *American Journal of Therapeutics,* 5(3): 181-187.

40. Ernst, E., 2000, Mistletoe therapy: An alternative cancer treatment, *MMW Fortschr Med.* 142(45): 52-53 (in German); Hutt, N., Kopferschmitt-Kubler, M., Cabalion, J., Purohit, A., Alt, M., and Pauli, G., 2001, Anaphylactic reactions after therapeutic injection of mistletoe (*Viscum album* L.), *Allergology Immunopathology (Madr).,* 29(5): 201-203.

41. Attele, A.S., Wu, J.A., and Yuan, C.S., 1999, Ginseng pharmacology: Multiple constituents and multiple actions, *Biochemical Pharmacology,* 58(11): 1685-1693.

42. Ota, T., Maeda, M., Odashima, S., Ninomiya-Tsuji, J., and Tatsuka, M., 1997, G1 phase-specific suppression of the Cdk2 activity by ginsenoside Rh2 in cultured murine cells, *Life Science,* 60(2): 39-44.

43. Sohn, J., Lee, C.H., Chung, D.J., Park, S.H., Kim, I., and Hwang, W.I., 1998, Effect of petroleum ether extract of *Panax ginseng* roots on proliferation and cell cycle progression of human renal cell carcinoma cells, *Experimental Molecular Medicine,* 30(1): 47-51.

44. Oh, M., Choi, Y.H., Choi, S., Chung, H., Kim, K., Kim, S.I., Kim, D.K., and Kim, N.D., 1999, Anti-proliferating effects of ginsenoside Rh2 on MCF-7 human breast cancer cells, *International Journal of Oncology,* 14(5): 869-875.

45. Duda, R.B., Zhong, Y., Navas, V., Li, M.Z., Toy, B.R., and Alavarez, J., 1999, American ginseng and breast cancer therapeutic agents synergistically inhibit MCF-7 breast cancer cell growth, *Journal of Surgery Oncology,* 72(4): 230-239.

46. Duda, R.B., Kang, S.S., Archer, S.Y., Meng, S., and Hodin, R.A., 2001, American ginseng transcriptionally activates p21 mRNA in breast cancer cell lines, *Journal of Korean Medical Science,* 16(Suppl.): S54-S60.

47. Kim, Y.S., Jin, S.H., Lee, Y.H., Kim, S.I., and Park, J.D., 1999, Ginsenoside Rh2 induces apoptosis independently of Bcl-2, Bcl-xL, or Bax in C6Bu-1 cells, *Archives of Pharmaceutical Research,* 22(5): 448-453.

48. Park, I.H., Piao, L.Z., Kwon, S.W., Lee, Y.J., Cho, S.Y., Park, M.K., and Park, J.H., 2002, Cytotoxic dammarane glycosides from processed ginseng, *Chemical Pharmacy Bulletin* (Tokyo), 50(4): 538-540.

49. Lam, S.K. and Ng, T.B., 2001, Isolation of a novel thermolabile heterodimeric ribonuclease with antifungal and antiproliferative activities from roots of the sanchi ginseng *Panax notoginseng, Biochemical Biophysical Research Communications,* 285(2): 419-423.

50. Park, J.A., Kim, K.W., Kim, S.I., and Lee, S.K., 1998, Caspase 3 specifically cleaves p21WAF1/CIP1 in the earlier stage of apoptosis in SK-HEP-1 human hepatoma cells, *European Journal of Biochemistry,* 57(1): 242-248.

51. Wakabayashi, C., Murakami, K., Hasegawa, H., Murata, J., and Saiki, I., 1998, An intestinal bacterial metabolite of ginseng protopanaxadiol saponins has the ability to induce apoptosis in tumor cells, *Biochemistry and Biophysics Research Communications,* 246(3): 725-730.

52. Kim, S.E., Lee, Y.H., Park, J.H., and Lee, S.K., 1999, Ginsenoside-Rs4, a new type of ginseng saponin concurrently induces apoptosis and selectively elevates protein levels of p53 and p21WAF1 in human hepatoma SK-HEP-1 cells, *European Journal of Cancer,* 35(3): 507-511.

53. Kim, K.H., Lee, Y.S., Jung, I.S., Park, S.Y., Chung, H.Y., Lee, I.R., and Yun, Y.S., 1998, Acidic polysaccharide from *Panax ginseng,* ginsan, induces Th1 cell and macrophage cytokines and generates LAK cells in synergy with rIL-2, *Planta Medica,* 64(2): 110-115.

54. Liu, W.K., Xu, S.X., and Che, C.T., 2000, Anti-proliferative effect of ginseng saponins on human prostate cancer cell line, *Life Science,* 67(11): 1297-1306.

55. Lee, S.J., Ko, W.G., Kim, J.H., Sung, J.H., Moon, C.K., and Lee, B.H., 2000, Induction of apoptosis by a novel intestinal metabolite of ginseng saponin via cytochrome c-mediated activation of caspase-3 protease, *Biochemical Pharmacology,* 60(5): 677-685.

56. Park, M.T., Cha, H.J., Jeong, J.W., Kim, S.I., Chung, H.Y., Kim, N.D., Kim, O.H., and Kim K.W., 1999, Glucocorticoid receptor-induced down-regulation of MMP-9 by ginseng components, PD and PT contributes to inhibition of the invasive capacity of HT1080 human fibrosarcoma cells, *Molecules and Cells,* 9(5): 476-483.

57. Fujimoto, J., Sakaguchi, H., Aoki, I., Toyoki, H., Khatun, S., and Tamaya, T., 2001, Inhibitory effect of ginsenoside-Rb2 on invasiveness of uterine endometrial cancer cells to the basement membrane, *European Journal of Gynaecology and Oncology,* 22(5): 339-341.

58. Chang, M.S., Lee, S.G., and Rho, H.M., 1999, Transcriptional activation of Cu/Zn superoxide dismutase and catalase genes by panaxadiol ginsenosides extracted from *Panax ginseng, Phytotherapy Research,* 13(8): 641-644.

59. Kang, K.S., Kang, B.C., Lee, B.J., Che, J.H., Li, G.X., Trosko, J.E., and Lee, Y.S., 2000, Preventive effect of epicatechin and ginsenoside Rb(2) on the inhibition of gap junctional intercellular communication by TPA and H(2)O(2), *Cancer Letters,* 152(1): 97-106.

60. Lee, S.J., Sung, J.H., Lee, S.J., Moon, C.K., and Lee, B.H., 1999, Antitumor activity of a novel ginseng saponin metabolite in human pulmonary adenocarcinoma cells resistant to cisplatin, *Cancer Letters,* 144(1): 39-43; Gao, R., Jin, J., and Niu. Y., 1999, Potentiated effects of total saponins of *Panax Ginseng* on inhibition of leukemic cells by cytotoxic drugs, *Zhongguo Zhong Xi Yi Jie He Za Zhi,* 19(1): 17-19 (in Chinese).

61. Amato, P., Christophe, S., and Mellon, P.L., 2002, Estrogenic activity of herbs commonly used as remedies for menopausal symptoms, *Menopause,* 9(2): 145-150.

62. Xiaoguang, C., Hongyan, L., Xiaohong, L., Zhaodi, F., Yan, L., Lihua, T., and Rui, H., 1998, Cancer chemopreventive and therapeutic activities of red ginseng, *Journal of Ethnopharmacology,* 60(1): 71-78.

63. Konoshima, T., Takasaki, M., and Tokuda, H., 1999, Anti-carcinogenic activity of the roots of Panax notoginseng, II, *Biol Pharm Bull.,* 22(10): 1150-1152.

64. Konoshima, T., Takasaki, M, Tokuda, H., Nishino, H., Duc, N.M., Kasai, R., and Yamasaki, K., 1998, Anti-tumor-promoting activity of majonoside-R2 from Vietnamese ginseng, *Panax vietnamensis*, Ha et Grushv. (I), *Biology and Pharmacy Bulletin*, 21(8): 834-838.

65. Keum, Y.S., Park, K.K., Lee, J.M., Chun, K.S., Park, J.H., Lee, S.K., Kwon, H., and Surh, Y.J., 2000, Antioxidant and anti-tumor promoting activities of the methanol extract of heat-processed ginseng, *Cancer Letters*, 150(1): 41-48.

66. Surh, Y.J., Na, H.K., Lee, J.Y., and Keum, Y.S., 2001, Molecular mechanisms underlying anti-tumor promoting activities of heat-processed *Panax ginseng* C.A. Meyer, *Journal of Korean Medical Science*, 16(Suppl.): S38-S41.

67. Li, W., Wanibuchi, H., Salim, E.I., Wei, M., Yamamoto, S., Nishino, H., and Fukushima, S., 2000, Inhibition by ginseng of 1,2-dimethylhydrazine induction of aberrant crypt foci in the rat colon, *Nutrition and Cancer*, 36(1): 66-73.

68. Wargovich, M.J., 2001, Colon cancer chemoprevention with ginseng and other botanicals, *Journal of Korean Medical Science*, 16(Suppl.): S81-S86.

69. You, W.C., Chang, Y.S., Heinrich, J., Ma, J.L., Liu, W.D., Zhang, L., Brown, L.M., Yang, C.S., Gail, M.H., Fraumeni, J.F. Jr., and Xu, G.W., 2001, An intervention trial to inhibit the progression of precancerous gastric lesions: Compliance, serum micronutrients and S-allyl cysteine levels, and toxicity, *European Journal of Cancer Prevention*, 10(3): 257-263.

70. Yun, Y.S., Moon, H.S., Oh, Y.R., Jo, S.K., Kim, Y.J., and Yun, T.K., 1987, Effect of red ginseng on natural killer cell activity in mice with lung adenoma induced by urethane and benzo(a)pyrene, *Cancer Detection and Prevention Suppl.*, 1: 301-309.

71. Konoshima, T., Takasaki, M., Ichiishi, E., Murakami, T., Tokuda, H., Nishino, H., Duc, N.M., Kasai, R., and Yamasaki, K., 1999, Cancer chemopreventive activity of majonoside-R2 from Vietnamese ginseng, *Panax vietnamensis*, *Cancer Letters*, 147(1-2): 11-16.

72. Lee, Y.S., Chung, I.S., Lee, I.R., Kim, K.H., Hong, W.S., and Yun, Y.S., 1997, Activation of multiple effector pathways of immune system by the antineoplastic immunostimulator acidic polysaccharide ginsan isolated from *Panax ginseng*, *Anticancer Research*, 17(1A): 323-331.

73. Yun, T.K., Lee, Y.S., Lee, Y.H., Kim, S.I., and Yun, H.Y., 2001, Anticarcinogenic effect of Panax ginseng C.A. Meyer and identification of active compounds, *Journal of Korean Medical Science*, 16(Suppl.): S6-S18.

74. Yun, T.K., Yun, Y.S., and Han, I.W., 1983, Anticarcinogenic effect of long-term oral administration of red ginseng on newborn mice exposed to various chemical carcinogens, *Cancer Detection Previews*, 6(6): 515-525.

75. Wu, X.G. and Zhu, D.H., 1990, Influence of ginseng upon the development of liver cancer induced by diethylnitrosamine in rats, *Journal of Tongji Medical University*, 10(3): 141-145, 133.

76. Wu, X.G., Zhu, D.H., and Li, X., 2001, Anticarcinogenic effect of red ginseng on the development of liver cancer induced by diethylnitrosamine in rats, *Journal of Korean Medical Science*, 16: 561-565.

77. Konoshima et al., 1999, Cancer chemopreventive activity.

78. Nishino, H., Tokuda, H., Ii, T., Takemura, M., Kuchide, M., Kanazawa, M., Mou, X.Y., Bu, P., Takayasu, J., Onozuka, M., et al., 2001, Cancer chemoprevention

by ginseng in mouse liver and other organs, *Journal of Korean Medical Science,* 16(Suppl.): S66-S69.

79. Watanabe, T., Sugie, S., Okamoto, K., Rahman, K.M., Ushida, J., and Mori, H., 2001, Chemopreventive effects of scordinin on diethylnitrosamine and phenobarbital-induced hepatocarcinogenesis in male F344 rats, *Japan Journal of Cancer Research,* 92(6): 6.

80. Bespalov, V.G., Alexandrov, V.A., Limarenko, A.Y., Voytenkov, B.O., Okulov, V.B., Kabulov, M.K., Peresunko, A.P., Slepyan, L.I., and Davydov, V.V., 2001, Chemoprevention of mammary, cervix and nervous system carcinogenesis in animals using cultured *Panax ginseng* drugs and preliminary clinical trials in patients with precancerous lesions of the esophagus and endometrium, *Journal of Korean Medical Science,* 16(Suppl.): S42-S53.

81. Nakata, H., Kikuchi, Y., Tode, T., Hirata, J., Kita, T., Ishii, K., Kudoh, K., Nagata, I., and Shinomiya, N., 1998, Inhibitory effects of ginsenoside Rh2 on tumor growth in nude mice bearing human ovarian cancer cells, *Japan Journal of Cancer Research,* 89(7): 733-740.

82. Yun, T.K. and Choi, S.Y., 1995, Preventive effect of ginseng intake against various human cancers: A case-control study on 1987 pairs, *Cancer Epidemiology Biomarkers Prev,* 4(4): 401-408.

83. Yun, T.K. and Choi, S.Y., 1998, Non-organ specific cancer prevention of ginseng: A prospective study in Korea, *International Journal of Epidemiology,* 27(3): 359-364.

84. de la Taille, A., Hayek, O.R., Burchardt, M., Burchardt, T., and Katz, A.E., 2000, Role of herbal compounds (PC-SPES) in hormone-refractory prostate cancer: Two case reports, *Journal of Alternative and Complementary Medicine,* 6(5): 449-451.

85. Ibid.

86. Bespalov et al., 2001, Chemoprevention of mammary, cervix, and nervous system carcinogenesis.

87. See, D.M., Broumand, N., Sahl, L., and Tilles, J.G., 1997, In vitro effects of echinacea and ginseng on natural killer and antibody-dependent cell cytotoxicity in healthy subjects and chronic fatigue syndrome or acquired immunodeficiency syndrome patients, *Immunopharmacology,* 35(3): 229-235; Scaglione, F., Ferrara, F., Dugnani, S., Falchi, M., Santoro, G., and Fraschini, F., 1990, Immunomodulatory effects of two extracts of *Panax ginseng* C.A. Meyer, *Drugs Experimental and Clinical Research,* 16(10): 537-542.

88. Coon, J.T. and Ernst, E., 2002, *Panax ginseng:* A systematic review of adverse effects and drug interactions, *Drug Safety,* 25(5): 323-344.

89. Li, X., Xie, J., and Li, W., 1998, Garlic oil induces differentiation and apoptosis of human gastric cancer cell line, *Zhonghua Zhong Liu Za Zhi,* 20(5): 325-327 (in Chinese).

90. Frantz, D.J., Hughes, B.G., Nelson, D.R., Murray, B.K., and Christensen, M.J., 2000, Cell cycle arrest and differential gene expression in HT-29 cells exposed to an aqueous garlic extract, *Nutrition and Cancer,* 38(2): 255-264.

91. Tang, F., Zhou, J., and Gu, L., 2001, In vivo and in vitro effects of selenium-enriched garlic on growth of human gastric carcinoma cells, *Zhonghua Zhong Liu Za Zhi,* 23(6): 461-464 (in Chinese).

92. Soni, K.B., Lahiri, M., Chackradeo, P., Bhide, S.V., and Kuttan, R., 1997, Protective effect of food additives on aflatoxin-induced mutagenicity and hepatocarcinogenicity, *Cancer Letters,* 115(2): 129-133; Sivam, G.P., Lampe, J.W., Ulness, B., Swanzy, S.R., and Potter, J.D., 1997, *Helicobacter pylori*—in vitro susceptibility to garlic *(Allium sativum)* extract, *Nutrition and Cancer,* 27(2): 118-121; Sivam, G.P., 2001, Protection against *Helicobacter pylori* and other bacterial infections by garlic, *Journal of Nutrition,* 131: 1106-1108.

93. Hu, X., Cao, B.N., Hu, G., He, J., Yang, D.Q., and Wan, Y.S., 2002, Attenuation of cell migration and induction of cell death by aged garlic extract in rat sarcoma cells, *International Journal of Molecular Medicine,* 9(6): 641-643.

94. Yamasaki, T. and Lau, B.H., 1997, Garlic compounds protect vascular endothelial cells from oxidant injury, *Nippon Yakurigaku Zasshi,* 110(Suppl. 1): 138P-141P (in Japanese); Ide, N., Nelson, A.B., and Lau, B.H., 1997, Aged garlic extract and its constituents inhibit Cu(2+)-induced oxidative modification of low density lipoprotein, *Planta Medica,* 63(3): 263-264.

95. Li, Y., Yang, L., Cui, J.T., Li, W.M., Guo, R.F., and Lu, Y.Y., 2002, Construction of cDNA representational difference analysis based on two cDNA libraries and identification of garlic inducible expression genes in human gastric cancer cells, *World Journal of Gastroenterology,* 8(2): 208-212; Li, Y. and Lu, Y.Y., 2002, Applying a highly specific and reproducible cDNA RDA method to clone garlic upregulated genes in human gastric cancer cells, *World Journal of Gastroenterology,* 8(2): 213-216.

96. Knowles, L.M. and Milner, J.A., 2000, Allyl sulfides modify cell growth, *Drug Metabolism and Drug Interactions,* 17(1-4): 81-107; Knowles, L.M. and Milner, J.A., 2001, Possible mechanism by which allyl sulfides suppress neoplastic cell proliferation, *Journal of Nutrition,* 131(3s): 1061S-1066S.

97. Sakamoto, K., Lawson, L.D., and Milner, J.A., 1997, Allyl sulfides from garlic suppress the in vitro proliferation of human A549 lung tumor cells, *Nutrition and Cancer,* 292: 152-156; Chen, G.W., Chung, J.G., Hsieh, C.L., and Lin, J.G., 1998, Effects of the garlic components diallyl sulfide and diallyl disulfide on arylamine N-acetyltransferase activity in human colon tumour cells, *Food Chemistry and Toxicology,* 36(9-10): 761-770; Chung, J.G., 1999, Effects of garlic components diallyl sulfide and diallyl disulfide on arylamine N-acetyltransferase activity in human bladder tumor cells, *Drug Chemistry and Toxicology,* 22(2): 343-358; Iciek, M.B., Rokita, H.B., and Wlodek, L.B., 2001, Effects of diallyl disulfide and other donors of sulfane sulfur on the proliferation of human hepatoma cell line (HepG2), *Neoplasma,* 48(4): 307-312.

98. Hong, Y.S., Ham, Y.A., Choi, J.H., and Kim, J., 2000, Effects of allyl sulfur compounds and garlic extract on the expression of Bcl-2, Bax, and p53 in non small cell lung cancer cell lines, *Experimental and Molecular Medicine,* 32(3): 127-134.

99. Robert, V., Mouille, B., Mayeur, C., Michaud, M., and Blachier, F., 2001, Effects of the garlic compound diallyl disulfide on the metabolism, adherence and cell cycle of HT-29 colon carcinoma cells: Evidence of sensitive and resistant subpopulations, *Carcinogenesis,* 22(8): 1155-1161.

100. Nakagawa, H., Tsuta, K., Kiuchi, K., Senzaki, H., Tanaka, K., Hioki, K., and Tsubura, A., 2001, Growth inhibitory effects of diallyl disulfide on human breast cancer cell lines, *Carcinogenesis,* 22(6): 891-897.

101. Bottone, F.G. Jr, Baek, S.J., Nixon, J.B., and Eling, T.E., 2002, Diallyl disulfide (DADS) induces the antitumorigenic NSAID-activated gene (NAG-1) by a p53-dependent mechanism in human colorectal HCT 116 cells, *Journal of Nutrition,* 132(4): 773-778.

102. Kwon, K.B., Yoo, S.J., Ryu, D.G., Yang, J.Y., Rho, H.W., Kim, J.S., Park, J.W., Kim, H.R., and Park, B.H., 2002, Induction of apoptosis by diallyl disulfide through activation of caspase-3 in human leukemia HL-60 cells, *Biochemical Pharmacology,* 63(1): 41-47.

103. Song, K. and Milner, J.A., 2001, The influence of heating on the anticancer properties of garlic, *Journal of Nutrition,* 131(3s): 1054S-1057S.

104. Sigounas, G., Hooker, J., Anagnostou, A., and Steiner, M., 1997, S-allylmercaptocysteine inhibits cell proliferation and reduces the viability of erythroleukemia, breast, and prostate cancer cell lines, *Nutrition and Cancer,* 27(2): 186-191.

105. Sigounas, G., Hooker, J.L., Li, W., Anagnostou, A., and Steiner, M., 1997, S-allylmercaptocysteine, a stable thioallyl compound, induces apoptosis in erythroleukemia cell lines, *Nutrition and Cancer,* 28(2): 153-159.

106. Pinto, J.T., Qiao, C., Xing, J., Rivlin, R.S., Protomastro, M.L., Weissler, M.L., Tao, Y., Thaler, H., and Heston, W.D., 1997, Effects of garlic thioallyl derivatives on growth, glutathione concentration, and polyamine formation of human prostate carcinoma cells in culture, *American Journal of Clinical Nutrition,* 66(2): 398-405.

107. Pinto, J.T., Qiao, C., Xing, J., Suffoletto, B.P., Schubert, K.B., Rivlin, R.S., Huryk, R.F., Bacich, D.J., and Heston, W.D., 2000, Alterations of prostate biomarker expression and testosterone utilization in human LNCaP prostatic carcinoma cells by garlic-derived S-allylmercaptocysteine, *Prostate,* 45(4): 304-314.

108. Shirin, H., Pinto, J.T., Kawabata, Y., Soh, J.W., Delohery, T., Moss, S.F., Murty, V., Rivlin, R.S., Holt, P.R., and Weinstein, I.B., 2001, Antiproliferative effects of S-allylmercaptocysteine on colon cancer cells when tested alone or in combination with sulindac sulfide, *Cancer Research,* 61(2): 725-731.

109. Dirsch, V.M., Gerbes, A.L., and Vollmar, A.M., 1998, Ajoene, a compound of garlic, induces apoptosis in human promyeloleukemic cells, accompanied by generation of reactive oxygen species and activation of nuclear factor kappa B, *Molecular Pharmacology,* 53(3): 402-407.

110. Ahmed, N., Laverick, L., Sammons, J., Zhang, H., Maslin, D.J., and Hassan H.T., 2001, Ajoene, a garlic-derived natural compound, enhances chemotherapy-induced apoptosis in human myeloid leukaemia CD34-positive resistant cells, *Anticancer Research,* 21(5): 3519-3523.

111. Dirsch, V.M. and Vollmar, A.M., 2001, Ajoene, a natural product with non-steroidal anti-inflammatory drug (NSAID)-like properties? *Biochemical Pharmacology,* 61(5): 587-593.

112. Zheng, S., Yang, H., Zhang, S., Wang, X., Yu, L., Lu, J., and Li, J., 1997, Initial study on naturally occurring products from traditional Chinese herbs and vegetables for chemoprevention, *Journal of Cell Biochemistry,* Suppl. 27: 106-112; Thatte, U., Bagadey, S., and Dahanukar, S., 2000, Modulation of programmed cell death by medicinal plants, *Cell and Molecular Biology,* 46(1): 199-214.

113. Hirsch, K., Danilenko, M., Giat, J., Miron, T., Rabinkov, A., Wilchek, M., Mirelman, D., Levy, J., and Sharoni, Y., 2000, Effect of purified allicin, the major in-

gredient of freshly crushed garlic, on cancer cell proliferation, *Nutrition and Cancer,* 38(2): 245-254.

114. Riggs, D.R., DeHaven, J.I., and Lamm, D.L., 1997, *Allium sativum* (garlic) treatment for murine transitional cell carcinoma, *Cancer,* 79(10): 1987-1994.

115. Samaranayake, M.D., Wickramasinghe, S.M., Angunawela, P., Jayasekera, S., Iwai, S., and Fukushima, S., 2000, Inhibition of chemically induced liver carcinogenesis in Wistar rats by garlic *(Allium sativum), Phytotherapy Research,* 14(7): 564-567.

116. Ip, C., Birringer, M., Block, E., Kotrebai, M., Tyson, J.F., Uden, P.C., and Lisk, D.J., 2000, Chemical speciation influences comparative activity of selenium-enriched garlic and yeast in mammary cancer prevention, *Journal of Agricultural and Food Chemistry,* 48(9): 4452; Dong, Y., Lisk, D., Block, E., and Ip, C., 2001, Characterization of the biological activity of gamma-glutamyl-Se-methylseleno-cysteine: A novel, naturally occurring anticancer agent from garlic, *Cancer Research,* 61(7): 2923-2928.

117. Moon, D.G., Cheon, J., Yoon, D.H., Park, H.S., Kim, H.K., Kim, J.J., and Koh, S.K., 2000, *Allium sativum* potentiates suicide gene therapy for murine transitional cell carcinoma, *Nutrition and Cancer,* 38(1): 98-105.

118. Singh, A. and Shukla, Y., 1998, Antitumour activity of diallyl sulfide on polycyclic aromatic hydrocarbon-induced mouse skin carcinogenesis, *Cancer Letters,* 131(2): 209-214.

119. Shukla, Y., Singh, A., and Srivastava, B., 1999, Inhibition of carcinogen-induced activity of gamma-glutamyl transpeptidase by certain dietary constituents in mouse skin, *Biomedical and Environmental Science,* 12(2): 110-115.

120. Singh, A. and Shukla, Y., 1998, Antitumor activity of diallyl sulfide in two-stage mouse skin model of carcinogenesis, *Biomedical and Environmental Science,* 11(3): 258-263.

121. Lea, M.A. and Randolph, V.M., 2001, Induction of histone acetylation in rat liver and hepatoma by organosulfur compounds including diallyl disulfide, *Anticancer Research,* 21(4A): 2841-2845.

122. Watanabe, S., Kitade, Y., Masaki, T., Nishioka, M., Satoh, K., and Nishino, H., 2001, Effects of lycopene and Sho-saiko-to on hepatocarcinogenesis in a rat model of spontaneous liver cancer, *Nutrition and Cancer,* 39(1): 96-101.

123. Balasenthil, S., Arivazhagan, S., Ramachandran, C.R., and Nagini, S., 1999, Effects of garlic on 7,12-dimethylbenz[a]anthracene-induced hamster buccal pouch carcinogenesis, *Cancer Detection Preview,* 23(6): 534-538.

124. Balasenthil, S. and Nagini, S., 2000, Inhibition of 7,12-dimethyl-benz[a]anthra-cene-induced hamster buccal pouch carcinogenesis by S-allyl-cysteine, *Oral Oncology,* 36(4): 382-386.

125. Balasenthil, S., Ramachandran, C.R., and Nagini, S., 2001, Prevention of 4-nitroquinoline 1-oxide-induced rat tongue carcinogenesis by garlic, *Fitoterapia,* 72(5): 524-531.

126. Arivazhagan, S., Balasenthil, S., and Nagini, S., 2000, Garlic and neem leaf extracts enhance hepatic glutathione and glutathione dependent enzymes during N-methyl-N'-nitro-N-nitrosoguanidine (MNNG)-induced gastric carcinogenesis in rats, *Phytotherapy Research,* 14(4): 291-293; Arivazhagan, S., Nagini, S., Santhiya, S.T., and Ramesh, A., 2001, Protection of N-methyl-N'-nitro-N-nitrosoguanidine-

induced in vivo clastogenicity by aqueous garlic extract, *Asia Pacific Journal of Clinical Nutrition,* 10(3): 238-241.

127. Mostafa, M.G., Mima, T., Ohnishi, S.T., and Mori, K., 2000, S-allylcysteine ameliorates doxorubicin toxicity in the heart and liver in mice, *Planta Medica,* 66(2): 148-151.

128. Munday, R. and Munday, C.M., 1999, Low doses of diallyl disulfide, a compound derived from garlic, increase tissue activities of quinone reductase and glutathione transferase in the gastrointestinal tract of the rat, *Nutrition and Cancer,* 34(1): 42-48; Munday, R. and Munday, C.M., 2001, Relative activities of organosulfur compounds derived from onions and garlic in increasing tissue activities of quinone reductase and glutathione transferase in rat tissues, *Nutrition and Cancer,* 40(2): 205-210.

129. Hu, X. and Singh, S.V., 1997, Glutathione S-transferases of female A/J mouse lung and their induction by anticarcinogenic organosulfides from garlic, *Archives of Biochemistry and Biophysics Apr,* 340(2): 279-286; Srivastava, S.K., Hu, X., Xia, H., Zaren, H.A., Chatterjee, M.L., Agarwal, R., and Singh, S.V., 1997, Mechanism of differential efficacy of garlic organosulfides in preventing benzo(a)pyrene-induced cancer in mice, *Cancer Letters,* 118(1): 61-67.

130. Singh, S.V., Pan, S.S., Srivastava, S.K., Xia, H., Hu, X., Zaren, H.A., and Orchard, J.L., 1998, Differential induction of NAD(P)H:quinone oxidoreductase by anti-carcinogenic organosulfides from garlic, *Biochemical and Biophysical Research Communications,* 244(3): 917-920.

131. Yang, C.S., Chhabra, S.K., Hong, J.Y., and Smith, T.J., 2001, Mechanisms of inhibition of chemical toxicity and carcinogenesis by diallyl sulfide (DAS) and related compounds from garlic, *Journal of Nutrition,* 131(3s): 1041S-1045S.

132. Balasenthil, S., Rao, K.S., and Nagini, S., 2002, Apoptosis induction by S-allylcysteine, a garlic constituent, during 7,12-dimethylbenz[a]anthracene-induced hamster buccal pouch carcinogenesis, *Cell Biochemistry and Function,* 20(3): 263-268; Tang, Z., Sheng, Z., Liu, S., Jian, X., Sun, K., and Yan, M., 1997, [The preventing function of garlic on experimental oral precancer and its effect on natural killer cells, T-lymphocytes and interleukin-2], *Hunan Yi Ke Da Xue Xue Bao,* 22(3): 246-248. Chinese; Kyo, E., Uda, N., Kasuga, S., and Itakura, Y., 2001, Immunomodulatory effects of aged garlic extract, *Journal of Nutrition,* 131(3s): 1075S-1079S.

133. Horie, T., Awazu, S., Itakura, Y., and Fuwa, T., 2001, Alleviation by garlic of antitumor drug-induced damage to the intestine, *Journal of Nutrition,* 131(3s): 1071S-1074S.

134. Key, T.J., Silcocks, P.B., Davey, G.K., Appleby, P.N., and Bishop, D.T., 1997, A case-control study of diet and prostate cancer, *British Journal of Cancer,* 76(5): 678-687.

135. Challier, B., Perarnau, J.M., and Viel, J.F., 1998, Garlic, onion and cereal fibre as protective factors for breast cancer: A French case-control study, *European Journal of Epidemiology,* 14(8): 737-747.

136. Le Marchand, L., Hankin, J.H., Wilkens, L.R., Kolonel, L.N., Englyst, H.N., and Lyu, L.C., 1997, Dietary fiber and colorectal cancer risk, *Epidemiology,* 8(6): 658-665.

137. Levi, F., Pasche, C., La Vecchia, C., Lucchini, F., and Franceschi, S., 1999, Food groups and colorectal cancer risk, *British Journal of Cancer,* 79(7-8): 1283-1287.

138. Kim, H.J., Chang, W.K., Kim, M.K., Lee, S.S., and Choi, B.Y., 2002, Dietary factors and gastric cancer in Korea: A case-control study, *International Journal of Cancer,* 97(4): 531-535.

139. Gao, C.M., Takezaki, T., Ding, J.H., Li, M.S., and Tajima, K., 1999, Protective effect of allium vegetables against both esophageal and stomach cancer: A simultaneous case-referent study of a high-epidemic area in Jiangsu Province, China, *Japanese Journal of Cancer Research,* 90(6): 614-621; Takezaki, T., Gao, C.M., Wu, J.Z., Ding, J.H., Liu, Y.T., Zhang, Y., Li, S.P., Su, P., Liu, T.K., and Tajima, K., 2001, Dietary protective and risk factors for esophageal and stomach cancers in a low-epidemic area for stomach cancer in Jiangsu Province, China: Comparison with those in a high-epidemic area, *Japan Journal of Cancer Research,* 92(11): 1157-1165.

140. Palli, D., Russo, A., Ottini, L., Masala, G., Saieva, C., Amorosi, A., Cama, A., D'Amico, C., Falchetti, M., Palmirotta, R., et al., 2001, Red meat, family history, and increased risk of gastric cancer with microsatellite instability, *Cancer Research,* 61(14): 5415-5419; Yun et al., 1987, Effect of red ginseng.

141. Fleischauer, A.T., Poole, C., and Arab, L., 2000, Garlic consumption and cancer prevention: Meta-analyses of colorectal and stomach cancers, *American Journal of Clinical Nutrition,* 72(4): 1047-1052.

142. Gail, M.H., You, W.C., Chang, Y.S., Zhang, L., Blot, W.J., Brown, L.M., Groves, F.D., Heinrich, J.P., Hu, J., Jin, M.L., et al., 1998, Factorial trial of three interventions to reduce the progression of precancerous gastric lesions in Shandong, China: Design issues and initial data, *Control Clinical Trials,* 19(4): 352-369; You et al., 2001, An intervention trial to inhibit the progression of precancerous lesions.

143. Ishikawa, H., 2002, Cancer prevention in familial cancer, *Gan To Kagaku Ryoho,* 29(4): 545-549 (in Japanese).

Chapter 15

Phytochemicals and Prevention of Coronary Heart Disease

D. Francesco Visioli
Simona Grande
Claudio Galli

INTRODUCTION

There are now many epidemiological and clinical studies that demonstrate how a proper diet, in which plant foods provide the major portion of caloric intake, reduce the development of certain diseases, in particular cancer and cardiovascular disease.[1] In the past few years there has been a steadily growing interest in the role that oligonutrients commonly found in plant foods might play in the maintenance of human health. Although plant oligonutrient deficiencies in diets are yet to be described, the recent discovery that several phenolic molecules, namely catechols, exert interesting biological activities led to a reevaluation of the contribution of these compounds to human health.[2] This is also leading to increasing marketing of "functional foods" or "nutraceuticals," in other words, fortified or enriched foodstuff that is attributed health-promoting effects. The rationale that certain phytochemicals might contribute to lowering cardiovascular disease is based on the hypothesis that oxidative and inflammatory processes play a role in the onset and development of this pathology.[3] Several phytochemicals (notably catechols) are endowed with antioxidant and enzyme-modulating activities, hence expressing the potential to lower cardiovascular risk. It is noteworthy that a diet that derives a large proportion of its calories from plant food allows the intake of several grams of phytochemicals per day. Although the absorption, metabolism, and excretion of most of these compounds are far from being elucidated, such high intake makes this group of molecules particularly interesting from the preventive medicine point of view.[4] Also, it is important to note that, apart from phytochemicals, the human body cannot synthesize lipid- and water-soluble vitamins that, therefore, must also be taken up from the diet or provided through functional foods or pharmaceutical preparations.

In this chapter we examine the role of such oligonutrients, namely those present in high amounts in the most widely consumed diets, and their potential contribution to the prevention of atherosclerotic cardiovascular disease.

PLANT-DERIVED PHENOLS

Phenolics are derivatives of benzene with one or more hydroxyl groups associated with their ring, and they can be conveniently classified into at least ten different classes according to their chemical structure.

Two main synthetic pathways lead to the synthesis of phenolic compounds in plants: the acetate and the shikimate pathways (Figure 15.1). Phenolics are important to plant physiology, contributing to resistance to microorganisms and insects, pigmentation, and organoleptic characteristics (odor and flavor).

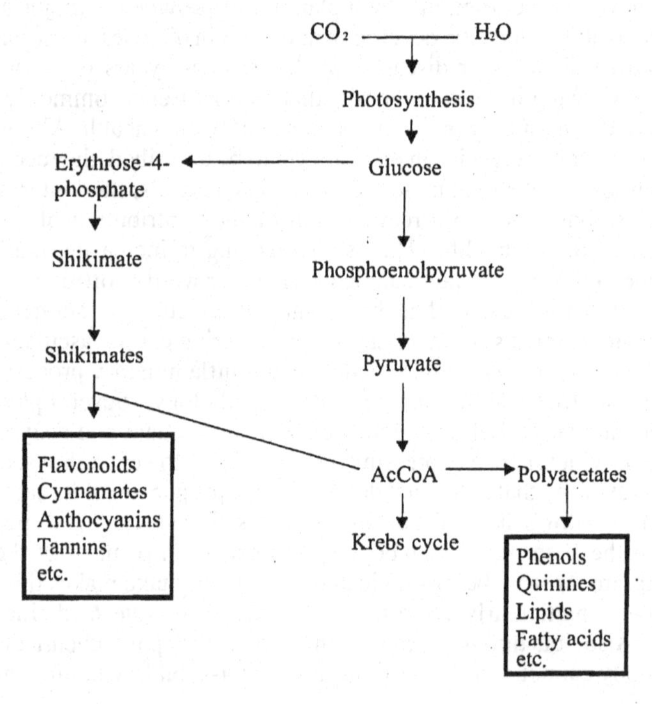

FIGURE 15.1. Schematic representation of the two major pathways involved in the production of phenolic compounds in plants.

It is noteworthy that fruits and vegetables require a variety of antioxidant compounds of different origin that may protect plants from environmental stress, including ultraviolet (UV) light radiation and relatively high temperatures. Accordingly, phenolics-rich crops such as grapes, olives, and heavily pigmented vegetables are particularly abundant in the Mediterranean area, where the combination of heat and light radiation stimulates plant antioxidant defenses. In addition to phenolic compounds, legumes (soybeans, peanuts, beans, and peas) contain flavonoids, isoflavonoids, isoflavones, coumestans, lignans, and other molecules, some of which also act as estrogenic agonists/antagonists; such compounds have been suggested to contribute to reducing the risk of atherosclerosis.

SOY

Soy is a major source of protein worldwide. Soybeans contain variable (normally around 5 $mg{\cdot}kg^{-1}$ protein) amounts of isoflavones that have been hypothesized to play a relevant role in the health effects on the cardiovascular system attributed to soy.

In particular, genistein and daidzein are endowed with weak proestrogenic or antiestrogenic activities and thus have the potential to interact with estrogen receptors and decrease serum cholesterol concentration.[5] In vitro studies have shown that genistein (a specific inhibitor for tyrosine kinases) might prevent the development of atheroma by inhibiting cell adhesion and proliferation, by inhibiting low-density lipoproteins (LDL) oxidation, and by altering growth factor activity.

Alas, it is difficult to discern the individual effects of soy components; thus the mechanisms responsible for the observed reduction of cholesterol concentration by soy are still unclear.[6] Moreover, the effects of the other soy constituents such as soluble fibers, saponins, and oligosaccharides should not be ignored; the up-regulation of the LDL receptor by soy proteins has also been demonstrated in vitro and ex vivo but, as the inhibition of tyrosine kinase by soy isoflavones should result in a hypercholesterolemic effect, this may not be the exclusive mechanism of action of soy components. To further spark debate on the healthful properties of soy isoflavones, recent data questioned their bioavailability in humans.[7] Finally, it should be stated that the proteic components of soy exhibit remarkable cholesterol-lowering properties, more evident in hypercholesterolemic subjects, that render a soy-based hypocholesterolemic diet very cost-effective. Based on current research data, most of the cardioprotective effects of a diet in which soy plays a relevant role are to be attributed to its proteic component.[8]

COCOA

Cocoa (*Theobroma cacao* L.) contains appreciable amounts (~20 mg total phenolic gallic acid equivalent per gram of cocoa powder) of phenolic substances, the most representative being (–)-epicatechin. Such phenolics confer chocolate fat a particular resistance to peroxidation; furthermore, accumulating in vivo and ex vivo studies indicate that they are endowed with biological activities, including antioxidant capacity and vasorelaxant and immunoregulatory effects.[9]

Although chocolate as a source of bioactive compounds might appear superfluous, it is important to note that, in a representative sample of Dutch population of young age, chocolate contributes 20 percent of the total catechin intake, suggesting that in younger age groups chocolate provides a significant portion of the daily antioxidant intake and that this contribution should be taken into account in epidemiological studies.[10]

TEA

Tea is a very rich source of oligonutrients. In fact, tea may very well be the major source of antioxidant worldwide, due to the very high concentration of these compounds—for example, quercetin, kaempferol, myrecitin, and epigallocatechin gallate—in tea, combined with the frequent consumption of this beverage. Tea is second to water as the most widely consumed beverage worldwide.

Several in vitro studies have shown the potent antioxidant activities of tea extracts, which also possess interesting tumor-inhibiting properties.[11] Also, some epidemiological studies correlated a higher flavonoid intake from tea with a lower incidence of coronary heart disease (CHD) although others failed to record protective effects.[12] Controlled studies show that tea components are taken up from the diet and that tea consumption has beneficial effects on the cardiovascular system, for example, it improves vasomotion.[13] As controlled, randomized long-term studies with tea (and other food items) are very unpractical and unlikely to be ever carried out, we must rely on animal data and short-term human trials. Yet based on the available data, it appears that tea oligonutrients do contribute to human health.

WINE

The observed inverse correlation between moderate alcohol intake and mortality from cardiovascular disease has promoted considerable interest in the components of alcoholic beverages. The most popular example followed the American television program *60 Minutes* (CBS Television, November 17, 1991). During this program, the so-called French paradox (i.e., low mortality rate from cardiovascular disorders recorded in France despite a high saturated fat intake and the presence of other established risk factor for cardiovascular disease, such as cigarette smoke and high cholesterol levels) was revealed to the lay public.[14] The rationale on which the French paradox is based is that red wine, at difference with other alcoholic beverages, contains substantial amounts of phytochemicals, most of which are phenolic in nature and are synthesized by red grapes as a form of self-protection from relatively high temperatures known to exist within a dark grape.[15]

Many of such wine components exhibit in vitro and in vivo antioxidant activities, including increased plasma antioxidant capacity and increased resistance of LDL to chemically induced oxidation that have been recorded following wine or wine-phenolic intake. Other cardioprotective properties suggested for red wine and grape juice include improved endothelial function and inhibition of platelet activity, suggesting that the antiplatelet activities of wine phenolics, thus far exhibited in vitro, are retained in the bloodstream.[16]

Thus far, human volunteer studies have yielded controversial results, leaving an open question as to whether the decreased mortality from coronary heart disease observed in moderate drinkers is attributable to wine pytochemicals, to ethanol itself, or even the diet of wine consumers.[17] Indeed, the wine hypothesis is currently not supported by the literature, leaving an open question whether it is "the drink or the drinker."[18] The contribution of important confounders such as higher socioeconomic status of moderate drinkers (eminently in the United States), the fact that wine consumption is usually associated with meals (possibly favoring the absorption of the other nutrients or limiting the formation and absorption of products of antioxidant from the gut), or discrepancies in the coroner systems of various countries is yet to be fully elucidated. Last but not least, the fact that cardiovascular mortality is extremely low in Muslim Mediterranean countries, e.g., in the Maghreb, where wine and alcohol consumption is forbidden by religious beliefs, should not be overlooked, as it partially questions the true contribution of alcoholic beverages to coronary pathology. Finally, even though research on wine components is widespread and certainly important to understand the contribution of this important diet component to human health, it should not be forgotten that alcohol has serious noxious side effects, including damage to the gastrointestinal (GI) tract. Furthermore, alcoholism is to be regarded as a very deleterious social phenomenon that is responsible for a high number of deaths, including those associated with violent crime or motor-vehicle accidents. Hence, all anecdotal as well as experimental data regarding the potentially positive effects of drinking any form of alcohol, including red wine, should be accompanied by such a warning/disclaimer, and the form in which messages concerning the health effects of wine is to be conveyed to the lay public deserves careful attention.[19]

OLIVE OIL

Besides its dietary use, olive oil has been employed for thousands of years for its putative medicinal properties. Indeed, olives contain high amounts of phenolic compounds that confer raw fruits a peculiar bitterness.[20] During olive oil production, and depending on their partition coefficient, most of these compounds are transferred to either olive oil (which, at difference from seed oils, is not obtained by chemical extraction but rather by physical means) or its waste water.[21] Indeed, olive mill waste waters contain high amounts of phenolics which are currently discarded but can be recovered by a newly developed technology and employed in preservative chemistry or as nutritional antioxidant supplements.[22]

Depending on its degree of acidity, olive oil is classified into different grades that also serve as guidelines for the consumer in the choice of the preferred kind of oil. From this classification, it can be concluded that the most valuable kind of olive oil is the extra virgin variety, obtained from intact olives that are quickly processed and cold-pressed. In particular, the phenolic fraction (see Box 15.1) responsible for the stability and flavor of extra virgin olive oil and endowed with pharmacological properties should be an additional, valuable marker of olive oil quality. As discussed next, the choice of good quality, in other words, phenol-rich extra virgin olive oil, albeit more expensive, provides considerable amounts of bioavailable compounds that might play a role in human health.

During the refining process, which is necessary when the free acidity of an oil exceeds the legal limit, almost all the minor components, phenols in particular, are destroyed, thus reducing the sharpness of the flavor.[23] This type of oil is marketed as olive oil and can usually be distinguished from the extra virgin oil by its blander taste. In fact, a high phenolic content confers a very bitter and pungent zest to olive oil. In particular, phenolic acids such as phenol and cinnamic acid are responsible for the bitter sensation that can be detected on the lateral and posterior areas of the tongue while secoiridoids confer the oil's peculiar pungency. In brief, intact olives that have been hand-picked at the best time, that are immediately brought to the mill, that are processed right away in a clean plant, and that are crushed and pressed at

BOX 15.1. Olive oil constituents.

Triacylglycerols:	*Polyphenols* (50-800 mg·kg^{-1}):
Hydrocarbons	Hydroxytyrosol
Nonglyceric esters	Oleuropeine
Tocopherols	Tyrosol
Sterols	Caffeic acid
Terpenic acids	Ligstroside
Pigments	Vanillic acid
Chlorophylls	Hydroxytyrosol esters
Carotenoids	Synaptic acid
	Syringic acid

temperatures lower than 25 to 30°C yield a high-quality oil that is also rich in phenolic constituents.

In the past few years, the availability of pure compounds, either from commercial sources or extracted and purified from olive oil, prompted investigations on the antioxidant and other biological properties of olive oil phenols. In fact, a large body of epidemiological studies shows how the incidence of CDH is lowest in the Mediterranean countries, possibly due to the relatively safe and maybe protective dietary habits of this area where olive oil constitutes the principal source of fat.

The antioxidant activities of olive oil phenolics, particularly those of the complex phenol oleuropein and its derivative hydroxytyrosol, have been revealed through a variety of experimental cellular and animal models, including transition metal ion- and chemically induced oxidation of LDL and the generation of reactive oxygen species such as superoxide, trichloromethylperoxyl radicals and hypochlorous acid.[24]

Further antioxidant activities of olive oil phenolics are their ability to interfere with enzymatic activities relevant to human pathology. For example, they inhibit platelet aggregation, reduce the formation of proinflammatory molecule thromboxane B2 and leukotriene B4 by activated human leukocytes, and inhibit human neutrophils' respiratory burst.[25]

In addition to the demonstration of dose-dependent absorption of olive oil phenolics, human evidence of the antioxidant activity of these compounds, namely hydroxytyrosol, is also accumulating.[26] For example, we have been able to demonstrate that hydroxytyrosol, administered to rats as the only bioactive component of an olive mill waste water extract, is dose-dependently absorbed and is able to increase their plasma antioxidant capacity.[27] Furthermore, very low doses of hydroxytyrosol, i.e., 414 g per rat, are able to inhibit passive-smoking-induced oxidative stress in rats, as demonstrated by a reduced urinary excretion of the F_2-isoprostane 8-*iso*-PGF_2 (iPF_2-III).[28] Finally, a dose-dependent inhibition of the rate of 8-*iso*-PGF_2 excretion was observed in human volunteers who ingested olive oils added with increasing amounts of phenolics; interestingly, the urinary levels of 8-*iso*-PGF_2 inversely correlated with those of homovanillyl alcohol, i.e., a catechol-*O*-methyl-transferase (COMT)-derived metabolite of hydroxytyrosol, suggesting that this phenol enters into cellular compartments where it exerts its antioxidant activity.[29]

CONCLUSION

The excessive and unmatched production of free radicals and reactive oxygen species contributes to the onset of certain pathologies such as atherosclerotic heart disease and cancer. This leads to the recommendation of a diet high in fruits and vegetables, in other words, plant food with a large proportion of antioxidant vitamins, flavonoids, and polyphenols. As an example, the observation that in the Mediterranean area there is a lower incidence of CHD and certain types of cancers indicates that the Mediterranean diet, rich in grains, legumes, fresh fruits and vegetables, wine in moderate amounts, and olive oil, has beneficial effects on human health, notwithstanding the presence of traditional risk factors for coronary heart disease (e.g., blood pressure and plasma cholesterol concentrations do not differ much between the Mediterranean basin and other areas of the Western world).[30] A large amount of research points to the pharmacological, cardioprotective properties of many phytochemicals. Current limitations to the full confirmation of the cardioprotective role of phytochemicals include the scarcity of pure, standard compounds and the insufficiency of appropriate methodology to evaluate their concentrations in food—and, consequently, their average consumption—and body fluids after ingestion.[31] Yet based on present data, the adoption of diet rich in plant foods abundant in bioactive compounds is desirable.

NOTES

1. Keys, A., 1995, Mediterranean diet and public health: Personal reflections, *Am J Clin Nutr,* 61: 1321S-1323S.

2. Bravo, L., 1998, Polyphenols: Chemistry, dietary sources, metabolism, and nutritional significance, *Nutr Rev,* 56: 317-333.

3. Ross, R., 1999, Atherosclerosis: An inflammatory disease, *N Engl J Med,* 340: 115-126.

4. Bravo, 1998, Polyphenols.

5. Anderson, J.W., Smith, B.M., and Washnock, C.S., 1999, Cardiovascular and renal benefits of dry bean and soybean intake, *Am J Clin Nutr,* 70: 464S-474S.

6. Sirtori, C.R., 2000, Dubious benefits and potential risk of soy phyto-oestrogens, *Lancet,* 355: 849; Sirtori, C.R., 2001, Risks and benefits of soy phytoestrogens in cardiovascular diseases, cancer, climacteric symptoms and osteoporosis, *Drug Saf,* 24: 665-682.

7. Setchell, K.D., Brown, N.M., Zimmer-Nechemias, L., Brashear, W.T., Wolfe, B.E., Kirschner, A.S., and Heubi, J.E., 2002, Evidence for lack of absorption of soy isoflavone glycosides in humans, supporting the crucial role of intestinal metabolism for bioavailability, *Am J Clin Nutr,* 76: 447-453.

8. Sirtori, 2000, Dubious benefits of soy phytoestrogens; Sirtori, 2001, Risks and benefits of soy phytoestrogens.

9. Sanbongi, C., Suzuki, N., and Sakane, T., 1997, Polyphenols in chocolate, which have antioxidant activity, modulate immune functions in humans in vitro, *Cell Immunol.,* 177: 129-136; Wan, Y., Vinson, J.A., Etherton, T.D., Proch, J., Lazarus, S.A., and Kris-Etherton, P.M., 2001, Effects of cocoa powder and dark chocolate on LDL oxidative susceptibility and prostaglandin concentrations in humans, *Am J Clin Nutr,* 74: 596-602.

10. Arts, I.C., Hollman, P.C., and Kromhout, D., 1999, Chocolate as a source of tea flavonoids, *Lancet,* 354: 488.

11. McKay, D.L. and Blumberg, J.B., 2002, The role of tea in human health: An update, *J Am Coll Nutr,* 21: 1-13.

12. Ibid.

13. Duffy, S.J., Keaney, J.F.J., Holbrook, M., Gokce, N., Swerdloff, P.L., Frei, B., and Vita, J.A., 2001, Short- and long-term black tea consumption reverses endothelial dysfunction in patients with coronary artery disease, *Circulation,* 104: 151-156.

14. de Lorgeril, M., Salen, P., Paillard, F., Laporte, F., Boucher, F., and de Leiris, J., 2002, Mediterranean diet and the French paradox: Two distinct biogeographic concepts for one consolidated scientific theory on the role of nutrition in coronary heart disease, *Cardiovasc Res,* 54: 503-515.

15. Soleas, G.J., Diamandis, E.P., and Goldberg, D.M., 1997, Wine as a biological fluid: History, production, and role in disease prevention, *J Clin Lab Anal,* 11: 287-313.

16. Folts, J.D., 2002, Potential health benefits from the flavonoids in grape products on vascular disease, *Adv Exp Med Biol,* 505: 95-111; Freedman, J.E., Parker, C., Li, L., Perlman, J.A., Frei, B., Ivanov, V., Deak, L.R., Iafrati, M.D., and Folts, J.D., 2001, Select flavonoids and whole juice from purple grapes inhibit platelet function and enhance nitric oxide release, *Circulation,* 103: 2792-2798.

17. Barefoot, J.C., Gronbaek, M., Feaganes, J.R., McPherson, R.S., Williams, R.B., and Siegler, I.C., 2002, Alcoholic beverage preference, diet, and health habits in the UNC Alumni Heart Study, *Am J Clin Nutr,* 76: 466-472.

18. Rimm, E.B., 1996, Alcohol consumption and coronary heart disease: Good habits may be more important than just good wine, *Am J Epidemiol,* 143: 1094-1098; Rimm, E.B., Klatsky, A., Grobbee, D., and Stampfer, M.J., 1996, Review of moderate alcohol consumption and reduced risk of coronary heart disease: Is the effect due to beer, wine, or spirits? *Brit Med J,* 312: 731-736.

19. Gaziano, J.M. and Hennekens, C., 1995, Royal colleges' advice on alcohol consumption, *Brit Med J,* 311: 3-4.

20. Blekas, G., Vassilakis, C., Harizanis, C., Tsimidou, M., and Boskou, D.G., 2002, Biophenols in table olives, *J Agric Food Chem,* 50: 3688-3692.

21. Boskou, D., 2000, Olive oil, *World Rev Nutr Diet,* 87: 56-77.

22. Visioli, F. and Galli, C., 2002, Phenolics from olive oil and its waste products. Biological activities in in vitro and in vivo studies, *World Rev Nutr Diet,* 88: 233-237; Visioli, F. and Galli, C., 2002, Olives and their production waste products as sources of bioactive compounds, *Curr Topics Nutr Research,* 1: 85-88.

23. Boskou, 2000, Olive oil.

24. Visioli, F. and Galli, C., 1998, The effect of minor constituents of olive oil on cardiovascular disease: New findings, *Nutr Rev,* 56: 142-147.

25. Visioli, F. and Galli, C., 2002, Biological properties of olive oil phytochemicals, *Crit Rev Food Sci Nutr,* 42: 209-221.

26. Visioli, F., Galli, C., Bornet, F., Mattei, A., Patelli, R., Galli, G., and Caruso, D., 2000, Olive oil phenolics are dose-dependently absorbed in humans, *FEBS Lett,* 468: 159-160; Vissers, M.N., Zock, P.L., Roodenburg, A.J., Leenen, R., and Katan, M.B., 2002, Olive oil phenols are absorbed in humans, *J Nutr,* 132: 409-417.

27. Visioli, F., Caruso, D., Plasmati, E., Patelli, R., Mulinacci, N., Romani, A., Galli, G., and Galli, C., 2001, Hydroxytyrosol, as a component of olive mill waste water, is dose-dependently absorbed and increases the antioxidant capacity of rat plasma, *Free Rad Res,* 34: 301-305.

28. Visioli, F., Galli, C., Plasmati, E., Viappiani, S., Hernandez, A., Colombo, C., and Sala, A., 2000, Olive phenol hydroxytyrosol prevents passive smoking-induced oxidative stress, *Circulation,* 102: 2169-2171.

29. Visioli, F., Caruso, D., Galli, C., Viappiani, S., Galli, G., and Sala, A., 2000, Olive oils rich in natural catecholic phenols decrease isoprostane excretion in humans, *Biochem Biophys Res Commun,* 278: 797-799; Caruso, D., Visioli, F., Patelli, R., Galli, C., and Galli, G., 2001, Urinary excretion of olive oil phenols and their metabolites in humans, *Metabolism,* 50: 1426-1428.

30. Mancini, M. and Rubba, P., The Mediterranean diet in Italy, in A.P. Simopoulos and F. Visioli (eds.), *Mediterranean diets* (Basel: Karger, 2000), pp. 114-126.

31. Bravo, 1998, Polyphenols.

Chapter 16

A Modern Look at Folkloric Use of Anti-Infective Agents

Lester A. Mitscher

INTRODUCTION

Plants and animals, including humans, have at least a few things in common. Each supports a thriving population of microbial life that is ordinarily relatively benign. Metabolic or immune imbalance, wounding, or exposure to unusual microbial concentrations or species can lead to infections of plants or animals that are debilitating if not lethal. This has been an unfortunate characteristic of daily life for ages.

Whereas animals rely primarily on a functioning immune system for warding off infection, the very different construction of plant organs requires a different approach. Anti-infective plant macromolecules include protease inhibitors, glycosidase inhibitors, lectins, phytohemagglutinins, and tannins. These agents have not found significant usage in the treatment of infected animals. Of greater promise are the constitutive antimicrobial small molecules present before infection and the postinfectional agents elicited in response to infection (phytoalexins). A wide variety of such agents are present in plants and, in recent years, a number of them have been isolated and evaluated.

By and large the human or animal physiological response to infection makes comparatively little use of endogenous small molecules of this type. Instead, small molecules made by soil microorganisms, in other words, antibiotics, are used for this purpose. Higher plant agents with antimicrobial activity are plentiful but by and large have not found much contemporary use for treatment of bacterial or fungal infections. On the other hand, they have proven value in the treatment of protozoal and viral infections.

The human search for therapy or crop protection based on rational considerations was hindered until approximately 150 years ago by a virtual total lack of knowledge of microbes, and thus the origin and treatment of infectious disease were mysteries. Previously such little useful therapy that became available was found entirely empirically. Very early usages are recorded in the accounts of medical practice written by Pen Tsao about 3000 B.C.E., the Ebers Papyrus of about 1500 B.C.E. and Celsius' *De Medicina*. From today's perspective, much of the lore recorded in early works seems to be fanciful nonsense. Even after it became clear that microbes cause infectious diseases and even after scientists learned how to cultivate and study them in the laboratory, progress in finding selective poisons that would cure infections without serious harm to the host was slow. The comparatively few useful treatments that were available did not point to a generally productive means of finding more and better agents.

This depressing picture changed dramatically in the 1940s when a number of safe and effective antimicrobial agents were found among the metabolites of certain soil microbes and in the products of organic chemical synthesis. Rapidly thereafter, many of today's mainstay weapons against microbial disease were introduced, including penicillins, cephalosporins, macrolides, tetracyclines, sulfonamides, aminoglycosides, fluoroquinolones, the 'conazoles, and the like. The dramatic successes of antimicrobial chemotherapy are widely known and celebrated. Mortality from infectious diseases fell at an astonishing rate. Unfortunately, however, the need of all species to propagate their genes is universal. The discovery and application of antimicrobial agents by humans was an ecological disaster for microbes and for half a century they were unable to overcome their use. Microbial fecundity and genetic versatility and their very short generation times ultimately led to an unacceptable level of bacterial resistance to antimicrobial drugs. The positive selecting pressure of widespread and often careless antimicrobial use coupled with the appearance of bacterial resistance has now led to dramatic losses in the effectiveness of these "miracle drugs." One might draw a parallel with the bubonic plague of the Middle Ages or the influenza pandemic of 1918 in which human mortality at the hands of the microbes was similarly devastating. Nonetheless, eventually a new equilibrium point was reached and the human race continues. The parallel with resistance mechanisms employed by bacteria to ensure the continuation of their genera is unmistakable. Clearly the contest between plants and animals on one hand and microbes on the other is a war that will never cease. That is not to say that one should not press every advantage one has in order to overcome preventable mortality.

It is in this context that the material in this chapter should be considered. Of the present approaches chemists employ to extend the utility of antimicrobial agents, one notes the search for new structures, novel modes of action, inhibitors of microbial enzymes that inactivate or export antimicrobials, use of combinations of antimicrobials, use of immunostimulants, attack against virulence mechanisms, and so on. Antimicrobial agents from higher plants can be examined in this context.

The use of folkloric agents has become very popular in recent years and is a return to the age-old method of finding chemotherapy in natural materials. The fundamental difference is that one can now evaluate in the laboratory the range of utility of such agents in comparison with existing treatments. Study of the multitude of antimicrobial agents elaborated by higher plants to combat microbial predation has revealed that their chemical structures differ widely from those employed by soil microbes, that their biochemical modes of action are often different or usually unknown, that they are small molecules readily synthesized and modified, and that their potencies in infected animals are comparatively unimpressive. In economically advantaged parts of the world, the vogue for the use of these agents is often a matter of personal preference and belief. In cases of mild infections one can posit that this is not a particularly harmful practice even though resorting to powerful antimicrobials is more rational. Still, the rich have choices, and one should hesitate to interfere with their right to exercise their rights. In many parts of the world, however, where economic resources are often dramatically less, resorting to such agents is of necessity all that can be afforded. Although this is unfortunate, it is useful here to establish at least which preparations have the capacity to be effective when resorting to them is a necessity.

In considering the voluminous literature on folk medicines a number of important questions should form the basis of an informed opinion and recommendation.

1. Is there a significant history of human use?
2. What claims are made?
3. What science backs up these claims?
4. Is the evidence credible?
5. Are there reliable sources of material?
6. Are the active constituents known?
7. Are there suitable assays?
8. Is it safe?

Only rarely today can one answer all of these questions convincingly with respect to the majority of herbal medicines. Nonetheless a significant body of evidence is building up relating to these and related materials.

In this context, ultimately one can divide the agents that are potentially available into four broad classes. First are those for which scientific investigation has proved beyond a reasonable doubt that they are safe and reasonably effective. Next is a group in which some credible evidence suggests that they could be useful. Then there is a group for which the evidence is still equivocal and more study is needed. Finally are those whose efficacy has not been demonstrated and for which there may even be evidence of harm; use of these agents should be discouraged.

Of the ancient anti-infective remedies of higher plant origin for which credible modern evidence has been obtained, only berberine, emetine, quinine, and sanguinarine find much use today and these uses, with the exception of quinine, are mostly restricted to minor indications. Other materials for which significant evidence is available validating anti-infective properties include allyl isothiocyanate, chelidonine, conessine, cepharanthine, harmine, kawain, pinosylvine, thujaplicins, and thymol.

BERBERINE

Berberine, a protoberberine alkaloid, is the main antimicrobial substance present in barberry (*Berberis vulgaris* L. and *B. aquifolium* Nutt. [Syn. *Mahonia aquifolium* Nutt.], Berberidaceae), goldenseal (*Hydrastis canadensis* L., Ranunculaceae), and yellow root (*Xanthorhiza simplicissima* Marsh., Ranunculaceae) and many other plants as well (see Figure 16.1). It may, in fact, be the most widely distributed alkaloid of all. Berberine was first isolated in 1826. Barberry has been used for medicinal purposes from at least the Middle Ages and other berberine-containing plants (found in at least 7 different plant families and 23 genera) have been used in the Orient for treating diarrhea for many centuries. Use in gastrointestinal distress including bowel looseness is the primary present use. Berberine is poorly absorbed from the gastrointestinal (GI) tract following oral administration so it is mostly used topically or for gastrointestinal complaints. Berberine it-

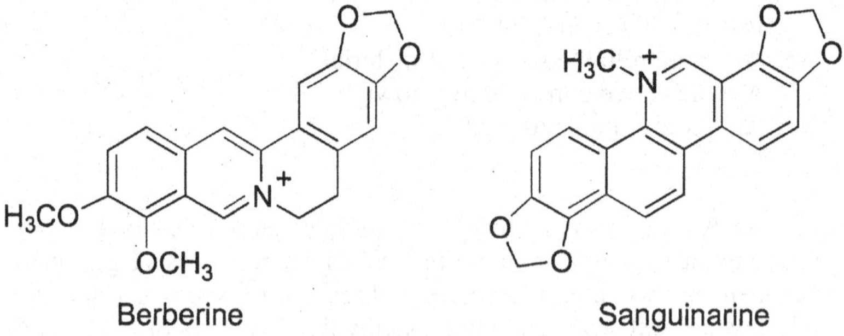

Berberine

Sanguinarine

FIGURE 16.1. Chemical structures of berberine and sanguinarine.

self has been used as an astringent in eye drops, but this use is now archaic. The broad-spectrum bactericidal activity of berberine has been well validated in the laboratory.[1]

Berberine's molecular mode of action appears to be a consequence of its ability to intercalate into DNA.[2] Interestingly, berberine's effect on whole cells seems to be opposed by active efflux due to a multidrug resistance pump. This effect is opposed by the presence of 5'-methoxyhydnocarpin in plant extracts which inhibits the efflux pump.[3] Berberine has been used in combination with tetracycline for the treatment of cholera with significant improvement being noted.[4] The alkaloid's antimicrobial action in diarrhea is possibly also assisted by antisecretory and antimotility properties. Berberine, however, has other pharmacological effects as well (anticonvulsant, immunostimulant, anti-inflammatory, antioxidant, sedative, hypotensive, uterine stimulant, etc.).[5] Its action as an uterotrophic agent would preclude use during pregnancy and its ability to inhibit the anticoagulant effect of heparin and its cardiostimulant action are significant contraindications.[6] Side effects clearly inhibit its acceptance, as does the general availability of more powerful and safer antibiotics. It is perhaps fortunate that berberine does not produce high blood levels following oral administration.

In conclusion, berberine is unquestionably active as an antimicrobial agent but is poorly absorbed on oral administration and has too many side effects to find systemic use following injection. Few reports of toxicity have been published other than accounts of allergy to the plants from which it comes. Use in topical or GI infections appear supportable, but the agent is inferior to clinically significant antibiotics.

SANGUINARINE

Sanguinarine is an alkaloid found in bloodroot (*Sanguinaria canadensis* L., Papaveraceae) and other plants of the Papaveraceae (see Figure 16.1). It has long been used in many folkloric medicinal applications, most persistently in treatment of tumors. It finds commercial use today in dentifrices wherein it inhibits plaque formation partly as a consequence of its bacteriostatic action against bacteria believed to play a role in plaque formation aided by its ability to cling to plaque itself.[7] The antimicrobial activity of sanguinarine and its analogs is well established in laboratory studies and is likely to be a consequence of its ability to intercalate into DNA.[8] Sanguinarine forms reversible adducts with a number of nucleophiles, including alcohols in basic solutions. These are more lipophilic by far than sanguinarine itself but revert readily on attempted storage. This property enhances the antimicrobial effect of the agent in vitro but does not significantly enhance the possibility of central absorption and toxicity because of the ready reversibility of this reaction.[9] Sanguinarine is poorly absorbed following oral administration enhancing its safety as a dentifrice. Central effects of sanguinarine include some central nervous system depression and narcosis, a papaverine-like effect on heart and smooth muscle, and inhibition of glycolysis by disruption of the oxidative decarboxylation of pyruvate.[10]

In conclusion, sanguinarine definitely possesses antimicrobial activity and has shown proven utility in preventing dental plaque. It is not very toxic, but oral administration in quantity is discouraged. The comparatively small amounts swallowed following use in dentifrices appear to be harmless.

QUININE

Quinine, a quinoline alkaloid, is found in Jesuit's bark or Peruvian bark (*Cinchona ledgeriana* Moens ex. Trim, *C. succirubra* Pav. ex Klotsch, *C. calisaya* Wedd., and a variety of related *Cinchona* species in the family Rubiaceae) (see Figure 16.2). Its value in treating the mosquito-borne protozoan disease malaria was apparently revealed to Europeans in Peru around 1600, and plant extracts rapidly entered medicine in Europe. Quinine was isolated from the medicament in 1820 and usage switched largely to the pure alkaloid. The story of its cultivation in what is now Indonesia and Africa as well as its native South American range is lurid and makes interesting reading.[11] Quinine suppresses but does not cure the vivax form of malaria and was once nearly abandoned in the United States in favor of more potent chloroquine and related synthetics. Its use recovered when the

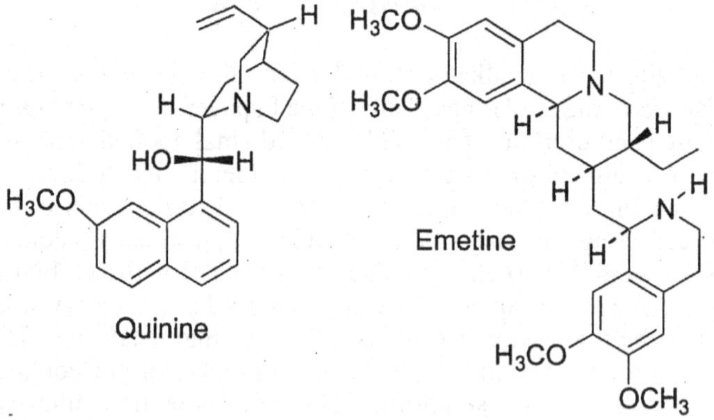

FIGURE 16.2. Chemical structures of quinine and emetine.

falciparum form of malaria became chloroquine resistant but retained sensitivity to quinine. Malaria is one of the most devastating of the microbial diseases afflicting humans with approximately 500 million people infected with it and an annual mortality of about 2 million—largely young children, pregnant females, and nonimmune persons. The protozoa causing malaria go through a variety of growth phases in their life cycle and the erythrocytic nonsexual schizontal form is the most sensitive to the action of quinine and related alkaloids.

Quinine, as a weak base, accumulates in the acidic food vacuoles of susceptible protozoa and inhibits the nonenzymatic polymerization of toxic heme into a nontoxic mixed polymer. Interruption of this process is injurious to the plasmodia. The heme is a digestion product of erythrocytes and hemoglobin and, ordinarily, generates toxic reactive oxygen species that would kill the parasites. When this is polymerized to insoluble, nonreactive hemozoin, however, the plasmodia are able to survive.[12]

Quinine also finds occasional use in prevention of nocturnal leg cramps caused by vascular spasms.[13] It is a cardiodepressant and was historically used in the treatment of cardiac arrhythmias, although its optical isomer quinidine is preferred for this use.

Quinine itself is well absorbed when taken orally, significantly metabolized in the liver by oxidation catalyzed by CYP3A4, leading to significant drug-drug interactions, and is excreted in the urine.[14] The drug is comparatively toxic, causing tinnitis (ringing of the ears), high-tone deafness, severe headaches, convulsions, visual disturbances, vertigo, nausea, vomiting, postural hypotension (known collectively as cinchonism), as well as hypoglycemia and hypotension. Acute hemolytic anemia has also occurred during the course of antimalarial therapy in individuals deficient in glucose-6-phosphate dehydrogenase.[15] It is extremely bitter-tasting and one of its more popular uses is as a bitter in gin beverages.

In summary, despite its long history of use, significant toxicities, and the development of resistance, quinine remains a valuable antimalarial drug for certain patients. It is doubtless the most successful, in modern terms, of all of the antimicrobial drugs of antiquity.

EMETINE

Emetine is an isoquinoline alkaloid isolated from ipecac [*Cephaelis ipecacuanha* (Brotero) A. Richard (Brazilian ipecac) and *C. acuminata* Karsten (Cartagena ipecac), family Rubiaceae] (see Figure 16.2). Centuries ago Portuguese missionaries brought news from the New World of the utility of ipecac in the treatment of dysentery to Europe. Emetine itself was first isolated in 1817 but not obtained in pure form until 1894. The ancient use of the agent was for emesis and as an expectorant, as its name suggests. Its use as an antiamoebic agent in Western medicine dates from about 1912.[16] Today, syrup of ipecac is still used in first-aid kits as an emetic in poisonings. Care must be exercised not to confuse it with the fluid extract, which contains much more ipecac than the syrup, and its accidental substitution can lead to death. Its other common uses are in the treatment of intestinal amoebic infections and in some extraintestinal amoebic conditions. Severe cumulative cardiotoxicity limits the desirability of emetine in the treatment of chronic amoebic dysentery.[17] The molecular mode of action of emetine appears to be consequences of protein biosynthesis inhibition.[18] The powdered plant is also a powerful respiratory irritant, so care must be exercised in using it. Metronidazole, a synthetic agent, is regarded as safer for the treatment of amebic dysentery and is preferred in modern medicine.

INDIVIDUAL ANTIMICROBIAL MEDICINAL PLANTS OF LESSER PROMINENCE

In the following cases, the plants and agents considered are those for which there are a lesser history of human consumption but some modern laboratory evidence providing a plausible basis for their possible use.

Allspice

Allspice [*Pimenta dioica* (L.) Merr., Myrtaceae family] is a well-known spice plant that contains a volatile oil. The content of eugenol in the oil is usually regarded as responsible for its useful properties, including its antimicrobial action.[19] Eugenol is indisputably a weak antimicrobial agent.

Aloe

Aloe species (*Aloe vera* L., *A. perryi* Baker, *A. barbadensis* Miller, *A. ferox* Miller, Liliaceae family) have been used as a cathartic and an antibacterial for several thousand years, although the lay public today knows it best as a remedy for burns and as a component in cosmetics. The evidence on its use as an antibacterial is contradictory, with some species showing activity and others not but, a number of recent studies are supportive.[20] The systemic toxicity of aloe would seem to preclude its anti-infective use other than as a topical agent where its other wound-healing attributes would assist. The structure of the anthraquinone constituents suggests the possibility of genotoxicity via intercalation, also suggesting that internal use is to be avoided.

Alpinia

Alpinia (*Alpinia officinarum* Hance.) has been used for a long time as a flavoring agent and in the treatment of GI complaints. Recent studies confirm antifungal properties as well.[21] These properties elicit little interest today outside of the culinary arts.

Artemisia

Artemisia, or sweet wormwood (*Artemisia annua* L.), has been used in the traditional Chinese medical system for at least 2,000 years for treatment of fevers. Systematic investigation of its extracts resulted in isolation of the active antimalarial agent, artemesinin (qinghaosu).[22] This interesting peroxide-containing terpenoid is active in vivo against chloroquine-resistant *Plasmodium falciparum* strains and cerebral malaria following reaction with hemin, and detailed molecular modeling studies have resulted in docking models to rationalize the action of this agent.[23] Extensive chemical manipulations of artemesinin have resulted in the production of numerous active analogs for further exploitation.[24] It appears likely that novel antimalarial chemotherapy will emerge from analogous studies.

Astragalus

Astragalus (*Astragalus membranaceus* Bunge, Fabaceae family) is a very well-known and venerable drug in Chinese traditional medicine in which it is used to stimulate the immune system. In Western folk medicine it is used to counteract the immune suppression often associated with cancer chemotherapy. It contains cycloartane saponins that have been shown to stimulate human lymphocytes ex vivo.[25] Astragalans I and II are polysaccharides that stimulate immunological responses when administered intraperitoneally (i.p.) in mice (but not orally).[26] Studies in immune-suppressed and immune-competent human patients demonstrated restoration or augmentation of local graft-versus-host rejection using astragalus extract.[27] It has been suggested that *astragalus* administration improves symptomology in HIV-infected patients, but these studies are not strong. The rapid turnover rate of T-cells in HIV-infected patients makes immunostimulation a doubtful therapy, but any improvement in immune status for these patients would be desirable.[28] The extracts of *astragalus* appear to be safe, but suggestions of mutagenicity need to be examined further.

Bitter Melon

Bitter melon (*Momordica charantia* L., Cucurbitaceae family) has been confirmed to possess antiviral properties (against polio, herpes simplex 1, and HIV). The active constituent appears to be a 28.6 kD protein named MRK29 that inhibits HIV-1 reverse transcriptase.[29] The bitter taste of this plant would be a disincentive for oral use, and there has been insufficient examination of MRK29 to establish whether its use would lead to hypersensitivity.

Black Cumin

Black cumin *(Nigella sativa)* seed extract inhibits gram-positive and -negative bacteria and fungi in vitro and cures mice infected with nonlethal staphylococci.[30] The active constituent has not been identified. Further work is apparently necessary.

Burdock

Burdock (*Arctium lappa* L., Asteraceae family) is generally considered to be a food, but folkloric records of its use in the treatment of common viral infections exist. Indeed, recent work suggests activity against HIV, but the evidence for this is unconvincing.[31]

Calophyllum

Calophyllum (*Calophyllum lanigerum* Linn.) is a tree growing in the South Seas. From this tree has been isolated (+)-calanolide A, a naturally occurring coumarin. Calanolide A inhibits the AIDS virus by interfering with RNA-dependant DNA polymerase (reverse transcriptase).[32] Interestingly, the giant African snail, *Achatina fulica,* which is known to eat from this tree, also produces calophyllolide, which also is found in *C. lanigerum.* The snail also produces the inophyllums B and P that also inhibit the HIV-1 virus.[33] The suggestion is that these substances may result via the food chain or through modification of these dietary substances following consumption by the snail.

From the latex of *Calophyllum teysmannii* was also isolated souratrolide, a related coumarin, also found to be active against HIV-1 reverse transcriptase.[34] *Calophyllum cordato-oblongum,* a Sri Lankan species, was found to contain cordatolide A and B, which also inhibit HIV-1.[35] Another Sri Lankan plant, *C. moonii,* contained calozeyloxanthone, a natural product active against vancomycin-resistant *Staphylococcus aureus.*[36]

Collectively these findings with various *Calophyllum* species suggest that further study of this genus might lead to practical results.

Calendula

Calendula (marigold) (*Calendula officinalis* L., Compositae family) has been used topically in folk medicine since the early Middle Ages to promote wound healing. Recent studies support antibacterial activity and the sesquiterpene glycosides and triterpenoid saponins are reportedly antiviral.[37] The plant and its extracts appear not to be toxic, but the history of allergy to members of the Compositae family must be borne in mind.

Cat's Claw (Una de Gato)

Cat's claw [*Uncaria tomentosa* (Willd.) DC and *U. guianensis* (Aubl.) (Gmel.), Rubiaceae family] has a long history of use in Latin America for wound healing and for treatment of a variety of other complaints, including gastric ulcers. Some evidence exists of efficacy in cancer and immunodeficiency states, and some anecdotal evidence of use in Europe in supplementation of azidothymidine (AZT) in AIDS treatment is found.[38] The oxindole alkaloids have been shown to stimulate phagocytosis, which may account for the folkloric uses.[39]

Citronella

Citronella [*Cymbopogon nardus* (L.) Rendle and *C. winterianus* Jowitt, Gramineae family] produces an oil best known as an insect repellant, although recent work indicates that some in vitro antibacterial and antifungal activity is also present.[40]

Clove

Clove [*Syzgium aromaticum* (L.) Merr. et L. M. Perry, Myrtaceae family] has an ancient history for culinary and medicinal use. Most interesting from the perspective of this chapter is its use in dentistry where it has a long reputation and current use as an anodyne and antiseptic. Eugenol, kaempferol, and myricetin, in particular, are known to be active against oral bacteria.[41]

Use of cloves in cigarettes, however, is to be discouraged because of the irritation that accompanies this use.

Cranberry

Cranberry (*Vaccinium macrocarpon* Ait., *V. oxycoccos* L., *V. erythrocarpum* Michx., *V. vitis*, and *V. edule*), a popular relish and juice component, has found common use in acidifying the urine for about a century and a half, although this use is not strongly supported by scientific evidence. A direct antibacterial effect is not supported in the literature, but it seems more likely that the components of the juice are useful in inhibiting urinary tract infections because they interfere with bacterial adherence to mucous membranes.[42] Clinical trials are under way.

Echinacea

Echinacea [*Echinacea angustifolia* DC, *E. purpurea* (L) Moench, and *E. pallida* (Nutt.) Briton, Compositae family] are very popular herbals in the United States and Europe. Teas were introduced rapidly following Western contact with the High Plains Indians of North America. The various *Echinacea* preparations are popular for their reputed immunostimulant properties, although experts disagree with respect to responsible constituents and the clinical information distilled from numerous trials is contradictory. There appears to be no direct antimicrobial action by extracts but rather general stimulation of the cells of the passive immune system is the most likely causative factor.[43] Lack of identification of a convincing active component hinders standardization of *Echinacea*. This may well contribute strongly to confusion about its efficacy.

Garlic

Garlic (*Allium sativum* L., Liliaceae family) is mostly used as a condiment, but it also has a long history as a potential medicament. Garlic's broad spectrum antimicrobial activity has been known for centuries.[44] In the 1800s it was commonly used by inhalation for the treatment of tuberculosis, but modern studies indicate that the potency is low against mycobacteria. Certainly many more effective antitubercular drugs are in use today. Several more recent studies confirm antibacterial and antifungal activity, especially sulfur-containing substances in onion volatile oil.[45] Of these, ajoene and related compounds are prominent.[46]

Ginger

Ginger's (*Zingiber officinale* Roscoe, Zingiberaceae family) medicinal and flavoring use in Asia goes back for centuries, and its medicinal and culinary use in Western medicine goes back to the early Middle Ages. Its primary contemporary use is in the treatment of motion sickness and various gastrointestinal (GI) distress syndromes. Quite recently, however, a cyclic sesquiterpene named zerumbone has been found to possess antibacterial properties.[47]

Hydrangea

Hydrangea (*Hydrangea macrophylla* Syringe) was closely investigated during the World War II era as a potential replacement for quinine, which was no longer readily available to the Western allies, for the treatment of malaria. Although not leading as yet to marketed products, this work continues. Recently the dihydroisocoumarin glycosides, the thunberginols, have been proven to be orally active against a variety of bacteria as well as in murine malaria due to *Plasmodium yoelii,* although cures were not obtained.[48]

Hyssop

Hyssop (*Hyssopus officinalis* L., Labiatae family) has been used in herbal medicine for centuries. Recently a polysaccharide has been isolated from it that demonstrated strong activity against HIV-1.[49] The plant is generally regarded as safe, but convulsions have been reported so caution should be exercised until this is sorted out.

Larch

Larch [*Larix dahurica* L., *L. decidua* Mill., *L. eurolepis* Gord., *L. gmelinii, L. kaempferi, L. laricina* Koch., *L. leptolepis* (Sieb. et Zucc.) Gord., *L. occidentalis* (Nutt.), and *L. sibirica Ledeb.,* Pinaceae family] is primarily a timber plant but contains arabinogalactans that are immunostimulatory in the same sense as echinacea. Thus its medicinal applications are based upon this indirect antimicrobial effect rather than a direct antiseptic action.[50]

Lemon Verbena

Lemon verbena [*Aloysia triphylla* (L'Her.) Britt., Verbenaceae family] is a venerable medicinal plant now primarily restricted to uses in perfumery, teas, and as an ornamental. The volatile oil is obtained by steam distillation of the leaves and, as with many volatile oils, shows some in vitro antibacterial properties.[51] Recently a study of Argentinian medicinal plants, including lemon verbena, identified some flavonoids in the plant as also possessing antimicrobial activity.[52] Many flavonoids possess antimicrobial activity, but these are usually too weak to find medicinal use for this indication.

Licorice

Licorice root (*Glycyrrhiza glabra* L., Leguminoseae family) is used extensively as a flavoring agent and finds medical applications as a demulcent, expectorant, antiulcer, anti-inflammatory, and laxative agent. A significant number of flavonoid constituents have been identified that possess antimicrobial properties.[53]

Mace

Mace (*Myristica fragrans* Houtt., Myristicaceae family) has a venerable history of culinary and medicinal use in the East Indies and India. Particular reference to the treatment of diarrhea is found. Quite recently the resorcinols malabaracone B and C were isolated from the plant and shown to have strong antibacterial and antifungal activity.[54] This has not apparently been pursued.

Mastic

Mastic (*Pistachia lentiscus* L., Anacardiaceae family) has been used since prerecorded time as incense and as part of the embalming treatments. Today it finds use as a flavoring agent in some gums and alcoholic beverages (Greek retsina, for example). It finds a variety of other uses as well, some of which are antimicrobial in nature. Perhaps the most interesting of these is its in vitro activity (albeit weak) against clinical strains of *Helicobacter pylori,* the microbe responsible in many cases of gastric ulcers.[55] Mastic has been used for some time in the treatment of benign gastric ulcers, so these findings are suggestive of a cause-and-effect relationship.

Meadowsweet

Meadowsweet (*Filipendula ulmaria* L. Maxim., *Spiraea ulmaria* L., Rosaceae family) has a long reputation based upon its pleasant odor and has medicinal significance primarily as the progenitor of aspirin. Its mild antibacterial activity is attributed to the presence of certain flavonoids and phenolics.[56]

Neem

Neem (*Azadirachta indica* A. Juss., Meliaceae family), an important Indian medicinal plant, is one of those plants with a wide variety of active constituents that consequently finds many folkloric uses. Some of the more important of these are as a directly applied vaginal contraceptive and as an insect and pest repellant. Its antibacterial properties have been most explored against *Staphylococcus mutans* and lactobacilli for use in tooth preservative preparations.[57] Mahmoodin, a tetranor triterpenoid, has been isolated from neem and demonstrated to be antibacterial.[58]

Onion

Onion (*Allium cepa*, Liliaceae family) has been used for at least 5,000 years for a variety of purposes including, particularly, as a food. It has an ancient and confirmed reputation for antifungal and antibacterial activity.[59] This is believed to be associated with sulfur-containing constituents. This activity has been suggested as being useful for oral pathogenic bacteria.[60] Interestingly this activity was stable in the presence of cysteine suggesting that electrophilicity is not a major component of the mode of action. Another recent study reaffirms activity of onion oil against gram-positive bacteria and fungi but generally not against gram-negative bacteria.[61]

Passionflower

Passionflower (*Passiflora incarnata,* Passifloraceae family) was found by Spanish explorers in Peru about 1600. The name of the plant has religious rather than carnal associations. It finds present folkloric use in treatment of sleep disorders. There have been reports of antimicrobial activity against staphylococcus, streptococcus, molds, and yeasts. The activity appears to be unstable, disappearing on storage, and possibly attributed to 4-hydroxy-2-cyclopentenone (passicol).[62]

Peppermint

Peppermint (*Mentha piperita* L., Labiatae, family) is a well-known condiment and flavoring agent that finds many uses, especially for gastrointestinal complaints. It contains a great many constituents of which menthol and menthone in particular show some antimicrobial activity. Prominent among these is activity against *Escherichia coli* 0157:H7, a particularly dangerous foodborne toxogenic microorganism.[63] Peppermint has a number of unpleasant side effects if taken chronically or intensively and also shows a number of drug-drug interactions, so more than the usual caution should be employed if it is to be used.

Quassia

Quassia (*Picrasma excelsa* and *Quassia amara* L., Simaroubaceae family) is an intensely bitter-tasting plant with a history of use as an antimalarial that has recently been conformed in murine studies in vivo.[64] It has also a history of use in measles, as an antibacterial and antifungal, in birth control, and as an insecticide. It is not recommended for use in pregnant females.

Rue

Rue (*Ruta graveolens* L., *R. montana* L., and *R. bracteosa,* Rutaceae family) has been used since ancient times as an insect repellant and as an antispasmodic, sedative, and to stimulate the onset of menses. It contains a comparatively large number (more than a dozen) of components that display antimicrobial activity. In particular the acridone alkaloids are active whereas the coumarins are less active.[65] The potential for genotoxicity of the rue acridones due to intercalation is worrisome.

Sarsaparilla

Sarsaparilla (*Smilax aristolochiafolia* Mill., *S. officinalis* Kunth, *S. rigelii,* Killip et Morton, *S. febrifuga, S. sarsaparilla, S. ornata,* Liliaceae family) was introduced in medicine to treat syphilis in the late fifteenth century. This use, although minor compared to its other uses, seems to have persisted to this day. It is believed that the molecular mode of action justifying this use involves the ability of sarsaparilla constituents to bind to endotoxins. This effect leads to its adjunctive use in a number of disease conditions. The direct antimicrobial effects of sarsaparilla appear not to be strong.[66] Obviously much more efficacious treatments for syphilis are available today.

A recent study of plants used in Guatemala for intestinal complaints involving enterobacteria was able to rationalize the folkloric use of *S. lundelii* for treatment of such diseases in man.[67]

Savory

Savory is a collective name for a group of condiment plants (*Satureja hortensis* L. and *S. montana* L., Labiatae family) frequently used in cooking. These plants are comparatively rich in volatile oil and, as with many volatile oils, therefore possess some antimicrobial activity.[68]

Skullcap

Skullcap (*Scutellaria lateriflora* L., Lamiaceae family) has been used in North American folk medicine since about 1800 when it was one of the few treatments available at that time for hydrophobia. Whatever the basis for this, it has generally fallen into disuse for any medicinal purpose. Recent studies, however, have identified the flavonoid baicalin as an inhibitor of HIV-1 infection by apparently interfering with viral entry by blocking the cell surface chemokine coreceptors.[69]

St. John's Wort

St. John's wort (*Hypericum perforatum* L., Hypericaceae family) has been used as a medicinal since at least the Middle Ages. It was especially prized for its wound-healing properties. Today it has become an object of special interest in the treatment of anxiety and depression disorders. Of special interest in this review is its potential utility in the treatment of AIDS. Hypericin and pseudohypericin show activity against a variety of viruses and are in clinical trial for this indication.[70] The mode of action of these agents is thought to be associated with a light-activated membrane effect. Broad-spectrum antibacterial activity is also recorded.[71] The side effects of the herb are generally mild.

Stevia

Stevia (*Stevia rebaudiana,* Astraceae family) has an ancient value as a sweetener and indeed is available as a nonnutritive sweetener based apparently upon its content of glycosides of *ent*-kaurene. In addition, extracts have shown significant broad-spectrum in vitro antibacterial activity based at least in part upon their content of the flavanol glycoside, ombuoside.[72]

Styrax/Storax

Styrax, also known as storax (*Ipomoea orizabensis* L., Convolvulaceae family) produces an antimicrobially active resin containing the cyclic glycolipid orazabin (scammonin 1). This compound was narrowly anti-gram-positive active with best activity against certain *Staphylococcus aureus* strains.[73] A related plant, *Styrax ferrugineus,* has produced nor-lignans with antibacterial and antifungal activities.[74]

Terminalia

Terminalia, also known as arjuna (*Terminalia arjuna,* Retz., *T. bellirica,* and *T. chebula*), is a medicinal plant revered in Indian and Tibetan medical systems with a usage going back in ayurvedic medicine for about 3,000 years. It has a demonstrated cardiotonic activity and is used also for a plethora of other complaints. Antifungal and antiviral properties are also recorded.[75] The fruiting bodies of the plant contain significant quantities of gallic acid and methyl gallate, both of which showed activity against methicillin-resistant *Staphylococcus aureus.*[76] The tannin content of *Terminalia* extracts go a long way toward rationalizing these anti-infective properties and making it comparatively unattractive to pursue further.

Yarrow

Yarrow (*Achillea millefolium* L., Compositae family) is a medicinal plant whose original uses (in wound healing and hemostasis) go back to legendary times well before the Christian era. It has many present folkloric uses and is included in contemporary beverages and food as a flavoring agent. The oil is antibacterial and antifungal, although specific components responsible for this have not been identified. The most recent relevant study (of *A. alpina*) refers to its use in Chinese traditional medicine and confirms antimicrobial action.[77]

CONCLUSION

It is amply clear that antimicrobial agents are plentiful in higher plants. In many of these cases there has been insufficient chemical study to identify the active constituents and to provide detailed comparative antimicrobial and pharmacological evidence in favor of their emergence as regular medications. Except where noted, however, the history of human consumption suggests that when used carefully and not to excess, they may do some good in mild cases and are unlikely to do much harm. Where possible, however, resorting to more powerful and specific antibiotics is preferred when infection takes place.

The literature is replete with reports of higher plants that have antimicrobial constituents but where no in vivo studies have been performed or where the results have been unconvincing. These reports are not reviewed in this chapter because there is no clear folkloric association with the majority of this work. Unless or until such detailed studies are performed, one cannot recommend use of such agents under any circumstances.

Whether a new and useful antibacterial, antifungal, antiprotozol, or antiviral agent rests obscurely in this literature or among the agents covered in this review is conjectural. It is clearly, however, not impossible.

NOTES

1. Maiti, M., Molecular aspects on the interaction of sanguinarine with B-form, Z-form and HL-form DNA structures. *Indian J Biochem Biophys* 2001, 38, (1-2), 20-26; Leung, A. Y., *Encyclopedia of Common Natural Ingredients Used in Food, Drugs and Cosmetics*. J. Wiley and Sons: New York, 1980; Jampani, H. B.; Keschavarz-Shokri, A.; Morton, M. D.; Mitscher, L. A.; Shawar, R.; Baker, W. R., Novel antitubercular natural products: Berberine, related alkaloids, hycandam, a new protoberberine lactam and two new quinic acid esters, hycandinic acid esters-1 and -2, from the roots of *Hydrastis canadensis*. *Journal of Natural Products* 1998, 61, 1187-1193; Okunade, A. L.; Hufford, C. D.; Richardson, M. D.; Peterson, J. R.; Clark, A. M., Antimicrobial properties of alkaloids from Xanthorhiza simplicissima. *J Pharm Sci* 1994, 83, (3), 404-406; McCutcheon, A. R.; Ellis, S. M.; Hancock, R. E.; Towers, G. H., Antifungal screening of medicinal plants of British Columbian native peoples. *J Ethnopharmacol* 1994, 44, (3), 157-169; Scazzocchio, F.; Cometa, M. F.; Tomassini, L.; Palmery, M., Antibacterial activity of Hydrastis canadensis extract and its major isolated alkaloids. *Planta Med* 2001, 67, (6), 561-564; Iwasa, K.; Nanba, H.; Lee, D. U.; Kang, S. I., Structure-activity relationships of protoberberines having antimicrobial activity. *Planta Med* 1998, 64, (8), 748-751.

2. Pepeljnjak, S.; Petricic, J., The antimicrobic effect of berberine and tinctura berberidis. *Pharmazie* 1992, 47, (4), 307-308.

3. Amin, A. H.; Subbaiah, T. V.; Abbasi, K. M., Berberine sulfate: Antimicrobial activity, bioassay, and mode of action. *Can J Microbiol* 1969, 15, (9), 1067-1076.

4. Stermitz, F. R.; Lorenz, P.; Tawara, J. N.; Zenewicz, L. A.; Lewis, K., Synergy in a medicinal plant: Antimicrobial action of berberine potentiated by 5'-methoxyhydnocarpin, a multidrug pump inhibitor. *Proc Natl Acad Sci U S A* 2000, 97, (4), 1433-1437; Khin Maung, U.; Myo, K.; Nyunt Nyunt, W.; Aye, K.; Tin, U., Clinical trial of berberine in acute watery diarrhoea. *Br Med J (Clin Res Ed)* 1985, 291, (6509), 1601-1605; Khin Maung, U.; Myo, K.; Nyunt Nyunt, W.; Tin, U., Clinical trial of high-dose berberine and tetracycline in cholera. *J Diarrhoeal Dis Res* 1987, 5, (3), 184-187.

5. Newall, C., *Herbal Medicines*. Pharmaceutical Press: London, 1996, pp. 151-152.

6. Bonesvoll, P.; Gjermo, P., A comparision between chlorhexidine and some quaternary ammonium compounds with regard to retention, salivary concentration and plaque-inhibiting effect in the human mouth after mouth rinses. *Arch Oral Biol* 1978, 23, (4), 289-294; Dzink, J. L.; Socransky, S. S., Comparative in vitro activity of sanguinarine against oral microbial isolates. *Antimicrob Agents Chemother* 1985, 27, (4), 663-665; Wolinsky, L. E.; Cuomo, J.;

Quesada, K.; Bato, T.; Camargo, P. M., A comparative pilot study of the effects of a dentifrice containing green tea bioflavonoids, sanguinarine or triclosan on oral bacterial biofilm formation. *J Clin Dent* 2000, 11, (2), 53-59; Eley, B. M., Antibacterial agents in the control of supragingival plaque—A review. *Br Dent J* 1999, 186, (6), 286-296; Godowski, K. C., Antimicrobial action of sanguinarine. *J Clin Dent* 1989, 1, (4), 96-101.

7. Newton, S. M.; Lau, C.; Gurcha, S. S.; Besra, G. S.; Wright, C. W., The evaluation of forty-three plant species for in vitro antimycobacterial activities; isolation of active constituents from Psoralea corylifolia and Sanguinaria canadensis. *J Ethnopharmacol* 2002, 79, (1), 57-67; Stiborova, M.; Simanek, V.; Frei, E.; Hobza, P.; Ulrichova, J., DNA adduct formation from quaternary benzo[c]phenanthridine alkaloids sanguinarine and chelerythrine as revealed by the 32P-postlabeling technique. *Chem Biol Interact* 2002, 140, (3), 231-242; Saran, A.; Srivastava, S.; Coutinho, E.; Maiti, M., 1H NMR investigation of the interaction of berberine and sanguinarine with DNA. *Indian J Biochem Biophys* 1995, 32, (2), 74-77.

8. Mitscher, L. A.; Park, Y. H.; Clark, D.; Clark, G. W., 3rd; Hammesfahr, P. D.; Wu, W. N.; Beal, J. L., Antimicrobial agents from higher plants. An investigation of Hunnemannia fumariaefolia pseudoalcoholates of sanguinarine and chelerythrine. *Lloydia* 1978, 41, (2), 145-150.

9. Yankell, S. L., Saliva glycolysis and plaque. *Compend Contin Educ Dent* 1984, Suppl 5, S57-60.

10. Lee, M. R., Plants against malaria. Part 1: Cinchona or the Peruvian bark. *J R Coll Physicians Edinb* 2002, 32, (3), 189-196; Basilico, N.; Tognazioli, C.; Picot, S.; Ravagnani, F.; Taramelli, D., Synergistic and antagonistic interactions between haemozoin and bacterial endotoxin on human and mouse macrophages. *Parassitologia* 2003, 45, (3-4), 135-140.

11. Fitch, C. D.; Chou, A. C., Regulation of heme polymerizing activity and the antimalarial action of chloroquine. *Antimicrob Agents Chemother* 1997, 41, (11), 2461-2465.

12. Butler, J. V.; Mulkerrin, E. C.; O'Keeffe, S. T., Nocturnal leg cramps in older people. *Postgrad Med J* 2002, 78, (924), 596-598.

13. Mirghani, R. A.; Hellgren, U.; Westerberg, P. A.; Ericsson, O.; Bertilsson, L.; Gustafsson, L. L., The roles of cytochrome P450 3A4 and 1A2 in the 3-hydroxylation of quinine in vivo. *Clin Pharmacol Ther* 1999, 66, (5), 454-460.

14. Olin, B. R.; Hebel, S. K., *Drug Facts and Comparisons*. ed.; Drug Facts and Comparisons: St. Louis, MO, 1993.

15. Morton, J. F., *Major Medicinal Plants*. ed.; Charles C Thomas, Inc.: Springfield, IL, 1977.

16. Lemmens-Gruber, R.; Studenik, C.; Karkhaneh, A.; Heistracher, P., Mechanism of sodium channel blockade in the cardiotoxic action of emetine dihydrochloride in isolated cardiac preparations and ventricular myocytes of guinea pigs. *J Cardiovasc Pharmacol* 1997, 30, (5), 554-561.

17. Flickinger, C. J., Ribosomal aggregates in amebae exposed to the protein synthesis inhibitor emetine. *Exp Cell Res* 1972, 74, (2), 541-546.

18. Conner, D. E.; Beuchat, L. R., Sensitivity of heat-stressed yeasts to essential oils of plants. *Appl Environ Microbiol* 1984, 47, (2), 229-233.

19. Klein, A. D.; Penneys, N. S., Aloe vera. *J Am Acad Dermatol* 1988, 18, (4 Pt 1), 714-20; Bruce, W. G., Investigations of antibacterial activity in the aloe. *S Afr Med J* 1967, 41, (38), 984; Ali, N. A.; Julich, W. D.; Kusnick, C.; Lindequist, U., Screening of Yemeni medicinal plants for antibacterial and cytotoxic activities. *J Ethnopharmacol* 2001, 74, (2), 173-179; Hatano, T.; Uebayashi, H.; Ito, H.; Shiota, S.; Tsuchiya, T.; Yoshida, T., Phenolic constituents of Cassia seeds and antibacterial effect of some naphthalenes and anthraquinones on methicillin-resistant Staphylococcus aureus. *Chem Pharm Bull (Tokyo)* 1999, 47, (8), 1121-1127.

20. Janssen, A. M.; Scheffer, J. J., Acetoxychavicol acetate, an antifungal component of Alpinia galanga. *Planta Med* 1985, (6), 507-511; Morita, H.; Itokawa, H., Cytotoxic and antifungal diterpenes from the seeds of Alpinia galanga. *Planta Med* 1988, 54, (2), 117-120; Ray, P. G.; Majumdar, S. K., Antifungal flavonoid from Alpinia officinarum Hance. *Indian J Exp Biol* 1976, 14, (6), 712-714; Haraguchi, H.; Kuwata, Y.; Inada, K.; Shingu, K.; Miyahara, K.; Nagao, M.; Yagi, A., Antifungal activity from Alpinia galanga and the competition for incorporation of unsaturated fatty acids in cell growth. *Planta Med* 1996, 62, (4), 308-313.

21. Trevett, A.; Lalloo, D., A new look at an old drug: Artemesinin and qinghaosu. *P N G Med J* 1992, 35, (4), 264-269.

22. Cheng, F.; Shen, J.; Luo, X.; Zhu, W.; Gu, J.; Ji, R.; Jiang, H.; Chen, K., Molecular docking and 3-D-QSAR studies on the possible antimalarial mechanism of artemisinin analogues. *Bioorg Med Chem* 2002, 10, (9), 2883-2891.

23. Avery, M. A.; Alvim-Gaston, M.; Rodrigues, C. R.; Barreiro, E. J.; Cohen, F. E.; Sabnis, Y. A.; Woolfrey, J. R., Structure-activity relationships of the antimalarial agent artemisinin. 6. The development of predictive in vitro potency models using CoMFA and HQSAR methodologies. *J Med Chem* 2002, 45, (2), 292-303.

24. Bedir, E.; Pugh, N.; Calis, I.; Pasco, D. S.; Khan, I. A., Immunostimulatory effects of cycloartane-type triterpene glycosides from astragalus species. *Biol Pharm Bull* 2000, 23, (7), 834-837.

25. Zhao, K. W.; Kong, H. Y., [Effect of Astragalan on secretion of tumor necrosis factors in human peripheral blood mononuclear cells]. *Zhongguo Zhong Xi Yi Jie He Za Zhi* 1993, 13, (5), 263-265, 259.

26. Chu, D. T.; Wong, W. L.; Mavligit, G. M., Immunotherapy with Chinese medicinal herbs. I. Immune restoration of local xenogeneic graft-versus-host reaction in cancer patients by fractionated Astragalus membranaceus in vitro. *J Clin Lab Immunol* 1988, 25, (3), 119-123.

27. Burack, J. H.; Cohen, M. R.; Hahn, J. A.; Abrams, D. I., Pilot randomized controlled trial of Chinese herbal treatment for HIV-associated symptoms. *J Acquir Immune Defic Syndr Hum Retrovirol* 1996, 12, (4), 386-393.

28. Jiratchariyakul, W.; Wiwat, C.; Vongsakul, M.; Somanabandhu, A.; Leelamanit, W.; Fujii, I.; Suwannaroj, N.; Ebizuka, Y., HIV inhibitor from Thai bitter gourd. *Planta Med* 2001, 67, (4), 350-353.

29. Hanafy, M. S.; Hatem, M. E., Studies on the antimicrobial activity of Nigella sativa seed (black cumin). *J Ethnopharmacol* 1991, 34, (2-3), 275-278.

30. Yao, X. J.; Wainberg, M. A.; Parniak, M. A., Mechanism of inhibition of HIV-1 infection in vitro by purified extract of Prunella vulgaris. *Virology* 1992, 187, (1), 56-62.

31. Kashman, Y.; Gustafson, K. R.; Fuller, R. W.; Cardellina, J. H., 2nd; McMahon, J. B.; Currens, M. J.; Buckheit, R. W., Jr.; Hughes, S. H.; Cragg, G. M.; Boyd, M. R., The calanolides, a novel HIV-inhibitory class of coumarin derivatives from the tropical rainforest tree, Calophyllum lanigerum. *J Med Chem* 1992, 35, (15), 2735-2743.

32. Patil, A. D.; Freyer, A. J.; Eggleston, D. S.; Haltiwanger, R. C.; Bean, M. F.; Taylor, P. B.; Caranfa, M. J.; Breen, A. L.; Bartus, H. R.; Johnson, R. K.; et al., The inophyllums, novel inhibitors of HIV-1 reverse transcriptase isolated from the Malaysian tree, Calophyllum inophyllum Linn. *J Med Chem* 1993, 36, (26), 4131-4138.

33. Pengsuparp, T.; Serit, M.; Hughes, S. H.; Soejarto, D. D.; Pezzuto, J. M., Specific inhibition of human immunodeficiency virus type 1 reverse transcriptase mediated by soulattrolide, a coumarin isolated from the latex of calophyllum teysmannii. *J Nat Prod* 1996, 59, (9), 839-842.

34. Dharmaratne, H. R.; Wanigasekera, W. M.; Mata-Greenwood, E.; Pezzuto, J. M., Inhibition of human immunodeficiency virus type 1 reverse transcriptase activity by cordatolides isolated from Calophyllum cordato-oblongum. *Planta Med* 1998, 64, (5), 460-461.

35. Sakagami, Y.; Kajimura, K.; Wijesinghe, W. M.; Dharmaratne, H. R., Antibacterial activity of calozeyloxanthone isolated from Calophyllum species against vancomycin-resistant Enterococci (VRE) and synergism with antibiotics. *Planta Med* 2002, 68, (6), 541-543.

36. Dumenil, G.; Chemli, R.; Balansard, C.; Guiraud, H.; Lallemand, M., Evaluation of antibacterial properties of marigold flowers (Calendula officinalis L.) and mother homeopathic tinctures of C. officinalis L. and C. arvensis L. *Ann Pharm Fr* 1980, 38, (6), 493-499.

37. De Tommasi, N.; Conti, C.; Stein, M. L.; Pizza, C., Structure and in vitro antiviral activity of triterpenoid saponins from Calendula arvensis. *Planta Med* 1991, 57, (3), 250-253; De Tommasi, N.; Pizza, C.; Conti, C.; Orsi, N.; Stein, M. L., Structure and in vitro antiviral activity of sesquiterpene glycosides from Calendula arvensis. *J Nat Prod* 1990, 53, (4), 830-835.

38. Wagner, H.; Kreutzkamp, B.; Jurcic, K., [The alkaloids of Uncaria tomentosa and their phagocytosis-stimulating action]. *Planta Med* 1985, (5), 419-423.

39. Pattnaik, S.; Subramanyam, V. R.; Kole, C., Antibacterial and antifungal activity of ten essential oils in vitro. *Microbios* 1996, 86, (349), 237-246.

40. Cai, L.; Wu, C. D., Compounds from Syzygium aromaticum possessing growth inhibitory activity against oral pathogens. *J Nat Prod* 1996, 59, (10), 987-990; Dorman, H. J.; Deans, S. G., Antimicrobial agents from plants: Antibacterial activity of plant volatile oils. *J Appl Microbiol* 2000, 88, (2), 308-316; Shafi, P. M.; Rosamma, M. K.; Jamil, K.; Reddy, P. S., Antibacterial activity of the essential oil from Aristolochia indica. *Fitoterapia* 2002, 73, (5), 439-441.

41. Lee, Y. L.; Owens, J.; Thrupp, L.; Cesario, T. C., Does cranberry juice have antibacterial activity? *Jama* 2000, 283, (13), 1691.

42. Bauer, R.; Wagner, H., Echinacea-species as potential immunostimulatory drugs. In *Phytomedicines of Europe: Chemistry and Biologic Activity*, ed.; Farnsworth, N. R.; Wagner, H. Eds. Academic Press: New York, 1991.

43. Kabelik, J., [Antimicrobial properties of garlic]. *Pharmazie* 1970, 25, (4), 266-270; Didry, N.; Pinkas, M.; Dubreuil, L., [Antibacterial activity of species of the genus Allium]. *Pharmazie* 1987, 42, (10), 687-688.

44. Avato, P.; Tursil, E.; Vitali, C.; Miccolis, V.; Candido, V., Allylsulfide constituents of garlic volatile oil as antimicrobial agents. *Phytomedicine* 2000, 7, (3), 239-243; Tsao, S. M.; Yin, M. C., In-vitro antimicrobial activity of four diallyl sulphides occurring naturally in garlic and Chinese leek oils. *J Med Microbiol* 2001, 50, (7), 646-649; Harris, J. C.; Cottrell, S. L.; Plummer, S.; Lloyd, D., Antimicrobial properties of Allium sativum (garlic). *Appl Microbiol Biotechnol* 2001, 57, (3), 282-286.

45. Yoshida, S.; Kasuga, S.; Hayashi, N.; Ushiroguchi, T.; Matsuura, H.; Nakagawa, S., Antifungal activity of ajoene derived from garlic. *Appl Environ Microbiol* 1987, 53, (3), 615-617; Naganawa, R.; Iwata, N.; Ishikawa, K.; Fukuda, H.; Fujino, T.; Suzuki, A., Inhibition of microbial growth by ajoene, a sulfur-containing compound derived from garlic. *Appl Environ Microbiol* 1996, 62, (11), 4238-4242.

46. Yoshida, H.; Katsuzaki, H.; Ohta, R.; Ishikawa, K.; Fukuda, H.; Fujino, T.; Suzuki, A., An organosulfur compound isolated from oil-macerated garlic extract, and its antimicrobial effect. *Biosci Biotechnol Biochem* 1999, 63, (3), 588-590.

47. Kitayama, T.; Yamamoto, K.; Utsumi, R.; Takatani, M.; Hill, R. K.; Kawai, Y.; Sawada, S.; Okamoto, T., Chemistry of zerumbone. 2. Regulation of ring bond cleavage and unique antibacterial activities of zerumbone derivatives. *Biosci Biotechnol Biochem* 2001, 65, (10), 2193-2199.

48. Yoshikawa, M.; Matsuda, H.; Shimoda, H.; Shimada, H.; Harada, E.; Naitoh, Y.; Miki, A.; Yamahara, J.; Murakami, N., Development of bioactive functions in hydrangeae dulcis folium. V. On the antiallergic and antimicrobial principles of hydrangeae dulcis folium. (2). Thunberginols C, D, and E, thunberginol G 3'-O-glucoside, (-)-hydrangenol 4'-o-glucoside, and (+)-hydrangenol 4'-O-glucoside. *Chem Pharm Bull (Tokyo)* 1996, 44, (8), 1440-1447; Yoshikawa, M.; Uchida, E.; Chatani, N.; Kobayashi, H.; Naitoh, Y.; Okuno, Y.; Matsuda, H.; Yamahara, J.; Murakami, N., Thunberginols C, D, and E, new antiallergic and antimicrobial dihydroisocoumarins, and thunberginol G 3'-O-glucoside and (-)-hydrangenol 4'-O-glucoside, new dihydroisocoumarin glycosides, from Hydrangeae Dulcis Folium. *Chem Pharm Bull (Tokyo)* 1992, 40, (12), 3352-3354; Yoshikawa, M.; Uchida, E.; Chatani, N.; Murakami, N.; Yamahara, J., Thunberginols A, B, and F, new antiallergic and antimicrobial principles from hydrangeae dulcis folium. *Chem Pharm Bull (Tokyo)* 1992, 40, (11), 3121-3123; Ishih, A.; Ikeya, C.; Yanoh, M.; Takezoe, H.; Miyase, T.; Terada, M., A potent antimalarial activity of Hydrangea macrophylla var. Otaksa leaf extract against Plasmodium yoelii 17XL in mice. *Parasitol Int* 2001, 50, (1), 33-39.

49. Gollapudi, S.; Sharma, H. A.; Aggarwal, S.; Byers, L. D.; Ensley, H. E.; Gupta, S., Isolation of a previously unidentified polysaccharide (MAR-10) from Hyssop officinalis that exhibits strong activity against human immunodeficiency virus type 1. *Biochem Biophys Res Commun* 1995, 210, (1), 145-151; Kreis, W.; Kaplan, M. H.; Freeman, J.; Sun, D. K.; Sarin, P. S., Inhibition of HIV replication by Hyssop officinalis extracts. *Antiviral Res* 1990, 14, (6), 323-337.

50. Currier, N. L.; Lejtenyi, D.; Miller, S. C., Effect over time of in-vivo administration of the polysaccharide arabinogalactan on immune and hemopoietic cell lineages in murine

spleen and bone marrow. *Phytomedicine* 2003, 10, (2-3), 145-153; Kim, L. S.; Waters, R. F.; Burkholder, P. M., Immunological activity of larch arabinogalactan and Echinacea: A preliminary, randomized, double-blind, placebo-controlled trial. *Altern Med Rev* 2002, 7, (2), 138-149.

51. Duke, J. A., *Handbook of Medicinal Herbs*. ed.; CRC Press: Boca Raton, FL, 1985.

52. Hernandez, N. E.; Tereschuk, M. L.; Abdala, L. R., Antimicrobial activity of flavonoids in medicinal plants from Tafi del Valle (Tucuman, Argentina). *J Ethnopharmacol* 2000, 73, (1-2), 317-322.

53. Fukai, T.; Marumo, A.; Kaitou, K.; Kanda, T.; Terada, S.; Nomura, T., Antimicrobial activity of licorice flavonoids against methicillin-resistant Staphylococcus aureus. *Fitoterapia* 2002, 73, (6), 536-539; Fukui, H.; Goto, K.; Tabata, M., Two antimicrobial flavanones from the leaves of Glycyrrhiza glabra. *Chem Pharm Bull (Tokyo)* 1988, 36, (10), 4174-4176; Tsukiyama, R.; Katsura, H.; Tokuriki, N.; Kobayashi, M., Antibacterial activity of licochalcone A against spore-forming bacteria. *Antimicrob Agents Chemother* 2002, 46, (5), 1226-1230; Mitscher, L. A.; Park, Y. H.; Clark, D.; Beal, J. L., Antimicrobial agents from higher plants. Antimicrobial isoflavanoids and related substances from Glycyrrhiza glabra L. var. typica. *J Nat Prod* 1980, 43, (2), 259-269.

54. Orabi, K. Y.; Mossa, J. S.; el-Feraly, F. S., Isolation and characterization of two antimicrobial agents from mace (Myristica fragrans). *J Nat Prod* 1991, 54, (3), 856-859.

55. Marone, P.; Bono, L.; Leone, E.; Bona, S.; Carretto, E.; Perversi, L., Bactericidal activity of Pistacia lentiscus mastic gum against Helicobacter pylori. *J Chemother* 2001, 13, (6), 611-614.

56. Rauha, J. P.; Remes, S.; Heinonen, M.; Hopia, A.; Kahkonen, M.; Kujala, T.; Pihlaja, K.; Vuorela, H.; Vuorela, P., Antimicrobial effects of Finnish plant extracts containing flavonoids and other phenolic compounds. *Int J Food Microbiol* 2000, 56, (1), 3-12.

57. Vanka, A.; Tandon, S.; Rao, S. R.; Udupa, N.; Ramkumar, P., The effect of indigenous Neem Azadirachta indica [correction of (Adirachta indica)] mouth wash on Streptococcus mutans and lactobacilli growth. *Indian J Dent Res* 2001, 12, (3), 133-144.

58. Siddiqui, S.; Faizi, S.; Siddiqui, B. S.; Ghiasuddin, Constituents of Azadirachta indica: Isolation and structure elucidation of a new antibacterial tetranortriterpenoid, mahmoodin, and a new protolimonoid, naheedin. *J Nat Prod* 1992, 55, (3), 303-310.

59. Guarrera, P. M., Traditional antihelmintic, antiparasitic and repellent uses of plants in Central Italy. *J Ethnopharmacol* 1999, 68, (1-3), 183-192.

60. Kim, J. W.; Kim, Y. S.; Kyung, K. H., Inhibitory activity of essential oils of garlic and onion against bacteria and yeasts. *J Food Prot* 2004, 67, (3), 499-504.

61. Zohri, A. N.; Abdel-Gawad, K.; Saber, S., Antibacterial, antidermatophytic and antitoxigenic activities of onion (Allium cepa L.) oil. *Microbiol Res* 1995, 150, (2), 167-172.

62. Nicolls, J. M.; Birner, J.; Forsell, P., Passicol, an antibacterial and antifungal agent produced by Passiflora plant species: Qualitative and quantitative range of activity. *Antimicrob Agents Chemother* 1973, 3, (1), 110-117; Birner, J.; Nicolls, J. M., Passicol, an antibacterial and antifungal agent produced by Passiflora plant species: Preparation and physicochemical characteristics. *Antimicrob Agents Chemother* 1973, 3, (1), 105-109.

63. Imai, H.; Osawa, K.; Yasuda, H.; Hamashima, H.; Arai, T.; Sasatsu, M., Inhibition by the essential oils of peppermint and spearmint of the growth of pathogenic bacteria. *Microbios* 2001, 106 Suppl 1, 31-39.

64. Ajaiyeoba, E. O.; Abalogu, U. I.; Krebs, H. C.; Oduola, A. M., In vivo antimalarial activities of Quassia amara and Quassia undulata plant extracts in mice. *J Ethnopharmacol* 1999, 67, (3), 321-325.

65. Wolters, B.; Eilert, U., Antimicrobial substances in callus cultures of Ruta graveolens. *Planta Med* 1981, 43, (2), 166-174.

66. Murray, M., In *The healing power of herbs*, 2nd ed. Prima Publishing: Rocklin, CA, 1995; pp 302-305.

67. Caceres, A.; Cano, O.; Samayoa, B.; Aguilar, L., Plants used in Guatemala for the treatment of gastrointestinal disorders. 1. Screening of 84 plants against enterobacteria. *J Ethnopharmacol* 1990, 30, (1), 55-73.

68. Sonboli, A.; Fakhari, A.; Kanani, M. R.; Yousefzadi, M., Antimicrobial activity, essential oil composition and micromorphology of trichomes of Satureja laxiflora C. Koch from Iran. *Z Naturforsch [C]* 2004, 59, (11-12), 777-781.

69. Li, B. Q.; Fu, T.; Dongyan, Y.; Mikovits, J. A.; Ruscetti, F. W.; Wang, J. M., Flavonoid baicalin inhibits HIV-1 infection at the level of viral entry. *Biochem Biophys Res Commun* 2000, 276, (2), 534-538.

70. Gulick, R. M.; McAuliffe, V.; Holden-Wiltse, J.; Crumpacker, C.; Liebes, L.; Stein, D. S.; Meehan, P.; Hussey, S.; Forcht, J.; Valentine, F. T., Phase I studies of hypericin, the active compound in St. John's Wort, as an antiretroviral agent in HIV-infected adults. AIDS Clinical Trials Group Protocols 150 and 258. *Ann Intern Med* 1999, 130, (6), 510-514.

71. Kizil, G.; Toker, Z.; Ozen, H. C.; Aytekin, C., The antimicrobial activity of essential oils of Hypericum scabrum, Hypericum scabroides and Hypericum triquetrifolium. *Phytother Res* 2004, 18, (4), 339-341; Schempp, C. M.; Pelz, K.; Wittmer, A.; Schopf, E.; Simon, J. C., Antibacterial activity of hyperforin from St John's wort, against multiresistant Staphylococcus aureus and gram-positive bacteria. *Lancet* 1999, 353, (9170), 2129.

72. Amaro-Luis, J. M.; Adrian, M.; Diaz, C., Isolation, identification and antimicrobial activity of ombuoside from Stevia triflora. *Ann Pharm Fr* 1997, 55, (6), 262-268.

73. Mitscher, L. A.; Telikepalli, H., Bioassay-directed discovery of natural product leads. Antibacterials and antifungals from unusual sources. In *Advances in natural product chemistry*, ed. Atta-ur-Rahman. Harwood Academic Publishers: Basel, 1992, pp. 281-310.

74. Pauletti, P. M.; Araujo, A. R.; Young, M. C.; Giesbrecht, A. M.; Bolzani, V. D., nor-Lignans from the leaves of Styrax ferrugineus (Styracaceae) with antibacterial and antifungal activity. *Phytochemistry* 2000, 55, (6), 597-601.

75.Gupta, M.; Mazumder, U. K.; Manikandan, L.; Bhattacharya, S.; Haldar, P. K.; Roy, S., Antibacterial activity of Terminalia pallida. *Fitoterapia* 2002, 73, (2), 165-167; 114.Suguna, L.; Singh, S.; Sivakumar, P.; Sampath, P.; Chandrakasan, G., Influence of Terminalia chebula on dermal wound healing in rats. *Phytother Res* 2002, 16, (3), 227-31; .Ahmad, I.; Mehmood, Z.; Mohammad, F., Screening of some Indian medicinal plants for their antimicrobial properties. *J Ethnopharmacol* 1998, 62, (2), 183-193.

76. Sato, Y.; Oketani, H.; Singyouchi, K.; Ohtsubo, T.; Kihara, M.; Shibata, H.; Higuti, T., Extraction and purification of effective antimicrobial constituents of Terminalia chebula RETS. against methicillin-resistant Staphylococcus aureus. *Biol Pharm Bull* 1997, 20, (4), 401-404.

77. Liu, M. L.; Li, M. L.; Hu, S. K., Study of antibacterial and antiinflammatory components of Achillea alpina. *J Tradit Chin Med* 1983, 3, (3), 213-216.

Chapter 17

Use of Medicinal Plants in CNS Disorders

Peter John Houghton

INTRODUCTION

The Central Nervous System (CNS)

The CNS comprises the brain and spinal cord and is an amazingly complex part of the mammalian body. In human beings it has developed to such an extent that it occupies a much greater proportion of body mass than it does any other creature. Although thought processes and reason are most commonly associated with the CNS, it should not be forgotten that almost every aspect of physiological function is affected by CNS activity and that "brain death" is widely accepted as the definition of the end of human life.

The CNS consists of an interconnected series of cells that communicate with one another by a series of electrical and chemical impulses. The actual link between one cell and another commonly consists of the release of chemical compounds, called neurotransmitters, from the donor cell, which bind to receptors on the acceptor cell. Homeostasis of levels of chemical mediators is controlled by complex feedback systems involving breakdown of the mediators, reuptake into the donor cell, and competition from chemicals acting on different receptors that produce an opposing effect or that block the usual receptor.

Neurotransmitters and CNS Function

Neurotransmitters are of several chemical types, and some of them are important in neurotransmission and neuromuscular transmission in the peripheral nervous system. Some neurotransmitters occur throughout the CNS while others are more localized, and they may stimulate or inhibit information transmission. The particular activity of a neurotransmitter varies according to the type of receptor upon which it acts. An increasing number of subtypes within the CNS are being discovered. The chemical structures of many drugs acting on the CNS show similarities to particular transmitters, but it should be emphasised that agreement about the chemical basis of many CNS disorders is still not resolved and thus a fully rational approach to treatment of the majority of conditions is not yet possible.

Scope of This Chapter

Malfunction or disease of the CNS may be shown by symptoms in other parts of the body but, for the purposes of this chapter, only those CNS disorders that are seen by an effect on activities mainly associated with the CNS, for example, cognition, alertness, sedation, will be considered as targets for treatment with plants and their constituents. Since the activity of a plant or its extract relies fundamentally on its constituent chemicals, most of the plants mentioned here affect the CNS by providing compounds that act in the same, or opposite, way as the chemical transmitters found in the brain, which accelerates or inhibits the chemical transformations in the CNS or which directly or indirectly protects the CNS against harmful chemicals or processes.

Almost every culture in the world has used plants in medicine or as toxins that affect the CNS, but often it is difficult to discern medical treatment of CNS disorders. This is partly due to the confusion, until fairly recently even in Western thought, between the physical and spiritual aspects of manifestations of disorders of the CNS. Plant materials alone were very rarely used; it was much more common for prayers, exorcism, divination, and other religious practices to be used concurrently or exclusively. In addition, CNS disorders could be categorized only crudely by their symptoms and not in biochemical or clinical terms such as is done in orthodox medicine today. Nevertheless, a corpus of information has accumulated in most cultures about plants that have an effect on the CNS which have consequently been applied to the treatment of disease. A detailed treatment of such plants and their constituents is not possible in this chapter but can be found in several recent texts.[1]

Some aspects of the interaction between plants and the CNS will not be considered in any direct way. These include the use of plants as hallucinogens or as drugs of abuse. In the former case, although these are used in religious, superstitious, social, and medical contexts in many, if not all, societies, there is, as yet, little place for hallucinogens in the treatment of patients in either orthodox or herbal medicine in Western culture. The other category not considered, drugs of abuse, is now practically synonymous with the term *drug* in present Western culture and is taken by large numbers of people not as preventatives or cures for disease but as "recreational" drugs to relieve the pressures or boredom of ordinary existence. Although the majority of those widely used are plant material, or are extracted from plants, their use is not considered to be medicinal—although some substances are mentioned because they are also used therapeutically.

It should also be pointed out that many drugs act on the CNS but are used for relief of symptoms in other parts of the body. Thus vomiting, associated with the gastrointestinal system, is frequently controlled by drugs, such as scopolamine, that suppress the CNS, and coughing, a respiratory tract complaint, by suppressants of the cough center in the CNS, such as codeine. These substances will be considered briefly at the end of the chapter.

DISORDERS OF THE CNS

General Considerations

The CNS is so complex, both in structure and function, that it is extremely vulnerable to factors causing it to malfunction. It is actually surprising that, in most individuals, it appears to function at acceptable levels most of the time! Disorders can be categorized in various ways, such as their causes, but it is generally most convenient to classify them either in terms of their reversal of a state not considered to be a reflection of good health (e.g., antidepressants) or for the treatment of a general disease state (e.g., neurodegeneration).

Precise diagnosis of a CNS disorder is often difficult and is often defined by symptoms rather than changes in histology or brain chemistry. Obtaining material from the patient for analyzing these latter features presents many problems and in some cases, for example, dementia, a precise identification of the type can only be made postmortem.

The disorders that can occur are categorized as follows:

- Neurodegenerative diseases
- Diseases associated with electrophysiological malfunction
- Suppression of the CNS
- Overexcitement of the CNS
- Conditions displaying excessive aspects of normal function, for example, algesia, migraine, nausea, and vomiting

Neurodegenerative Diseases

Diseases in which the physical and chemical structure of the brain degenerates in a progressive manner have received much attention in recent years in industrially developed countries. The major classes are (1) the various dementias associated with loss of cognitive function and memory and (2) Parkinson's disease. Alzheimer's disease is the best-known example of the first type, and the tragic loss of memory and changes in personality over time is well known due to common experience and attention from the national media.

Diseases Associated with Electrophysiological Malfunction

The flow of electrical currents and impulses in the brain follows regular patterns and these can be recorded on an electroencephalograph (EEG). Patterns observed in healthy individuals are different or disturbed in several CNS disorders. The link between abnormal electrical activity and levels of chemical substances is still largely unexplored, but abnormal "firing" of electrical impulses is best known because of its association with epilepsy. Although some plants are used for the treatment of epilepsy in other cultures, there is no substantial use of plants or their constituents in either Western herbal or orthodox medicine.

Suppression of the CNS

Suppression of the CNS may be due to a variety of factors and is usually associated with lack of motor activity, inability to concentrate, apathy toward external factors, drowsiness, and sleep. Subclinical conditions are part of normal life for most people due to physical or mental exertion, loss of sleep, emotional disturbance, changes in body chemistry (especially hormone profiles), or a poor diet. Plants or their extracts are used in most cultures to give a mild stimulant effect to the CNS, which eliminates the suppression and helps concentration for tasks such as driving and social interactions.

More severe suppression of the CNS is associated with depressive disorders that have a debilitating effect on the normal social activities of the sufferer and may even affect his or her physical health. Depression has been said to be a major health problem for the industrialized world for the twenty-first century. Several classes of synthetic drugs have been developed as antidepressants, but it is only recently that much attention has been paid to plant products that achieve the same ends.

Overexcitement of the CNS

Overexcitement of the CNS results in frenzied physical activity, inability to concentrate, and irrational behavior and thoughts. Mania is the collective name given to various mixtures of these symptoms. In many cases the patient uses large amounts of energy and eventually suffers exhaustion. Another aspect of overexcitement of the CNS is increased anxiety, often accompanied by sleeplessness and difficulty in carrying out the activities of normal life.

In the past, such conditions were treated with sedatives and hypnotics, which tended to suppress all CNS activity so that the patient was reduced to a drowsy state most of the time. During the past fifty years, this approach has been largely abandoned with the introduction of tranquilizers and anxiolytics. These drugs have a more selective suppressant effect and enable patients to remain relatively alert and to be available for nonchemical means of treating their conditions, such as psychological counseling. Their widespread use has stimulated much debate about habit-forming and psychological dependence, as well as concern over the negative effects of long-term use.

Conditions Displaying Unwanted Aspects of Normal Function

Algesia

Algesia, the sensation of pain, is a necessary defensive mechanism of the body as a protection from or limitation of physical damage from external sources. It also serves as a signal of many types of dysfunction within the body. The sensation of pain, often called nociception, is connected with CNS activity, and the feeling of pain can be reduced or abolished by suppression of the appropriate centers in the CNS. Although this may be undesirable in some situations, since harm may occur to the subject, in many acute and chronic disease states pain relief is an important part of treatment with regard to the patient's quality of life.

Migraine

Migraine is a particular type of pain, also incorporating nausea and giddiness, focused in the head. It is obvious from some of the symptoms that CNS activity is affected. Its etiology is still not clear since different types of substance and stress situations appear to act as triggers according to the individual patient. Blood flow and levels of the neurotransmitter 5-hydroxytryptamine (5HT) appear to be important in the occurrence and treatment of migraine attacks.

Nausea and Vomiting

Nausea and vomiting are similar to algesia in the sense that they are important aspects of normal body function in protecting the body from harmful ingested substances. However, they can be inconvenient in some situations, such as travel, pregnancy, radiation therapy, and medication, and it may desirable to suppress them, one method being the suppression of the center in the brain that triggers vomiting.

PLANTS AND DERIVED COMPOUNDS USED TO TREAT NEURODEGENERATIVE DISEASES

The two most common neurodegenerative diseases encountered are Alzheimer's disease (AD) and Parkinson's disease (PD)—although several diseases of the Alzheimer's type exist with similar presenting symptoms and a precise diagnosis can only be made postmortem in many cases by chemical and histological analysis of the brain. Both of these diseases are common aspects of aging and, with the increasing proportion of older people in many societies, the care of those suffering from these conditions is an increasing strain on resources of time, personnel, and money.

Alzheimer's Disease

Alzheimer's disease is characterized by increasing loss of memory and cognitive function and, at an anatomical level, by plaques of a protein called amyloid and by neurofibrillary tangles. The main target of current chemotherapy is the fact that levels of the neurotransmitter acetylcholine are low and can be enhanced by inhibiting the breakdown of this compound by inhibition of acetylcholinesterase, the enzyme responsible. Although symptoms similar to those of AD have been known in elderly patients for several centuries, the disease was first described in 1907 and seems to be much more prevalent now than historically, although the reasons for this are debated. Possibly because of a lower incidence in the past, there are comparatively little ethnopharmacological data for treatment of AD but more symptoms, such as loss of memory. Cholinesterase-inhibiting drugs based on natural sources have recently been introduced for symptomatic relief of AD in its early stages, but these were not primarily selected for ethnopharmacological reasons.

Alkaloids in Clinical Use

The classic cholinesterase inhibitor is physostigmine (see Figure 17.1), or eserine, originally isolated from the Nigerian ordeal poison Calabar bean *Physostigma venenosum* Bal. and for years used mainly to treat myasthenia gravis or in ophthalmology to decrease the size of the pupil. When the clinical potential of the cholinesterase inhibitors in AD became realized, it was noted that physostigmine itself was not very suitable since it had unfavorable pharmacokinetics in having a short half-life. Many analogous compounds were designed to overcome this and the one which has been marketed is rivastigmine (see Figure 17.2) (Exelon), which was introduced in 2000 with substantial clinical support.[2]

More recently, a naturally occurring alkaloid, galantamine (see Figure 17.3), has also been licensed in Europe and the United States. This alkaloid was originally isolated from the bulbs of snowdrops (*Galanthus* spp.) by Bulgarian workers. Interest in its cholinesterase inhibitory properties and

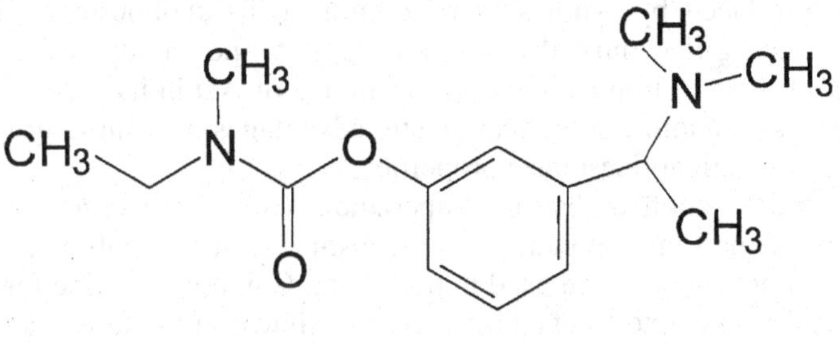

FIGURE 17.1. The chemical structure of physostigmine.

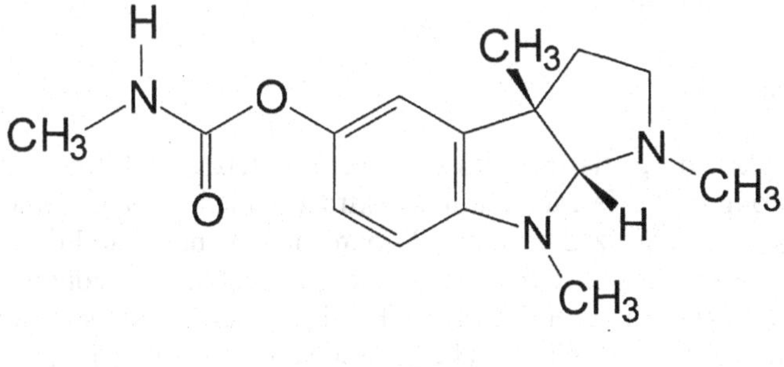

FIGURE 17.2. The chemical structure of rivastigmine.

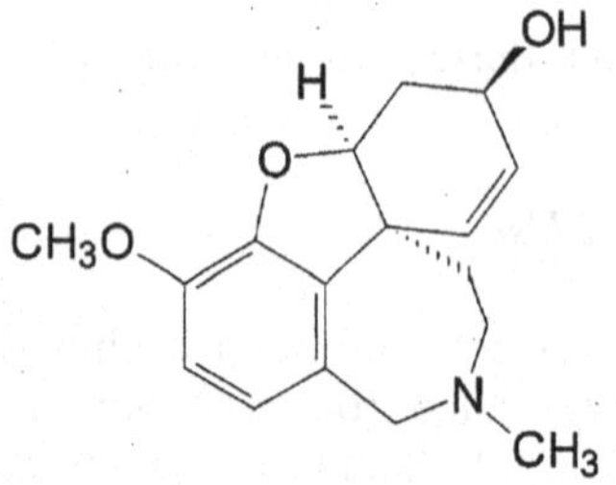

FIGURE 17.3. The chemical structure of galantamine.

potential in treating AD led to commercial development in the United Kingdom, particularly since it was found in the much larger bulbs of species of *Narcissus,* commonly called daffodils. These were already grown on a large scale in the United Kingdom for horticultural purposes, so it was comparatively easy to produce large amounts of the alkaloid for the clinical trials that led to the licensing of the drug. An important clinical study showed that it was well tolerated and significantly improved cognitive function.[3] Although galantamine was originally extracted from the harvested bulbs, it is now produced by synthesis. In addition to its cholinesterase inhibitory properties, galantamine also possesses nicotinergic activity, which is thought to contribute to improvement in symptoms of AD in its early stages, thus giving galantamine some therapeutic advantages over those drugs relying solely on anticholinesterase properties.[4]

In China, another alkaloid, huperzine A (see Figure 17.4), is used clinically for the same purpose. It is a constituent of the club moss *Huperzia serrata,* which has been used in traditional Chinese medicine for hundreds of years for symptoms of aging. A recent clinical study showed a significant improvement in memory and behavior in AD patients when huperzine A was given. It appears to be less toxic than the synthetic compounds donepezil and tacrine.[5]

Ginkgo

A plant drug that has attracted some interest as a treatment of early symptoms of AD is a standardized extract of the leaves of *Ginkgo biloba* L. Extracts standardized to about 24 percent flavonoids and 6 percent ginkgolide terpenoids have been shown to be beneficial in reducing the symptoms of aging in several studies. The improvements noted have been ascribed to a range of relevant effects associated with the compounds present. Those of major importance include the vasorelaxant and antioxidant effects

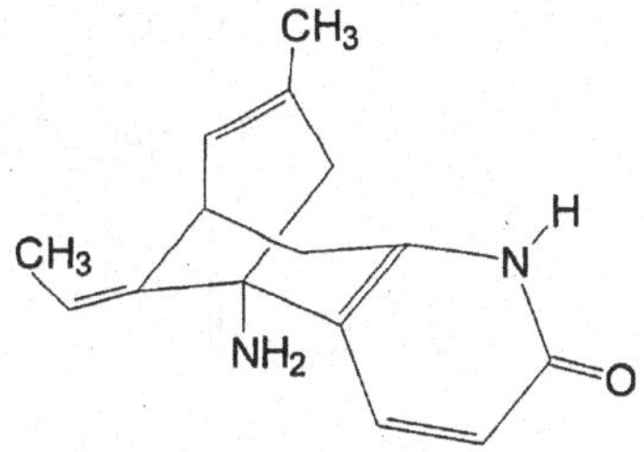

FIGURE 17.4. The chemical structure of huperzine A.

of the flavonoids and the antithrombotic (via inhibition of platelet-aggregating factor [PAF]) and antioxidant effects of the ginkgolides and bilobalide. Two recent studies specifically aimed at patients displaying characteristic symptoms of AD have shown some beneficial effects, although a recently published study in healthy volunteers showed no significant improvements in the group administered ginkgo extract.[6]

Sage

Sage species, *Salvia officinalis* L. and *S. lavandulaefolia* Vahl., are traditional herbal plants that have attracted recent interest. Sixteenth- and seventeenth-century herbals mentioned the reputation of sage as being good for the memory. The oil has been found to inhibit cholinesterase in vitro and in vivo in the CNS, and the major compounds responsible have been shown to be the monoterpenes 1,8-cineole and related compounds.[7] Other activities such as an ability to bind to estrogen receptors and antioxidant effects, which could reduce the likelihood of AD occurring, have also been found in *S. lavandulaefolia* oil and leaves.[8]

Parkinson's Disease

Parkinson's disease is another common neurodegenerative condition found in elderly people. It is characterized by tremor and shuffling gait and by decreased congitive processes and memory, and it is linked to low levels of the neurotransmitter L-dopamine in the striatum in the CNS.

The major therapeutic approach is to elevate the levels of dopamine by either inhibition of monoamine oxidase (MAO), which metabolizes dopamine to less active compounds, or by increasing the concentration of the precursor of dopamine by administering L-DOPA (see Figure 17.5). This compound is found in amounts in various species of bean, notably *Mucuna*

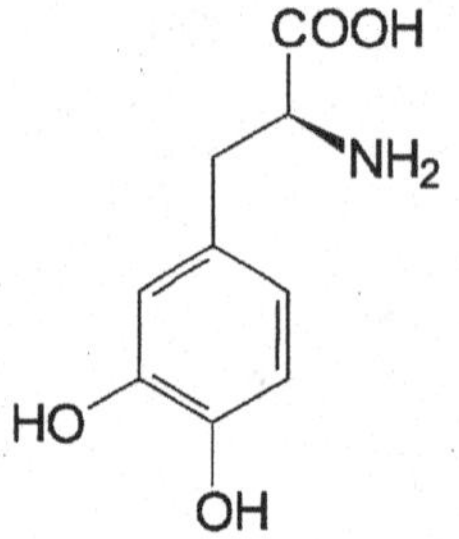

FIGURE 17.5. The chemical structure of L-DOPA.

spp., in sufficient concentrations to make their growing and harvesting economically worthwhile, although the drug is now mainly obtained by synthesis.

The tropane alkaloids, especially hyoscine (scopolamine) (see Figure 17.6), have been used to treat PD since they antagonize cholinergic activity at the muscarinic receptors in the striatum and so increase dopamine activity. The naturally occurring alkaloids are not much used for this purpose, but synthetic derivatives and analogs such as benzatropine are used which also inhibit dopamine reuptake.

Another group of compounds used for the treatment of PD are derivatives of the ergot indole alkaloids. These act on a variety of receptors in the CNS and have a history of causing hallucinations and psychiatric disturbances in populations in which rye contaminated with ergot has been ingested. The naturally occurring alkaloids are not specific enough as stimulators of the dopamine receptors but several semisynthetic derivatives, for example, bromocryptine, pergolide are used clinically since they have a more specific dopaminergic action.[9]

PLANTS AND DERIVED COMPOUNDS USED AS ANXIOLYTICS, TRANQUILIZERS, AND SEDATIVES

Reserpine

Most traditions of medicine include plants that are used as tranquilizers and sedatives. In orthodox Western medicine, most of the drugs used are synthetic compounds, and historically chlorpromazine, the barbiturates, and benzodiazepines have been widely used since the 1950s. However, it is not always realized that the first true tranquilizer introduced into Western medicine was the natural product reserpine (see Figure 17.7), derived from

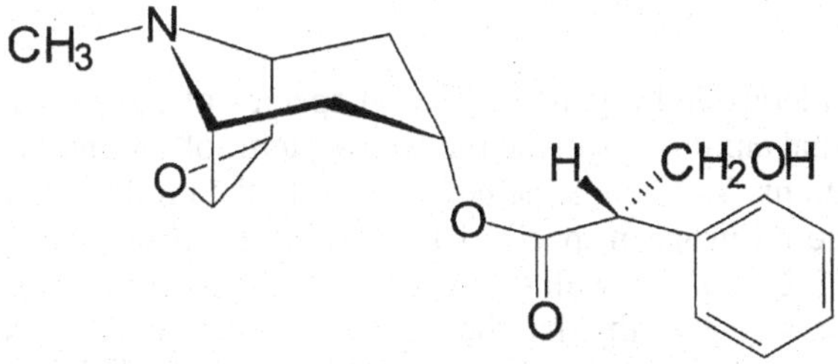

FIGURE 17.6. The chemical structure of hyoscine.

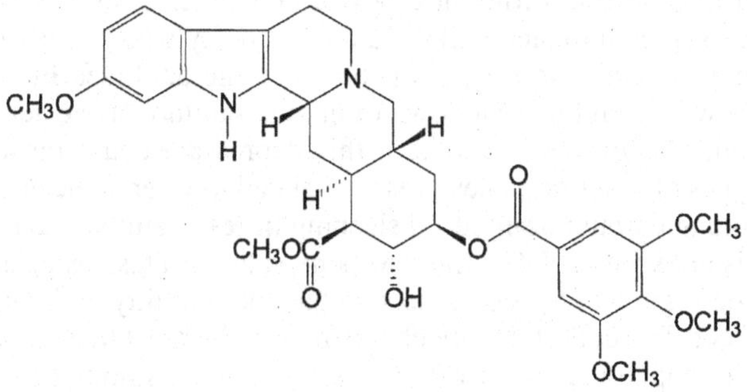

FIGURE 17.7. The chemical structure of reserpine.

the ancient ayurvedic drug sarpagandha, the roots of *Rauwolfia serpentina* (L.) Benth. ex Kurz, which was traditionally used to treat psychiatric illness as well as snakebite and diseases due to hypertension. Reserpine was isolated in 1952 from the complex mixture of alkaloids present in the plant and for some time, before the introduction of chlorpromazine, was the only drug available for treating psychotic patients. It acts by depleting the stores of serotonin (5HT) and noradrenaline in the synaptic terminals of the CNS, by thus producing a calming effect without inducing sleep. It was also widely used for treating hypertension but is now seldom used because of the danger of precipitating suicidal events in some patients due to the depression produced by depletion of the central monamines.[10]

Although reserpine is the only isolated compound to have achieved prominence as an anxiolytic, tranquilizer, or sedative, several herbal drugs are widely used for this purpose.

Valerian

The underground organs of several species of the genus *Valeriana* are major herbal anxiolytics in the medical systems of different cultures in temperate regions. Valerian is the name given to the product obtained from the northwestern European species *Valeriana officinalis* L., but Indian valerian *V. wallichii* DC. is also well known, and other species are used in Japan and China.[11] It is noteworthy that the different species are all used for much the same purpose. Valerian is a classic example of an herbal drug whose therapeutic qualities as an anxiolytic and to aid the onset of sleep have been demonstrated clinically but in which no single active ingredient can be identified. In fact, several different types of constituents are present for which there are some pharmacological data for activity which could contribute to the overall effect. The sesquiterpene valerenic acid (see Figure 17.8) has been shown to inhibit breakdown of gamma-aminobutyric acid (GABA) in the brain, thus elevating levels of this depressant neurotransmitter, while other types of sesquiterpenes present, related to valeranone and kessyl alcohol, prolong barbiturate-induced sleeping times in animals, but the mode of action is not known.[12] The monoterpene valepotriates, especially didrovaltrate (see Figure 17.9), decrease spontaneous motility and have a spasmolytic effect. The underlying mechanism is unclear, but they have been found to bind to dopamine receptors and may inhibit the stimulatory effect of endogenous dopamine in the CNS.[13] A large amount of GABA is present in roots of *V. officinalis* which, if it is absorbed from the gastrointestinal (GI) tract and crosses the blood-brain barrier, could exert a depressant effect. Recently, lignans present have been shown to inhibit 5HT binding to its receptors, which would have a net depressant effect on the CNS.[14]

Kava

Another herbal drug that has spread from its local use in Polynesia to widespread use in Europe and North America is kava (sometimes called

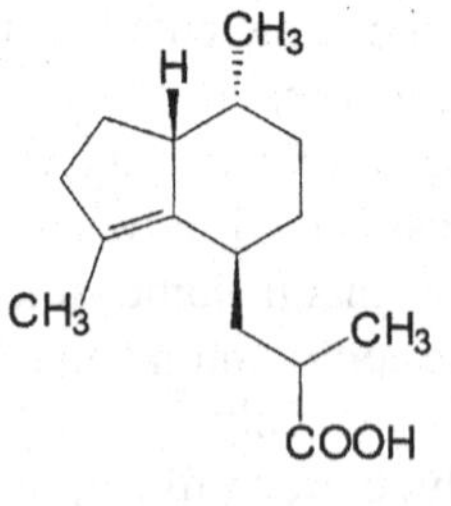

FIGURE 17.8. The chemical structure of valerenic acid.

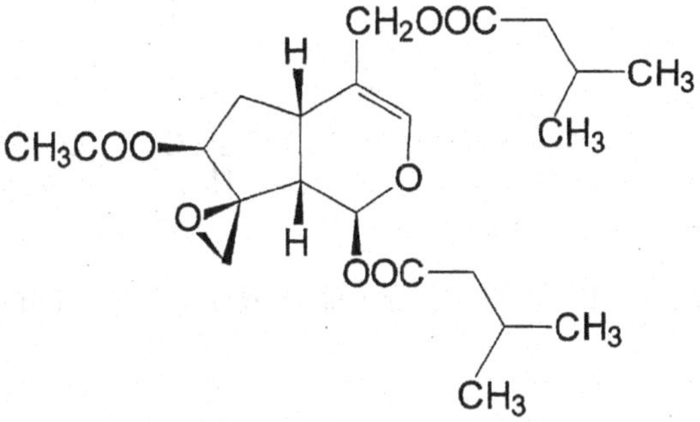

FIGURE 17.9. The chemical structure of didrovaltrate.

kava-kava), the roots of *Piper methysticum* Forst. An aqueous infusion of the roots is used widely in a social context in Fiji and Tonga and induces a relaxed state, and, if sufficiently large doses are taken, the subject falls asleep. Dried kava was marketed and has extensively been employed as a tranquilizing herbal medicine until recently when, due to some deaths from liver toxicity associated with its use, its sale and supply was banned in some countries and severely restricted in others. The underlying cause of the toxicity is not clear and reflects a similar confusion regarding the compounds responsible for activity. The root contains a series of unusual compounds known as kavapyrones. The most active compound at antagonizing strychnine-induced convulsions was dihydromethysticin (see Figure 17.10) but this constitutes only 5 percent of the total extract and related compounds present appear to contribute synergistically to produce a much greater activity for the total extract. The pharmacology and clinical efficacy of kava have been described in detail elsewhere.[15]

Hops

Hops, the female fruits of *Humulus lupulus* L., are used as an infusion and in pillows to aid sleep. The plant is also used as a bitter flavoring in beer. In several well-established herbal preparations a combination of hops and valerian is used to reduce anxiety and promote sleep, and several animal and clinical studies have been carried out that support such a use. 2-Methyl-3-buten-2-ol, which is volatile and appears to be a degradation product of other compounds present in hops, has been identified as a sedative when inhaled but not when given orally.[16]

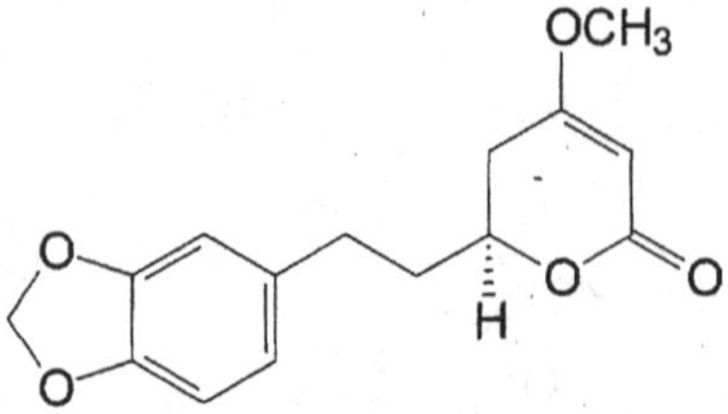

FIGURE 17.10. The chemical structure of dihydromethysticin.

Cannabis

Possibly the best-known and most-used herbal drug of abuse is cannabis, *Cannabis sativa* L. which is known by a large number of names, including marijuana, Indian hemp, and ganja, and has been used for thousands of years, in a variety of forms, and for many different purposes in traditional medicine in the Middle East and India, from where it is thought to originate. The herb, leaves, and flowering tops of the female plant are used, but the resin secreted by glandular trichomes on the flower bracts contains a high concentration of the psychoactive substances, so is sometimes used in this form or dissolved in edible oil. It is chiefly employed, albeit illegally in most countries, by inhalation through smoking or by ingestion in the form of a tincture.

The chemistry of cannabis is complex, and therefore so is the pharmacology, and it has a number of potential medical applications that are currently being investigated, especially its ability to reduce the pain and spasm associated with multiple sclerosis and cancer pain. A major effect, which is probably one reason why it is used as a leisure drug, is its a depressant effect on the CNS leading to relaxation and a state of detachment, although at higher doses it may cause hallucinations and euphoria. Δ9-Tetrahydrocannabinol (Δ9-THC) (see Figure 17.11), a member of a unique group of compounds called the cannabinoids found only in the *Cannabis* genus, is considered to be the major compound that affects the CNS, but related compounds also contribute to the effect. Considerable interest was generated by the discoveries of cannabinoid receptors in the CNS and of endogenous compounds that bound to these receptors such as anandamide. Agonist stimulation of the receptor affects many body systems including behavior, motor function, memory, and analgesia, and accounts for the complex effects of cannabis. Many different varieties of cannabis exist with a corresponding variety of total and relative amounts of the individual constituents and, thereby, different spectrums of activity. Recent reviews give a good overview of current work on cannabis.[17]

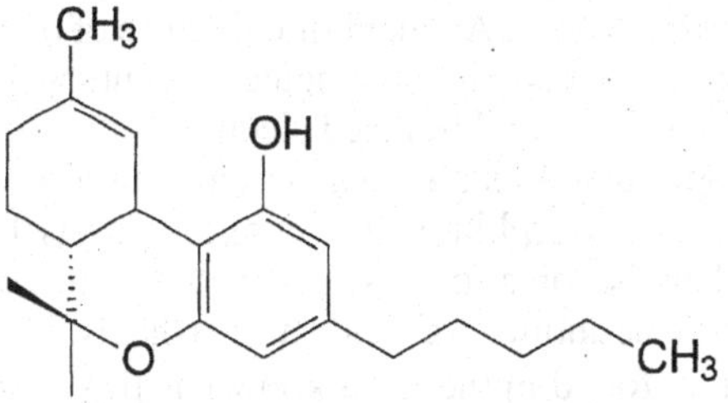

FIGURE 17.11. The chemical structure of Δ9-tetrahydrocannabinol.

Other Herbals in Use

Teas made from a variety of herbs such as chamomile [*Chamaemelum nobile* (L.) All. and *Matricaria recucita* L. flowers], lime flowers (*Tilia cordata* Mill., *T. platyphyllos* Scop., *T.* × *europea*) and passionflower (*Passiflora incarnata* L.) have a reputation for a mild sedative effect. The chemical reason for any activity is not certain, but recent work has shown that some flavonoids found in these plants bind to benzodiazepine receptors, so this may explain their reputed activity.[18]

PLANTS AND DERIVED COMPOUNDS WITH NARCOTIC EFFECTS

A narcotic effect is demonstrated by the ability of a compound or extract to send a subject into sleep and anaesthesia. Until the discovery of the narcotic effects of volatile solvents such as chloroform and ether, plant extracts were extensively used for this purpose in surgery.

Opium, Alkaloids, and Derivatives

Opium and its derived alkaloids occupy an important place in this respect. Opium is the dried latex of the poppy *Papaver somniferum* L., whose pharmacological effects have been known in the Mediterranean region for a very long time. The opium poppy capsule is featured in various artifacts from Minoan, Egyptian, and Greek civilizations of that area. The specific epithet *somniferum* indicates its ability to induce drowsiness and sleep, a theme reflected in the label *morphine,* derived from Morpheus, the Greek god of sleep, given to the major alkaloid present in opium. Opium contains about 30 different alkaloids and they can constitute up to 25 percent of the weight of the crude opium. An alcoholic tincture of opium, known as laudanum, was a popular remedy in the nineteenth century for babies since it was analgesic, antidiarrheal, and induced sleep.

Medically, opium and its alkaloids are now used primarily as analgesics rather than narcotics. Morphine (see Figure 17.12) has the strongest analgesic action while other alkaloids such as codeine have a much less addictive effect and are used more as antitussives and antidiarrheals. Morphine binds to three types of receptors found in the CNS known as μ, κ, and δ opioid receptors, which are distributed in different proportions according to the particular part of the CNS. Endogenous compounds known as the endorphins bind to these receptors. The overall effect of agonism varies according to the particular dominant receptor and whether the receptor is pre- or postsynaptic. Reduction of the amount of acetylcholine released is a feature of opioid receptor stimulation as is a decrease in central norepinephrine. Dopamine transmission in the mesolimbic system is thought to be increased by μ agonists, but κ agonists are seen to inhibit release of the same transmitter.

Many drugs have been developed by semisynthesis from the naturally occurring alkaloids or synthesised completely using them as templates. Not many of these derivatives have been made explicitly as improvements on the narcotic effects of morphine. The most commonly used derivative with this property, etorphine, is used to sedate large animals.

Tropane Alkaloids

The solanaceous plants containing tropane alkaloids are also of historical significance, since they were widely used as anaesthetics for operations and are still used in preoperative medication, partly for their ability to dry bronchial and salivary secretions as well induce a degree of amnesia and sedation. A large number of species from the Solanaceae family contain hyoscyamine and hyoscine (scopolamine) (see Figure 17.6), and these in-

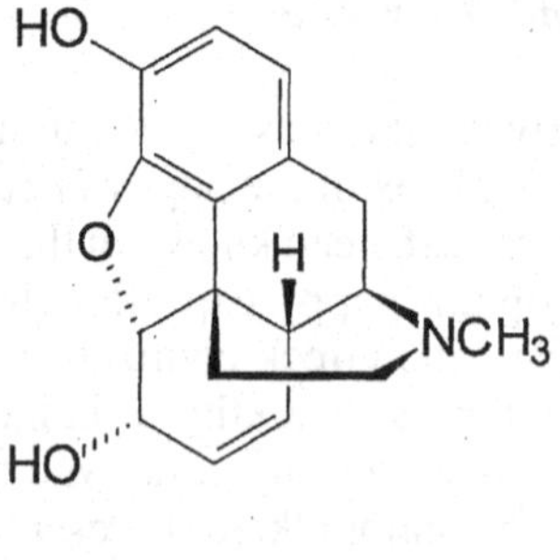

FIGURE 17.12. The chemical structure of morphine.

clude *Datura stramonium* L. (thornapple), *Atropa belladonna* L. (deadly nightshade), *Hyoscyamus niger* L. (henbane), *Mandragora officinarum* L. (mandrake), and *Brugmansia* spp. Both of these alkaloids have the same effect of acting as antagonists to acetylcholine at the muscarinic cholinergic receptors. A multitude of effects throughout the body are seen because of this activity, including loss of memory, hallucinations in larger doses, and a depressant effect on the CNS resulting in sleep. These plants have a long and fascinating history of use in "witches' brews," hallucinatory preparations, and in shamanistic divination. Hyoscine is the compound that has a greater effect on the CNS since, unlike hyoscyamine, it crosses the blood-brain barrier; therefore, it is the tropane alkaloid that is largely responsible for the CNS effects.

PLANTS AND DERIVED COMPOUNDS USED AS CNS STIMULANTS

Stimulant Plants Used in a Social Context

It is important to remember the maxim first coined by Paracelsus in the late fifteenth century, that "only the dose determines that a substance is not a poison" when these products are considered. To a greater or lesser extent overdosage can lead to adverse symptoms and chronic health problems, and in some cases psychological, if not physical dependence. However, in their native context, traditional methods of preparation and social conventions often curtail excessive use, and often it is only when they are used in alien cultures that serious problems arise.

Caffeine-Containing Plants

It is interesting to note that many beverages that originate from a local tradition but that are now used worldwide contain the xanthine derivative caffeine (see Figure 17.13), one of the purine alkaloids. Caffeine occurs in appreciable amounts in a variety of otherwise unrelated species of plants, and the more common ones used in beverages are listed in Table 17.1. Caffeine has a complex pharmacology and its stimulant effect on the CNS is only one of its activities. Caffeine is appreciably soluble in hot water so, beverages made by infusing the plant material contain appreciable amounts. However, the amount absorbed and the rate of absorption from the GI tract are determined by other constituents present in the extract and result in different effects. Thus the caffeine from coffee is quickly absorbed and gives a rapid CNS stimulatory effect, while the effect due to a similar dose in a cup

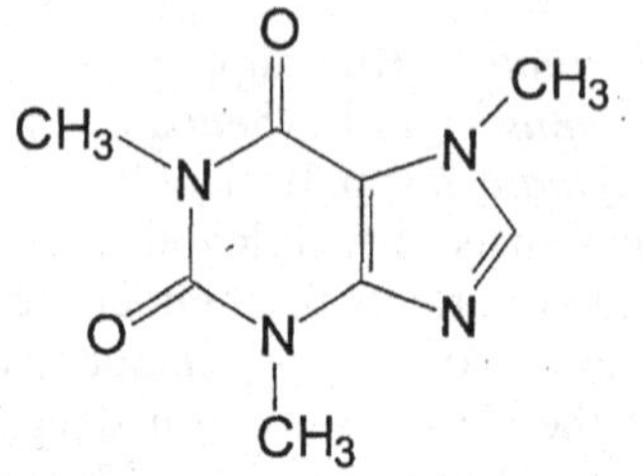

FIGURE 17.13. The chemical structure of caffeine.

of tea has a more gentle effect because the tannins present reduce the rate of absorption across the gut wall.

Nicotine

Nicotine (see Figure 17.14) is not uncommon in small amounts in plant tissue, but it is industrially obtained from the tobacco plant *Nicotiana tabacum* L. The dried leaves of this species are most commonly smoked in the form of cigarettes or cigars by hundreds of millions of people worldwide despite strong evidence of the serious effects on health. The persistence in use can partly be ascribed to the addictive properties of nicotine but also partly to the benefits perceived by those who smoke in providing mental stimulation or aiding relaxation. The pharmacology of nicotine is complex and many body systems are affected by it. As far as the CNS is concerned, the effect is largely stimulatory, caused by increased dopamine levels which causes cortical activation, as demonstrated by EEG measurements, but it is also anxiolytic. Nicotine also appears to improve learning and memory, and this is thought to be associated with its ability to increase the theta rhythm in the hippocampus. An increase in cerebral blood flow and in glucose metabolism, which have been shown to take place when nicotine is administered in the brain, are also associated with improved CNS performance.

Areca Nut

Another plant material used by millions of people in southern Asia, and by communities of their diaspora in other parts of the world, is the shredded seeds of the palm species *Areca catechu* L., named betel nut or pan. These are often soaked in an extract made from spices, or mixed with the spices themselves, mixed with lime, and wrapped in the leaves of *Piper betel* L.

TABLE 17.1. Plants containing caffeine used for preparation of beverages.

Common English name	Species	Part of plant used	Geographical origin	Area of major present use	Amount of caffeine present in dried plant material (percent w/w)
Coffee	*Coffea arabica*, *C. robusta* (Rubiaceae)	Seed	Ethiopia	Worldwide	1.0-2.0
Tea	*Camellia sinensis* (Theaceae)	Young leaves	China	Worldwide but especially Indian subcontinent, China, and where British influence is present	1.0-5.0
Cocoa and chocolate	*Theobroma cacao* (Sterculiaceae)	Fermented seed kernels	Mexico	Worldwide	0.05-0.36
Maté	*Ilex paraguensis* (Aquifoliaceae)	Leaves	Paraguay, Brazil, Argentina	Southern South America	0.2-2.0
Kola, cola	*Cola acuminata*, *C. astrophora* and other *Cola* spp. (Sterculiaceae)	Seed kernels	West Africa	Worldwide	1.0-2.5
Guarana	*Paullinia cupana* (Sapindaceae)	Seeds	Upper Amazon	Brazil	2.5-7.0

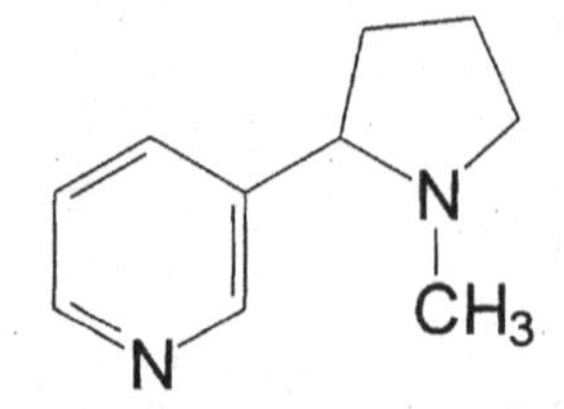

FIGURE 17.14. The chemical structure of nicotine.

The quid thus formed is masticated and has a stimulatory effect thought to be due to the alkaloids present, of which arecoline (see Figure 17.15) is the chief one. Arecoline acts in a stimulatory fashion at the muscarinic acetylcholine receptors and increases cerebral glucose metabolism and blood flow. It has been demonstrated to provide some improvement in cognitive function.

Ephedra, Khat, and Contained Compounds

Several natural compounds are very similar in chemical structure to the catechol monoamine neurotransmitters, for example, epinephrine (adrenaline), which perform a stimulatory role in the peripheral and central nervous systems. These compounds act in similar ways and are known generally as sympathomimetics. The best-known isolated compound is ephedrine (see Figure 17.16), originally isolated from the Chinese plant *Ephedra sinica* Stapf, which is not employed for its CNS stimulatory properties but for those activities useful in treating respiratory ailments, such as drying up secretions and bronchorelaxation. The stimulation of the CNS is a useful bonus when the drug is given in such situations, but ephedra has achieved some notoriety due to toxicity when abused in high doses as a mental stimulant, since it gives much the same effect in high doses as MMDA (Ecstasy).

The young leafy shoots of *Catha edulis* (Vahl) Forsskal. ex Endl., known as khat, are chewed extensively in Ethiopia and Yemen for their ability to improve mental concentration and prevent sleep. The major active compound is generally thought to be cathinone (see Figure 17.17), a phenylephrine analog similar in structure to ephedrine, and this has been subjected to a fairly thorough study of its effects. Cathinone is similar in structure to the group of synthetic CNS stimulants known as the amphetamines and it is likely to act in much the same way by inducing release of the monoamine neurotransmitters, such as dopamine, and by decreasing firing of substantia nigra neurons. Khat has not entered Western medicine as a medicinal plant

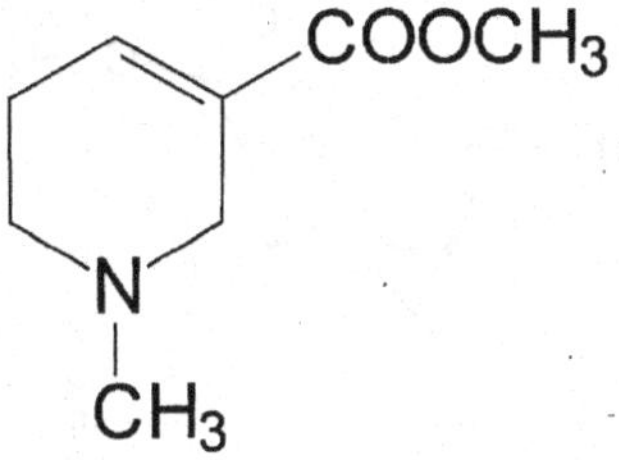

FIGURE 17.15. The chemical structure of arecoline.

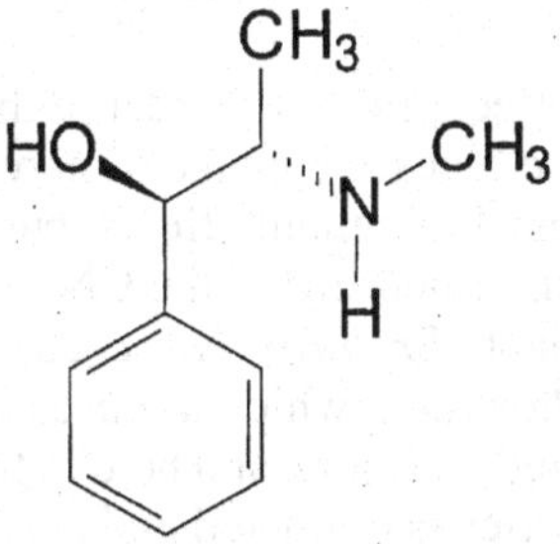

FIGURE 17.16. The chemical structure of ephedrine.

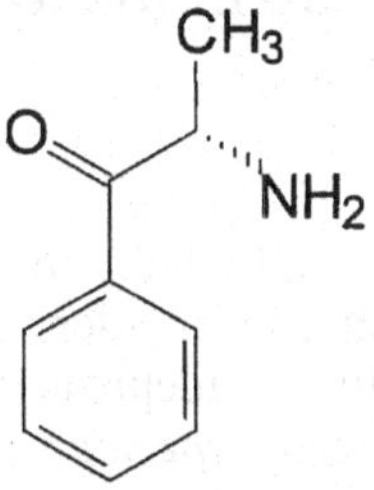

FIGURE 17.17. The chemical structure of cathinone.

and its use has not spread very far outside the cultural and ethnic groups who have used it traditionally.

Coca and Cocaine

The most well-known strongly stimulant plant material known is probably cocaine (see Figure 17.18), which is not used medicinally to any appre-

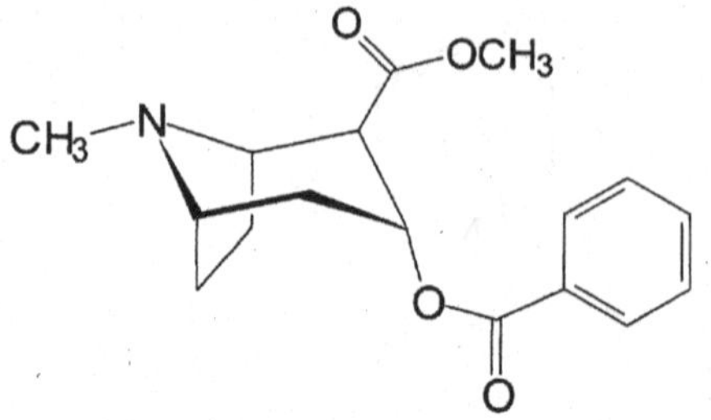

FIGURE 17.18. The chemical structure of cocaine.

ciable extent for its effects in this respect, although it is widely used as a drug of abuse. Cocaine blocks reuptake of norepinephrine (noradrenaline), thus exaggerating typical stimulant effects, but some thinking associates it with direct stimulation of neurons in the CNS. Cocaine is derived from the leaves of the coca plant, *Erythroxylum coca* Lam., *E. novogratense* var. *truxillense* (Rusby) Plowman, which are used in their native environment, of South America mixed with lime and held in the mouth, where a slow supply of the contained bases is extracted and absorbed.

PLANTS AND DERIVED COMPOUNDS USED IN DEPRESSION

St. John's Wort

Depression is a widespread condition and, although the condition has been known since antiquity, its incidence appears to be much higher in more affluent societies. Most antidepressants are synthetic compounds, but the use of St. John's wort herb *Hypericum perforatum* L. has grown considerably in recent years as a herbal treatment for mild to moderate depression. Although the herb has been known for centuries, it was mainly used for other purposes such as wound healing, and its current popular use does not have a very strong traditional base, although the plant was reputed to ward off evil spirits—a property which might indicate some psychological effects.

The clinical evidence for its usefulness in mild depression has a solid basis, and it has been demonstrated to be as effective as the traditional synthetic tricyclic antidepressants such as imipramine, with fewer and weaker adverse events in those taking standardized commercial preparations.[19] A recent three-way study claimed that St. John's wort has no effect in major depression but it has been pointed out that the standard drug used in the study also showed no significant effect.[20] The chemical basis for the observed clinical effects is still not completely elucidated, although the monoamine oxidase inhibitory effects of the flavonoids, and possibly the naphthodione hypericin, and the serotonin reuptake inhibitor phloridzin, are considered to play major parts in the overall effect. Because of its widespread use, reports of interactions with several other drugs are now appearing, based on the ability of the extract to up-regulate cytochrome P450 enzymes in the liver.[21]

PLANTS USED FOR CONDITIONS THAT DISPLAY EXCESSIVE ASPECTS OF NORMAL FUNCTION

Algesia

The sensation of pain is a useful protective mechanism at preventing harm to the body, but in some conditions the intensity of pain has to be controlled since it adversely affects work and other aspects of normal life. Algesia is felt in any part of the body, but it can be suppressed by analgesics that act on the CNS. Mild to moderate analgesia in all parts of the body can be achieved by the salicylates and their many semisynthetic derivatives and analogs, most notably aspirin. Aspirin is a derivative of salicylic acid, obtained originally from plant material due to the hydrolysis of glycosides such as salicin from *Salix* spp. (willow) and *Filipendula ulmaria* L. (meadowsweet). Analgesic activity is primarily local, rather than in the CNS, due to inhibition of prostaglandin synthesis.

The opiates have been mentioned previously. Their analgesic effect depends on their stimulation of presynaptic opioid receptors on the terminals of sensory neurons in the spinal cord, resulting in reduced amounts of substance P being released. A similarity in structure between the opiates and the endogenous analgesic substances known as endorphins can be seen, which explains the activity of morphine and related compounds.

Migraine

A particularly severe pain which is usually accompanied by severe CNS symptoms is encountered in the series of clinical conditions known collectively as migraine. Migraine attacks can be precipitated by a variety of internal and external factors and are caused by alteration in several aspects of physiology. The use of ergot alkaloids such as ergotamine, which was considered to improve the condition by constriction of blood vessels, has now largely disappeared, but more selective ergot-derived compounds such as methysergide are still employed.

Feverfew

In the United Kingdom, Canada, and some other countries, the dried leaves of feverfew *Tanacetum parthenium* (L.) Schultz Bip (Asteraceae) are used extensively as an herbal prophylactic for migraine headaches. This use is based on folklore but has been underpinned by several positive clinical studies.[22] The nature of the compounds responsible is not very clear. At first it was thought that the sesquiterpene lactones, especially parthenolide (see Figure 17.19), exerted their effect through inhibition of eicosanoid synthesis and also by inhibiting the expression of proinflammatory cytokines. However, a clinical study that showed no significant improvement with a preparation containing a standardized amount of parthenolide has suggested that other compounds might also need to be present for useful effects to be demonstrated.[23]

Antinausea Plants

Hyoscine (see Figure 17.6) has been used extensively as a treatment for the nausea and vomiting associated with motion sickness. The activity is due partly to its depressant effect on the motor and vegetative centers in the CNS, although the reduction in gut motility due to its peripheral anticholinergic effects also plays a part.

Powdered ginger rhizomes (*Zingiber officinale* L.) have attracted some attention in recent years because of their antiemetic effect. The effect appears to be based more on peripheral than CNS activity, and clinical studies suggest that it is better for sickness associated with pregnancy and postoperative nausea than for motion sickness.

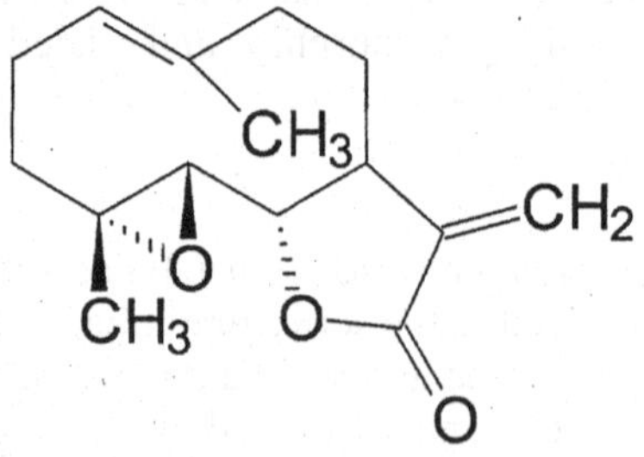

CH_3 CH_2 O CH_3 O O

FIGURE 17.19. The chemical structure of parthenolide.

ADAPTOGENS

An adaptogen enables an animal to cope with stresses by returning its general physiology and metabolism to the "normal" state. The organism is therefore able to tolerate conditions better and becomes less fatigued. The concept of balance, which provides the philosophical basis for Chinese and ayurvedic medicine, is amenable to the concept of this type of medicine, so it is not surprising that several herbal products for which adaptogenic properties are claimed originate from these cultures. Although the plant extracts are claimed to affect many different systems in the body, the CNS is also affected such that depressed patients are stimulated while those suffering agitation become more relaxed.

Ginseng

The major plants encountered worldwide that belong to this category are ginseng and eleutherococcus, also known as Siberian ginseng. These are the roots of several species of the Araliaceae, Korean ginseng coming from *Panax ginseng* Meyer, American ginseng from *P. quinquefolium* L., and eleutherococcus from *Eleutherococcus senticosus* (Rupr. & Maxim.) Maxim. The *Panax* species contain the triterpenoid ginsenosides based on (20 S)-protopaxadiol and (20 S)-protopanaxatriol (see Figures 17.20 and 17.21). The diols tend to have sedative and tranquilizing effects and the triols exert more of a stimulant action. These compounds were thought to be primarily responsible for the effects seen when extracts were administered, but more recently the polysaccharide components have also been shown to play a part. Numerous studies on animals and clinical trials have demonstrated that extracts of *Panax* species are adaptogenic with a predominant stimulant effect and ability to reduce fatigue and raise threshold levels for fatigue, thus enabling a greater amount of exercise to be undertaken. Memory, learning, and resistance to infection are also improved on long-term use of ginseng. *Eleutherococcus* contains different types of constituents consisting of lignans, coumarins, and sterols, but the overall effects are very similar to those shown by *Panax* species. The research literature on ginseng and *Eleutherococcus* is large, and much is literature from the Far East, which is not very accessible, but recent comprehensive reviews are readily available.[24]

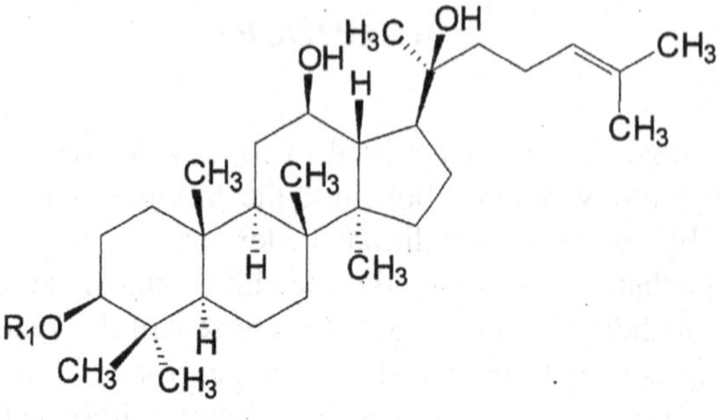

FIGURE 17.20. The chemical structure of protopanaxadiol.

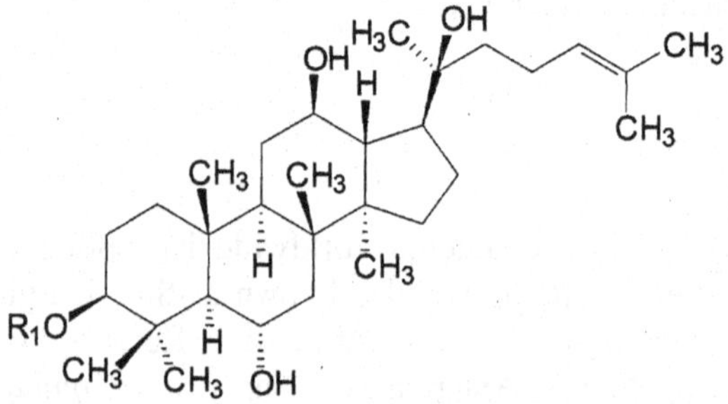

FIGURE 17.21. The chemical structure of protopanaxatriol.

Ashwagandha

Withania somnifera (L.) Dunal (Solanaceae) is sometimes known as Indian ginseng or ashwagandha and is an important plant in ayurvedic medicine which is used for a variety of purposes, including debility.[25] It has less of a stimulant effect than ginseng and often is observed to have a mild sedative activity. The root is most frequently used, although other parts of the plant are mentioned in the literature.

The major compounds of interest are unusual steroid derivatives known as withanolides, which occur as the free compounds, for example, withaferin A or as glycosides (see Figure 17.22). Alkaloids, including those of the tropane type, have also been isolated and tested for activity relevant to CNS performance. Extracts of *Withania* demonstrate adaptogenic activity in animals and also enhance cognitive performance, in spite of their sedative properties. The alkaloids appear to be responsible for some of the sedative effect, and the sitoindosides and withaferins have been shown to be

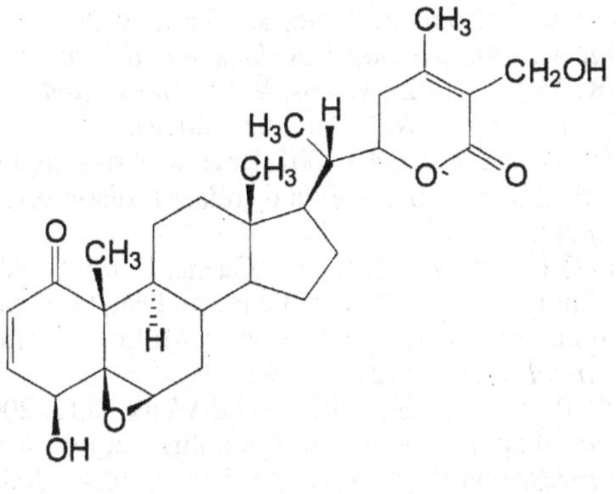

FIGURE 17.22. The chemical structure of withaferin A.

associated with immunostimulant activity, GABAergic effects, and the ability to improve CNS-cholinergic performance in animals.

CONCLUSION

It can be seen that the treatment of CNS disorders in orthodox medicine is aided by many plant-derived compounds. There is no reason why advances in knowledge of neurochemistry and psychopharmacology, coupled with investigations into plant extracts, should not continue to provide more molecules of interest for development of new drugs. However, the revival of interest in herbal medicines, particularly those used in other cultures, has provided some fascinating observations in recent years. The greater acceptance of the enhanced efficacy of mixtures compared to individual compounds, as well as the scientific validation of traditional uses, may result in some of these products entering mainstream medicine rather than being confined to the complementary or alternative fringe. In either case, more investigation needs to be carried out into the chemical constituents of many plants and the effects that they have in the CNS.

NOTES

1. Spinella, M., *The psychopharmacology of herbal medicine* (Cambridge, MA: MIT Press, 2001); Mills, S.Y. and Bone, K., *Principles and practice of phytotherapy* (Edinburgh: Churchill Livingstone, 2000); Barnes, J., Anderson, L.A., and Phillipson, J.D. *Herbal medicines: A guide for healthcare professionals* (Second edition) (London: Pharmaceutical Press, 2002); Sandberg, F. and Corrigan, D., *Natural remedies:*

Their origins and uses (London: Taylor and Francis, 2001), pp. 102-116; Dewick, P.M., *Medicinal natural products: A biosynthetic approach* (Second edition) (Chichester, UK: Wiley, 2002); Evans, W.C., *Trease and Evans' pharmacognosy* (Fifteenth edition) (London: W.B. Saunders, 2002).

2. Grossberg, G. and Desai, A., 2001, Review of rivastigmine and its clinical applications in Alzheimer's disease and related disorders, *Expert Opinion in Pharmacotherapeutics,* 2(4): 653-666.

3. Wilcock, G.K., Lilienfeld, S., and Gaens, E. [on behalf of the Galantamine International-1 Study Group], 2000, Efficacy and safety of galantamine in patients with mild to moderate Alzheimer's disease: Multicentre randomised controlled trial, *British Medical Journal,* 321: 1445-1449.

4. Woodruff-Pak, D.S., Vogel, R.W., and Wenk, G.L., 2001, Galantamine: Effect on nicotinic receptor binding, acetylcholinesterase inhibition, and learning, *Proceedings of the National Academy of Sciences,* 98(4): 2089-2094.

5. Shu, Y-Z. 1998, Recent natural products based drug development: A pharmaceutical industry perspective, *Journal of Natural Products,* 61(6): 1053-1071.

6. Barnes, J., 2002, Herbal therapeutics: Cognitive deficiency and dementia, *Pharmaceutical Journal,* 269: 160-162; Mix, J.A., Crews, W.D., 2002, A double-blind, placebo-controlled, randomized trial of *Ginkgo biloba* extract EGb 761 in a sample of cognitively intact older adults: Neuropsychological findings, *Human Psychopharmacology—Clinical and Experimental,* 17: 267-277.

7. Perry, N.S.L., Houghton, P.J., Theobald, A.E., Jenner, P., and Perry, E.K., 2000, In-vitro inhibition of human erythrocyte acetylcholine esterase by *Salvia lavandulaefolia* essential oil and constituent terpenes, *Journal of Pharmacy and Pharmacology,* 52: 895-902.

8. Perry, N.S.L., Houghton, P.J., Sampson, J., Theobald, A.E., Hart, S., Lis-Balchin, M., Hoult, J.R.S., Evans, P., Jenner, P., Milligan, S., and Perry, E.K., 2001, In-vitro activity of *S. lavandulaefolia* (Spanish sage) relevant to treatment of Alzheimer's disease, *Journal of Pharmacy and Pharmacology,* 53: 1347-1356.

9. Groger, D. and Floss, H., 1998, Biochemistry of ergot alkaloids: Achievements and challenges, *The Alkaloids, Chemistry and Pharmacology,* 50: 171-218.

10. Charney, D.S., 1998, Monoamine dysfunction and the pathophysiology and treatment of depression, *Journal of Clinical Psychiatry,* 59(Suppl. 14): 11-14.

11. Dweck, A.C., Valerian: The genus *Valeriana,* in P.J. Houghton (ed.), *Medicinal and aromatic plants—Industrial profiles* (London: Harwood Academic Publishers, 1997), pp. 1-21.

12. Hölzl, J., The genus valeriana medicinal and aromatic plants—Industrial profiles, in P.J. Houghton (ed.), *Medicinal and aromatic plants—Industrial profiles* (London: Harwood Academic Publishers, 1997), pp. 55-75.

13. Houghton, P.J., 1999, The scientific basis for the reputed activity of valerian, *Journal of Pharmacy and Pharmacology,* 51: 505-512.

14. Ibid.

15. Singh, Y.N. and Blumenthal, M., 1997, Valerian, *Herbalgram,* 39:33-56; Pittler, M.H. and Ernst, E., 2000, Efficacy of kava extract for treating anxiety: Systematic review and meta-analysis, *Journal of Clinical Psychopharmacology,* 20(1): 84-89.

16. Barnes, Anderson, and Phillipson, 2002, *Herbal medicines.*

17. Spinella, 2001, *The psychopharmacology of herbal medicine;* British Medical Association, *Therapeutic uses of cannabis* (London: Harwood Academic, 1997); Brown, D.T., Cannabis: The genus *Cannabis,* in P.J. Houghton (ed.), *Medicinal and aromatic plants—Industrial profiles* (London: Harwood Academic Publishers, 1997).

18. Paladini, A.C., Marder, M., Viola, H., Wolfman, C., Wasowski, C., and Medina, J.H., 1999, Flavonoids and the central nervous system: From forgotten factors to potent anxiolytic compounds, *Journal of Pharmacy and Pharmacology,* 51(5): 519-526.

19. Linde, K., Ramirez, G., Mulrow, C.D., Pauls, A., Widenhammer, W., and Melchart, D., 1996, St. John's wort for depression—An overview and meta-analysis of controlled clinical trials, *British Medical Journal,* 313: 253-258; Di Carlo, G., Borrelli, F., Ernst, E., and Izzo, A.A., 2001, St John's wort: Prozac from the plant kingdom, *Trends in Pharmacological Sciences,* 22: 292-297; Barnes, J., Anderson, L.A., and Phillipson, J.D., 2001, St John's wort (*Hypericum perforatum* L.): A review of its chemistry, pharmacology and clinical properties, *Journal of Pharmacy and Pharmacology,* 53: 583-600.

20. Hypericum Depression Trial Study Group, 2002, Effect of *Hypericum perforatum* (St. John's wort) in major depressive disorder, *Journal of the American Medical Association,* 287: 1807-1814.

21. Di Carlo et al., 2001, St. John's wort.

22. Palevitch. D., 1997, Feverfew *(Tanacetum parthenium)* as a prophylactic treatment for migraine: A double-blind, placebo-controlled study, *Phytotherapy Research,* 11: 508-511; Murphy, J.J., 1988, Randomised, double-blind, placebo-controlled trial of feverfew in migraine prevention, *Lancet,* 2: 189-192.

23. De Weerdt, C.J., Bootsma, H.P.R., and Hendriks, H., 1996, Herbal medicines in migraine prevention: Randomized double-blind placebo-controlled crossover trial of a feverfew preparation, *Phytomedicine,* 3: 225-230.

24. Spinella, 2001, *The psychopharmacology of herbal medicine;* Mills and Bone, 2000, *Principles and practice of phytotherapy;* Barnest, Anderson, and Phillipson, 2002, *Herbal medicines;* Farnsworth, N.R., Kinghorn, A.D., Soejarto, D.D., and Waller, D.P., Siberian ginseng *(Eleutherococcus senticosus):* Current status as an adaptogen, in H. Wagner, H. Hikino, and N.R. Farnsworth (eds.), *Economic and medicinal plant research,* Volume 1 (London: Academic Press, 1985), pp. 155-209.

25. Mills and Bone, 2000, *Principles and practice of phytotherapy.*

Chapter 18

Herbal Medicine in Endocrinology and Metabolic Disease

Graham Pinn

INTRODUCTION

This chapter reviews herbal options in treating metabolic and endocrine disease. Conditions such as diabetes and gout have been known since antiquity and have a wide range of natural treatments. The twentieth century heart-disease epidemic has focused attention on the cholesterol-lowering properties of plants; these claims have to be viewed with greater care as there is no tradition of usage. Increased obesity results in a wide range of medical problems. A simple solution is perhaps the most important challenge in treating metabolic disease and herbal options are being explored. Concern about the side effects of hormone replacement therapy during menopause has encouraged a review of traditional plant therapy. However, long-term plant usage may also be associated with side effects.

GOUT

In past times the typical symptoms of severe pain and swelling, particularly in the big toe, were attributed to overindulgent living. We now know that the primary problem is not usually an excessive intake of purine-containing foods but an enzyme defect in urate metabolism. This inherited defect causes a buildup of uric acid which, when it crystallizes in joints, causes the acute inflammation we know as gout.

Gout attacks three to five people per thousand. The frequency increases with age so that more than 1 percent of the over-30 population are troubled with attacks. With a disease this common, it is no surprise that many traditional remedies are found in different parts of the world. Four centuries ago Culpepper listed more than 40 different herbal treatments.[1] Other subsequent herbal textbooks offer similar numbers of usually differing treatment options. Treatment is often advised with plants which are mild diuretics (remove excess water) with the theory that the toxic substance (uric acid) will also be removed through the kidneys.[2] It is, however, well known that, with orthodox diuretic therapy and removal of fluid, there is in fact an increase in the level of uric acid and an increase in frequency of attacks of gout. Treatments such as dandelion *(Taraxacum officinale)*, celery *(Apium graveolens)*, parsley *(Petroselinium crispum)*, nettle *(Urtica dioica)*, and carrot *(Daucus carota)* are all mild diuretics, but perhaps some other mode of action is present in gout treatment. No reliable clinical studies have been done to confirm their benefit in acute attacks.[3]

If attacks are frequent, orthodox medicine prescribes a tablet called allopurinol which is a xanthine oxidase inhibitor that interferes with the metabolism of purine and stops the buildup of uric acid. In nature, some fruits have weak xanthine-oxidase-inhibiting properties due to their flavonoid content; there has been recent interest in chiso *(Perilla frutescens)*, skullcap *(Scutellaria baicalensis)*, and yarrow *(Achillea millefolium)* have also been suggested as preventative treatments. Cyanidins in some fruits are also weak xanthine oxidase inhibitors, and it has been suggested that eating half a pound of cherries per day would be a therapeutic option![4]

For acute attacks, the most rapidly effective conventional treatment is colchicines (see Photo 18.1). This alkaloid was first isolated from the meadow saffron *(Colchicum autumnale)* in 1814. Colchicine suppresses the release of inflammation-causing chemicals from the white blood cells; unfortunately, many people are unable to tolerate it because of gastrointestinal side effects.[5] Another traditional European treatment is gout weed *(Aegopodium podagraria)*; despite its name and traditional use, it has not been shown to have any significant effect.[6]

Many other plants have been found to contain a natural aspirin-like substance. The original sources from which aspirin was developed were willow bark *(Salix alba)* and meadow saffron *(Spirea ulmaria)*, and theoretically other plants containing aspirin-like substances might have a similar effect in reducing inflammation. In Europe plants such as poplar (*Populus* spp.), in the Middle East plants such as turmeric *(Curcuma longa)* and boswellia *(Boswellia serata)*, from the Far East ginger *(Zingiber officinale)*, and in North America black cohosh *(Cicimifuga racemosa)* have all traditionally been used to treat inflamed joints. Despite their aspirin content their effect is much slower than orthodox therapies. Devil's claw *(Harpagophytum procumbens)* is a native of South Africa. It has also traditionally been used and has been found to have analgesic effects, although little anti-inflammatory potential.[7] No clinical trials have adequately demonstrated the effectiveness of these and many other plants used to treat gout.

PHOTO 18.1. Meadow saffron *(Colchicum autumnale).*

DIABETES MELLITUS

Diabetes was first described in ancient Egypt and Greece more than 3,000 years ago. With the increasing age and obesity of the world's population, an epidemic of non-insulin-requiring diabetes is developing. In some Western populations two-thirds of males and half of females are now overweight—this in combination with more than one in five of the population now being over sixty years of age has caused the increased incidence. Currently there are an estimated 100 million diabetics worldwide. This is expected to grow to 350 million by 2025. About half of these people are living in countries without access to orthodox medicines, and herbal alternatives are widely used.

In the Indian subcontinent, genetic factors have produced a high incidence of diabetes and many traditional ayurvedic medicines are used. As plant metabolism is based on carbohydrates, much research has explored the sugar-lowering potential of plants. Plants that have been reviewed include the Madagascar periwinkle *(Catharanthus roseus)* (see Photo 18.2), bitter gourd *(Mormordica charantia),* and fenugreek *(Trigonella feonum-*

PHOTO 18.2. Madagascar periwinkle *(Catharanthus roseus).*

graecum).[8] Jambul *(Sygyium cumini)* has a long been used in ayurvedic medicine, but there are no satisfactory clinical studies. Gymnema *(Gymnema sylvestre)* is another plant that has extensive animal studies but no good human studies. Long-term evaluation has suggested an effect both on sugar and cholesterol levels.[9]

Diabetes in Europe was relatively uncommon in the past but has now become a major problem. The most useful medication for treating diabetes in the obese is the biguanide metformin. This drug has been developed from a commonly used herbal remedy: goat's rue *(Galega officinalis).* This drug makes insulin more effective, both in the liver and at the muscle receptor, and tends to cause weight loss. Goat's rue was originally used in the 1500s to treat the plague. It was first found to lower blood sugar in 1927 and extensive research was carried out in the 1930s to produce a synthetic compound.

A number of other potential mechanisms may be used by which the carbohydrate metabolism might be improved by plants (see Table 18.1). High fiber intake reduces and slows sugar absorption, and a number of bulking agents have been suggested helpful in diabetes.[10] One in particular, guar gum *(Cyamopsis tetragonoloba),* had a vogue as a diabetic dietary supplement in biscuit form. *Aloe vera* may also work in the same way to reduce sugar

TABLE 18.1. Herbal plants and diabetes proposed mode of action.

Benefit	**Herbal**
Reduce/slow sugar absorption	aloe *(Aloe vera)*
	guar *(Cyamopsis tetragonoloba)*
	wild oats *(Avena sativa)*
	barley *(Hordeum disticum)*
Stimulate Insulin production	Indian ipecac *(Gymnema sylvestre)*
	fenugreek *(Trigonella foenum-graecum)*
Stimulate liver function	chicory *(Cichorium intybus)*
	ginseng *(Panax ginseng)*
Reduce Insulin resistance	goat's rue *(Galega officinalis)*
Unknown	periwinkle *(Catharanthus roseus)*
	chinese foxglove *(Rehmannia glutinosa)*

and triglyceride levels.[11] A related plant, *Aloe dhufarensis,* has traditionally been used in the Middle East to treat diabetes.[12]

Plants that contain high levels of chromium reduce insulin resistance. Human studies also suggest a benefit in lipid lowering, and animal studies suggest a benefit in hypertension. Barley *(Hordeum disticum),* as well as slowing sugar absorption, has a high chromium content which has been shown to lower blood sugars in rats.[13] The sulphonamide-containing antibiotics were found by serendipity to have a hypoglycemic effect when they were first used as antibiotics. Sulphonylurea hypoglycaemic drugs were subsequently developed. Sulphur-containing herbs such as garlic *(Allium sativum)* and onion *(Allium cepa)* have been found to have some effect in stimulating insulin production.[14] Garlic has been traditionally used in Arabic medicine to treat diabetes.[15]

Other herbs that also have anecdotal evidence of benefit include those which are thought to improve liver function and hepatic glucose utilization in the liver, including chicory *(Cichorium intybus)* and ginseng *(Panax ginseng).* A brief study initially suggested ginseng may be an effective hypoglycemic agent, but despite widespread use and numerous studies there is little evidence of its effectiveness in diabetes or hyperlipidemia.[16] Diabetes is another condition in which antioxidants have been suggested beneficial, the U.S. Physician Study showed no reduction in incidence of Type II diabetes mellitus with long term B-carotene supplementation.[17]

Other herbs that have been traditionally used include carrot, nettle, dandelion, sunflower *(Helianthus annuus),* life root *(Senecio neurorensis),* bilberry *(Vaccinium myrtillus),* St. Mary's thistle *(Silybum marianum),* licorice *(Glycyrrhiza glabra),* and andrographis *(Andrographis paniculata).*

More than 1,200 hypoglycemic plants, half with traditional and half with experimental evidence, are awaiting further research with promise for the development of future treatments.[18]

Herbal treatments have also been used to treat diabetic complications involving the nerve and eye. Evening primrose *(Oenothera biennis)* and baical skullcap *(Scutellaria baicalensis)* have been found to reduce toxic metabolite buildup in nerves by aldose reductase inhibition. This enzyme pathway results in sorbitol accumulation, a cause of chronic complications.[19] Topical cayenne *(Capsicum)* has been found to reduce pain in diabetic peripheral neuropathy.[20]

An extensive overview by Ernst in 1997 concluded that the lack of adequate studies of hypoglycemic plant remedies makes their use difficult to support.[21]

OBESITY

As already discussed, adult (and, worryingly, childhood) obesity is increasing, which predisposes sufferers to a whole range of health problems including diabetes, hypertension, and heart disease. Obesity also affects the body more generally to cause problems with indigestion, breathing difficulty such as sleep apnea, arthritis of the spine and legs, and a general reduction in quality of life. The appropriate management is to restrict calorie intake to less than that expended by the individual. This is very simple but difficult to maintain on a long-term basis and why so many "wonder cures" for weight reduction are on the market. Orthodox medicine has currently produced some short-term treatments, but the problem is a lifelong one, and as such a lifelong treatment without side effects is required.

Hydroxy citric acid is a compound derived from the Indian plant *Garcinia cambogia,* and it is incorporated into many weight-reducing products.[22] In vitro it has beneficial effects on fat metabolism, and animal studies have demonstrated appetite suppression and weight reduction. In human study, however, no benefit has been shown. *Gymnema* has also been used in ayurvedic medicine as a means of reducing appetite and losing weight. Some animal experiments suggest benefit, but no human studies have been conducted. A South African cactus *Hoodia annulata* has traditionally been used by the Kalahari bushmen to suppress appetite when food is short. This plant will soon be studied in a clinical trial as a weight loss aid.

Other nonproven herbal remedies include guarana *(Paullinia cupana)* with its high caffeine content, bladderwrack *(Fucus vesiculosus)* with its high iodine content, and chili *(Capsicum annuum)*. Ma huang *(Ephedra sinica)* contains ephedrine; it has been used in weight-reducing regimes but has now been banned in many countries because of side effects and abuse potential. Side effects relate primarily to the nervous system and cardiovascular system, with psychosis, seizures, strokes, and heart attacks being reported.[23]

HYPERLIPIDEMIA

It is only in the past 40 or so years that the importance of cholesterol as a cause of disease in heart and circulation has been recognized, and perhaps only in the past ten years its importance in diabetes management. There is no historical basis for lipid-lowering herbal therapy; therefore herbal remedies need closer scientific evaluation. Orthodox treatment is mainly with a group of drugs called the statins; several of those very effective agents (e.g., simvastatin and pravastatin) are produced by fungi.

As is the case with diabetes, a high-fiber diet reduces cholesterol absorption and simple bulking agents such as psyllium *(Plantago isphagula)* have been found to have small effectives. Reduction in intake of saturated fats and substitution with soya bean *(Glycine soja)* is also an effective dietary remedy, and a meta-analysis of 38 controlled clinical trials confirmed benefit.[24] The first clinical studies attempting to lower cholesterol levels were carried out with soy in the 1960s, prior to the availability of the first manufactured lipid-lowering agent (see Photo 18.3).

Garlic has been extensively researched, and fresh garlic has been found effective in lowering cholesterol (see Photo 18.4). Tablet and capsule form appear less effective. Garlic also seems to have additional cardiovascular benefits including antiplatelet action, reducing blood viscosity, and improving endothelial function. Controversy about the lipid-lowering potential of garlic has occurred, but a recent meta-analysis has confirmed modest benefit.[25]

Plant sitosterols and sitostanols have been known for 40 years to reduce cholesterol absorption. Vegetarian diets contain up to ten times the level of these chemicals as omnivorous diets, so their addition to the diet significantly reduces cholesterol absorption. Soy products containing these compounds are now available in a variety of spreads, and regular use produces significant improvement in the lipid profile and is recommended by the Australian Heart Foundation. More recently, pine tree–derived sterols have become available as additives to spreads and have confirmed low-density lipoprotein cholesterol-lowering properties.[26] Another dietary approach is to increase consumption of nuts. A study of increased walnut intake has shown beneficial effects on total cholesterol levels.[27] The beneficial effects

PHOTO 18.3. Soybean (*Glycine soja*).

PHOTO 18.4. Garlic (*Allium sativum*).

of olive oil and fish oils are well known, and evening primrose *(Oenothera biennis)* also may have some benefit.

A study of the effect of red clover *(Trifolium pratense)*, a treatment of menopausal symptoms, has shown no cholesterol-lowering effect in post-menopausal women (see Photo 18.5), but a recent small placebo-controlled trial suggested an increase in high-density lipoprotein (HDL), the "good" cholesterol.[28] A number of in vitro and animal studies of globe artichoke *(Cynara scolymus)* have been carried out confirming reduced hepatic cholesterol production. Several small clinical studies have also confirmed this response with falling total cholesterol and an increase in HDL.[29] Anecdotal reports suggest rhubarb *(Rheum palmatum)*, devil's claw *(Harpagophytum procumbens)*, gulgul *(Commiphora mukul)*, and turmeric *(Curcuma longa)* also may have some effect. Many herbs may have lipid-lowering potential, but adequate studies have not as yet been carried out to confirm their effectiveness in humans.

THYROID DISORDER

Thyroid overactivity (thyrotoxicosis) is an uncommon condition affecting mainly women. Underactivity (myxoedema) is common, particularly in older women, and is often undiagnosed. It is estimated that as many as 2 or 3

PHOTO 18.5. Red clover *(Trifolium pratense)*.

percent of the elderly female population may in fact have an underactive thyroid gland.

Bugleweed *(Lycopus virginicus)* and gypsy wort *(Lycopus europaeus)* are European plants with many traditional uses including treating palpitations, goiter (thyroid swelling), and Grave's disease.[30] Experimental work has suggested an effect on output of various pituitary hormones including thyroid stimulating hormone (TSH), gonadotropins, and prolactin. Work in humans has suggested effects on iodine metabolism and thyroxine hormone release. The clinical effect, however, seems to be mild. Laboratory work showed that extracts of both *Lithospermum officinale* and lemon balm *(Melissa officinalis)* react with the antibody produced in thyrotoxicosis to block its effect.[31] No clinical studies have been done to confirm the potential benefit.

An underactive thyroid gland may be caused by iodine deficiency in some parts of the world. This can be corrected by increasing one's dietary intake. Addition of iodine to salt is one option in iodine-deficient areas. Kelp *(Fucus vesiculosys)* is a type of seaweed that contains high levels of iodine; this is also of use for people with iodine deficiency and may have some effect in those who have an underactive thyroid gland. Other natural therapies which have been suggested include radish *(Raphanus sativus)*, used in Russian folk medicine, walnuts (*Juglans* spp.), in Turkish folk medicine, and mustard *(Brassica nigra)* in Europe. No clinical studies support the claims of effectiveness. The German Commission E supports the use of motherwort *(Leonurus cardiaca)* to treat thyroid overactivity. Its benefit may be by controlling the resulting palpitations through its direct effect of the heart.[32]

MENOPAUSE

Many women suffer from menopausal symptoms in middle age, with 70 percent estimated to experience flushes and 40 percent mood swings. There are increasing concerns about the side effects of hormone replacement therapy and natural alternatives are commonly used.[33] A Scandinavian survey found 21 percent of women taking hormone replacement and 45 percent alternative therapies for their symptoms.[34]

Plant estrogens are found in many fruits, vegetables, and cereals. The addition of soy to the diet has been promoted as an alternative, but large amounts need to be consumed to produce a biological effect. These plants and extracts do show evidence of benefit in treating high cholesterol, as well as osteoporosis, and they perhaps reduce the risk of some cancers. Conversely, concern has been expressed about the side effect of high levels of phytoestrogens; epidemiological evidence suggests an adverse effect on male sexual development in pregnancy.[35] Similarly, the use of soy-based infant foods has been reviewed with concerns about high exposure to isoflavones and possible biological effects.[36]

Black cohosh *(Cimicifuga racemosa)* was traditionally used by the Algonquin Indians of North America to treat premenstrual problems (see Photo 18.6). It is approved by the German Commission E for treating premenstrual syndrome, dysmenorrhea, and menopausal symptoms.[37] A small short-term Australian study has shown benefit, with a 70 percent improvement in menopausal symptoms after three months, and a review of eight human studies found black cohosh to be a safe, effective alternative to estrogen therapy.[38] Wild yam *(Dioscorea villosa)* has been widely promoted as an alternative approach to hormone therapy, yet it does not seem to have any benefit related to phytoestrogen intake. It contains dioscin, a progesterone precursor which the body is unable to convert to progesterone. It may possibly work because it also contains other precursors which are convertible, but limited studies show no benefit.[39] Other herbs which may have a similar action include ginseng *(Panax ginseng)*. Evening primrose oil *(Oenothera biennis)* (see Photo 18.7) and angelica (dong quai).[40] Despite their popularity, they have no proven benefit. A recent Australian study of a traditional Chinese formula containing 12 herbs showed no symptom improvement over three months.[41]

PHOTO 18.6. Black cohosh *(Cimicifuga racemosa)*.

PHOTO 18.7. Evening primrose *(Oenothera biennis)*.

Motherwort *(Leonurus cardiaca)* is an old-fashioned European herb used to treat menopausal symptoms, in *Culpepper's Herbal* it was noted that it "makes mothers joyful and settles the womb, also promotes women's courses."[42] St. John's wort *(Hypericum perforatum)* is also traditionally used to help with menopausal depression and premenstrual mood lability. It is effective, but increasing concerns from orthodox therapy about possible drug interactions exist.[43] It seems as effective as commonly used orthodox treatments for mood change.[44]

TOXICITY

In any condition requiring long-term maintenance treatment the potential for toxicity increases whether treatment is orthodox or herbal. With herbal treatment few long-term studies have been conducted by which to confirm efficacy or toxicity.

Periwinkle (used to treat diabetes) is known to contain toxic substances (such as the *Vinca* alkaloids, used to treat cancer) and has the potential to affect the bone marrow when used long term. When first investigated, animal experimentation showed little hypoglycemic effect but did demonstrate leukopenia. This finding led to the use of a derivative as an anticancer agent.

Case reports have suggested a possible interaction between replacement therapy with thyroxine and the use of celery seed *(Apium graveolens).*

CONCLUSION

Our diet has changed from that of the hunter-gatherer, but our metabolism has not. People are reluctant to change their eating habits, but herbal supplementation may reduce the resulting metabolic disorders. This change should not, however, be based on laboratory experimentation, pseudoscience, or testimonials. Until such time as medical benefit is proven in proper clinical trials we should continue to encourage the use of the effective orthodox treatments with known effects and side effects. In the long term it is likely that many traditional plant therapies will be proven effective and incorporated into orthodox treatment regimes.

NOTES

1. Culpepper, N., *Culpeppers complete herbal and English Physician* (Manchester: J. Gleave & Son, 1826 [reprint]).
2. Weiss, R.F., *Herbal medicine* (Beaconsfield: AB Arcaman, 1996 [reprint]), p. 272.
3. Mills, S. and Bone, K., *Principles of practice of phytotherapy* (Churchill Livingstone, 2000), p. 249.
4. Duke, J.A., *The green pharmacy* (Emmaus, PA: Rodale Press, 1997), p. 224.
5. Wallace, S.L. and Singer, J.L., 1988, Therapy of gout, *Rheumatic Disease Clinics of North America,* 14: 441.
6. Grieve, M., *A modern herbal* (Twickenham: Tiger Books, 1998 [reprint]), p. 368.
7. Chrubasik, S., Junck, H., Breitschwerdet, H., et al., 1999, Effectiveness of *Harpagophytum* extract WAS 1531 in the treatment of exacerbation of low back pain, *European Journal of Anaesthesiology,* 16: 118-129.
8. Swanston-Flatt, S.K., Day, C., Flatt, P.R., Gould, B.J., and Bailey, C.J., 1989, Glycaemic affects of traditional European plant treatments for diabetes: Studies in normal and diabetic mice, *Diabetes Res,* 10: 69-73; Leatherdale, B.A., Panesar, P.K., Singh, G., Atkins, T.W., Bailey, C.J., and Bignell, A.H., 1981, Improvement in glucose tolerance to *Mormordica charantia* (Karela), *British Medical Journal,* 282: 1823-1824; Abdel-Barry, J.A., Abdel Hassan, I.A., and Al-Kakeim, M.H., 1997, Hypoglycaemic and anti-hyperglycaemic effects of *Trigonella foenum-graecum* leaf in normal and alloxan induced diabetic rats, *Journal of Ethnopharmacology,* 58: 149-155.
9. Okabayashi, Y., Taui, S., Fijisaiwa, T., Koide, M., Hasegawa, H., Nakamusa, T., Fujii, M., and Otsuki, M., 1990, Effect of *Gymnema sylvestre* on glucose homeostasis in rats, *Diabetes Research Clinic Practice,* 9: 143.

10. Chandalia, M., Garg, A., Lutjohan, D., et al., 2000, Beneficial effects of high dietary fibre in patients with type 2 diabetes mellitus, *New England Journal of Medicine,* 342: 1392-1398.

11. Yongchaiyuda, S., Rungpitarangsi, V., Bunyapraphatsara, N., and Chokechaijareoporn, O., 1996, Anti diabetic activity of *Aloe vera* juice: Clinical trials in new cases of diabetes mellitus, *Phytomedicine,* 3: 241-243.

12. Gazanfar, S., *Arabian medicinal plants* (Boca Raton, FL: CRC Publications, 1994).

13. Naismith, B.T., 1993, Therapeutic value of barley in the management of diabetes, *Annals of Nutritional Metabolism,* 35: 61-64.

14. Day, G., 1984, The allium alliance, *Nutrition Food Science,* 90: 20-21.

15. Gazanfar, 1994, *Arabian medicinal plants.*

16. Vogler, B.K., Pittler, M.H., and Ernst, E., 1999, The efficacy of ginseng: A systematic review of randomised clinical trials, *European Journal of Clinical Pharmacology,* 55: 567-575.

17. Lui, S., Ajam, V., and Chae, L., 1999, Long term beta carotene supplementation and risk of type 2 diabetes mellitus, *JAMA,* 282: 1073-1075.

18. Marles, R.J. and Farnsworth, N.R., 1995, Anti-diabetic plants and their active constituents, *Phytomedicine,* 2: 13-189.

19. Keen, H., Payan, J., Allawi, J., et al., 1993, Treatment of diabetic neuropathy with gamma-linolenic acid: The Gamma-Linolenic Acid Multicentre Trial Group, *Diabetes Care,* 16: 8-15.

20. Tandan, R., Lewis, G.A., Krusinski, P.B., et al., 1992, Topical capsaicin in painful diabetic neuropathy: A controlled study with long term follow up, *Diabetes Care,* 15: 8-14.

21. Ernst, E., 1997, Plants with hypoglycaemic activity in humans, *Phytomedicine,* 4: 73-78.

22. Heymsfield, S.B., Allison, D.B., and Vasselli, J.R., 1998, *Garcinia cambogia* (hydroxycitric acid) on a potential anti obesity agent, *JAMA,* 280: 1596-1600.

23. Wooltorton, E. and Sibbald, B., 2002, Ephedra/ephedrine: Cardiovascular and CNS effects, *Canadian Medical Association Journal,* 166: 633.

24. Anderson, J.W., Johnstone, B.M., and Cook-Newell, M.E., 1995, Meta-analysis of the effects of soy protein intake on serum lipids, *New England Journal of Medicine,* 333: 276-282.

25. Stevinson, C., Pittler, M.H., and Ernst, E., 2000, Garlic for treating hypercholesterolaemia, *Annals of Internal Medicine,* 133: 420-429.

26. Blair, S.N., 2000, Incremental reduction of serum total cholesterol and low density lipoprotein cholesterol with the addition of plant sterol ester containing spread to statin therapy, *American Journal of Cardiology,* 86: 46-52.

27. Sabate, J., Fraser, G.E., Burke, K., Knutsen, S.F., Bennett, H., and Linsted, K.D., 1993, Effects of walnuts on serum lipid levels and blood pressure in normal men, *New England Journal of Medicine,* 328: 613-607.

28. Campbell, M., 2001, Isoflavone supplements and lipid lowering, International Congress of Endocrinology, Sydney.

29. Kraft, K., 1997, Artichoke leaf extract: Recent findings reflecting effect on lipid metabolism, liver and gastro-intestinal tracts, *Phytomedicine,* 4: 369-378.

30. Harvey, R., 1995-1996, *Lycopus europatus* and *Lycopus virginalis*: A review of the scientific literature, *British Journal of Phytotherapy,* 4: 55-65.

31. Auf'molk, M., Ingbar, J.C., and Kubota, K., 1985, Extracts and auto oxidised constituents of certain plants inhibit receptor binding and biological activity of Graves immunoglobulins, *Endocrinology,* 116: 1687-1693.

32. Blumenthal, M., Goldberg, A., and Brinckmann, J., *Herbal medicine expanded Commission E monographs, 2000* (Austin, TX: American Botanical Council, 2000).

33. Vastag, B., 2002, Hormone replacement therapy falls out of favour with expert committee, *JAMA,* 287: 24-25.

34. Stadberg, E., Mattson, L., and Mulson, I., 1997, The prevalence and severity of climacteric symptoms and the use of different treatment regimens in a Swedish population, *Acta Obstetrica et Gynecologica, Scandinavia,* 76: 442-448.

35. Whitten, P.L., Lewis, C., Russell, E., Naftolin, F., 1995, Potential adverse effects of phytoestrogens, *Journal of Nutrition,* 125(Suppl. 3): 715-765.

36. Irvine, C.H.G., Fitzpatrick, M.G., and Alexander, S.L., 1998, Phytooestrogen in soy based infant foods: Concentrations, daily intake and possible biological effects. *Proceedings of the Society of Experimental Biology and Medicine,* 217: 247-253.

37. Vastag, 2002, Hormone replacement therapy falls out of favour.

38. Lieberman, S., 1998, A review of the effectiveness of *Cimicifuga racemosa* (black cohosh) for the symptoms of menopause, *Journal of Women's Health,* 7:525-529.

39. Komesaroff, P.A, Black, C.V.S., and Cable, V., 1998, Effects of wild yam extract on menopausal symptoms and hormonal and biochemical parameters (abstract). Australasian Menopause Conference, Auckland.

40. Chenoy, R., Hussain, S., Tayob, Y., O'Brien, P.M., Moss, M.Y., and Morse, P.F., 1994, Effect of gamolenic acid from evening primrose oil on menopausal flushes, *British Medical Journal,* 308: 501-503; Hirata, J.D., Swiersz, L.M., Zell, B., Small, R., Ettinger, B., 1997, Does dong quai have oestrogenic effects in postmenopausal women? A double blind placebo controlled trial, *Fertility and Sterility,* 68: 981-986.

41. Davis, S.R., Briganti, E.M., and Chen, R.Q., 2001, The effects of Chinese medicinal herbs on postmenopausal vasomotor symptoms of Australian women, *Medical Journal of Australia,* 174: 68-71.

42. Culpepper, 1826, *Culpeppers complete herbal,* p. 239.

43. Ernst, E., 1999, Second thoughts about the safety of St. John's wort, *Lancet,* 354: 2014-2015.

44. Larkin, M., 2000, St. John's wort included in US depression guidelines, *Lancet,* 55: 1619.

Chapter 19

Bioactive Saponins from Plants: Recent Developments

Marie-Aleth Lacaille-Dubois

INTRODUCTION

Saponins are a heterogenous group of natural products both with respect to structure and properties. They are found in many plant-derived foods and medicinal plants. They are used as anti-inflammatory (glycyrrhizin, aescin), wound-healing (asiaticoside), veinotonic (aescin, ruscogenin glycosides), expectorant (senegosides), and spasmolytic (hederasaponin) products. These compounds are structurally composed of a triterpenoid or steroid aglycone moiety and quite complex oligosaccharidic substituents. The hydrophilic properties of the glycoside part and the more or less lipophilic properties of the aglycone part give saponins their amphiphilic or surfactant properties. They have the ability to form stable aqueous foams, to form complexes with membrane steroids and lipid compounds, often resulting in partial destruction of the membrane structure (e.g., ability to hemolyze red blood cells). There has been an increase in the interest of the biological effects of saponins which were evaluated by many new in vitro and in vivo test systems. They are often related to their membrane-interacting properties, resulting in potential toxic properties or specific biological effects which have been reviewed.[1]

This chapter summarizes the most relevant results obtained in the past five years concerning the broad spectrum of biological and pharmacological activities of saponins, such as anti-inflammatory, antihepatotoxic, hypoglycemic, antimicrobial, and antiviral activities. Also, the important role of saponins on the cardiovascular system and the central nervous system is presented, whereas the antitumor and immunomodulating activitites are only briefly reported because they are detailed elsewhere. The discussion also focuses on the significant achievements in the understanding of their mechanism of action and structure-activity relationships.

ANTI-INFLAMMATORY ACTIVITY

The inhibition of the enzymatic formation and release of inflammatory mediators (prostaglandins [PG] and thromboxanes by cyclooxygenases [COX] and leukotrienes [LT] by 5-lipoxygenases) are considered good models for in vitro bioassays.

Platycodin D (see Figure 19.1), from the roots of *Platycodon grandiflorum,* inhibited TPA-induced PG E2 production in a concentration-dependent manner at 10 and 30 μM in rat peritoneal macrophages by inhibiting the induction of COX-2 protein but not by the direct inhibition of COX-1 and COX-2.[2] Furthermore, platycodin D and D3 increased airway mucin release from rat hamster both in vitro and in vivo.[3] The strong mucin-releasing effect and lower toxicity of platycodin D3 would be suitable for the development of a new type of expectorant drug for the treatment of various airway diseases. Kalopanax saponin A (see Figure 19.2, hederagenin glycoside from *Kalopanax pictus*) was shown to inhibit lipopolysaccharide (LPS)-induced inducible NO synthase (iNOS) and COX-2 protein expression in a dose-dependent manner and consequently NO, PG E2, and tumor necrosis factor-α (TNF-α) production by the macrophage cell line RAW 264.7.[4] The antirheumatoidal effect of this saponin has been confirmed in vivo. Namely on Freund's complete adjuvant reagent-induced rheumatoidal arthritis in rats, the administration of kalopanax saponin A inhibited edema, agglutination, vascular permeability, and trypsin inhibitor.[5] Furthermore kalopanax saponins (50 $mg \cdot kg^{-1}$ p.o.) showed a significant inhibition of vascular permeability in mice.[6] Among several in vivo models of acute inflammation, the most frequently used are the carragenin-induced hind paw edema in mouse or in rats and the mouse ear edema induced by several irritants (12-*O*-tetradecanoylphorbol-13-acetate [TPA], arachidonic acid [AA], ethylphenylpropionate [EPP], phorbol myristate acetate [PMA]). All the saponins from *Bupleurum rotundifolium* (see Figure 19.3) were

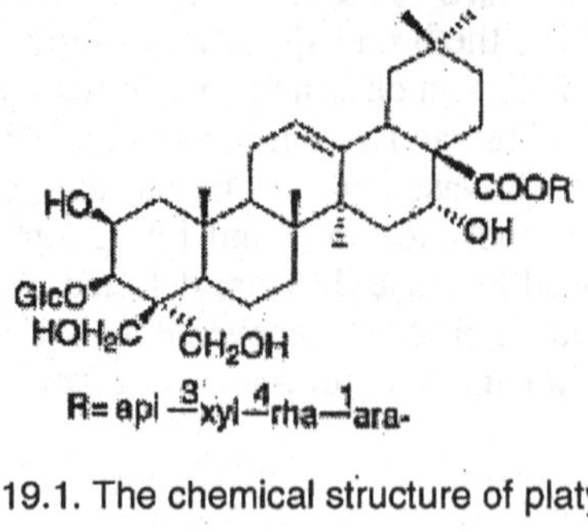

HO
COOR
OH
GlcO
HOH_2C CH_2OH
R= api $\overset{3}{—}$xyl$\overset{4}{—}$rha$\overset{1}{—}$ara-

FIGURE 19.1. The chemical structure of platycodin D.

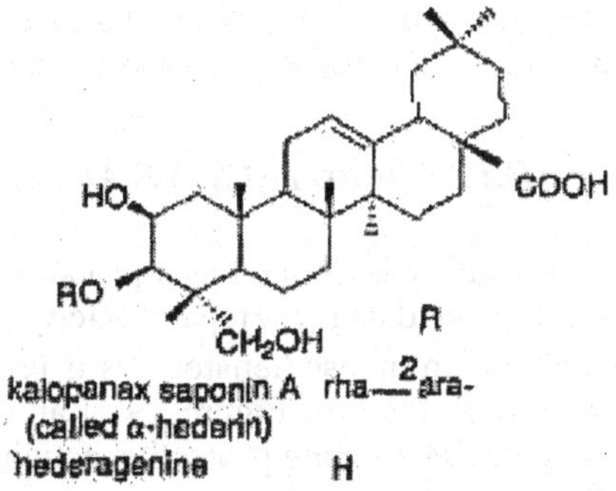

FIGURE 19.2. The chemical structure of hederagenin glycosides.

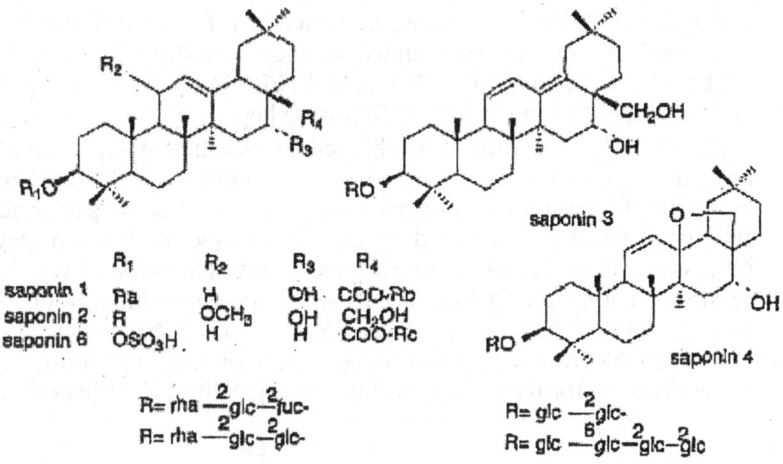

FIGURE 19.3. The chemical structure of *Bupleurum* saponins.

found to be effective against TPA-induced ear edema in mice (ID_{50} 248, 288, 128, 99, 297 nmol/ear for saponins 1, 2, 3, 4 and 6, respectively; indomethacin, 350 nmol/ear).[7] Saponins 3 and 6 were also active on a TPA multiple-dose model of skin chronic inflammation. Among hederagenin glycosides, loniceroside A, dipsacus saponins B and C having more than four sugar residues showed significant anti-inflammatory activity against both ear edema induced by AA or croton oil (20 to 37 percent).[8] A standardized preparation of 13 ginsenosides (G115) was reported to protect the muscles studied from injury and inflammation after eccentric exercise, although the exact mechanism of this protection needs to be further evaluated.[9]

HEPATOPROTECTIVE ACTIVITY

Antihepatotoxic activity is defined as a property exhibited by drugs or agents that prevents liver damage. Hepatotoxicity induced by *D*-galactosamine (Gal-N)/TNF-α in mouse hepatocytes is used as a model to study liver injury. Majonoside R2 (see Figure 19.4) an ocotillol saponin from *Panax vietnamensis* showed strong protective activity against Gal-N/TNF-α-induced cell death in primary cultured mouse hepatocytes (IC_{50} 82.4 μM, by comparison with the positive control silibinin, IC_{50} 14 μM).[10] These findings suggested that MR2 may have protected the hepatocytes from apoptosis via an inhibition of TNF-α production by activated macrophages and a direct inhibition of apoptosis induced by TNF-α.[11] These findings were confirmed in vivo. Pretreatment of mice with MR2 (50 $mg \cdot kg^{-1}$ or 10 $mg \cdot kg^{-1}$, i.p.) at 12 and 1 hour before Gal-N/LPS injection significantly inhibited apoptosis and suppressed following hepatic necrosis. Furthermore, MR2 (50 $mg \cdot kg^{-1}$) significantly inhibited the elevation of the serum TNF-α level, an important mediator of apoptosis.[12] In this study the furan ring of the ocotillol-type side chain seems to be important for the hepatoprotective activity, while a pyran ring does not. It was shown that a ruscogenin glycoside isolated from *Liriope muscari* improved the delayed-type hypersensitivity but not the CCl_4-induced liver injury which might be due to the selective dysfunction of immune cells that caused hepatocyte damage rather than the protection of hepatocyte membranes.[13] These results will be of significance for the therapy of immunologically related liver diseases.

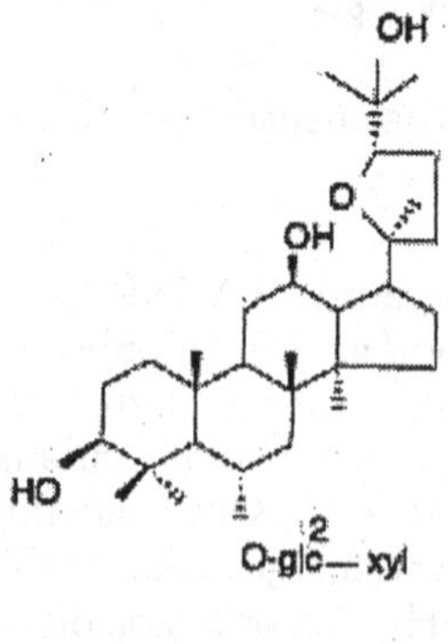

FIGURE 19.4. The chemical structure of *Ginseng vietnamesis* saponin MR2.

Three new triterpene glucosides (ursane and lupane-type) from *Combretum quadrangulare* (see 1, 2, and 5 on Figure 19.5) showed significant hepatoprotective effects in the Gal-N/TNF-α-induced cell death model at 25-200 μM in a concentration-dependent manner.[14]

They showed at 50 mM, 37.6, 40.9 and 67.5 percent inhibition against cell death, respectively, while a positive control, silibin, revealed 61.2 percent inhibition. Gypenoside, a saponin extract derived from *Gynostemma pentaphyllum,* was shown to significantly reduce the increase of SGOT and SGPT activities in liver injury induced by CCl_4 for eight weeks in rats.[15] Chikusetsusaponin IVa, an oleanolic acid glycoside (see Figure 19.6), showed hepatoprotective activity in an in vitro immunological liver injury of primary cultured rat hepatocytes.[16] Although a rhamnosyl derivative of this compound similarly showed hepatoprotective activity, its prosapogenin did not show any activity, but on the contrary exhibited cytotoxicity toward liver cells. Therefore, the activity of these types of saponins could represent a balance between hepatoprotective action and hepatotoxicity.[17]

A similar observation was obtained from experiments with some glucuronide analogs prepared from kaikosaponin III (see Figure 19.7) (an olean-12-ene type triterpene saponin, with a methyl group at C-28 and a glucuronic acid moiety at C-3) which were evaluated using immunologically induced liver injury in primary cultured rat hepatocytes.[18] In this study the sophoradiol monoglucuronide was less active than the kaikosaponin I at 30 to 200 μM. Then, the order of preventive effect was kaikosaponin I > kaikosaponin III > soyasaponin I. This information supported previously obtained structure-hepatoprotective relationship data, namely that the methyl group at C-24 enhances the hepatoprotective activity.

	R
1	CH_3
2	CH_2OH

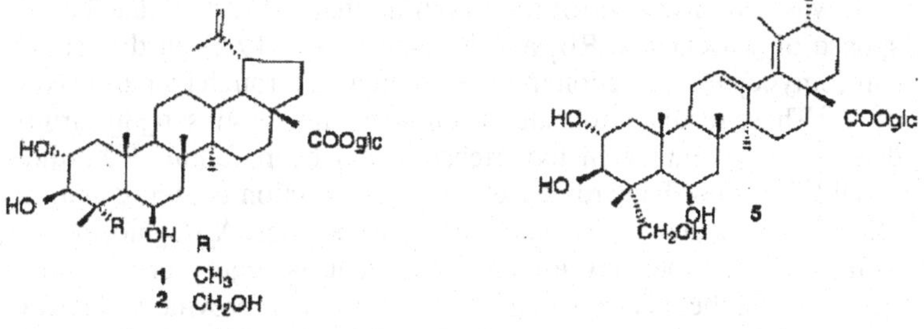

FIGURE 19.5. The chemical structure of triterpene-glucosides from *Combretum quadrangulare.*

20-S-Protopanaxadiol

	R	R_1
G-Rb_1	glc —2 glc-	glc —6 glc-
G-Rb_2	glc —2 glc-	ara —6 glc-
G-Rb_3	glc —2 glc-	xyl —6 glc
G-Rc	glc —2 glc-	ara(f) —6 glc-
G-Rd	glc —2 glc-	glc-
G-Rh_2	glc-	H
G-Rg_3	glc —2 glc-	H
G-Rs_1	6Acglc —2 glc-	xyl —6 glc-
G-Ra_1	glc —2 glc-	xyl —4 ara —6 glc-

G = ginsenoside; C = chikusetsusaponin

20-S-Protopanaxatriol

	R	R_1
G-Re	rha —2 glc-	glc-
G-Rf	glc —2 glc-	H
G-Rh_1	glc-	H
G-Rg_1	glc-	glc-
G-Rg_2 = C-I	rha —2 glc-	H

	R	R_1
C IVa	glc A-	glc-
oleanolic acid	H	H

FIGURE 19.6. The chemical structure of the ginsenosides of *Panax ginseng*.

CARDIOVASCULAR ACTIVITY

It was demonstrated for the first time that a direct cardiac depressive response of ginsenoside Rb_1 and Re (see Figure 19.6) on the rat ventricular cardiomyocyte contraction might be mediated through increased NO production.[19] These results implicate the clinical value of ginseng in cardiovascular disease in patients with hypertension and heart failure.[20] Further studies should focus on the cardiac excitation-contraction coupling and membrane ion channels with Rb_1, Re, and other ginsenosides. Astragaloside was shown to have some protecting effects on myocardial injury of viral myocarditis mice. One of the mechanism may be due to the improvment of the activity of myocardial sarco/endoplasmic Ca(2+)-ATPase (SERCA).[21]

Ischemia/reperfusion (I/R, in vivo) or hypoxia/reoxygenation (H/R, in vitro) is usually observed in various diseases such as myocardium infarc-

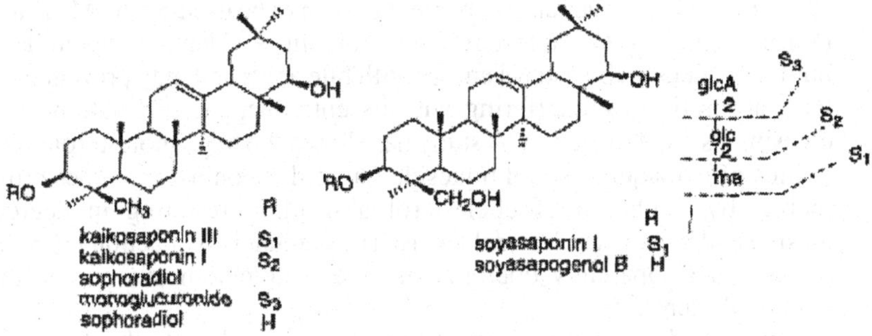

FIGURE 19.7. The chemical structure of saponins from leguminous plants.

tion. Protein tyrosine kinase (PTK) signaling pathways play important roles in such injuries. Inhibition of PTK activation can protect against I/R- or H/R-induced damages. Ginsenoside Rb_1, Rd, Ra_1, and Ro (see Figure 19.6) showed significant inhibitory effects on PKT activation induced by a H/R model in cultured human umbilical vein endothelials cells.[22] Dose-response experiments revealed that ginsenoside Rb_1 was the most active compound and it completely blocked PTK activation at a wide range of concentrations. A saponin mixture of *Gynostemma pentaphyllum* at 0.1 to 100 $\mu g \cdot mL^{-1}$ elicited a concentration-dependent vasorelaxation of porcine coronary rings through the direct release of endothelium-derived NO which was antagonized by the nitric oxide synthase inhibitor N-nitro-L-arginine methyl ester.[23] The endothelium-dependent relaxation elicited by ginsenoside Rg_3 in isolated rat aorta is due to an increased formation of NO by endothelial cells and seems to involve the activation of tetraethylammonium-sensitive K^+ channel.[24] In addition to this, ginsenoside Rg_3 was able to inhibit directly vascular smooth muscle tone possibly by inhibition of Ca^{2+} influx and stimulation of K^+ efflux, via activation of tetraethylammonium-sensitive K^+ channels.[25]

A saponin mixture of *Herniaria glabra* at a dose of 200 $mg \cdot kg^{-1}$ significantly decreased systolic and diastolic blood pressure in spontaneously hypertensive rats from 187.60 ± 5.94 mm Hg 119.10 ± 7.79 mm Hg at day 0 to 141.60 ± 7.51/ 90.40 ± 7.68 mm Hg at day 30 in comparison with furosemide treated rats ($p < 0.001$).[26]

Reports of plants dietary additives in East Africa support a role for hypocholesterolemic agents in the diet, and provide explanation of the low incidence of cardiovascular disease of populations that traditionally consume high levels of dietary fat and cholesterol. It was shown that 81 percent of the Batemi additives and 82 percent known to be used by the Maasai contain potentially hypocholesterolemic saponins.[27] These compounds may bind with dietary cholesterol and/or with bile acids and may prevent cholesterol absorption by interfering with its enterohepatic circulation and increasing its fecal excretion. A study has shown the hypocholesterolemic activities of pamaqueside and tiqueside (steroid saponins) in cholesterol fed rabbits by inhibition of cholesterol absorption resulting in decreased plasma and hepatic cholesterol levels. The mechanism of action is probably lumenal but apparently does not involve a stoichiometric complexation with cholesterol.[28]

Six steroidal saponins from *Anemarrhena asphodeloides* (see Figure 19.8) showed remarkable inhibiting effect on platelet aggregation. Five of them were not hemolytic, suggesting that such compounds might be used as a novel antithrombotic agent in postmyocardial infarction.[29] On the contrary, new noroleanane-type triterpene saponins from *Camellia japonica* showed a concentration-dependent aggregation of platelets in the concentration range (10 to 100μg·mL^{-1}).[30]

The antioxidant effect of various individual ginsenosides was evaluated on the experimental system of free-radical-initiated peroxidation such as hemolysis of human erythrocytes induced thermally by 2,2'-azobis(2-amidinopropanehydrochloride) (AAPH).[31] The IC_{50} of ginsenosides (see Figure 19.6) in AAPH-induced hemolysis of the erythrocyte was found to be in the order $Rb_3 < Rb_1 < Rg_2 < Re < Rg_1–Rc < Rh_1 < R1, Rb_1, Rc$, and Rg_2, prolonging the lag time of hemolysis. Contrarily, Rg3, Rd, and Rh1, and high concentrations of Rb3, Rg1, and Rh2 function as prooxidants to accelerate AAPH-induced hemolysis. Although the mechanism of action will be further studied in detail, this information may be useful in the clinical usage of ginsenosides.[32]

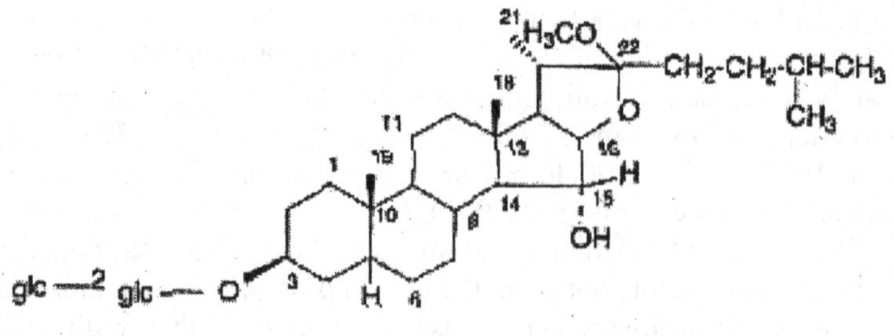

FIGURE 19.8. The chemical structure of anemarrhenasaponin.

ACTIVITY ON THE CENTRAL NERVOUS SYSTEM

Most works published in recent years concern the saponins from *Panax ginseng;* however, some papers are related to other plants. Dipsacus saponin C (DSC) from *Dipsacus afer* administered intrathecally (i.t.) produced antinociception in a dose-dependent-manner (from 3.75 to 30 μg) as measured by the tail-flick assay.[33] This effect may be mediated by GABAergic, serotoninergic, and noradrenergic receptors located at the spinal cord level. Furthermore, cholecystokinin (CCK) has an antagonistic action against DSC-induced antinociception.[34] Platycodin D from *Platycodon grandiflorum* (see Figure 19.1) administered intracerebroventricularly (i.c.v.) showed an antinociceptive effect in a dose-dependent manner as measured by the tail-flick assay which was maintained at least one hour.[35]

Ginsenosides Rb_1, Rb_2, Rc, Rd, and Rg_1 administered spinally appear to be responsible for blocking the antinociception induced by U50 (a kappa opioid receptor agonist) administered supraspinally, whereas ginsenosides Rb_2 and Re administered supraspinally appear to be responsible for blocking the antinociception induced by U50 administered supraspinally.[36] Another study using the same bioassay suggested that the ginseng total saponin (100 to 200 $mg \cdot kg^{-1}$, i.p.) dose-dependently attenuates U50 (40 $mg \cdot kg^{-1}$ i.p.)-induced analgesia and inhibits tolerance to its analgesia by acting on benzodiazepine receptors and $GABA_A$-gated chloride channels.[37] The effects of *Panax notoginseng* roots on spontaneous behavior have been evaluated in different groups of Wistar rats (ten per group) through gastric tube for ten days in two doses (43 and 86 $mg \cdot kg^{-1}$ in a volume of 5 $ml \cdot kg^{-1}$ per day, versus animals treated with the same dose of saline).[38] A significant increase in spontaneous motility was found ($p < 0.05$) whereas the feeding behavior was not affected by the treatment. Observed effects did not seem dose related. Vietnamese ginseng saponin (15 to 25 $mg \cdot kg^{-1}$ p.o.) and its major component MR2 (see Figure 19.4) (1 to 10 $mg \cdot kg^{-1}$ i.p.) were shown to have preventive effects on the psychological stress-induced enhancement of lipid peroxidation in the mouse brain which was partly due to enhancement of $GABA_A$-ergic systems in the brain.[39] On the other hand, ginseng has been used as an antistress agent and ginsenosides might be responsible for its antistress efficacy by selective regulation of voltage-dependent Ca^{2+} channel subtypes, which are mainly involved in catecholamine secretion in chromaffin cells.[40] A study has shown that the possible mechanism through which ginsenoside Rb_1 ameliorates affects the impairment of learning performance observed in Alzheimer's disease (AD) was to minimize the toxic effect of β-amyloid peptides.[41]

HYPOGLYCEMIC ACTIVITY

Many natural drugs are used in traditional medicine as antidiabetics, and several active compounds have been identified by in vivo bioassay using experimental diabetic animals induced by alloxan or streptozotocin (STZ) or by using the oral sucrose tolerance test (an animal model of insulin-dependent diabetes mellitus) and by in vitro bioassay testing glucose transport activity.

Prevention of hyperglycemia and hyperinsulinemia by retardation of glucose uptake by the small intestine is one successful approach to improving insulin resistance in diabetes mellitus. Several approaches may be taken to retard glucose uptake in the small intestine: delaying the gastric emptying rate of the gastrointestinal contents, inhibiting the digestive enzyme, inhibiting active glucose transport, and so on.[42] The speed of gastric emptying is important in the regulation of glucose homeostasis. The usual situation in patients with diabetes is delayed gastric emptying. It was reported that glycyrrhizin (GZ) reduced the postprandial blood glucose rise in normal mice. A study showed that the long-term GZ treatment (2.7 to 4.1 $g \cdot kg^{-1}$ diet) has an antidiabetic effect in non-insulin-dependent diabetic model mice.[43] Investigation of the mode of action for the hypoglycemic activity of some representative saponins such as momordin Ic (oleanolic acid-3-O-β-D-glucopyranosyl-(1->3)-β-D-glucuronopyranoside) (12.5 to 50 $mg \cdot kg^{-1}$), escins Ia and IIb (50 $mg \cdot kg^{-1}$) (see Figure 19.9) calendasaponins (see Figure 19.10) revealed that they inhibited gastric emptying in mice loaded with a nonnutrient or nutrient meal.[44] Furthermore, calendasaponins exhibited potent inhibitory effects on an increase in serum glucose levels in glucose-loaded rats.[45]

It was shown that escin Ib (see Figure 19.9) inhibits gastric emptying in mice, at least in part, mediated by capsaicin-sensitive sensory nerves to stimulate the synthesis and/or the release of dopamine, to act through central $dopamine_2$ receptor, which in turn causes the release of prostaglandins.[46] Similar conclusions were obtained with oleanolic acid glycosides obtained from several medicinal foodstuffs.[47] Furthermore, by examination of the structure-activity relationships, the 3-O-glucuronide moiety and the 28-carboxyl group in oleanolic acid glycosides were required to exert the inhibitory activity on the increase of serum glucose levels in oral glucose-loaded rats.[48] The antihyperglycemic and antiobese effects of the extract of *Panax ginseng* berries in obese diabetic mice have been evaluated after a daily intraperitoneal injection of 150 $mg \cdot kg^{-1}$ for 12 days.[49] On day 12, the extract-treated mice became normoglycemic (137 ± 6.7 $mg \cdot dL^{-1}$) and had significantly improved glucose tolerance. The improvement in blood glu-

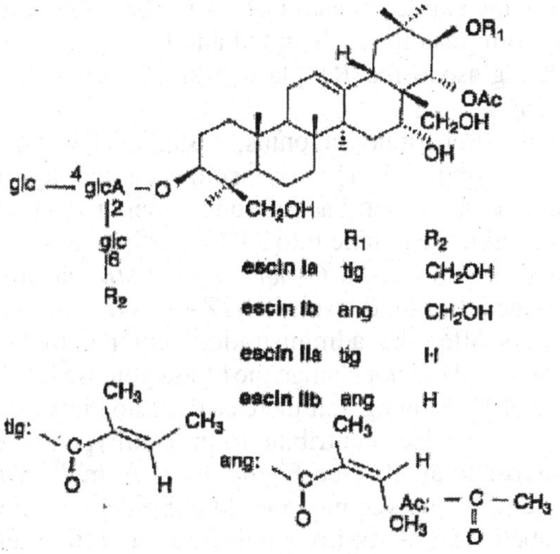
OR1
H
OAc
CH2OH
OH
glc —4 glcA — O
2
glc
6
R2
CH2OH
R1 R2
escin Ia tig CH2OH
escin Ib ang CH2OH
escin IIa tig H
escin IIb ang H
CH3
tig:
C
O
CH3
H
CH3
ang:
C
O
H
CH3
Ac: — C — CH3
O

FIGURE 19.9. The chemical structures of escins Ia, Ib, IIa, IIb.

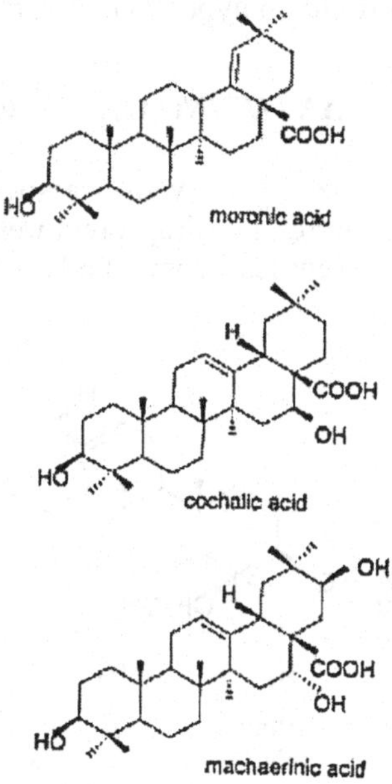
COOH
HO
moronic acid
H
COOH
OH
HO
cochalic acid
OH
H
COOH
OH
HO
machaerinic acid

FIGURE 19.10. The chemical structures of the genins of the calendasaponins.

cose levels in the extract-treated mice was associated with a significant reduction in serum insulin levels in fed and fasting mice. Additional studies revealed that ginsenoside Re plays a significant role in the antihyperglycemic action.

Two novel triterpenoid saponins, assamicins I and II from *Aesculus assamica* (see Figure 19.11) with insulin-like activity, were shown to inhibit the release of free fatty acids from epinephrine-treated rat adipocytes and enhanced glucose uptake into 3T3-L1 adipocytes.[50]

Gymnemic acid IV from *Gymnema sylvestre,* at doses of 3.4 to 13.4 $mg \cdot kg^{-1}$ reduced the blood levels in STZ-diabetic mice by 13.5 to 60.0 percent six hours after the administration, comparable to the potency of glibenclamide, and did not change the blood glucose levels on normal mice. Furthermore at 13.4 $mg \cdot kg^{-1}$, it increased plasma insulin levels in STZ diabetic mice, which might contribute to the antihyperglycemic effect of the plant.[51] Kaikosaponin III (see Figure 19.7) from *Pueraria thunbergiana* showed potent hypoglycemic and hypolipidemic effects in the STZ-induced diabetic rats.[52] An i.p. administration with 5 and 10 $mg \cdot kg^{-1}$ for seven days to the rats reduced blood glucose, total cholesterol, LDL- and VLDL-cholesterol and triglyceride levels when compared to control group. It has been shown that the antioxidant action of kaikosaponin III may alleviate toxicity and contribute to hypoglycemic effects.

ANTIFUNGAL ACTIVITY

The increasing incidence of mycoses associated with AIDS and treatment with immunosuppressive drugs requires the search for new antimycotic compounds. New lead compounds are also required for agricul-

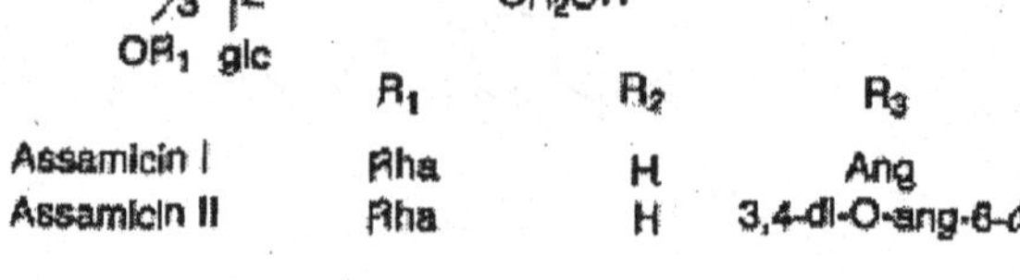

	R_1	R_2	R_3
Assamicin I	Rha	H	Ang
Assamicin II	Rha	H	3,4-di-O-ang-6-deoxy-glc

FIGURE 19.11. The chemical structures of the assamicins.

tural use. Saponins are an important source of antifungal compounds.[53] The antifungal properties of saponins are generally ascribed to the ability of these molecules to bind with sterols in fungal membranes, so causing pore formation and loss of membrane integrity. Bioactivity-guided fractionation of the active extracts led to the isolation and characterization of many antifungal saponins. A large panel of fungal species (human pathogenic yeasts, dermatophytes) was used to assess the minimal inhibitory concentration (MI) and the potential minimum fungicidal concentration (MFC) of saponins. Recent studies demonstrating significant antifungal activities of various saponins are summarized in Table 19.1.[54]

ANTIVIRAL ACTIVITY

The ever-increasing resistance of human pathogens to current anti-infective agents is a serious medical problem. It is evident that the continued search for new drugs and the investigation of new targets are essential for the development of more effective antiviral chemotherapy. We present here the most relevant in vitro activities obtained in recent years against some important enveloped viruses, *Herpes simplex* virus (HSV), human immunodeficiency virus (HIV), vesicular stomatis virus (VSV), and influenza viruses. Often, such results do not present a counterpart in vivo. Some triterpene saponins were reported with anti-HSV-1 activity.[55] They did not show evidence of cytotoxicity under antiviral test conditions. The oleanane-type saponin inhibited HSV-1 DNA synthesis, whereas the ursane type seemed to inhibit viral capsid protein synthesis. Oleanane triterpene saponins from some fabaceous plants (see Figure 19.7) were tested for their anti-HSV-1 activity by the plaque reduction assay.[56] They were classified into three groups soyasapogenol B-, sophoradiol- and soyasapogenol E-glycosides. Among sophoradiol glycosides, the order of potency was kaikasaponins III > I > sophoradiol monoglucuronide. It was suggested that the trisaccharide group showed greater action than the disaccharide group. Neither the monoglucuronide of sophoradiol nor that of soyasapogenol B showed activity. Concerning the structure/activity relationships it seems that a glucosyl unit in the central sugar moiety showed greater action. In comparison with the activities for a group having the same trisaccharide, it seems that the carbonyl group at C-22 would be more effective than the hydroxyl group in anti-HSV-1 activity, while the hydroxyl group at C-24 could reduce the activity.[57]

The effects of a mixture of tea-seed saponins from *Camellia sinensis* var. *sinensis* were evaluated on human influenza viruses types A and B in MDCK cells.[58] At the concentrations of 60, 80, and 100 $\mu g \cdot mL^{-1}$,

TABLE 19.1. Antifungal saponins discovered in recent years.

Compound	Source	Fungus	Concentration ($\mu g \cdot mL^{-1}$)
Hederacolchiside A[a]	*Hedera colchica*	*Candida* spp.	MFC 6.25-25
		Trichosporon cutaneum	MFC 12.5
		Cryptococcus neoformans	MFC 12.5
		Trichophyton spp.	MFC 12.5
α-hederin[a]	*Hedera colchica*	*Candida* spp.	MFC 6.25-100
		Trichosporon cutaneum	MFC 100
		Cryptococcus neoformans	MFC 100
		Trichophyton spp.	MFC 12.5
Polygalacic acid glycosides[b]	*Solidago virgaurea*	*Candida albicans* strains	100 percent inh 100-200
	Heteropappus biennis	*C. tropicalis, C. krusei, C. glabrata*	100 percent inh 100-400
Jujubogenin glycosides[c]	*Colubrina retusa*	*C. albicans*	MIC 50
		C. neoformans	MIC 50
		A. fumigatus	MIC 50
Astraverrucin II[d]	*Astragalus verrucosus*	*Aspergillus niger*	Inh z. (21/22 mm)
		Botrytis cinerea	Inh z. (22/23 mm)
Phytolaccoside B[e]	*Phytolacca tetramera*	*C. albicans,* *C. tropicalis,* *C. neoformans,* *S. cerevesiae,* *Aspergillus fumigatus, A.* *flavus, A. niger, T.* *mentagrophytes,* *T. rubrum*	MIC 50-125 MIC 50 MIC 50-125 MIC 125 MIC 25-50
Tetrahydroxyolean-ene-glycoside[f]	*Lupinus angustifolius*	*C. albicans*	MIC 25-30
Cycloartane glycosides[g]	*Thalictrum minus*	*C. albicans*	78.7 percent inhibition at 1 $mg \cdot ml^{-1}$
Hederagenin glycosides[h]	*Kalopanax pictus* var. *chinense*	*Microsporum canis, Trichophyton mentagrophytes, Cryptococcus neoformans, C. albicans, Coccidioides immitis*	MIC 6.25-25

Compound	Source	Fungus	Concentration ($\mu g \cdot mL^{-1}$)
Spirostane glycoside[i]	*Dioscorea futschauensis*	*Pyricularia orizae*	MMDC 2-12
CAY-1, spirostan saponin (see Figure 19.12)[j]	*Capsicum frutescens*	*Aspergillus flavus, fumigatus, parasiticus, niger*	LD_{90} 3-20μM
Spirostanol saponins[k]	*Allium porrum*	*Fusarium culmorum*	ED_{50} 30
Spirostanol saponins[l]	*Tribulus terrestris*	*C. albicans, Cryptococcus neoformans*	MIC 1.5-6.25
Spirostanol saponins[m]	*Yucca schidigera*	Yeast strains	MIC 3.13-100
SC-1, spirostan saponin[n]	*Solanum chrysotrichum*	*Trichophyton mentagrophytes*	MIC 40

[a] Mshvildadze, V., Favel, A., Delmas, F., Elias, R., Faure, R., Decanosidze, G., Kemertelidze, E., and Balansard, G., 2000, Antifungal and antiprotozoal activities of saponins from *Hedera colchica, Pharmazie,* 55: 325-326.
[b] Bader, G., Seibold, M., Tintelnot, K., and Hiller, K., 2000, Cytotoxicity of triterpenoid saponins. Part 2: Relationships between the structures of glycosides of polygalacic acid and their activities against pathogenic *Candida* species, *Pharmazie,* 55(1): 72-74.
[c] Li, X.C., ElSohly, H.N., Nimrod, A.C., and Clark, A.M., 1999, Antifungal jujubogenin saponins from *Colubrina retusa, Journal of Natural Products,* 62(5): 674-677.
[d] Pistelli, L., Bertoli, A., Lepori, E., Morelli, I., and Panizzi, L. 2002, Antimicrobial and antifungal activity of crude extracts and isolated saponins from *Astragalus verrucosus, Fitoterapia,* 73(4): 336-339.
[e] Escalanta, A.M., Santecchia, C.B., Lopez, S.N., Gattuso, M.A., Ravelo, A.G., Monache, F.D., Sierra, M.G., and Zacchino, A.S., 2002, Isolation of antifungal saponins from *Phytolacca tetramera,* an Argentinean species in critic risk, *Journal of Ethnopharmacology,* 82: 29-34.
[f] Woldemichael, G.M. and Wink, M., 2002, Triterpene glycosides of *Lupinus angustifolius, Phytochemistry,* 60: 323-327.
[g] Gromova, A.S., Lutsky, V.I., Li, D., Wood, S.G., Owen, N.L., Semenov, A.A., and Grant, D.M., 2000, Thalicosides A1-A3, minor cycloartane bisdesmosides from Thalictrum minus, *Journal of Natural Products,* 63: 911-914.
[h] Lee, M.W., Kim, S.U., and Hahn, D.R., 2001, Antifungal activity of modified hederagenin glycosides from the leaves of *Kalopanax pictum* var. *chinense, Biological and Pharmaceutical Bulletin,* 24: 718-719.
[i] Liu, H.W., Hu., K., Zhao, Q.C., Cui, C.B., Kobayashi, H., and Yao, X.S., 2002, Bioactive saponins from *Dioscorea futschauensis, Pharmazie,* 57: 570-572.
[j] De Lucca, A.J., Bland, J.M., Vigo, C.B., Cushion, M., Selitrennikoff, C.P., Peter, J., and Walsh, T.J., 2002, CAY-1, a fungicidal saponin from *Capsicum* spp. fruit, *Medical Mycology,* 40: 131-137.
[k] Carotenuto, A., Fattorusso, E., Lanzotti, V., and Magno, S., 1999, Spirostanol saponins of *Allium porrum, Phytochemistry,* 51: 1077-1082.
[l] Bedir, E., Khan, I.A., and Walker, L.A., 2002, Biologically active steroidal glycosides from *Tribulus terrestris, Pharmazie,* 57(7): 491-493.

TABLE 19.1 *(continued)*

[m] Miyakoshi, M., Tamura, Y., Masuda, H., Mizutani, K., Tanaka, O., Ikeda, T., Ohtani, K., Kasai, R., and Yamasaki, K., 2000, Antiyeast steroidal saponins from *Yucca schidigera* (Mohave Yucca), a new anti-food-deteriorating agent, *Journal of Natural Products,* 63: 332-338.

[n] Alvarez, L., Perez, M.C., Gonzalez, J.L., Navarro, V., Villarreal, M.L., and Olson, J.O., 2001, SC-1, an antimycotic spirostan saponin from *Solanum chrysotrichum, Planta Medica,* 67(4): 372-374.

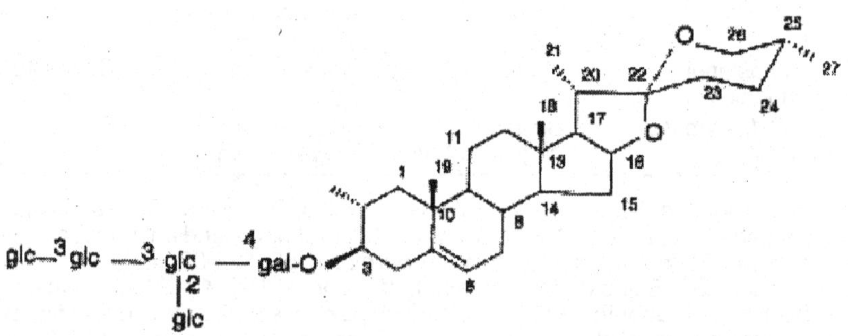

FIGURE 19.12. The chemical structure of CAY-1, a fungicidal saponin from *Capsicum.*

respectively, the mixture inactivated viruses A (H3N2) and (H1N1) and B/Lee/40 almost completely. The mixture also inactivated type A virus A after inoculation at concentrations of 1-30 $\mu g \cdot mL^{-1}$ dose-dependently (IC_{50} 24 $\mu g \cdot mL^{-1}$). Moreover, mechanism of action and safety studies are needed.

Among saponins from *Maesa lanceolata,* maesasaponin V_3 (see Figure 19.13) exhibited the strongest extracellular activity against HSV-1 virus with a viral titer reduction factor (RF) of 10^3 at 10 $\mu g \cdot mL^{-1}$, this being the ratio of the viral titers in the virus control and in the presence of the maximal nontoxic dose of the test substance. Cetrimonium bromide was used as positive control (RF = 10^3 at 50 $\mu g \cdot mL^{-1}$).[59]

The influence of the substitution pattern of several acylated triterpenoid saponins from *M. lanceolata* on their biological activities was investigated and structure-activity relationships were established. In general, 21,22-diacylation with a free 16-OH appeared to be associated with a virucidal activity (RF = 10^3 at 50 $\mu g \cdot mL^{-1}$). In the anti-HIV-1 and anti-HIV-2 screening, the mixture as well as the maesasaponins showed no activity. A saikosaponin, buddlejasaponin IV (see Figure 19.14) from *Bupleurum rigidum,*

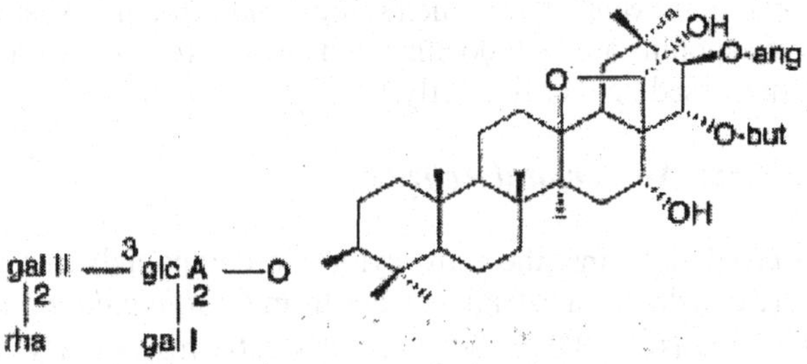

FIGURE 19.13. The chemical structure of maesasaponin V_3.

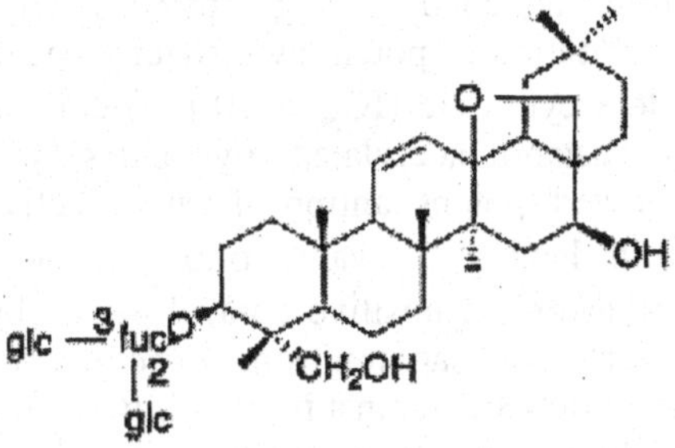

FIGURE 19.14. The chemical structure of the saikosaponin buddlejasaponin IV.

was tested for its in vitro activity against HSV-1, VSV, and poliovirus type 1. It showed a virucidal activity against VSV, an RNA virus at concentrations ranging from 20 to 25 $\mu g \cdot mL^{-1}$ without any cytotoxic effect at this concentration range, being as effective as the reference drug, dextran sulphate.[60] Two steroid saponins from *Solanum torvum* exhibited antiviral activity (HSV-1) (IC_{50} 23.2 and 17.4 $\mu g \cdot mL^{-1}$, respectively).[61]

In the search for new anti-HIV agents, escins from *Aesculus chinensis* were found to show moderate anti-HIV-1 protease activity with an inhibition of 86.1 ± 0.2 percent at 100 μM whereas escin Ia and Ib (see Figure 19.9) displayed this activity with IC_{50} of 35 and 50 μM, respectively.[62]

ANTICANCER AND IMMUNE SYSTEM–RELATED ACTIVITY

Advances in the treatment and prevention of cancer will require the continued development of novel and improved chemotherapeutic and chemopreventive agents. Since cytotoxic, apoptosis-inducing, antitumor, antimutagenic, chemopreventive, antimetastatic, and immune system enhancing effects of saponins are well documented elsewhere, the most relevant results will be discussed here only briefly.[63]

Cytotoxic and Antitumoral Activity

Of steroid saponins, the most cytotoxic compounds that have been reported are two cholestane triglycosides from *Ornithogalum saundersiae* (see Figure 19.15) (IC_{50} 0.016 μM and 0.014 μM, respectively, against HL60 cells) compared with the positive control drugs, etoposide (IC_{50} 0.025 μM), adriamycin (IC_{50} 0.0072 μM), and methotrexate (IC_{50} 0.012 μM).[64]

The ester groups attached to the glycoside moiety were found to be essential for the exhibition of potent cytotoxicity because the deacyl derivatives were far less cytotoxic (IC_{50} > 10 μM) compared with the original compounds. Spirostanol acetylated glycosides isolated from *Dracaena angustifolia* showed potent antiproliferative activity against HT-1080 fibrosarcoma cells, having IC_{50} values of 0.2, 0.3, and 0.6 μM, respectively, comparable to that of the positive control doxorubicin (IC_{50} 0.2 μM).[65] Protodioscin, a furostanol saponin from *Dioscorea colletii* var. *hypoglauca,* was shown to be cytotoxic against most cell lines from leukemia and solid tumor in the NCI's human cancer panel, especially selectively against one leukemia line (MOLT 4), two colon cancer lines (HCT 116 and SW-620), one CNS cancer line (SNB-75), one melanoma line (LOX IMVI), and one renal cancer line (786-0), with GI_{50} < 2 μM.[66] Methyl protoneodioscin, giving similar results, is in the list for the NCI's in vivo assay in nude mice after evaluation of nontoxicity (maximum tolerated dose, MTD > 600 mg·kg^{-1}).[67] Among triterpene glycosides from *Kalopanax pictus*, kalopanax saponin A (commonly called α-hederin) (see Figure 19.2) showed significant cytotoxicity on several types of tumor cells. In the in vivo experiments, kalopanax saponin A (15 mg·kg^{-1}, i.p.) apparently increased the life span of mice bearing colon 26 and 3LL Lewis lung carcinoma, as well as cisplatin

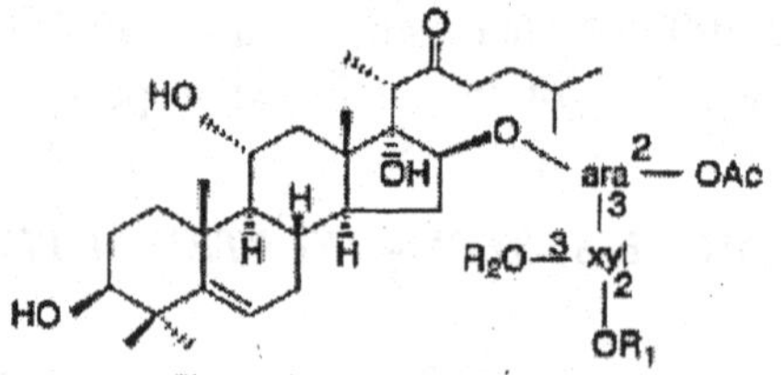

	R1	R2
1	3,4-dimethoxybenzoyl	glc
2	3,4,5-trimethoxybenzoyl	glc
3	p-methoxybenzoyl	H

FIGURE 19.15. The chemical structures of cholestane glycosides from *Ornithogalum saundersiae.*

(3 mg·kg^{-1}, i.p.).[68] A comparison of the cytotoxicity of hederacolchisid A1, a new oleanolic acid monodesmoside from *Hedera colchica* (IC_{50} 4.5 μM in malignant melanoma versus normal cells IC_{50} 7.5 μM) with that of five other saponins from this plant offers some new information about structure/activity relationships. It was observed that for the same sugar sequence, monodesmosides with oleanolic acid as aglycone exhibit higher cytotoxicity than those containing hederagenin.[69] It was recently reported that a fraction CC-5 from the seeds of *Nigella sativa* caused dose-dependent inhibition of tumor induction and inhibited tumor growth time and dose dependently when given before tumor transplantation. α-Hederin, isolated from this bioactive fraction, was found to inhibit tumor growth more significantly than the alkilating agent, cyclophosphamide. These results were confirmed in vivo. After administration to mice with formed tumors, α-hederin produced significant dose-dependent tumor inhibition rate (TIR) values in comparison with the cyclophosphamide-treated group.[70] A study has demonstrated that this molecule stimulates NO release and is able to upregulate iNOS expression through NF-kB transactivation, which may be a mechanism whereby α-hederin elicits its biological effects.[71] Apoptosis could be a major mechanism by which α-hederin is preventing tumor growth. Apoptosis is characterized by cell shrinkage, membrane bleeding, chromatin condensation, and formation of a DNA ladder of multiples of 180 base pairs caused by internucleosomal DNA cleavage. Since it has recently been reported that cancer chemotherapeutics exert part of their pharmacological effect by triggering apoptotic cell death, apoptosis-inducing compounds in tumor cells have become useful as leading compounds for the development of anticancer drugs. Ginsenoside Rh-2 (p.o., s.c.) inhibited growth of human ovarian cells transplanted into nude mice and significantly prolonged the survival times of mice.[72] Apoptosis-inducing activity and NK-cell enhancing activity were two approaches of mechanism of action. The number of reports on saponins having apoptotic activity is increasing, among them, ginsenosides, furostanol saponins from Liliaceae, avicins from *Acacia victoriae* (see Figure 19.16), momordin I from *Ampelopsis japonica,* and so on.[73]

It was observed in the author's group that the combined use of some acylsaponins from *Silene fortunei* and cisplatin increased the intracellular platinum concentration and potentiated the in vitro cytotoxic effect of cisplatin in human colon cancer, HT-29 cells.[74] The absence of potentiating activity of deacylated saponins suggested the importance of the p-methoxycinnamoyl group for activity. Such effect was also observed in vivo with a combination of digitonin and cisplatin during isolated lung perfusion.[75]

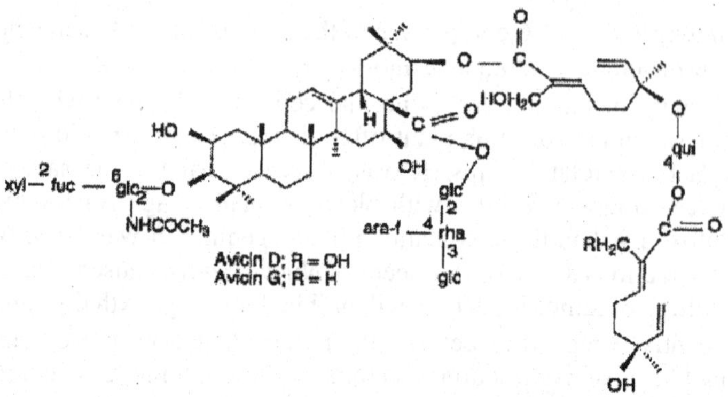

FIGURE 19.16. The chemical structure of saponins from *Acacia victoriae*.

Antimutagenic and Chemopreventive Activity

In carcinogenesis, the initiation (mutation) involves the direct action of the carcinogen on target cells (or after its metabolic activation) yielding initiated cells, whereas the promotion/progression means that the initiated cells are stimulated to proliferate yielding tumor cells. Inhibition of mutagenesis and inhibition of the tumor promotion/progression have been used as a screen in working models for the discovery of agents that may be useful for preventing cancer. A fraction of soybean saponins containing group B soyasaponins and 2,3-dihydro-2,5-dihydroxy-6-methyl-4H-pyran-4-one (DDMP) repressed 2-acetoxyacetylaminofluorene (2AAAF)-induced DNA damage in Chinese hamster ovary cells as measured by alkaline Comet assay.[76] Majonoside-R2 (see Figure 19.4) exhibited a potent anti-tumor-promoting activity in a two-stage carcinogenesis test of mouse hepatic tumor induced by *N*-nitrosodiethylamine as an initiator and phenobarbital as a non-TPA-type promotor. Furthermore, MR-2 exhibited a remarkable inhibitory effect on a two-stage carcinogenesis test of mouse skin tumors induced by a nitric oxide donor (an hexenamide derivative) as an initiator and TPA as a promoter.[77] Some reports described in a recent review have suggested that fewer glycosylated protopanaxadiol derivatives are effective in cancer-prevention activities.[78] Ginsenoside Rh_2 and Rg_3 possess the most promising anticancer activities. A triterpenoid saponin mixture from *Acacia victoriae,* having acacic acid as aglycone (see Figure 19.16), was shown to be active in an initiation/promotion model in mouse skin carcinogenesis.[79] It was also observed that a 62 percent and 74 percent reduction occurred by avicins in H-ras mutations at codon 61 in the 7,12-dimethylbenz[*a*]anthracene (DMBA) and DMBA/TPA models.

Activity on the Immune System

The immunological properties of saponins could be divided into two groups such as adjuvants and immunostimulants. The administration of an antigen in combination with saponins considered to be immunoadjuvants yields an enhanced immune response against the antigen resulting in increased production of antibodies and cytotoxic T lymphocytes. Nonspecific imunostimulants are given on their own in order to elicit a generalized state of resistance to pathogens or tumors.[80]

The astragalosides, cycloartane-type triterpene glycosides from Turkish species, have been tested for their immunostimulant activity by using a transcription-based bioassay for NF-κB activation in a human macrophage/monocyte cell line. Astragaloside I (see Figure 19.17) at 10 $\mu g \cdot mL^{-1}$, increased NF-κB directed luciferase expression to levels about 65 percent as compared with maximal stimulation by *E. coli* LPS.[81]

Two cycloartane-type triterpene saponins from *Astragalus peregrinus* were shown to stimulate the proliferation of mouse splenocytes at a dose of 0.1 $\mu g \cdot mL^{-1}$ in the presence or absence of Con A.[82] A series of endotoxin free samples were examined at subtoxic concentrations for immunological properties in the lymphocyte proliferation test (LPT) and for phagocytosis activity. Among them, songarosaponin C from *Verbascum songaricum* possesses the highest immunosuppressive activity and is more active than cyclosporin A.[83]

QS 21, a quillaic acid-based triterpene saponin (see Figure 19.18), has been shown to be an effective immunological adjuvant enhancing the antibody responses and $CD8^+$ cytotoxic T lymphocytes with a wide variety of

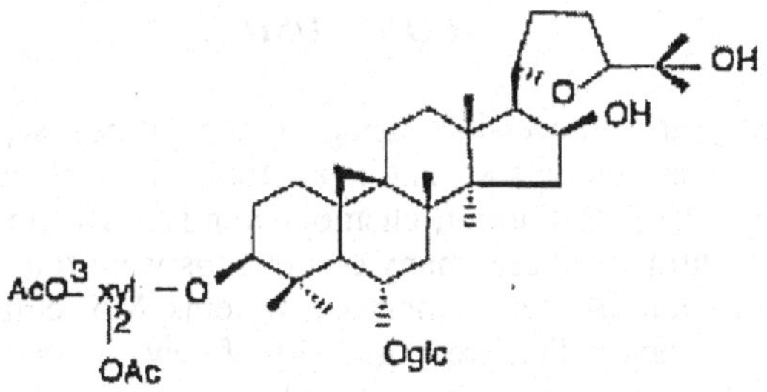

FIGURE 19.17. The chemical structure of astragaloside I from *Astragalus oleifolius*.

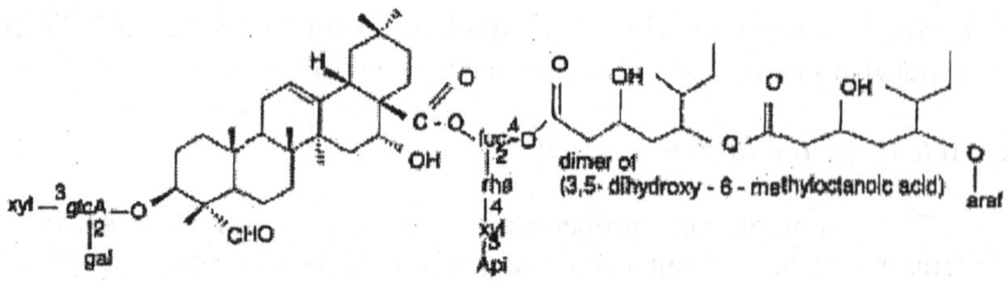

FIGURE 19.18. The chemical structure of QS 21 from *Quillaja saponaria.*

antigens, and to have a relatively low toxicity in preclinical studies. QS 21 has been evaluated in a large number of vaccines in Phase I and II human clinical trials, including cancer immunotherapeutics, HIV recombinant envelopes, and malarial antigens.[84] QS 21 at a dose of 50 to 100 μg has been tested in more than 2,600 individuals to date. Since QS 21 formulations caused immediate injection-site pain, a recent study has shown the improvement of their acceptability through reformulation with certain excipients.[85] To understand the importance of the acylation to the adjuvant activity, some derivatives were prepared, such as the deacylated QS 21 (DS-1). In contrast to QS 21, DS-1 did not stimulate a strong level of antibody and CTL response.[86]

Semisynthetic triterpene saponins known as GPI-0100 were prepared by deacylation of a crude mixture of *Q. saponaria* saponins and then coupling dodecylamine with the carboxyl group of the glucuronic acid residue of the deacylated saponin through an amide bond.[87] RDS-1 is the HPLC-purified analog of QS 21. These studies suggested that reacylation at a different site than the native compound with a synthetic fatty acid (dodecylamine) does not substantially improve the diminished adjuvant activity of the deacylated compound.[88]

CONCLUSION

In recent years, an ever-increasing number of new saponins have been discovered as antiphlogistic, antihepatotoxic, hypoglycemic, antifungal, antiviral, cytotoxic, antitumor, chemopreventive, and immunomodulating agents. In addition to these, many compounds were found to be active on the cardiovascular and central nervous systems. Molecular biological data have been obtained with a great diversity of new bioassays using enzymes and cells, allowing structure-activity relationship studies. Explanations of the mechanism of action sometimes remain in dark and further investigations are needed. Most of the saponins known as amphiphilic substances seem to accomplish their activity through the ability to intercalate into the plasma membrane. This leads to changes in membrane fluidity, and thus affects membrane functions, eliciting a cellular response. The broad potent immunoadjuvant activity led to the selection of QS 21 for clinical studies. It seems that the presence of acyl groups in many natural saponins improved the activity. It is the case, for example, of cholestane glycosides *(Ornithogalum saundersiae),* Avicins D and G *(Acacia victoriae),* jenissensosides C, D *(Silene jeniseensis),* QS 21 *(Quillaja saponaria),* Maesasaponin V_3 *(Maesa lanceolata),* and assamicins I and II *(Aesculus assamica).* According to the promising results obtained with acylated saponins, the development of semisynthetic compounds by chemical modification on the aglycone, or in the sugar chain of natural saponins, might improve considerably their pharmacological activities.

NOTES

1. Hostettmann, K. and Marston, A., Triterpine saponins: Pharmacological and biological properties, in Phillipson, J.D., Ayres, D.C., and Baxter, H. (eds.), *Saponins* (Cambridge, UK: Cambridge University Press, 1995), pp. 232-306; Lacaille-Dubois, M.A. and Wagner, H., Bioactive saponins from plants: An update, in Atta-ur-Rahman (ed.), *Studies in natural products chemistry,* Volume 21 (Amsterdam, The Netherlands: Elsevier, 2000), pp. 633-634; Rao, A.V. and Gurfinkel, D.M., 2000, The bioactivity of saponins: Triterpenoid and steroidal glycosides, *Drug Metabolism and Drug Interactions,* 17: 211-215.

2. Kim, Y.P., Lee, E.B., Kim, S.Y., Li, D., Ban, H.S., Lim, S.S., Shin, K.H., and Okuchi, K., 2001, Inhibition of prostaglandin D isolated from the root of *Platycodon grandiflorum, Planta Medica,* 67: 362-364.

3. Shin, C.Y., Lee, W.J., Lee, E.B., Choi, E.Y., and Ko, K.H., 2002, Platycodin D and D3 increase airway mucin release in vivo and in vitro in rats and hamsters, *Planta Medica,* 68: 221-225.

4. Kim, Y.K., Kim, R.G., Park, S.J., Ha, J.H., Choi, J.W., Park, H.J., and Lee, K.T., 2002, In vitro antiinflammatory activity of kalopanaxsaponin A isolated from *Kalopanax pictus* in murine macrophage RAW 264.7 cells, *Biological and Pharmaceutical Bulletin,* 25(4): 472-476.

5. Choi, J., Huh, K., Kim, S.H., Lee, K.T., Park, H.J., and Han, Y.N., 2002, Antinociceptive and anti-rheumatoidal effects of *Kalopanax pictus* extract and its saponin components in experimental animals, *Journal of Ethnopharmacology,* 79(2): 199-204.

6. Li Da, W., Lee, E.B., Kang, S.S., Hyun, J.E., and Whang, W.K., 2002, Activity-guided isolation of saponins from *Kalopanax pictus* with anti-inflammatory activity, *Chemical and Pharmaceutical Bulletin,* 50(7): 900-903.

7. Navarro, P., Giner, R.M., Recio, M.C., Manez, S., Cerda-Nicolas, M., and Rios, J.L., 2001, In vivo anti-inflammatory activity of saponins from *Bupleurum rotundifolium, Life Sciences,* 68(10): 1199-1206.

8. Kim, S.Y., Son, K.H., Chang, H.W., Kang, S.S., and Kim, H.P., 1999, Inhibition of mouse ear edema by steroidal and triterpenoid saponins, *Archives of Pharmacal Research,* 22: 313-316.

9. Cabral de Oliveira, A.C., Perez, A.C., Merino, G., Prieto, J.G., and Alvarez, A.I., 2001, Protective effects of *Panax ginseng* on muscle injury and inflammation after eccentric exercise, *Comparative Biochemistry and Physiology,* Part C 130: 369-377.

10. Le Tran, Q., Adnyana, I.K., Tezuka, Y., Nagaoka, T., Tran, Q.K., and Kadota, S., 2001, Triterpene saponins from Vietnamese ginseng *(Panax vietnamensis),* and their hepatoprotective activity, *Journal of Natural Products,* 64: 456-461.

11. Le Tran, Q., Adnyana, I.K., Tezuka, Y., Harimaya, Y., Saiki, I., Kurashige, Y., Tran, Q.K., and Kadota, S., 2002, Hepatoprotective effect of majonoside R2, the major saponin from Vietnamese ginseng *(Panax vietnamensis), Planta Medica,* 68: 402-406.

12. Ibid.

13. Wu, F., Cao, J., Jiang, J., Yu, B., and Xu, Q., 2001, Ruscogenin glycoside (Lm3) isolated from *Liriope muscari* improves liver injury by disfunctioning liver-infiltrating lymphocytes, *Journal of Pharmacy and Pharmacology,* 53: 681-688.

14. Adnyana, I.K., Tezuka, Y., Banskota, A.H., Xiong, Q., Tran, K.Q., and Kadota, S., 2000, Quadranosides I-V, new triterpene glucosides from the seeds of *Combretum quadrangulare, Journal of Natural Products,* 63(4): 496-500.

15. Chen, J.C., Tsai, C.C., Chen, L.D., Chen, H.H., and Wang, W.C., 2000, Therapeutic effect of gypenoside on chronic liver injury and fibrosis induced by CCl_4 in rats, *American Journal of Chinese Medicine,* 28(2): 175-185.

16. Kinjo, J., Okawa, M., Udayama, M., Sohno, Y., Hirakawa, T., Shii, Y., and Nohara, T., 1999, Hepatoprotective and hepatotoxic actions of oleanolic acid-type triterpenoidal glucuronides on rat primary hepatocyte cultures, *Chemical and Pharmaceutical Bulletin,* 47(2): 290-292.

17. Ibid.

18. Kinjo, J., Ikeda, T., Okawa, M., Udayama, M., Hirakawa, T., Shii, Y., and Nohara, T., 2000, Hepatoprotective and hepatotoxic activities of sopharadiol analogs on rat primary liver cell cultures, *Biological and Pharmaceutical Bulletin,* 23(9): 1118-1121.

19. Scott, G.I., Colligan, P.B., Ren, B.H., and Ren, J., 2001, Ginsenosides Rb1 and Re decrease cardiac contraction in adult rat ventricular myocytes: Role of nitric oxide, *British Journal of Pharmacology,* 134(6): 1159-1165.

20. Sung, J., Han, K.H. , Zo, J.H., Park, H.J., Kim, C.H., and Oh, B.H., 2000, Effects of red ginseng upon vascular endothelial function in patients with essential hypertension, *American Journal of Chinese Medicine,* 28: 205-216.

21. Lu, S., Zhang, J., and Yang, D., 1999, Effects of Astragaloside in treating myocardial injury and myocardial Sarco/Endoplasmic Ca(2+)-ATPase of viral myocarditis mice, *Zhongguo Zhong Xi Yi Jie He Za Zhi,* 19(11): 672-674.

22. Dou, D.Q., Zhang, Y.W., Zhang, L., Chen, Y.J., and Yao, X.S., 2001, The inhibitory effect of ginsenosides on PTK activated by hypoxia/reoxygenation in cultured human umbilical vein endothelial cells, *Planta Medica,* 67: 19-23.

23. Tanner, M.A., Bu, X., Steimle, J.A., and Myers, P.R., 1999, The direct release of nitric oxide by gypenosides derived from the herb *Gynostemma pentaphyllum, Nitric Oxide,* 3(5): 359-365.

24. Kim, N.D., Kang, S.Y., Park, J.H., and Schini-Kerth, V.B., 1999, Ginsenoside Rg3 mediates endothelium-dependent relaxation in response to ginsenosides in rat aorta: Role of K^+ channels, *European Journal of Pharmacology,* 367: 41-49.

25. Kim, N.D., Kang, S.Y., Kim, M.J., Park, J.H., and Schini-Kerth, V.B., 1999, The ginsenoside Rg3 evokes endothelium-independent relaxation in rat aortic rings: Role of K^+ channels, *European Journal of Pharmacology,* 367: 51-57.

26. Rhiouani, H., Settaf, A., Lyoussi, B., Cherrah, Y., Lacaille-Dubois, M.A., and Hassar, M., 1999, Effect of *Herniaria glabra* saponins in the spontaneously hypertensive rat, *Thérapie,* 54: 735-739.

27. Johns, T., Mahunnah, R.L.A., Sanaya, P., Chapman, L., and Ticktin, T., 1999, Saponins and phenolic content in plant dietary additives of a traditional subsistence community, the Batemi of Ngorongoro District, Tanzania, *Journal of Ethnopharmacology,* 66(1): 1-10.

28. Morehouse, L.A., Bangerter, F.W., DeNinno, M.P., Inskeep, P.B., McCarthy, P.A., Pettini, J.L., Savoy, Y.E., Sugarman, E.D., Wilkins, R.W., Wilson, T.C. et al., 1999, Comparison of synthetic saponin cholesterol absorption inhibitors in rabbits: Evidence for a non-stoichiometric, intestinal mechanism of action, *Journal of Lipid Research,* 40(3): 464-474.

29. Zhang, J., Meng, Z., Zhang, M., Ma, D., Xu, S., and Kodama, H., 1999, Effects of six steroidal saponins from anemarrhenae rhizoma on platelet aggregation and hemolysis in human blood, *Clinica Chimica Acta,* 289: 79-88.

30. Yoshikawa, M., Morikawa, T., Jujiwara, E., Ohgushi, T., Asao, Y., and Matsuda, H., 2001, New oleanane-type triterpene saponins with gastroprotective effect and platelet aggregation activity from the flowers of *Camellia japonica*: Revised structures of camellenodiol and camelledionol, *Heterocycles,* 55: 1653-1657.

31. Liu, Z.Q., Luo, X.Y., Sun, Y.X., Chen, Y.P., and Wang, Z.C., 2002, Can ginsenosides protect human erythrocytes against free-radical-induced hemolysis? *Biochimica et Biophysica Acta,* 1572(1): 58-66.

32. Ibid.

33. Suh, H.W., Song, D.K. Huh, S.O. Son, K.H., and Kim, Y.H., 2000, Antinociceptive mechanisms of Dipsacus saponin C administered intrathecally in mice, *Journal of Ethnopharmacology,* 71: 211-218.

34. Choi, S.S., Han, E.J., Lee, T.H., Lee, J.K., Han, K.J., Lee, H.K., and Suh, H.W., 2002, Antinociceptive mechanisms of platycodin D administered intracerebroventricularly in the mouse, *Planta Medica,* 68: 794-798.

35. Ibid.

36. Suh, H.W., Song, D.K., Huh, S.O., and Kim, Y.H., 2000, Modulatory role of ginsenosides injected intrathecally or intracerebroventricularly in the production of antinociception induced by kappa-opioid receptor agonist administered intracerebroventricularly in the mouse, *Planta Medica,* 66: 412-417.

37. Nemmani, K.V.S. and Ramarao, P., 2002, Role of benzodiazepine-$GABA_A$ receptor complex in attenuation of U-50, 488H-induced analgesia and inhibition of tolerance to its analgesia by ginseng total saponin in mice, *Life Sciences,* 70: 1727-1740.

38. Cicero, A.F.G., Bandieri, E., and Arletti, R., 2000, Orally administered *Panax notoginseng* influence on rat spontaneous behaviour, *Journal of Ethnopharmacology,* 73: 387-391.

39. Yobimoto, K., Matsumoto, K., Huong, N.T.T., Kasai, R., Yamasaki, K., and Watanabe, H., 2000, Suppressive effects of vietnamese ginseng saponin and its major component majonoside-R2 on psychological stress-induced enhancement of lipid peroxidation in the mouse brain, *Pharmacology Biochemistry and Behavior,* 66(3): 661-665.

40. Choi, S., Jung, S.Y., Kim, C.H., Kim, H.S., Rhim, H., Kim, S.C., and Nah, S.Y., 2001, Effect of ginsenosides on voltage-dependent Ca^{2+} channel subtypes in bovine chromaffin cells, *Journal of Ethnopharmacology,* 74(1): 75-81.

41. Lee, T.F., Shiao, Y.J., Chen, C.F., and Wang, L.C., 2001, Effect of ginseng saponins on α-amyloid-suppressed acetylcholine release from rat hippocampal slices, *Planta Medica,* 67: 634-637.

42. Takh, H., Kometani, T., Nishimura, T., Nakae, T., Okada, S., and Fushiki, T., 2001, Antidiabetic effect of glycyrrhizin in genetically diabetic KK-A mice, *Biological and Pharmaceutical Bulletin,* 24(5): 484-487.

43. Ibid.

44. Matsuda, H., Li, Y., Yamahara, J., and Yoshikawa, M., 1999, Inhibition of gastric emptying by triterpene saponin momordin C, in mice: Roles of blood glucose, capsaicin-sensitive sensory nerves, and central nervous system, *The Journal of Pharmacology and Experimental Therapeutics,* 289(2): 729-734; Matsuda, H., Li, Y., Murakami, T., Yamahara, J., and Yoshikawa, M., 1999, Effects of escins Ia, Ib, IIa, IIb from horse chestnuts on gastric emptying in mice, *European Journal of Pharmacology,* 368: 237-243; Yoshikawa, M., Murakami, T., Kishi, A., Kageura, T., and Matsuda, H., 2001, Medicinal flowers. III. Marigold. (1): Hypoglycemic, gastric emptying inhibitory, and gastroprotective principles and new oleanane-type triterpene oligoglycosides, calendasaponins A, B, C, and D, from Egyptian *Calendula officinalis, Chemical and Pharmaceutical Bulletin,* 49(7): 863-870.

45. Yoshikawa et al., 2001, Medicinal flowers.

46. Matsuda, H., Li, Y., and Yoshikawa, M., 2000, Possible involvement of dopamine and $dopamine_2$ receptors in the inhibitions of gastric emptying by escin Ib in mice, *Life Sciences,* 67(24): 2921-2927.

47. Yoshikawa, M. and Matsuda, H., 2000, Antidiabetogenic activity of oleanolic acid glycosides from medicinal foostuffs, *Biofactors,* 13: 231-237.

48. Ibid.

49. Attele A.S., Zhou, Y.P., Xie, J.T., Wu, J.A., Zhang, L., Dey, L., Pugh, W., Rue, P.A., Polonsky, K.S., and Yuan, C.S., 2002, Antidiabetic effects of *Panax ginseng* berry extract and the identification of an effective component, *Diabetes,* 51(6): 1851-1858.

50. Sakurai T., Nishimura, T., Otake, N., Xinsheng, Y., Abe, K., Zeida, M., Nagasawa, H., and Sakuda, S., 2002, Assamicin I and II, novel triterpenoid saponins with insulin-like activity from *Aesculus assamica, Bioorganic and Medicinal Chemistry Letters,* 12: 807-810.

51. Sugihara, Y., Nojima, H., Matsuda, H., Murakami, T., Yoshikawa, M., and Kimura, I., 2000, Antihyperglycemic effects of gymnemic acid IV, a compound derived from *Gymnema sylvestre* leaves in streptozotocin-diabetic mice. *Journal of Asian Natural Products Research,* 2(4): 321-327.

52. Lee, K.T., Sohn, I.C., Kim, D.H., Choi, J.W., Kwon, S.H., and Park, H.J., 2000, Hypoglycemic and hypolipidemic effects of tectorigenin and kaikasaponin III

in the streptozotocin-Induced diabetic rat and their antioxidant activity in vitro. *Archives of Pharmacal Research,* 23(5): 461-466.

53. Lacaille-Dubois and Wagner, 2000, Bioactive saponins from plants: An update.

54. Mshvildadze, V., Favel, A., Delmas, F., Elias, R., Faure, R., Decanosidze, G., Kemertelidze, E., and Balansard, G., 2000, Antifungal and antiprotozoal activities of saponins from *Hedera colchica, Pharmazie,* 55: 325-326; Bader, G., Seibold, M., Tintelnot, K., and Hiller, K., 2000, Cytotoxicity of triterpenoid saponins. Part 2: Relationships between the structures of glycosides of polygalacic acid and their activities against pathogenic *Candida* species, *Pharmazie,* 55(1): 72-4; Li ,X.C., ElSohly, H.N., Nimrod, A.C., and Clark, A.M., 1999, Antifungal jujubogenin saponins from *Colubrina retusa, Journal of Natural Products,* 62(5): 674-677; Pistelli, L., Bertoli, A., Lepori, E., Morelli, I., and Panizzi, L. 2002, Antimicrobial and antifungal activity of crude extracts and isolated saponins from *Astragalus verrucosus, Fitoterapia,* 73(4): 336-339; Escalanta, A.M., Santecchia, C.B., Lopez, S.N., Gattuso, M.A., Ravelo, A.G., Monache, F.D., Sierra, M.G., and Zacchino, A.S., 2002, Isolation of antifungal saponins from *Phytolacca tetramera*, an Argentinean species in critic risk, *Journal of Ethnopharmacology,* 82: 29-34; Woldemichael, G.M. and Wink, M., 2002, Triterpene glycosides of *Lupinus angustifolius, Phytochemistry,* 60: 323-327; Gromova, A.S., Lutsky, V.I., Li, D., Wood, S.G., Owen, N.L., Semenov, A.A., and Grant, D.M., 2000, Thalicosides A1-A3, minor cycloartane bisdesmosides from *Thalictrum minus, Journal of Natural Products,* 63: 911-914; Lee, M.W., Kim, S.U., and Hahn, D.R., 2001, Antifungal activity of modified hederagenin glycosides from the leaves of *Kalopanax pictum* var. *chinense, Biological and Pharmaceutical Bulletin,* 24: 718-719; Liu, H.W., Hu., K., Zhao, Q.C., Cui, C.B., Kobayashi, H., and Yao, X.S., 2002, Bioactive saponins from *Dioscorea futschauensis, Pharmazie,* 57: 570-572; De Lucca, A.J., Bland, J.M., Vigo, C.B., Cushion, M., Selitrennikoff, C.P., Peter, J., and Walsh, T.J., 2002, CAY-1, a fungicidal saponin from *Capsicum* spp. fruit, *Medical Mycology,* 40: 131-137; Carotenuto, A., Fattorusso, E., Lanzotti, V., and Magno, S., 1999, Spirostanol saponins of *Allium porrum, Phytochemistry,* 51: 1077-1082; Bedir, E., Khan, I.A., and Walker, L.A., 2002, Biologically active steroidal glycosides from *Tribulus terrestris, Pharmazie,* 57(7): 491-493; Miyakoshi, M., Tamura, Y., Masuda, H., Mizutani, K., Tanaka, O., Ikeda, T., Ohtani, K., Kasai, R., and Yamasaki, K., 2000, Antiyeast steroidal saponins from *Yucca schidigera* (Mohave Yucca), a new anti-food-deteriorating agent, *Journal of Natural Products,* 63: 332-338; Alvarez, L., Perez, M.C., Gonzalez, J.L., Navarro, V., Villarreal, M.L., and Olson, J.O., 2001, SC-1, an antimycotic spirostan saponin from *Solanum chrysotrichum, Planta Medica,* 67(4): 372-374.

55. Simoes, C.M., Amoros, M., and Girre, L., 1999, Mechanism of antiviral activity of triterpenoid saponins, *Phytotherapy Research,* 13(4): 323-328.

56. Kinjo, J., Yokomizo, K., Hirakawa, T., Shii, Y., Nohara, T., and Uyeda, M., 2000, Anti-herpes virus activity of fabaceous triterpenoidal saponins, *Biological and Pharmaceutical Bulletin,* 23(7): 887-889.

57. Ibid.

58. Hayashi, K., Sagesaka, Y.M., Suzuki, T., and Suzuki, Y., 2000, Inactivation of human type A and B influenza viruses by tea-seed saponins, *Biosciences Biotechnology and Biochemistry,* 64: 184-186.

59. Apers, S., Baronikova, S., Sindambiwe, J.B., Witvrouw, M., De Clercq, E., Vanden Berghe, D., Van Marck, E., Vlietinck, A., and Pieters, L., 2001, Antiviral, haemolytic and molluscicidal activities of triterpenoid saponins from *Maesa lanceolata:* Establishment of structure-activity relationships, *Planta Medica,* 67(6): 528-532.

60. Bermejo, P., Abad, M.J., Diaz, A.M., Fernandez, L., De Santos, J., Sanchez, S., Villaescusa, L., Carrasco, L., and Irurzun, A., 2002, Antiviral activity of seven iridoids, three saikosaponins and one phenylpropanoid glycoside extracted from *Bupleurum rigidum* and *Scrophularia scorodonia, Planta Medica,* 68: 106-110.

61. Arthan, D., Svasti, J., Kittakoop, P., Pittayakhachonwut, D., Tanticharoen, M., and Thebtaranonth, Y., 2002, Antiviral isoflavonoid sulfate and steroid glycosides from the fruits of *Solanum torvum, Phytochemistry,* 59: 459-463.

62. Yang, X.W., Zhao, J., Cui, Y.X., Liu, X.H., Ma, C.M., Hattori, M., and Zhang, L.H., 1999, Anti-HIV-1 protease triterpenoid saponins from the seeds of *Aesculus chinensis, Journal of Natural Products,* 62: 1510-1513.

63. Lacaille-Dubois, M.A., Bioactive saponins with cancer-related and immunomodulatory activity: Recent advances, in Atta-ur-Rahman (ed.), *Studies in natural products chemistry* (Amsterdam: Elsevier, 2004), volume 30, in press.

64. Kuroda, M., Mimaki, Y., Yokosuka, A., Sashida, Y., and Beutler, J.A., 2001, Cytotoxic cholestane glycosides from the bulbs of *Ornithogalum saundersiae, Journal of Natural Products,* 64: 88-91.

65. Le Tran, Q., Tezuka, Y., Banskota, A.H., Tran, Q.K., Saiki, I., and Kadota, S., 2001, New spirostanol glycosides ans steroidal saponins from roots and rhizomes of *Dracaena angustifolia* and their antiproliferative activity, *Journal of Natural Products,* 64: 1127-1132.

66. Hu, K. and Yao, X.S., 2002, Protodioscin (NSC-698 796): Its spectrum of cytotoxicity against sixty human cancer cell lines in an anticancer drug screen panel, *Planta Medica,* 68: 297-301.

67. Hu, K. and Yao, X.S., 2002, The cytotoxicity of methyl protoneodioscin (NSC-698791) against human cancer cell lines in vitro, *Anticancer Research,* 22: 1001-1006.

68. Park, H.J., Kwon, S.H., Lee, J.H., Lee, K.H., Miyamoto, K., and Lee, K.T., 2001, Kalopanaxsaponin A is a basic saponin structure for the anti-tumor activity of hederagenin monodesmosides, *Planta Medica,* 67: 118-121.

69. Barthomeuf, C., Debiton, E., Mshvildadze, V., Kemertelidze, E., and Balansard, G., 2002), In vitro activity of hederacolchisid A1 compared with other saponins from *Hedera colchica* against proliferation of human carcinoma and melanoma cells, *Planta Medica,* 68: 672-675.

70. Kumara, S.S.M. and Huat, B.T.K., 2001, Extraction, isolation and characterisation of antitumor principle, α-hederin, from the seeds of *Nigella sativa, Planta Medica,* 67: 29-32.

71. Jeong, H.G. and Choi, C.Y., 2002, Expression of inducible nitric oxide synthase by α-hederin in macrophages, *Planta Medica,* 68: 392-396.

72. Shibata, S., 2001, Chemistry and cancer preventing activities of ginseng saponins and some related triterpenoid compounds, *Journal of Korean Medical Sciences,* 16(Suppl.): S28-S37.

73. Lee, S.J., Ko, W.G., Kim, J.H., Sung, J.H., Lee, S.J., Moon, C.K., and Lee, B.H., 2000, Induction of apoptosis by a novel intestinal metabolite of ginseng

saponin via cytochrome c-mediated activation of caspase-3 protease, *Biochemical Pharmacology,* 60: 677-685; Candra, E., Matsunaga, K., Fujiwara, H., Mimaki, Y., Kuroda, M., Sashida, Y., and Ohizumi, Y., 2002, Potent apoptotic effects of saponins from Liliaceae plants in L1210 cells, *Journal of Pharmacy and Pharmacology,* 54(2): 257-262; Haridas, V., Higuchi, M., Jayatilake, G.S., Bailey, D., Mujoo, K., Blake, M.E., Arntzen, C.J., and Gutterman, J.U., 2001, Avicins: Triterpenoid saponins from *Acacia victoriae* (Bentham) induce apoptosis by mitochondrial perturbation, *Proceedings of National Academy of Sciences USA,* 98(10): 5821-5826; Kim, J.H., Ju, E.M., Lee, D.K., and Hwang, H.J., 2002, Induction of apoptosis by momordin I in promyelocytic leukemia (HL-60) cells, *Anticancer Research,* 22(3): 1885-1889.

74. Gaidi, G., Correia, M., Chauffert, B., Beltramo, J.L., Wagner, H., and Lacaille-Dubois, M.A., 2002, Saponins–mediated potentiation of cisplatin accumulation and cytotoxicity in human colon cancer cells, *Planta Medica,* 68 : 70-72.

75. Tanaka, T., Kaneda, Y., Li, T.S., Matsuoka, T., Zempo, N., and Esato, K., 2001, Digitonin enhances the antitumor effect of cisplatin during isolated lung perfusion, *Annals of Thoracic Surgeons,* 72: 1173-1178.

76. Berhow, M.A., Wagner, E.D., Vaughn, S.F., and Plewa, M.J., 2000, Characterization and antimutagenic activity of soybean saponins, *Mutation Research,* 448(1): 11-22.

77. Konoshima, T., Takasaki, M., Ichiishi, E., Murakami, T., Tokuda, H., Nishino, H., Duc, N.M., Kasai, R., and Yamasaki, K., 1999, Cancer chemopreventive activity of majonoside-R2 from Vietnamese ginseng, *Panax vietnamensis, Cancer Letters,* 147: 11-16.

78. Shibata, 2001, Chemistry and cancer-preventing activities of ginseng saponins.

79. Hanausek, M., Ganesh P., Walaszek, Z., Arntzen, C.J., Slaga, T.J., and Gatterman, J.U., 2001, Avicins, a family of triterpenoid saponins from *Acacia victoriae* (Bentham), suppress H-ras mutations and aneupoidy in a murine skin carcinogenesis model, *Proceedings of National Academy of Sciences USA,* 98(20): 11551-11556.

80. Lacaille-Dubois and Wagner, 2000, Bioactive saponins from plants.

81. Bedir, E., Pugh, N., Calis, I., Pasco, D.S., and Khan, I., 2000, Immunostimulatory effects of cycloartane-type triterpene glycosides from *Astragalus* species, *Biological and Pharmaceutical Bulletin,* 23(7): 834-837.

82. Verotta, L., Guerrini, M., El-Sebakhy, N.A., Asaad, A.M., Toaima, S.M., Abou-Sheer, M.E., Luo, Y.D., and Pezzuto, J.M., 2001, Cycloartane saponins from *Astragalus peregrinus* as modulators of lymphocyte proliferation, *Fitoterapia,* 72(8): 894-905.

83. Bernhardt, M., Sturm, C., Shaker, K.H., Paper, D.H., Franz, G., and Seifert, K., 2001, Structurally related immunological effects of triterpenoid saponins, *Pharmazie,* 56(9): 741-743.

84. Press, J.B., Reynolds, R.C., May, R.D., and Marciani, D.J., 2000, Structure/function relationships of immunostimulating saponins, in Atta-ur-Rahman (ed.), *Studies in natural products chemistry,* Volume 24 (Amsterdam, The Netherlands: Elsevier, 2000), pp. 131-174.

85. Waite, D., Jacobson, E.W., Ennis, F.A., Edelman, R., White, B., Kammer, R., Anderson, C., and Kensil, C.R., 2001, Three double-blind, randomized trials evalu-

ating the safety and tolerance of different formulations of the saponin adjuvant QS-21, *Vaccine,* 19: 3957-3967.

86. Liu, G., Anderson, C., Scaltreto, H., Barbon, J., and Kensil, C.R., 2002, QS-21 structure/function studies: effect of acylation on adjuvant activity, *Vaccine,* 20: 2808-2815.

87. Marciani, D.J., Press, J.B., Reynolds, R.C., Pathak, A.K., Pathak, V., Gundy, L.E., Farmer, J.T., Koratich, M.S., and May, R.D., 2000, Development of semi-synthetic triterpenoid saponin derivatives with immune stimulating activity, *Vaccine,* 18: 3141-3151.

88. Liu et al., 2002, QS-21 structure/function studies.

Chapter 20

Natural Products and Herbal Medicines in the Gastrointestinal Tract

Zohar Kerem

INTRODUCTION

The prevention of degenerative diseases through dietary intervention has enormous potential as a simple, widely accepted, and inexpensive way of improving human health. It is only in recent times that we have begun to gain an understanding of what the dietary factors are and how they act at a mechanical level.[1] Edible plants, extracts, preparations, and medicines originating from plants form a very important part of the treatment of gastrointestinal (GI) disorders and diseases. The main disorders commonly encountered are dyspepsia, spasms of the intestine, ulceration, nausea and vomiting, constipation, and diarrhea. All have been extensively described and discussed in numerous publications. This chapter attempts to present the wide range of effects of natural products and medicinal plants in the GI tract.

For any drug, including herbal medicines, absorption or injection are required for the active compounds to reach their target organ or tissue. In many cases, the low absorption and fast metabolism of any foreign molecule is the limiting factor. Consequently, it is clear that the GI tract is the place where any ingested material will be present at its highest concentration. Therefore, it is suggested that the beneficial effects of herbal preparations may be pertinent and should be studied in the GI tract.

ANTIMICROBIAL ACTIVITIES OF HERBAL PRODUCTS

Conventional treatments of microbial infections in the GI tract are usually accomplished using one of several drugs. Due to the occurrence of unpleasant side effects and increasing resistance to the synthetic pharmaceuticals recommended for the treatment of microbial infections in the GI tract, interest in the quest for natural alternatives has been increasing. New wave antimicrobials appear to have their roots in natural herbal remedies. The currently available treatments, their efficacy, side effects, and different modes of action are being increasingly studied, and novel possible ways of advancing the use of natural products have been suggested.

Since the isolation of *Helicobacter pylori* from human gastric biopsy specimens by Warren and Marshall in 1983, many investigators have reported *H. pylori* in association with gastric disorders, e.g., chronic gastritis, duodenal ulcer, and gastric cancer.[2] Antibiotics contribute the main clinical strategy for the treatment of *H. pylori* infection through the eradication of this bacterium. However, its eradication can be difficult to achieve with conventional antibiotic therapies, requiring combinations of antibiotics, proton-pump inhibitors, and bismuth preparations. Moreover, adverse effects are regularly associated with these conventional treatments.[3] Garlic is one of the most extensively researched medicinal plants.[4] Its antibacterial action depends on allicin and is thought to be due to multiple inhibitory effects on various thiol-dependent enzymatic systems. Allicin is formed catalytically by crushing raw garlic or adding water to dried garlic, when the enzyme allicinase comes into contact with alliin. Steam distillation of mashed garlic produces garlic oil containing methyl and allyl sulfides of allicin, having the practical advantage of being more stable than allicin itself. Two controlled trials of garlic preparations used to eradicate *H. pylori* infection recorded failure.[5] Other trials reviewed by Martin and Ernst without control groups (thus not meeting the inclusion criteria of their review) similarly reported no significant results, although the garlic preparations used were different in all four trials.[6] Individual constituents of garlic oil and garlic powder have shown a range of potencies when tested in vitro against human enteric bacteria, including *H. pylori*.[7] Recently, compounds isolated from *Polygonum tinctorium* Lour., a plant commonly known as indigo, were shown to have many biological activities in vitro, including anti–*H. pylori* activity.[8] The substances showing anti–*H. pylori* activity were identified as tryptanthrin, kaempferol, 6-methoxykaempferol, and 3,5,4'- trihydroxy-6,7-methylenedioxy flavone. In addition, other investigators have reported that flavonoids (kaempferol) isolated from plants have an antiulcer effect, based on the inhibition of platelet-activating factor (PAF) formation by gastric mucosa. In a recent study, the anti–*H. pylori* effect of tryptanthrin and/or kaempferol on *H. pylori* was demonstrated in vivo, in an infected Mongolian gerbil model.[9] These results suggest the possibility of a new therapeutic strategy, using plant-derived compounds, such as tryptanthrin and kaempferol, for anti–*H. pylori* and other antimicrobial therapies.

Giardia intestinalis is yet another worldwide cause of intestinal infection. This tear-shaped flagellated protozoan lives in the small intestine and is transmitted primarily when the infective cysts are ingested in water. It is an example of what is known as a zoonosis, a parasite found in wild animals which can be transmitted to humans. This protozoan is regarded as the most common flagellate in the human digestive tract and is highly contagious. Treatment of this debilitating disease is usually accomplished using one of several drugs. Its conventional and alternative treatments have been thoroughly reviewed by Harris et al.[10] Metronidazole is the treatment of choice, but benzimidazoles are now being used more frequently. Other treatments include quinacrine, paromomycin, and furazolidone. Even though these drugs are used to treat the same disease, their modes of action differ in all cases.

Throughout history, garlic has often been used for gastrointestinal complaints.[11] Its antibacterial, antifungal, and antiviral properties are well established and have prompted an investigation into its possible use as an antiprotozoal against *Entamoeba histolytica.*[12] Inhibitory activity was noted with crude extract at 25 $\mu g \cdot mL^{-1}$ and lethal dosage was established as about 50 $\mu g \cdot mL^{-1}$. Encouraged by these results, a clinical trial was carried out on patients suffering from giardiasis.[13] Garlic was established as an antigiardial, resolving the symptoms in all patients within 24 hours (h) and completely removing any indication of giardiasis from the stool within 72 h, at a dosage of 1 $mg \cdot mL^{-1}$ twice daily. No in vitro calculations were possible as the workers could not culture the protozoa in vitro. Under certain conditions allicin degrades to diallyl trisulfide. This chemical is more stable than the extremely volatile allicin and is easily synthesized. In China, it is commercially available as a preparation called Dasuansu and has been prescribed for *E. histolytica* and *Trichomonas vaginalis* infections.

Several other medicinal botanicals, noted for their GI effects, have been screened for antigiardial activity. Mexico has a high number of catalogued medicinal plants, 14 of which are antidiarrhetics and antiparasitics.[14] Three species tested, *Justicia spicigera* (acanthus), *Lipia beriandieri* (oregano), and *Psidium guajava* (guava) proved more potent than the control tinidazole.[15] Further work with the Mayan communities of southern Mexico tested the six most important botanicals in the treatment of GI disturbances against the protozoa *E. histolytica* and *Giardia lamblia. Cuphea pinetorum* (loosestrife) was antiprotozoal in both cases, and *Rubus corifolius* (rose) and *Helianthemum glomeratum* (rockrose) were antigiardial. Methanol extracts were fractionated on a cellulose column to yield six fractions. Only one fraction was shown to maintain activity, and contained kaempferol and quercetin. Standards of these were tested for activity and it was established that kaempferol was responsible for the activity. Quercetin was also antigiardial but not antiamebic.[16]

These examples advocate the enormous potential of medicinal plants to act in the GI tract against microbial infections. The active substances in these plants may also present an array of novel chemical structures that may be used for the development of a new arsenal of synthetic drugs. Indeed, many hundreds of plants worldwide are used in traditional medicine as treatments for bacterial infections. Some of these have also been subjected to in vitro screening. Still, it is important to note that the efficacy of such herbal medicines has seldom been rigorously tested in controlled clinical trials. This was critically reviewed by Martin and Ernst.[17] They used four electronic databases to search for controlled clinical trials of antibacterial herbal medicines. Data were extracted and validated in a standardized fashion, according to predefined criteria, by two independent reviewers. Probably the most striking result of this review is the extreme paucity of controlled clinical trials testing herbal antibiotics. In light of the long history and present popularity of their use, it is surprising that so few trials have tested the efficacy of herbal antibiotics. The lack of patentable rights on herbal medicines probably is a major reason for this. Another reason could be that traditionally herbal medicine has been hesitant to embrace modern methods for efficacy testing. It could also be that herbal medicine has been less popular in developed countries during the last centuries.

The evidence accumulated to date tentatively suggests possible benefits from some herbal preparations with antibacterial activity. Further large-scale, well-designed clinical trials are required to provide more conclusive proof of their efficacy.

GI IMMUNOREGULATION

The GI tract constitutes one of the largest sites of exposure to the outside environment. The mucosae are bombarded immediately after birth by a large variety of microorganisms as well as by protein antigens from the environment, mainly in formula-fed infants. Therefore, the mucosal surface has to be well protected, probably almost 200 times better than that of the skin. The function of the GI tract in monitoring and sealing the host interior from intruders is called the gut barrier.

A variety of specific and nonspecific mechanisms operate to establish the host barrier; these include luminal mechanisms and digestive enzymes, the epithelial cells together with the tight junctions between them, and the gut immune system. Disruptions in the gut barrier follow injury from various causes including nonsteroidal anti-inflammatory drugs and oxidant stress. They involve mechanisms such as adenosine triphosphate depletion and damage to epithelial cell cytoskeletons that regulate tight junctions. Ample evidence links gut barrier dysfunction to multiorgan system failure in sepsis and immune dysregulation. In addition, the contribution of gut barrier dysfunction to GI disease is an evolving concept and is the focus of a thorough review by DeMeo et al.[18] Many responses, including immune system responses within the GI tract, may be induced or repressed through the use of medicinal herbs and the ingestion of active natural products. It is thus important to review the factors that may be activated or deactivated by dietary substances.

Tolerance to dietary antigens induced via the gut ("oral tolerance") appears to be a rather robust adaptive immune mechanism. During millions of years of evolution, the mucosal immune system has generated two arms of adaptive defense to handle these challenges: (1) antigen exclusion performed by secretory IgA (SIgA) and secretory IgM (SIgM) antibodies to modulate or inhibit colonization of microorganisms and dampen penetration of potentially dangerous soluble luminal agents, and (2) suppressive mechanisms to avoid local and peripheral overreaction (hypersensitivity) to innocuous substances bombarding the mucosal surfaces. The latter arm is referred to as oral tolerance when induced via the gut against dietary antigens, and it probably explains why overt and persistent immunological hypersensitivity, or allergy, to food proteins is relatively rare.

Lymphoid cells are located in three distinct compartments in the gut: organized gut-associated lymphoid tissue (GALT), the lamina propria, and the surface epithelium. GALT comprises the Peyer's patches, the appendix, and numerous solitary lymphoid follicles, especially in the large bowel. All of these lymphoid structures are believed to represent inductive sites for intestinal immune responses. The lamina propria and epithelial compartment constitute effector sites but are nevertheless important in terms of cellular expansion and differentiation within the mucosal immune system.[19]

An extract preparation (TJ-48) of a Japanese herbal (Kampo) medicine, Juzen-Taiho-To, has been found to show enhancing activity on proliferation of bone marrow cells mediated by Peyer's patch cells (intestinal immune system modulating activity). Later, when TJ-48 was fractionated by MeOH and water extractions, EtOH precipitation and dialysis, water-soluble dialyzable (F-3) and polysaccharide (F-5) fractions both showed significant intestinal immune system modulating activity in vitro; other fractions had no activity.[20] In the same work, oral administration of F-3 (150 $mg \cdot kg^{-1}$) also showed intestinal immune system-modulating activity in C3H/HeJ mice. The authors suggested that lignin-carbohydrate complexes are involved in intestinal immune system modulation by TJ-48.[21]

An important element of the immune system is the activation of the complement system, which initiates a number of defense mechanisms intended to protect the body from invading microorganisms and other insults. However, uncontrolled complement activation can lead to tissue damage and thereby be of immunopathological importance in acute and chronic inflammatory diseases of the GI tract. Little is known about the local protective measures operating against complement-induced damage along the human GI tract in health and disease.

A recent study evaluated the expression and distribution of complement-inhibitory molecules in the gastric and intestinal mucosa in normal controls and in chronic inflammatory diseases, including lesions associated with chronic *H. pylori* infection.[22] Sections of frozen tissue specimens were obtained from patients with *H. pylori* gastritis, celiac disease, Crohn's disease, or ulcerative colitis, and from histologically normal controls, and were examined by immunofluorescence with monoclonal antibodies to protein (CD59), decay-accelerating factor (DAF), and membrane cofactor protein (MCP). It was concluded by the study authors that epithelial complement-inhibitory molecules are expressed differently at various normal GI sites and also in association with mucosal disease, suggesting variable protective potential. Such molecules could play a role in the development of gastric atrophy by protecting areas of intestinal metaplasia. Conversely, parietal cells appeared to be potentially vulnerable targets for complement attack.

Induction of a full range of local and systemic immune responses to orally administered purified antigens requires nonliving adjuvant vectors. Medical botanicals and natural products present an exciting solution. Oral vaccines are easy and economical to administer and avoid the hazards of routes involving needles. More important, induction of effective immunity at a mucosal site can only be achieved by immunization via a mucosal route. The existence of this "common mucosal immune system" allows one surface to be primed by antigen given via another mucosal route, so that oral immunization can be effective in protecting against infections of the respiratory and urinary tracts, as well as inducing breast milk immunity.[23] Finally, the success of the oral polio vaccine demonstrates the applicability of this route for stimulating protective immunity against systemic infections.

Many difficulties have limited the development of oral subunit vaccines. Purified proteins are poorly immunogenic in general and are unable to enter the endogenous pathway of antigen processing required to generate MHC class I-restricted immune responses. Moreover, the usual result of feeding a soluble protein is profound systemic immunological unresponsiveness (oral tolerance), making it particularly difficult to induce primary immunity by this route.[24] For these reasons, orally administered antigens have to be given with an adjuvant which is active at mucosal surfaces. A number of such agents have been described, including live vectors such as mutant *Salmonella* and nonviable agents such as cholera toxin (CT) or the heat labile enterotoxin from *Escherichia coli* (LT).[25] In many cases, the use of these vectors is hampered by possible toxicity or a loss of attenuation and by the induction of potentially limiting immune responses to the adjuvant itself. For these reasons, it would be useful to have a nonliving adjuvant with no inherent immunogenicity.

Saponins, which are the active ingredient in many medicinal plants and are the topic of Chapter 19 in this book, show great potential as nonliving adjuvants, probably due to their surface-active nature. Quil A, a saponin produced from the bark of *Quillaja saponaria,* was integrated into lipophilic immune-stimulating complexes (ISCOMS).[26] ISCOMS are rigid cagelike particles that form spontaneously on mixing cholesterol, phosphatidyl choline, and saponin. Proteins containing hydrophobic groups incorporate readily into ISCOMS, and parenteral immunization with such vectors induces strong systemic antibody and T-cell-mediated immune responses.[27] Findings that indicated the potential of ISCOMS as vaccine vectors promoted studies exploring their usefulness in the induction of immunity by the oral route. ISCOMS were shown to increase the ability of the intestine to take up associated materials, as shown by the higher levels of circulating protein in mice fed ovalbumin after administration of ISCOMS.[28] The mechanisms underlying this are unknown but were suggested by the authors to reflect the detergent properties of saponins, which could increase the permeability of the epithelial cell membrane or have an effect on the junctions between enterocytes.[29] In either case, these effects are likely to enhance the access of ISCOMS and their associated antigens to the underlying immune apparatus.

It should be noted here that, in some cases, in vivo antimicrobial activity just described could be the outcome of induced immune response. One such case was demonstrated recently by Tripathi et al.[30] They tested *Piper longum* fruit for its efficacy against experimental infection of *G. lamblia* in mice. *Piper longum* fruit is used in traditional remedies as well as in the ayurvedic system of medicine against intestinal disorders. In an in vitro test, the ethanol extract of *P. longum* fruit powder showed 100 percent giardicidal activity. Low order activity was found in the n-butanol extract. Further fractionation resulted in a total loss of activity. The survival of trophozoites in mice at 900 $mg \cdot kg^{-1}$ body weight was 11 percent in powder, 8.5 percent in aqueous extract, and 5.8 percent in ethanol extract. The antigiardial activity of other fractions was comparable to the non-drug-treated control (47.6 percent). Interestingly, *P. longum* possessed demonstrable immunostimulatory activity, both specific and nonspecific, as evidenced by the standard test parameters such as hemagglutination titer, plaque-forming cell counts, macrophage migration index, and phagocytic index.

The role of herbal reactive substances in inducing elements of or the complete immune system through the GI tract is a fascinating issue. The important mechanisms of action differ along the digestive tract. Surface-active substances such as saponins and lignin-carbohydrate complexes appear to be the most active, alone or in combination with known antigens.

CHEMICAL CONSIDERATIONS IN THE GI TRACT

Lipophilicity is well known as a prime physicochemical descriptor of xenobiotics with relevance to their biological properties. In keeping with the outstanding importance of lipophilicity in biosciences, this topic is intensively treated by Pliska et al.[31] Still, the vague property of lipophilicity is not enough to predict permeability and transport of xenobiotics and drugs. The hydrophobic interactions of drugs with their receptors, the pharmacokinetic behavior of drug molecules, and the toxicological properties and pharmaceutical aspects such as solubility are examples of a steadily increasing number of topics in which lipophilicity plays an important role.

For instance, the determination of antioxidant capacity is largely dependent upon the lipid solubility of both oxidizing agents and antioxidants. This becomes important in determining the extent of oxidative damage, to, for instance, LDL.[32] In the case of protecting LDL, lipophilic antioxidants usually demonstrate potent scavenging activity.[33] We found great variability in partition coefficients among tested polyphenols, under various conditions representative of areas of the digestive tract (data to be published). Commonly, the partition coefficient of a compound is defined as the ratio of its solubility in octanol to its solubility in water. When the data are used as part of the characterization of biological systems, saline buffer is the proper aqueous model solution. Salinity, as well as the concentration of hydronium ions (pH), dramatically affects the partition coefficient of phenolics. These considerations suggest that the behavior of a polyphenol in the stomach will be different from its behavior in the intestine. When we assayed phenolic acids, we found that they tend to be more soluble in buffer than in octanol at intestinal pH. Nevertheless, the solubility of gallic acid in water increases with the addition of sodium chloride, while the water solubility of caffeic acid, like most organic substances, decreases with increasing salinity. When we assayed epicatechin and resveratrol, we found that their partition coefficients are hardly affected by the change of pH from the stomach to the intestine.

These data describe the complexity involved in predicting the distribution as well as the chemical behavior of a substance along the GI tract. The following sections detail enzyme and chemical interactions of natural products in the GI tract with regard to health-beneficial and medicinal activities.

HERBAL-INDUCED DRUG INTERACTIONS IN THE GI TRACT

St. John's wort *(Hypericum perforatum)* is one of the most commonly used herbal medicines for the treatment of depression in the United States and in the countries of the European Union. More than 50 products that contain St. John's wort are also consumed as dietary supplements in Japan. Recently, St. John's wort was reported to substantially decrease blood/plasma concentrations and efficacy of cyclosporine (INN), indinavir, and digoxin. Investigating the mechanisms of these St. John's wort–induced drug interactions, the administration of St. John's wort extract to rats for 14 days resulted in a 3.8-fold increase in intestinal P-glycoprotein/Mdr1 expression and in a 2.5-fold increase in hepatic CYP3A2 expression (Western blot analyses).[34] In a clinical study, the administration of St. John's wort extract to eight healthy male volunteers for 14 days resulted in an 18 percent decrease of digoxin exposure after a single digoxin dose (0.5 mg), a 1.4- and 1.5-fold increased expressions of duodenal P-glycoprotein/MDR1 and CYP3A4, respectively, and a 1.4-fold increase in the functional activity of hepatic CYP3A4 (C-14-erythromycin breath test). These results indicate direct inducing effects of St. John's wort on intestinal P-glycoprotein/MDR1 (in rats and humans), hepatic CYP3A2 (in rats), and intestinal and hepatic CYP3A4 (in humans). Therefore, the results of this study provide a mechanistic explanation for the previously observed drug interactions in patients and support the importance of intestinal P- glycoprotein/MDR1 in addition to intestinal and hepatic CYP3A4 for overall drug absorption and disposition in humans.

Accumulating data suggest that St. John's wort may reduce the plasma concentration of administered drugs, such as simvastatin, which is a widely used 3-hydroxy-3-methylglutaryl coenzyme A (HMG-CoA) reductase inhibitor and has beneficial effects on coronary disease and mortality rates in patients with hypercholesteremia. Simvastatin undergoes extensive first-pass metabolism-producing inactive metabolites in the intestinal wall and liver, which is primarily mediated by cytochrome P450 (CYP) 3A4. Thereafter, unchanged simvastatin, a lactone pro-drug, is converted to its active form, simvastatin hydroxy acid, in blood. The pharmacokinetic profile of simvastatin after repeated administration of St John's wort or its placebo in a double-blind, crossover study was reported by Sugimoto et al.[35] In this study, St. John's wort significantly decreased plasma simvastatin hydroxy acid. However, the therapeutic manifestation of coadministration of St. John's wort was not determined. St. John's wort may reduce the cholesterol-lowering effect of simvastatin, demonstrating the critical effects that herbal preparations may have on drug pharmacokinetics in the GI tract. Further study is needed to address this issue.

HIV-infected persons frequently use herbal remedies and dietary supplements along with conventional therapies.[36] Because garlic is one of the dietary supplements most commonly used by the HIV-positive population, the effect of garlic on the pharmacokinetics of saquinavir, a protease inhibitor, was studied, and the latter's plasma pharmacokinetic parameters were given.[37] In the presence of garlic, the plasma concentration of saquinavir during an eight-hour dosing interval decreased by 51 percent, and the mean maximum concentration (Cmax) decreased by 54 percent. After the ten-day washout period, Cmax values returned to 60 or 70 percent of their values at baseline. These data imply that patients should use caution when combining garlic supplements with saquinavir when it is used as a sole protease inhibitor.

The use of herbal products to treat a wide range of conditions is on the rise, leading to increased intake of phytochemicals. The data just presented represent recent studies that reveal potentially fatal interactions between herbal remedies and traditional drugs. A recent review by Ioannides summarizes many herbal remedies with the potential to modulate enzyme activities and thus participate in interactions with conventional drugs.[38] These include milk thistle, *Angelica dahurica,* ginseng, garlic preparations, danshen, and liquorice. Herbal products are currently not subject to the rigorous testing that is indispensable for conventional drugs. However, if potential drug interactions are to be predicted, it is essential that the ability of herbal products to interfere with drug-metabolizing enzyme systems be fully established.

ANTIOXIDANT AND REDUCING ACTIVITIES OF HERBAL PRODUCTS

The strong epidemiological evidence that diets rich in fruit and vegetables can promote health points to the role of natural antioxidants in foods.[39] Polyphenols, an integral part of food and feed, represent one of the most numerous and ubiquitous groups of plant metabolites. It is widely believed that traditional Mediterranean and Oriental diets, in contrast to northern European and American diets, include a significantly large amount of plant foods; this notable difference between the two eating styles, despite the similarities among other classic risk factors for cardiovascular disease

(CVD) such as high plasma cholesterol levels, has been associated with a lower risk of developing CVD and certain cancers.[40] Mediterranean diets are usually characterized by abundant plant foods (fruits, vegetables, breads, nuts, seeds, wine, and olive oil). Red wine has drawn particular interest due to the positive correlation between its consumption and the lower risk for CVD, known as the French paradox: in the 1990s, residents of Toulouse, France, who consume their alcohol largely in the form of red wine, were found to have a very low mortality rate from CVD, despite a fat consumption rate similar to that in the United States.[41] It was suggested that red wine impacts mortality more favorably than other alcoholic beverages, leading to the widespread belief that wine affords a protective effect.[42] The active substances unique to red wine are various polyphenols, including anthocyanins, flavonoids, and stilbenes.[43]

The wide range of biological effects exhibited by wine polyphenols are thought to be due to their powerful antioxidant properties in the body together with their ability to interact with redox-sensitive cell-signaling pathways.[44] The extent of the potency of polyphenols in vivo is dependent upon their absorption, metabolism and secretion.[45] Their antioxidant potential in the body may, however, be of importance only at sites where adverse oxidation occurs.[46]

The involvement of excessive free radical production and high exposure to reactive oxygen species (ROS), and the great number of epidemiological studies linking antioxidant intake with a reduced incidence of the aforementioned diseases, indicate that dietary antioxidants are likely to play a protective role. A balance between oxidant and antioxidant intracellular systems is therefore vital for cell function, regulation, and adaptation to diverse growth conditions.[47] The sources of ROS are numerous in modern life, among which food is a major route for introducing free radicals into the body.[48] Reactions between free radicals generated during oxidation and other food constituents dramatically alter the range of carcinogenic, cytotoxic, and atherogenic compounds produced; these include lipid hydroperoxides, oxycholesterols, malondialdehyde, and hydroxy alkanals.[49] The reactive species may be generated during processing and digestion, either in the gastric fluid, which contains absorbed oxygen and has a low pH, or in the gut environment, under neutral to basic pH.[50] Dietary antioxidants, such as plant polyphenols, reaching the digestive tract in the context of a whole meal, will affect overall oxidation and by-product formation. It was recently suggested that the stomach acts as a bioreactor in which many chemicals interact, and that the oxidation of high-fat, cholesterol-rich foods could be enhanced by endogenous metal catalysts in those foods.[51]

The antioxidant capacity of a substance is largely affected by the pH and polarity of the solvent: both factors change dramatically through the GI tract. Nevertheless, to the best of our knowledge, the capacity of dietary antioxidants during digestion has been poorly covered in the scientific literature. Gallic acid, for example, was found to exhibit better antioxidant behavior at pH 2 than at any higher pH (up to 8), where it may even act as a pro-oxidant.[52] *Trans*-resveratrol was found to effectively scavenge free radicals in the pH range of 2 to12.[53] The existence of lipophilic substances in the surrounding environment will also alter the antioxidant capacity, as well as the oxidative potential of the oxidant.[54]

Major emphasis has been placed in recent years on the potent in vitro antioxidant properties of herbal preparations as well as purified natural products from plants.[55] Indeed, the electron-donating properties of polyphenols and their structure-activity relationships have been intensively investigated and are well defined to explain their powerful antioxidant activities in vitro. However, as suggested here, their role as antioxidants may be of importance in the GI tract where their concentration is highest, while after absorption other mechanisms of action should be sought. Their precise mechanism of action in vivo will depend on the extent to which they are conjugated and metabolized—again, through the GI tract and during absorption across the small intestine.[56]

CONCLUSION

This chapter is an attempt to describe major reactions and interactions of natural products that take place in the digestive tract. Well-covered aspects, such as GI disorders and colon cancer, were intentionally not covered here. The digestive tract is a huge arena in which chemistry, biochemistry, and physiology meet, and the outcome may be beneficial at times but deleterious and even dangerous at others. Herbal medicine is becoming ever more appealing as synthetic drugs promote resistance and induce side effects, and as the raw materials for synthesis (oil) become scarce. Novel approaches are needed. One can only hope that intensive and well-controlled research will lead to the competent use of the vast data that has been collected throughout human history.

NOTES

1. Wolf, C.R., 2001, Chemoprevention: Increased potential to bear fruit, *Proc. Natl. Acad. Sci. USA*, 98: 2941-2943.

2. Warren, J.R. and Marshall, B.J.L.I., 1983, Unidentified curved bacilli on gastric epithelium in active gastritis, *Lancet,* 1: 1273-1275; Peek, R.M.J. and Blaser, M.J., 1993, *Helicobacter pylori* and gastrointestinal tract adenocarcinomas., *Nat.*

Rev. Cancer, 2: 28-37; Megraud, F. and Marshall, B.J., 2000, How to treat *Helicobacter pylori*. First-line, second-line, and future therapies, *Gastroenterol. Clin. North. Am.,* 29: 759-773; Nomura, A., Stemmermann, G.N., Chyou, P.H., Kato, I., Perez-Perez, G.I., and Blaser, M.J., 1991, *Helicobacter pylori* infection and gastric carcinoma among Japanese Americans in Hawaii, *N. Engl. J. Med.,* 325: 1132-1136; Parsonnet, J., Friedman, G.D., Vandersteen, D.P., Chang, Y., Vogelman, J.H., and Orentreich, N., 1991, *Helicobacter pylori* infection and the risk of gastric carcinoma, *N. Engl. J. Med.,* 325: 1127-1131.

3. Gaby, A.R., 2001, *Helicobacter pylori* eradication: Are there alternatives to antibiotics?, *Altern. Med. Rev.,* 6: 355-366.

4. Ankri, S. and Mirelman, D., 1999, Antimicrobial properties of allicin from garlic, *Microbes Infect.,* 1: 125-129.

5. Aydin, A., Ersoz, G., Tekesin, O., Akcicek, E., Tuncyurek, M., and Batur, Y., 1997, Does garlic oil have a role in the treatment of *Helicobacter pylori* infection? *Turkish J. Gastroenterol.,* 8: 181-184; Graham, D.Y., Anderson, S.-Y., and Lang, T., 1994, Garlic or jalapeno peppers for the treatment of *Helicobacter pylori* infection, *Am. J. Gastroenterol.,* 94: 1200-1202.

6. Martin, K.W. and Ernst, E., 2003, Herbal medicines for treatment of bacterial infections: A review of controlled clinical trials, *J. Antimicrob. Chemother.,* 51: 241-246.

7. Ross, Z.M., O'Gara, E.A., Hill, D.J., Sleightholme, H.V., and Maslin, D.J., 2001, Antimicrobial properties of garlic oil against human enteric bacteria: Evaluation of methodologies and comparisons with garlic oil sulfides and garlic powder, *Appl. Environ. Microbiol.,* 67: 475-480.

8. Hashimoto, T., Aga, H., Chaen, H., Fukuda, S., and Kurimoto, M., 1999, Isolation and identification of anti-*Helicobacter pylori* compounds from *Polygonum tinctorium* Lour., *Natural Medicines,* 53: 27-31.

9. Kataoka, M., Hirata, K., Kunikata, T., Ushio, S., Iwaki, K., Ohashi, K., Ikeda, M., and Kurimoto, M., 2001, Antibacterial action of tryptanthrin and kaempferol, isolated from the indigo plant (*Polygonum tinctorium* Lour.), against *Helicobacter pylori*-infected Mongolian gerbils., *J. Gastroenterol.,* 36: 5-9.

10. Harris, J.C., Plummer, S., and Lloyd, D., 2001, Antigiardial drugs, *Appl. Microbiol. Biotechnol.,* 57: 614-619.

11. Bolton, S., Null, G., and Troetel, W.M., 1982, The medicinal uses of garlic—Fact and fiction, *Am. Pharm.,* NS22: 40-43.

12. Ankri and Mirelman, 1999, Antimicrobial properties of allicin.

13. Soffar, S.A. and Mokhtar, G.M., 1991, Evaluation of the antiparasitic effect of aqueous garlic (*Allium sativum*) extract in hymenolepiasis nana and giardiasis, *J. Egypt Soc. Parasitol.,* 21: 497-502.

14. Harris, Plummer, and Lloyd, 2001, Antigiardial drugs.

15. Ponce-Macotela, M., Navarro-Alegria, I., Martinez-Gordillo, M.N., and Alvarez-Chacun, R., 1994, Efecto antigiardisico in vitro de 14 extractos de plantas, *Rev. Invest. Clin.,* 46: 343-347.

16. Calzada, F., Alanis, A.D., Meckes, M., Tapia-Contreras, A., and Cedillo-Rivera, R., 1998, In vitro susceptibility of *Entamoeba histolytica* and *Giardia lamblia* to some medicinal plants used by the people of Southern Mexico, *Phytother. Res.,* 12: 70-72.

17. Martin and Ernst, 2003, Herbal medicines for treatment of bacterial infections.

18. DeMeo, M.T., Mutlu, E.A., Keshavarzian, A., and Tobin, M.C., 2002, Intestinal permeation and gastrointestinal disease, *J. Clin. Gastroenterol.,* 34: 385-396.

19. Brandtzaeg, P., 2002, Current understanding of gastrointestinal immunoregulation and its relation to food allergy, *Ann. N.Y. Acad. Sci.,* 964: 13-45.

20. Kiyohara, H., Matsumoto, T., and Yamada, H., 2000, Lignin-carbohydrate complexes: Intestinal immune system modulating ingredients in Kampo (Japanese herbal) medicine, Juzen-Taiho-To, *Planta Medica,* 66: 20-24.

21. Ibid; Kiyohara, H., Matsumoto, T., and Yamada, H., 2002, Intestinal immune system modulating polysaccharides in a Japanese herbal (Kampo) medicine, Juzen-Taiho-To, *Phytomedicine,* 9: 614-624.

22. Berstad, A.E. and Brandtzaeg, P., 1998, Expression of cell membrane complement regulatory glycoproteins along the normal and diseased human gastrointestinal tract, *Gut,* 42: 522-529.

23. Ohtake, N., Nakai, Y., Yamamoto, M., Ishige, A., Sasaki, H., Fukuda, K., Hayashi, S., and Hayakawa, S., 2002, The herbal medicine Shosaiko-to exerts different modulating effects on lung local immune responses among mouse strains, *Int. Immunopharmacol.,* 2: 357-366; Reid, G., 1999, Potential preventive strategies and therapies in urinary tract infection, *World J. Urol.,* 17: 359-363; McGhee, J.R., Mestecky, J., Dertzbaugh, M.T., Eldridge, J.H., Hirasawa, M., and Kiyono, H., 1992, The mucosal immune system: From fundamental concepts to vaccine development, *Vaccine,* 10: 75-78.

24. Strobel, S. and Mowat, A.M., 1998, Immune responses to dietary antigens: Oral tolerance, *Immunol. Today,* 19: 173-181.

25. McGhee et al., 1992, The mucosal immune system.

26. Mowat, A.M., Smith, R.E., Donachie, A.M., Furrie, E., Grdic, D., and Lycke, N., 1999, Oral vaccination with immune stimulating complexes, *Immunol. Lett.,* 65: 133-140.

27. Morein, B., Lovgren, K., Ronnberg, B., Sjolander, A., and Villacres-Erikeson, M., 1995, Immunostimulating complexes: Clinical potential in vaccine development, *Clin. Immunother.,* 3: 461-475.

28. Mowat et al., 1999, Oral vaccination with immune stimulating complexes.

29. Ibid.

30. Tripathi, D.M., Gupta, N., Lakshmi, V., Saxena, K.C., and Agrawal, A.K., 1999, Antigiardial and immunostimulatory effect of *Piper longum* on giardiasis due to *Giardia lamblia, Phytotherapy Research,* 13: 561-565.

31. Pliska, V., Testa, B., and van de Waterbeemd, H. (eds.), *Lipophilicity in drug action and toxicology* (Weinheim, Germany: VCH, 1996).

32. Son, S. and Lewis, B.A., 2002, Free radical scavenging and antioxidative activity of caffeic acid amide and ester analogues: Structure-activity relationship, *J. Agric. Food Chem.,* 50: 468-472.

33. Massaeli, H., Sobrattee, S., and Pierce, G.N., 1999, The importance of lipid solubility in antioxidants and free radical generating systems for determining lipoprotein proxidation, *Free Radic. Biol. Med.,* 26: 1524-1530.

34. Durr, D., Stieger, B., Kullak-Ublick, G.A., Rentsch, K.M., Steinert, H.C., Meier, P J., and Fattinger, K., 2000, St John's wort induces intestinal P-glyco-

protein/MDR1 and intestinal and hepatic CYP3A4, *Clinical Pharmacology and Therapeutics,* 68: 598-604.

35. Sugimoto, K. et al., 2001, Different effects of St John's wort on the pharmacokinetics of simvastatin and pravastatin, *Clinical Pharmacology and Therapeutics,* 70: 518-524.

36. Ernst, E., 1997, Complementary AIDS therapies: the good, the bad, and the ugly, *Int. J. STD AIDS,* 8: 281-285.

37. Piscitelli, S.C., Burstein, A.H., Welden, N., Gallicano, K.D., and Falloon, J., 2002, The effect of garlic supplements on the pharmacokinetics of saquinavir, *Clin. Infect. Dis.,* 34: 234-238.

38. Ioannides, C., 2002, Topics in xenobiochemistry-pharmacokinetic interactions between herbal remedies and medicinal drugs, *Xenobiotica,* 32: 451-478.

39. Tikkanen, M.J., Wahala, K., Ojala, S., Vihma, V., and Adlercreutz, H., 1998, Effect of soybean phytoestrogen intake on low density lipoprotein oxidation resistance, *Proc. Natl. Acad. Sci. USA,* 95: 3106-3110; Caccetta, R.A.A., Burke, V., Mori, T.A., Beilin, L.J., Puddey, I.B., and Croft, K.D., 2001, Red wine polyphenols, in the absence of alcohol, reduce lipid peroxidative stress in smoking subjects, *Free Radical Biol. Med.,* 30: 636-642; Wolf, 2001, Chemoprevention.

40. Rice Evans, C., 2001, Flavonoid antioxidants, *Curr. Med. Chem.,* 8: 797-807; Halliwell, B., Zhao, K., and Whiteman, M., 2000, The gastrointestinal tract: A major site of antioxidant action? *Free Radical Res.,* 33: 819-830.

41. Seigneur, M. et al., 1990, Effect of the consumption of alcohol, white wine, and red wine on platelet function and serum lipids, *J. Appl. Cardiology,* 5: 215-222.

42. Gronbaek, M., 1999, Type of alcohol and mortality from cardiovascular disease, *Food Chem. Toxicol.,* 37: 921-924; Renaud, S. and de Lorgeril, M., 1992, Wine, alcohol, platelets, and the French paradox for coronary heart disease, *Lancet,* 339: 1523-1526.

43. Corder, R., Douthwaite, J.A., Lees, D.M., Khan, N.Q., Viseu Dos Santos, A.C., Wood, E.G., and Carrier, M.J., 2001, Endothelin-1 synthesis reduced by red wine, *Nature,* 414: 863-864.

44. Kuhnle, G., Spencer, J.P.E., Chowrimootoo, G., Schroeter, H., Debnam Edward, S., Srai, S.K.S., Rice Evans, C., and Hahn, U., 2000, Resveratrol is absorbed in the small intestine as resveratrol glucuronide, *Biochem. Biophys. Res. Comm.,* 272: 212-217; Jang, M.S., Jang, M.S., Udeani, G.O., Slowing, K.V., Thomas, C.F., Beecher, C.W., Fong, H.H., Farnsworth, N.R., Kinghorn, A.D., Mehta, R.G., Moon, R.C., et al., 1997, Cancer chemopreventive activity of resveratrol, a natural product derived from grapes, *Science,* 275: 218-220; Rice Evans, 2001, Flavonoid antioxidants, *Current Medical Chemistry,* 8: 797-807.

45. Soleas, G.J., Diamandis, E.P., and Goldberg, D.M., 2001, The world of resveratrol, *Adv. Exp. Med. Biol.,* 492: 159-182; Soleas, G.J., Yan, J., and Goldberg, D.M., 2001, Measurement of trans-resveratrol, (+)-catechin, and quercetin in rat and human blood and urine by gas chromatography with mass selective detection, *Methods Enzymol,* 335: 130-145; Scalbert, A. and Williamson, G., 2000, Dietary intake and bioavailability of polyphenols, *J. Nutr.,* 130: 2073-2085; Kuhnle et al., 2000, Resveratrol is absorbed in the small intesting as resveratrol glucoronide.

46. Tikkanen et al., 1998, Effect of soybean phytoestrogen intake on low density lipoprotein; Halliwell, Zhao, and Whiteman, 2000, The gastrointestinal tract.

47. Malins, D.C., Hellstrom, K.E., Anderson, K.M., Johnson, P.M., and Vinson, M.A., 2002, Antioxidant-induced changes in oxidized DNA, *Proc. Natl. Acad. Sci. USA,* 99: 5937-5941; Wolf, 2001, Chemoprevention.

48. Ames, B.N., Magaw, R., and Gold, L.S., 1987, Ranking possible carcinogenic hazards, *Science,* 236: 271-280.

49. Staprans, I., Pan, X. M., Rapp, J.H., and Feingold, K.R., 2003, Oxidized cholesterol in the diet is a source of oxidized lipoproteins in human serum, *J. Lipid Res.,* 44: 705-715.

50. Kanner, J. and Lapidot, T., 2001, The stomach as a bioreactor: Dietary lipid peroxidation in the gastric fluid and the effects of plant-derived antioxidants, *Free Radical Biol. Med.,* 31: 1388-1395.

51. Ibid.

52. Gunckel, S., Santander, P., Cordano, G., Ferreira, J., Munoz, S., Nunez Vergara, L.J., and Squella, J.A., 1998, Antioxidant activity of gallates: An electrochemical study in aqueous media, *Chem. Biol. Interact.,* 114: 45-59.

53. Stojanovic, S. and Brede, O., 2002, Elementary reaction of the antioxidant action of trans-stilbene derivatives: Resveratrol, pinosylvin and 4-hydroxystilbene, *Phys. Chem. Chem. Phys.,* 4: 757-764.

54. Aldini, G., Yeum, K.J., Carini, M., Krinsky, N.I., and Russell, M., 2003, (–)-Epigallocatechin-3-gallate prevents oxidative damage in both the aqueous and lipid compartments of human plasma, *Biochem. Biophys. Res. Commun.,* 302: 409-414.

55.Youdim, K.A., Spencer, J.P.E., Schroter, H., and Rice-Evans, C., 2002, Dietary flavonoids as potential neuroprotectants, *Biol. Chem.,* 383: 50-519.

56. Ibid.

PART V: HOPES AND DANGERS

Chapter 21

Challenges and Threats to Interdisciplinary Medicinal Plant Research

Michael Heinrich

INTRODUCTION: PATIENTS' DILEMMAS IN 2005

A scenario: Today, a mother in one of the poorer barrios of one of the major urban centers of the world has a sick child and is dissatisfied with the medical attention given to her by mainstream medical practitioners. She is keen on something natural and, like every parent, she is eager to give her child the best treatment. Many treatment options are available to her, including traditional Chinese medicine, ayurvedic medicine, herbalism based on European traditions, some alternative and complementary forms of therapy as they were developed over the past century (homeopathy, flower remedies, esoteric healing), and she may also use some of the elements of rational phytotherapy (see Chapter 1). But how is she going to come to a decision? As in Dan Moerman's example (see Chapter 6), availability of a medicine will be essential for selecting specific therapies. However, the woman 2,000 years ago had to rely on what she had learned from previous generations, or she could try a novel resource. In the twenty-first century, she will still rely on information she obtained from relatives or friends but also on information from the mass media (television, popular books, newspapers), and she now has a wide variety of potential resources at hand. It is important to note that scientific evidence has been accumulated for certain remedies used, while others are still not very well known. Some are little

Some data discussed in this chapter are part of the doctoral theses of Drs. B. Frei, M. Leonti, and C. Weimann. The concepts presented here have benefited from many fruitful discussions with the members of the Research Group on Ethnobiology at the Centre for Pharmacognosy and Phytotherapy, The School of Pharmacy, London (Jo Castle, Sarah Edwards, Johanna Kufer, Sabine Nebel, and Andrea Pieroni), as well as with Paul Bremner.

studied; others are well known, potent, and useful remedies with few side effects; and a last group includes those in which the risk-benefit analysis is clearly negative.

An example of the latter group are species of the genus *Aristolochia,* which are or were commonly used in many regions of the world. In November of 2002, I interviewed an herb seller in Mérida, Mexico, who recommended an unidentified species of *Aristolochia,* popularly known as *guaco,* for gastrointestinal problems and other conditions especially in children. With the Popoluca of Veracruz (Mexico) six *Aristolochia* species are used against stomach-ache and colic.[1] In Europe snake root or birthwort *(Aristolochia clematitis)* had long been used in medicine as an emmenagogue, abortifacient, diuretic, and for arthritis. Until the early 1980s several phytomedicines were widely sold in Europe. Its main active constituent, aristolochic acid, was formerly used in Germany against abscess, eczema, and other long-lasting skin diseases and as a nonspecific stimulant of the immune system, but in 1981 both the extract and the pure compound were withdrawn due to in vitro evidence for serious carcinogenic and nephrotoxic effects. More recently, a traditional Chinese medical herbal preparation—used for supposed weight-reducing effects—contained *Aristolochia fangchi* (Chinese snake root) and had been used extensively in a Belgian hospital during the early 1990s. This species, too, is known to possess large amounts of aristolochic acid. *Aristolochia fangchi* had inadvertently been substituted for another Chinese drug *(Stephania tetrandra).* Renal fibropathy (a progressive form of renal fibrosis) due to the use of *A. fangchi* was observed eight to ten years later in many of the patients. All *Aristolochia* species have been banned from the European markets for many years now. Overall, these data provide convincing evidence that the systemic use of *Aristolochia* species is very problematic and that, for example, in Mexico, appropriate programs for reducing the health risk posed by the use of *guaco* have to be launched.

This example is relevant in the context of a handbook of medicinal plants since it points to the global relevance of our (often local) knowledge about phytomedicines and highlights the need for making information about plants available on a worldwide basis. Therefore, we need not only a concerted research effort on all aspects of medicinal plants but also to synthesize such information appropriately (as it is done in this book). Three core areas that need our attention include the ones covered in this handbook:

- Medicinal plants and their (modern and historical) use
- New methodological developments
- Pharmacology and clinical use

In this chapter I highlight some areas that require specific attention in the immediate future and topics that are both challenging and promising. This is a forced ride through a huge field in which only some aspects can be addressed and many others were by necessity neglected.

MEDICINAL PLANTS AND THEIR MODERN AND HISTORICAL USES

Ethnobotany

The examples in Part II of this book show the richness and relevance of studies on indigenous plant use in such diverse regions as South America, Africa, China, and Europe. The rapid changes that many cultures undergo make such studies one of the top priorities in future medicinal plant research. Essential will be the use of appropriate quantitative methodologies and a truly *ethno*botanical approach.[2] Here I want to highlight three additional aspects: the history of plant use, especially in Europe; the relationship of medicinal plant research with biodiversity studies; and the impact of the Convention on Biological Diversity (CBD or Rio-Convention) and subsequent treatise on academic research.

History of Plant Use

From an ethnopharmacological point of view, Europe is one of the more neglected regions of the world. Very little is known about the current role of medicinal plants with the numerous European peoples and, just as important, little research has been conducted to integrate historical data on plant use with state-of-the-art biomedical research. Also, there is a need to integrate such historical data with modern ethnobotanical information.

Until the fifteenth century in the Christian parts of Europe the texts of the classic Greeks and Romans were copied and annotated, especially as they were recorded by the Arabs. The Italian monastery of Montecassino is one of the earliest examples of such a tradition; others developed around the monasteries of Chartres (France) and St. Gall (Switzerland). A common element of the monasteries was a medicinal plant garden, which was at the same time used for growing medicinal (and aromatic) herbs required for treating patients and for teaching the knowledge about medicinal plants to the younger generation. The species included in these gardens were common to practically all monasteries, and many of the species still are important medicinal plants of today.[3] Of particular interest in this context is the *Capitulare de Villis* of Charles the Great (747-814), who ordered that medicinal and other plants are to be grown in the king's gardens and in the monasteries. He specifically listed 24 species of medicinal plants. Walahfri(e)d Strabo (809-849), abbot of the monastery of Reichenau (Lake Constance), deserves mention because of his *Liber de Cultura Hortum* (Book on the growing of plants), the first "textbook" on (medical) botany, and the *Hortulus,* a Latin poem on the medical plants grown in the district. This poem is famous not only because of its poetry but also because of its vivid and excellent descriptions of the appearance and virtues of medicinal plants. Today many of the plants reported in the *Hortulus* are still important medicinal plants either as part of one of the European pharmacopoeias or as a species widely used by a people. However, less is known about species which were not that commonly used and a concerted effort is needed to identify these species botanically, as well as to study their uses over the centuries.

The period from the thirteenth to the seventeenth centuries saw great developments in the medical use of plants. A wide range of medical practitioners, including physicians, surgeons, and barber surgeons, used plants, minerals, and animals. One of the richest and yet least used resources (in a systematic and comparative manner) that we have about the history of basic medicine and pharmacy are the herbals, surgical and barber-surgical texts that summarize scholarly medical knowledge and offer clues as to the folk traditions of Europe. A comparative analysis of these sources is still lacking and may well provide novel ideas for further developing medicinal plant usage, but, just as important, it will provide exciting information about the development of medicinal plant usage in this crucial period. Similarly, the detailed study of other written traditions such as ayurvedic and Chinese medicine would be a great challenge.

Biodiversity and Phytomedicines

Modern medicinal plant research has to take into consideration the sustainable use of resources. Very often no immediate conflict exists between the local usage of a species and its conservation status. For example, the Orchidaceae are often rare and have relatively dispersed and often inaccessible populations, but they are also of very little importance as a medicinal resource. In our study with the Popoluca of Veracruz, Mexico, the Orchidaccae was one of the least utilized families in this ethnopharmacopoeia, while, for example, Asteraceae, Piperaceae, Fabaceae, Euphorbiaceae, and Lamiaceae, often with many weedy species, are the most frequently cited taxa.[4] Thus rareness or difficulty in obtaining the botanical drug will preclude the use of a species for medicinal purposes and this book.[5] This has also been very convincingly demonstrated in case of the Highlands of

Chiapas where weedy* species are far more like to be selected than other ones.[8] However, conflicts arise if such a drug becomes a more widely used commodity, especially if it is used in European phytomedicine (as in case of *Prunus africana*, red stinkwood) or as a source of an industrially produced drug (as, for example, *Taxus brevifolia*) (for an overview and further examples see Heinrich).[9]

Besides the view from the perspective of an individual species and its conservation status, one can also analyze the origin of a medicinal plant within the ecozones of a community. We have shown that both the Zapotec and the Mixe obtain most (59.2 percent and 71.6 percent, respectively) of their medicobotanical resources from their immediate environment and generally from disturbed habitats. These phytotherapeutic preparations are used to treat the common illnesses of the region. The comparatively lower number of medicinally important species collected from the secondary and primary forest vegetation contradicts the common popular assumption that these vegetation types are the principal sources of indigenously used medicinal plants.[10] Some of the species grown near the house originate from these forest habitats, but because they have been regarded as a useful remedy, they have been brought to the *solar* of the healer or its *corral* and grown there. Having easy access to the resources is therefore a factor with much influence on the diversity of the zones closer to the house.

Also, Voeks reported on the relative importance of primary as compared to secondary forest vegetation in Bahia, Brazil.[11] While primary vegetation is essential for obtaining timber, the secondary forests yielded a much larger number of medicinal plants. Since this author did not look at species from nonforest vegetation zones, a direct comparison is not possible. Comerford showed in the Petén region in Guatemala that regrown forests and intensively managed zones are more important for medicinal plant gatherings.[12] Overall, these data indicate that the indigenous use of these resources normally does not pose an threat to the species involved as long as the use is a locally restricted one and as long as the plant does not become a commodity.

The Convention on Biological Diversity

Much has been written about the impact of the Convention on Biological Diversity, and many excellent overviews also exist. It is, however, essential

* In this context weeds are defined as species "which grow entirely or predominantly in situations markedly disturbed by man (without, of course, a deliberately cultivated plant)"[6] especially in disturbed grounds and are considered as ecosustainable nutritional sources.[7]

to reiterate one of the main articles, which guides all research that potentially may lead to novel (phyto-)pharmaceutical or nutraceutical products:

> Each Contracting Party shall, as far as possible and as appropriate:
>
> Subject to its national legislation, respect, preserve and maintain knowledge, innovations and practices of indigenous and local communities embodying traditional lifestyles relevant for the conservation and sustainable use of biological diversity and promote their wider application with the approval and involvement of the holders of such knowledge, innovations and practices and encourage the equitable sharing of the benefits arising from the utilisation of such knowledge, innovations and practices.[13]

In recent years considerable emphasis has been put on developing appropriate legal frameworks for collaborative research between provider countries and users.[14,15] Sarah Laird's book provides excellent insights into this very complex topic and further stimulate the discussions in this field. However, the discussion has been hampered by a lack of emphasis on the potential *non*commercial benefits of the scientific study of medicinal plants for such "provider countries," and, also, on the role academic researchers in the field may play in such processes. In this situation ethnobotanists should take a new role as interlocutors between keepers of traditional knowledge and other potential uses of such resources. It is important that we take on the task of informing local communities about the outcome of bioscientific and other research. Of course, this requires culturally appropriate forms of transmitting such knowledge. Currently only very few funding bodies consider such an aspect of medicinal plant research as relevant. This area thus faces considerable challenge and needs novel and innovative approaches, especially with respect to research involving indigenous knowledge. In such novel approaches the native people should drive the research strategy and would—together with the researchers involved—(re)define the research strategy based on local needs. However, this is currently a vision for the future.

NEW METHODOLOGICAL DEVELOPMENTS

Biochemistry and Molecular Biology

Medicinal plant research has been greatly enriched by the recent developments in biochemistry and molecular biology, today often rephrased as metabolomics and genomics, respectively. However, the steps to be taken from initial biological activity to clinical application still remain the major challenge of drug development.

In our own research we have taken on this challenge by integrating the expertise of seven academic research groups and an industrial partner in an EU-sponsored consortium on anti-inflammatory natural products: AINP—new anti-inflammatory natural products from medicinal plants using inducible transcription factors and their signalling pathways as molecular targets. The consortium's aim is to identify inhibitors and/or modulators of the transcription factor nuclear factor kappa B (NF-κB). This transcription factor is ubiquitous and targets specific genes of the inflammatory response and apoptosis (i.e., interleukins, acute-phase genes, adhesion molecules, virus, and enzymes such as COX-2). The goal of the AINP consortium is to identify inhibitory extracts and compounds of the NF-κB cellular cascade from a targeted screening of European plants (mostly from Sardinia and southeastern Spain) with ethnobotanical uses in treating inflammatory conditions. It is based on previous research of our group, which identified sesquiterpene lactones from some Asteraceae and hypericin from St. John's wort as inhibitors of NF-κB.[16,17] No substantial screening of natural products with NF-κB inhibition activity has yet been undertaken. Contrary to previous approaches by groups in the field of medicinal plant research, some aspects are of interest in the context of this forward look:

- It draws together multidisciplinary expertise in ethnobotany, phytochemistry, and molecular biology among eight different research groups in Europe and Israel and specifically aims at drug development. The expertise available within the consortium includes natural product biology, molecular biology, biochemistry, immunology, neuropharmacology, and ethnobotany.
- A multitarget approach is far more promising than one which focuses just on a single target. A wide range of targets have been chosen to represent a number of key steps in the NF-κB activation pathway. The available screening targets within the EU consortium include, for example a *Luciferase* firefly luminescence assay to detect NF-κB inhibition based on interleukin (IL)-6 inhibition and electrophoretic mobility shift assays (EMSA) based on radiolabelled oligonucleotides of NF-κB, Oct-1, Ap-1; various interleukin (IL-6, IL1β) and PGE2 inhibition assays, IKK kinase inhibition assays, anti-ubiquitinilation assays; as well as assays relevant in the area of cell cycling and apoptosis. Included within these targets also is a GlaxoSmithKline group seeking to identify new targets within the kinase cascade re-

lated to NF-κB. At later stages of the project another partner will assess NF-κB dependent and independent processes in the brain—using both in vitro and animal whole organ in vivo models.

- Instead of limiting itself only to the early steps of analysis (primary pharmacological screens and phytochemistry), active substances and extracts are passed on to more advanced in vivo models and thus there is the potential to develop novel drug leads.
- The consortium is adding novel targets as research in biochemistry and molecular biology progresses.
- It focuses on plants from southern Europe, a region which has been neglected both from an ethnobotanical and a phytochemical perspective.
- In the event of successful lead compounds, 50 percent of all revenues will go back to the region of origin. In this case a trust fund with the specific aim to enable the sharing of benefits will be set up. It will make donations to the country or region which supplied the botanical material to support conservation, health training, and education at the community level or regional level when the source material originates from an uninhabited area.
- Clearly, a long-term goal of the project needs to be to develop appropriate mechanisms for returning information about the project and its results in a culturally appropriate form (see Chapter 7).

Although this final chapter cannot provide details on the molecular targets of relevance in the various therapeutic areas, as these can be found in Part III of the book, it demonstrates the relevance of novel strategies in medicinal plant research.

Quality Control and Chemical Analysis

Again the example of an adulterated Chinese traditional medicine can be cited to demonstrate the need for high levels of quality control. It is a multistep process that establishes the reproducible quality of a pharmaceutical product and covers all stages from the growing of the botanical material to the final control of the finished product and the evaluation of its stability and quality over time. Although the principles have remained the same, such novel techniques as DNA fingerprinting,[18] hyphenated chromatographic techniques,[19] and near-infrared spectroscopy have allowed for significant improvements in this field. However, such classical approaches as microscopic examination and TLC still remain the backbone of quality control.[20,21] More and more botanical drugs are now included in some of the main pharmacopoeias (e.g., the European pharmacopoeia), and the WHO has launched a series of monographs on important medicinal plants.[22] For each species information on the following aspects is given: drug's definition, synonyms, selected vernacular names, botanical and pharmacognostical description, purity tests (including ones on microbial contamination, pesticides and heavy metals), information on the plants chemistry, medical uses, pharmacology (experimental and clinical pharmacology, toxicology), adverse effects, and precautions. Of particular interest is the inclusion of clinical data, if such are available.[23] While this is an excellent starting point, the monographs have so far been biased toward plants prominent in Western, Chinese, and Asian Indian medicine, and much more effort needs to be put on plants commonly used, for example, in America south of the Rio Grande and in Africa.

MEDICAL APPLICATIONS: PHARMACOLOGY AND CLINICAL USE

Phytomedicines in Indigenous Medical Systems

As pointed out previously, there is an urgent need to study indigenous resources in order to identify potentially harmful species as well as ones which will provide a presumably save and efficacious means of therapy.[24] The following example from our projects in Mexico illustrates the need for research on indigenous phytomedicines.

In the medical ethnobotany of the Nahua of the Sierra de Zongolica in the Mexican state of Veracruz, *Baccharis conferta* (aerial parts, escobilla china) is used in the treatment of a variety of gastrointestinal illnesses especially diarrhea associated with gastrointestinal cramps. This species provides an interesting example of the role of indigenous phytomedicines in an indigenous culture and a classic ex vivo approach common in medicinal plant research. The aerial parts were investigated phytochemically and pharmacologically using the guinea pig ileum assay as a model.[25] The crude ethanolic extract showed a dose-dependant antispasmodic effect with the effect being particularly strong in flavonoid-rich fractions. Several flavonoids (apigenin-4',7-dimethylether, naringenin-4',7-dimethylether, pectolinarigenin, and cirsimaritin) were isolated, while others were identified in complex fractions by GC and MS. The flavonoids play an important role in the antispasmodic activity of this indigenous drug. In addition oleanolic acid and its methyl ester, as well as erythrodiol, were isolated. Oleanolic acid and methyl ester show weak antibacterial activity against *Micrococcus luteus* and *E. coli*. The phytochemical as well as the pharmacological data provide some evidence for the use of this plant in the treatment of gastrointestinal cramps. More important, it is not one compound which is responsible for the activity but a whole class. This makes *B. conferta* a prime example of a phytomedicine.

One often-neglected approach that has the potential of contributing significantly to primary health care (PHC) strategies is the use of plant extracts or phytomedicines and the clinical study of such preparations. Mueller, for example, reports clinical data of 48 semi-immune patients with malaria prior and after treatment with a tea prepared from *Artemisia annua*.[26] At the endpoint 77 percent were free of all symptoms, and in 92 percent of the cases no parasites were detectable in the blood.

New Phytomedicines for Europe and North America?

Recently ethnobotanical or ethnopharmaceutical approaches have come under serious criticism as being exploitive and not respecting indigenous traditions. As the previous example has shown, one of the main goals of ethnopharmaceutical research is the study of such indigenous preparations in order to contribute to a better understanding and ultimately to an improved use of such local resources. The transformation of a local food supplement to one used on a national or the international market has recently been exemplified by the transformation of Noni, a traditional Polynesian plant *(Morinada citrifolia)* into a Hawaiian and general U.S. health food supplement.[27] However, no scientific evidence is currently available to corroborate the claims by the companies that market noni products.

Turmeric or *Curcuma longa* (Zingiberaceae) is an example of a species that in recent years has been shown to have the potential for being developed either into a phytomedicine or one of its main constituents—curcumin—may well be a lead compound for novel anti-inflammatory pharmaceuticals. The species is endemic to peninsular India, especially the provinces of Tamil Nadu, West Bengal, and Maharashtra. It is the rhizome that is the source of turmeric, widely used in both Indian cuisine, the dyeing of cloth, and traditional medicine. Turmeric is produced by grinding the dried rhizome to a powder. The spice is used to dye cloth and confers on the robes of Buddhist monks their traditional bright orange color. At the close of the late ninteenth century turmeric was used in laboratories as "turmeric paper" to test for alkalinity before litmus paper was produced. Medicinally, turmeric is considered to be a strong antiseptic and is used to heal wounds, infections, jaundice, urinary diseases, ulcers, and also to reduce cholesterol levels. Turmeric in the form of a paste has been used to treat external conditions such as psoriasis (anti-inflammatory) and athlete's foot (antifungal).

A major medicinal compound from *C. longa* is the phenolic constituent curcumin. This compound has been shown to be effective against some forms of cancer and has been intensively examined as a possible anti-inflammatory drug.[28] In the United Kingdom an extract is currently developed as a veterinary medicine for use in canine arthritis.[29]

Other examples include *Harpagophytum procumbens* (devil's claw) and *Prunus africana* (synonym: *Pygeum africanum*; red stinkwood). The bark of *Prunus africana*—a tree from the higher altitude forests of Africa and Madagascar—has become an important European phytomedicine used for benign prostatic hyperplasia.[30] Within Africa, different decoctions of the plant are also used to treat various conditions including fevers, urinary tract infections, inflammation (bark tea), and wound dressings (leaves), and the leaf sap is drunk for insanity. Although it has been an established medicine for many years (i.e., has been a licensed drug in many European countries), recently the demand has increased substantially. In the regions of origin a high-quality extract is produced which—provided that sustainable forms of production are to be used—would give these regions a small but secure source of income. However, the botanical drug is still largely collected from the wild. Main importers include Italy, France, and Spain. Bark extracts are mostly sold in the two former countries as well as in Switzerland and Austria. Currently African countries export about 3,900 tons of bark/bark extract per year. Since the 1990s sales of *Prunus africana* containing botanical drugs have increase exponentially. For example, in 1995 Madagascar exported 1,200 tonnes (t) bark, while in previous years only around 200-600 t were exported.[31] Main producer countries are Cameroon and Madagascar and the overexploitation seems to have caused serious damage to the stinkwood populations. Consequently, conservationists have asked for a stop in the trade of this botanical drug. Alternatively, a sustainable production system would have to be developed, which would require a complex strategy for changing the expectations of producers, traders, pharmaceutical companies, and consumers, but also novel measures for assuring sustainable growth and appropriate forms of controlling these measures.[32]

Devil's claw or grapple plant is aptly named for its formidable "claw"—the dried hooked thorns of the fruit used in seed dispersal—which are a hazard to any passing clove-footed animal or careless human. The plant is native to southern and eastern Africa and it is collected in regions bordering the Kalahari Desert. It thrives in clay or sandy soils and is often found in the South African veldt of the west/north of Northern Province and the southern regions of South West Province. The species is used traditionally as a tonic, for "illnesses of the blood," fever, problems during pregnancy, and kidney and bladder problems. Starting in the mid-1980s, and with considerable research effort, African devil's claw has been developed into a very successful and relatively well-characterized phytomedicine.[33] The secondary storage roots are collected and, while they are still fresh, they are cut into small pieces and dried. The main exporters are South Africa and Namibia. Attempts are currently under way to cultivate the species and in this case there seems to be a better chance to develop sustainable forms of productions soon.

These three examples show the potential of further developing phytomedicines which may provide both improved treatment options for patients in Europe and income to producers in developing countries. However, these examples are scarce. Currently, the licensing requirements for new botanical pharmaceuticals are becoming more and more stringent and there is a definite retrograde step in research collaborations between, for example, Africa and Europe. This is in part due to novel research focus in Europe and consequently a reduced interest in natural product biology, but also due to a lack of appropriate research efforts in the area of phytomedical research.

New Pharmaceuticals from Plants

An example of a pharmaceutical newly developed on the basis of European natural resource is galanthamine, which has recently been licensed in European countries for the treatment of dementia, including Alzheimer's disease. It is derived from snowdrops (*Galanthus* spp. and *Leucojum* spp., especially *L. aestivum*) and *Narcissus* spp. (Amaryllidaceae).

Campothecin from the south and southeast Asian *Camptotheca acuminata* (Nyssaceae) is currently under development as a novel medication in cancer chemotherapy and acts as an inhibitor of the topoisomerase.

One current and topical example is that of the small Samoan tree *Homalanthus nutans*—the source of the anti-HIV-1 drug prostratin.[34] The tree had long been a staple medicinal plant of local peoples in the treatment of hepatitis. Ethnobotanical studies and cooperation with local communities have been conducted for many years and the drug is now classified by the National Cancer Institute to the AIDS Research Alliance as a candidate anti-AIDS drug. A significant portion of the license income is to be returned to the Samoan communities.[35]

Many other examples could have been selected, but these three examples demonstrate the opportunities for pharmaceutical drug development using plant-derived products, and this area surely will provide a continuous challenge and exciting results to researchers in the area of medicinal plant research. At this stage, only galanthamine has become a marketed pharmaceutical (and is no longer isolated from the plant but produced synthetically), and it remains to be seen which natural products will make it to the marked in the near future.

FUTURE NEEDS: PHYTOTHERAPY AND THE CONCEPT OF ETHNOPHARMACY

For many centuries, ethnobotany played an important role in the development of new useful products. As shown in this contribution and the other ones in this book, nature-derived drugs as well as phytomedicines are still important in pharmacy. Today the consumer demand for natural medicines is higher than it was ever before. Both in Europe and in the United States consumers spend at least $4 billion a year on herbal medicinal products. The study of indigenous medical systems and especially of medicinal plants (ethnopharmacy[36,37]) has been an essential basis both for the further development of autochthonous resources (indigenous phytotherapy) as well as for the discovery of new pharmaceuticals for the international market. Medicinal plants are not just a commodity but also an integral element of cultures all over the world. This and the importance of an interdisciplinary approach to ethnopharmacological or ethnopharmaceutical research is highlighted by the comparison with bioprospecting (see Table 21.1).

"Bioprospecting" focuses on the development of new single-entity drugs for the huge markets of the north. New potentially highly profitable pharmaceutical products are developed based on the biological and chemical diversity of the various ecosystems of the earth and the research requires an enormous financial input. The research goes from the collection of biogenic samples (plants, fungi, other microorganisms, and animals), to the subsequent analysis of the biological-pharmacological activities and to the study of the organisms' natural products to the development of drug templates or new drugs. Key in this search are high-throughput screening systems as they are established by the major international pharmaceutical companies. Huge libraries of compounds (and sometimes extracts) are screened for biological activity against specific targets. Biodiversity-derived products are only one of the many sources of material for these test batteries. This serves as a starting point for drug development. Currently some companies envision the screening of 500,000 samples a week against a single target. It thus becomes essential to have an enormous number of chemically diverse samples available. Clearly, the example given in Chapter 10 has a strong bioprospecting aspect, even though the starting point are plants used locally as medicine.

The other approach has mostly been termed an ethnopharmacological one. Ethnobotanical studies generally result in the documentation of a

TABLE 21.1. Ethnopharmacology compared to bioprospecting.

Ethnopharmacology	Bioprospecting
Overall goals	
(Herbal) drug development especially for local uses	Drug discovery for international market
Complex plant extracts (phytotherapy)	Pure natural products as drugs
Social importance of medicinal and other useful plants	—
Cultural meaning of resources and understanding of indigenous concepts about plant use and of the selection criteria for medicinal plants	—
Main disciplines involved	
Anthropology	—
Biology (ecology)	Biology including very prominently ecology
Pharmacology/molecular biology	Pharmacology/molecular biology
Pharmacognosy/phytochemistry	Phytochemistry
Number of samples collected	
Very few (up to several hundred)	As many as possible, preferably several thousand
Selected characteristics	
Detailed information on a small segment of the local flora (and fauna)	Limited information about many taxa
Database on ethnopharmaceutical uses of plants	Database on many taxa (including ecology)
Development of autochthonous resources (esp. local plant gardens, small-scale production of herbal preparations)	Inventory (→ expanded herbaria) economically sustainable alternative use to destructive exploitation (e.g., logging)
Pharmacological study	
Preferably using low-throughput screening assays which allow a detailed understanding of the local or indigenous uses	The assay is not selected based on local usage, instead high-throughput screening systems are used
Key problem	
Safety and efficacy of herbal preparations	Local agendas (rights) and compensation to access

Source: van Wyk, B-E., van Oudtshoorn, B., and Gericke, N., *Medicinal plants of South Africa* (Pretoria, South Africa: Briza Publications, 1997).

rather limited set of very well-documented useful plants (mostly medicinal, but also those known to be toxic or used in nutrition). In ethnopharmacology an important goal is the development of improved preparations for the use by local people. Thus it is essential to get information on the bioactive compounds from these plants, their relative contribution to the effects of the extract (including, for example, synergistic or antagonistic effects), the toxicological profile of the extract and its constituents and on improved galenic preparations to be prepared under local conditions. The examples discussed in Chapter 11 (*Baccharis conferta* and *Artemisia annua*) use an ethnopharmacological approach.

Ethnopharmacological approaches may lead to improved primary health care in marginal societies.[38] Thus a core task with respect to medicinal plants is to integrate the cultural history of (medicinal) plants and our current bioscientific knowledge of these resources and to use this information for a sustainable development of this exciting part of the interaction between people and plants.

To stress the breadth of such an approach the term *ethnopharmacy* may well be the most appropriate one, since this term encompasses all the relevant subdisciplines within the study of medicines:

- Pharmacognosy/natural product biology and chemistry
- Pharmacology
- Pharmaceutics (galenics)
- Drug delivery
- Toxicology
- Biopharmacy (bioavailability and metabolism studies)
- Pharmacy practice and policy

This will allow the development of selected local resources, especially medicinal plants into elements to be used, for example, in primary health care.

CONCLUSION

Discussion in this chapter first of all indicates that we need a patient-driven approach in the area of medicinal plant research. As the mother's dilemma in the introduction shows, patients request "something natural," and this challenge has to be taken on by society as a whole and specifically by pharmacy and researchers in the field of medicinal plant biology. Such research needs to be for the benefit of patients in poorer and richer countries alike.

In Europe we are fortunate in that our phytotherapeutical traditions and especially the more commonly used medicinal plants have been relatively well studied (see Chapter 1).[39] Thus, we are able to base our decisions on the scientific evidence about phytomedicines and especially on a benefit-risk analysis. Although some decisions by the regulatory authorities may well be seen and criticized as inadequate or too restrictive (e.g., the recent withdrawal of kava in many countries), only a pharmaceutical and science-based decision allows protection of the consumers and optimization of our phytopharmacopoeias. In conclusion, many important links exist between ethnopharmacy/ethnopharmacology and phytomedicine, and one of our core tasks of the future will be to further develop these links and to make such topics transparent to the general public.

NOTES

1. Leonti, M., Vibrans, H., Sticher, O., and Heinrich, M., 2001, Ethnopharmacology of the Popoluca, Mexico: An evaluation, *Journal of Pharmacy and Pharmacology,* 53: 1653-1669.
2. Iwu, M., Biodiversity and development of new phytomedicines and biodiversity conservation in Africa. Plenary lecture at the joint conference of the South African Association of Botanists/International Society for Ethnopharmacology, Pretoria, South Africa, January 8-11, 2003.
3. Heinrich, M., Barnes, J., Gibbons, S., and Williamson, E.M., *Fundamentals of pharmacognosy and phytotherapy* (London: Harcourt, 2004).
4. Leonti, M., Ramirez, F., Sticher, O., and Heinrich, M., 2003, Medicinal flora of the Popoluca, Mexico: A botanico-systematical perspective, *Economic Botany,* 57(2): 218-230.
5. Moerman, D., Pemberton, R.W., Keifer, D., and Berlin, B., 1999, A comparative analysis of five medicinal floras, *Journal of Ethnobiology,* 19: 49-67.
6. Cannel, R.J.P. (ed.), *Natural product isolation* (Totowa, NJ: Humana Press, 1998).
7. Comerford, S.C., 1996, Medicinal plants of two Mayan healers from San Andres, Peten, Guatemala, *Economic botany,* 50: 327-336.
8. Stepp, J.R. and Meorman, D.E., 2001, The importance of weeds in ethnopharmacology, *Journal of Ethnopharmacology,* 75: 19-23.
9. Heinrich, M., Pieroni, A., and Brenner, P., Medicinal plants and phytomedicines, in G. Prance and M. Nesbitt (eds.), *The Cultural History of Plants* (New York: Routledge, 2005).
10. Frei, B., Sticher, O., and Heinrich, M., 2000, Zapotec and Mixe use of tropical habitats for securing medicinal plants in Mexico, *Economic Botany,* 54(1): 73-81.
11. Voeks, R.A., 1996, Tropical forest healers and habitat preference, *Economic Botany,* 50: 381-400.
12. Comerford, 1996, Medicinal plants of two Mayan healers.
13. Secretariat of the Convention on Biological Diversity, *Handbook of the convention on biological diversity* (London: Earthscan, 2001).

14. Laird, S., (ed.), *Traditional knowledge: Equitable partnership in practice* (London: Earthscan, 2002).

15. Secretariat of the Convention on Biological Diversity, 2001, *Handbook of the convention on biological diversity.*

16. Bork, P.M., Schmitz, M.L., Kuhnt, M., Escher, C., and Heinrich, M., 1997, Sesquiterpene lactone containing Mexican Indian medicinal plants and pure sesquiterpene lactones as potent inhibitors of transcription factor NF-kB, *FEBS Lett.*, 402: 58-90.

17. Bork, P.M., Bacher, S., Schmitz, M.L., Kaspers, U., and Heinrich, M., 1999, Hypercin as a nonantioxident inhibitor of NF-kB, *Planta Medica,* 65: 297-300.

18. Shaw, P.C., Wang, J., and But, P.P.H., *Authentication of Chinese Medicinal Materials by DNA Technology* (Singapore: World Scientific Publishing Co., 2002).

19. Cannel, R.J. P. (ed.), 1998, *Natural Product Isolation.*

20. Wichtl, M. (ed.), *Herbal drugs and phytomedicines: A handbook for practise on a scientific basis* (third English edition translated from the fourth German edition, 2000) (Boca Raton: Medpharm Scientific Publishers, 2004).

21. Bisset, N.G. (ed.), *Herbal drugs and phytopharmaceuticals* (trans. from the second German edition) (Stuttgart: Medpharm, 1994).

22. World Health Organization, WHO monographs on selected medicinal plants, (volume one) (Geneva: WHO, 1999).

23. World Health Organization, 1999, WHO monographs.

24. Robineau, L. and Soejarto, D.D., TRAMIL: A research project on the medicinal plant resources of the Caribbean, in M.J. Balick, E. Elisabetsky, and S.A. Laird (eds.), *Medicinal resources of the tropical forest* (New York: Columbia University Press, 1996) pp. 317-325.

25. Weimann, U., Goransson, U., Pongprayoon-Claeson, P., Claeson, L., Bohlin, H., Rimpler, H., and Heinrich, M., 2002, Spasmolytic effects of *Baccharis conferta* and some of its constituents, *Journal of Pharmacy and Pharmacology,* 54: 99-104.

26. Mueller, M.S., Karhagomba, I.B., Hirt, H.M., and Wemakor, E., 2000, The potential of *Artemisia annua* L. as a locally produced remedy for malaria in the tropics: Agricultural, chemical, and clinical aspects, *Journal of Ethnopharmacology,* 73: 487-493.

27. Dixon, A.R., McMillen, H., and Etkin, N.L., 1999, Ferment this: The transformation of Noni, a traditional Polynesian medicine (*Morinada citrifola,* Rubiaceae), *Economic Botany,* 53(1): 51-68.

28. Surh, Y.J., Chun, K.S., Cha, H.H., Han, S.S., Keum, Y.S., Park, K.K., and Lee, S.S., 2001, Molecular mechanisms underlying chemopreventive activities of anti-inflammatory phytochemicals: Down regulation of COX-2 and iNOS through suppression of NF-kB activation, *Mutation Research,* 480-481: 243-268.

29. Heinrich, M., Pieroni, A., and Bremner, P., Medicinal plants and phytomedicines, in G. Prance and M. Nesbitt (eds.), *The Cultural History of Plants* (New York: Routledge, 2005).

30. Bombardelli, E. and Morazzoni, P., 1997, *Prunus africana* (Hook f.) Kalkm., *Fitoterpia,* 68(3): 205-217.

31. Cunningham, M., Cunningham, A.B., and Schippmann, U., *Trade in* Prunus africana *and implementation of CITES* (Bonn: Federal Agency for Nature Conservation, 1997).

32. Cunningham, Cunningham, and Schippmann, 1997, *Trade in* Prunus africana.

33. van Wyk, B.E., van Oudtshoorn, B., and Gericke, N., *Medicinal plants of South Africa* (South Africa: Briza Publications, 1997).

34. Cox, P.A. and Heinrich, M., 2001, Ethnobotanical drug discovery: Uncertainty and promise, *Pharmaceutical News,* 8(3): 55-59.

35. Cox and Heinrich, 2001, Ethnobotanical drug discovery.

36. Heinrich, M., *Ethnobotanik und ethnopharmazie: Eine einfuhrung* (Stuttgart: Wissenschaftliche Verlagsgesellschaft, 2001).

37. Leonti, Ramirez, Sticher, and Heinrich, 2003, Medicinal flora of the Popoluca, Mexico.

38. Mueller, Karhogomba, Hirt, and Wemakor, 2000, The potential of *Artemesia annua* L.

39. Schulz, Volker, Hansel, R., and Tyler, V.E., *Rational phytotherapy: A physician's guide to herbal medicine* (Berlin: Springer, 2001).

Index

Printed in the United States
by Baker & Taylor Publisher Services